T0184460

Marine Analytical Chemistry

Julián Blasco · Antonio Tovar-Sánchez
Editors

Marine Analytical Chemistry

 Springer

Editors
Julián Blasco (iD)
Department of Ecology and Coastal
Management
Instituto de Ciencias Marinas de Andalucía
(CSIC)
Puerto Real, Spain

Antonio Tovar-Sánchez
Department of Ecology and Coastal
Management
Instituto de Ciencias Marinas de Andalucía
(CSIC)
Puerto Real, Spain

ISBN 978-3-031-14488-2   ISBN 978-3-031-14486-8   (eBook)
https://doi.org/10.1007/978-3-031-14486-8

This Springer imprint is published by the registered company Springer Nature Switzerland AG
The registered company address is: Gewerbestrasse 11, 6330 Cham, Switzerland

Preface

Understanding the ocean functioning, chemical composition, biogeochemical processes, and dynamics, from coast to open ocean and from surface to depth, is a challenge for the marine community. The study of the physical and biogeochemical properties of the oceans requires the sampling of the water column and analysis of chemicals across the different oceanographic basins. Observation and sampling of the oceans involve the use of multiple and diverse platforms (e.g., oceanographic ship, satellite, RPAs, and ocean gliders) and equipment (e.g., sampling systems and analytical instruments) that are conditioned by the depth (from the sea surface microlayer to the deep ocean), the type of samples (e.g., water, suspended particles, biota, and sediments), and/or the parameters to be collected and/or analyzed (e.g., trace metals, radionuclides, organic pollutants, and emerging contaminants).

This book compiles information about the sampling, fate, behavior, and analytical methods and techniques currently used by the scientist community to study some chemicals linked to the functioning and pollution of our oceans. Chapter 1 "Carbonate system species and pH" and Chap. 2 "Dissolved organic matter" describe the methodology for the analysis, at marine environmental levels, of carbon, total alkalinity and pH, and dissolved organic matter, carbon, phosphorous, and nitrogen, respectively. Chapter 3 "Trace metals" describes the ultraclean techniques and procedures for collection of seawater and analysis of trace metals in the water column in order to produce reliable, accurate, and reproducible environmental trace metal data. Chapter 4 "Radionuclides as ocean tracers" describes the methods and techniques used to measure radionuclides and provides an overview of the use of radionuclides as tracers of processes occurring in the marine environment, such as biological pump (234Th, 238U), submarine groundwater discharge (226Ra, 228Ra, 223Ra, 224Ra, and 222Rn), or ocean circulation (129I).

However, the ocean is the main sink of many chemical substances, and the introduction of these can pose a danger to aquatic life and human beings. To know the fate, behavior, and effect of pollutants released to the marine environment is a challenge, which depends on physicochemical and biological variables. Some of these pollutants are well known, but others are emerging pollutants of recent concern. Chapter 5 "Persistent organic contaminants" shows the occurrence of many of these pollutants in environmental compartments - although many of them have been forbidden - and how they are distributed, transported, and biomagnified across the trophic chain. Chapter 6, "Emergent organic contaminants," points out the growing problem of the occurrence of these pollutants (e.g., pesticides, personal care products, flame retardants, plasticizers, and hormones) and how they can affect the flora and fauna. The drawbacks of traditional methods for the analysis are indicated and the use of promising techniques such as biosensors is mentioned. Chapter 7, "Nanoparticles in the marine environment," outlines processes that regulate the fate of nanoparticles in the marine environment and highlights the main parameters to be considered for sampling strategy. The advantages and disadvantages for nanoparticles analysis and the more promising methodologies are presented. Chapter 8, "Microplastics and nanoplastics," provides tools to approach the issue of microplastics including microfibers as a category. The chapter is focused on the sampling strategies and analytical approaches and standard requirements for reliable assessment of microplastics in marine matrices.

Nowadays, one of the promising approaches to understand the ocean functioning is the use of methodologies that allow us to get a global view of the ocean and coastal areas or to use system to monitor in a continuous way or in remote areas; thus, Chap. 9 "Satellite and RPAS (remotely piloted aircraft system)" and Chap. 10 "In situ sensing: Ocean glider" present remote (satellite and RPAs) and in situ (gliders) platforms to explore the ocean at high spatial and temporal resolutions.

Finally, Chap. 11 "Marine chemical (meta-)data management" provides best practices in preparing the generated chemical data from environmental studies to facilitate their dissemination and guarantee the reusability and long-lasting availability for the scientific community.

The chapters of this book include real examples derived from the experience acquired by their authors in their field of expertise, making this book especially useful for graduate/advanced graduated students. In addition, this book should be of interest to scientists from many disciplines such as oceanography, biology, chemistry, ecology, biogeochemistry, geosciences, fisheries, and climate change, among others.

Puerto Real, Spain   Julián Blasco
                                                                               Antonio Tovar-Sánchez

Contents

Carbonate System Species and pH

1

M. Fontela, A. Velo, P. J. Brown, and F. F. Pérez

Contents

Abstract

Accurate knowledge of the state of the marine carbon system is essential to constrain global carbon budgets. While analytical techniques for characterizing oceanic carbon chemistry are well established and diverse in approach, global monitoring still suffers from a lack of measurements being made both temporally and spatially. Here, we summarize the state of the art for the analysis of some of the main carbon parameters: total dissolved inorganic carbon, (C_T or DIC), pH, and total alkalinity (A_T). For each, a brief theoretical approach is followed by a description of technical methodology. All the selected techniques are detailed and

M. Fontela (✉)
Centre of Marine Sciences (CCMAR), Faro, Portugal

Centro de Investigación Mariña, Universidade de Vigo, Vigo, Spain
e-mail: mmfontela@ualg.pt

A. Velo · F. F. Pérez
Instituto Investigaciones Marinas (IIM-CSIC), Vigo, Spain

P. J. Brown
National Oceanography Centre (NOC), Southampton, UK

explained such that they can be used as standard methods aboard an oceano-graphic vessel. It is expected that by following the methods described in this chapter, independent researchers should be able to make comparable measurements of carbonate system species and pH.

1.1   Introduction

1.1.1   Global Carbon Cycle

Ocean circulation and marine biogeochemical processes drive and exchange large amounts of carbon, in what is called the marine carbon cycle. For billions of years, the ocean has incorporated large amounts of alkalinity from the continents, mostly in the form of bicarbonates, while continuously exchanging carbon dioxide (CO_2) with the atmosphere to reach a global dynamic equilibrium. Biological organisms in the ocean contribute to this by capturing CO_2 during primary production in the illuminated part of the ocean, before releasing it again during their decay as they sink into the deep ocean. Together these processes make the ocean a principal component of the global carbon cycle because of its capacity to accumulate carbon—while the atmosphere holds ~800 Gt C (gigatonnes of carbon) and the land biosphere ~2300 Gt C, the ocean carbon reservoir holds a huge 38,000 Gt C. In the present Anthropocene era, the ocean has absorbed around 30% of human-derived CO_2 emissions, which represents only 0.4% of the total carbon stored by the oceans. Therefore, accurate measurements of seawater carbon variables are crucial in order to be able to detect these small anthropogenic chemical changes caused by CO_2 increases in the ocean. The continued absorption of CO_2 causes a continuous decrease in pH and carbonate ion concentrations, a process known as Ocean Acidification (OA). OA has a wide range of ecological consequences, including impacts on calcareous organisms, changes in the speed of sound, effects on the speciation of trace elements, and changes in organoleptic properties of seawater. Furthermore, the evolution of ocean biogeochemical cycles are likely to be addition-ally modified by changes in OA rates. Understanding and modelling oceanic CO_2 uptake is thus essential to understand and predict how the carbon cycle will evolve and how the climate system will respond to the impact of human activities. This is possible by using observations to fully characterize the marine carbonate system (MCS), achieved through the determination of at least two of the four measurable variables of this system—dissolved inorganic carbon (C_T or DIC), pH, total alkalin-ity (A_T or TA), partial pressure of carbon dioxide (pCO_2)—as well as associated physical parameters (pressure, P, temperature, T, and salinity, S) and ancillary variables (nutrients).

1.1.2 Carbon Essential Ocean Variables

The United Nations Educational, Scientific and Cultural Organization (UNESCO) has defined a set of Sustainable Development Goals (SDGs), with SDG 14 focusing on "Life Below Water" and SDG target 14.3 being *minimizing and addressing the impacts of ocean acidification, including through enhanced scientific cooperation at all levels.*". The indicator 14.3.1 of this target is the measure of average marine acidity (pH) at representative places. Through various international oceanographic organizations, UNESCO provides guidance on undertaking OA observations to researchers of Member States and support on how to conduct annual reporting and OA data management to National Oceanographic Data Centers. Member States will be required to report annually on the collected data, following SDG indicator 14.3.1 methodology and through a dedicated data portal (https://oa.iode.org/). Furthermore, the Global Ocean Observation System (GOOS, https://goosocean.org/) considers Inorganic Carbon as an Essential Ocean Variable (EOV), where at least two of the four previously mentioned Sub-Variables (C_T, pH, A_T, and pCO_2) are needed. A GOOS Essential Ocean Variable is defined as a mandatory measurement to evaluate the state and change of the ocean at local and global level. As anthropogenic alteration of the MCS is relatively small (0.1% of C_T per year in the surface layer), high-quality observations are needed to track its progress. As a reference, the Global Ocean Acidification Observing Network (GOA-ON, http://www.goa-on.org/) has defined two uncertainty levels for marine carbon observations based on carbonate ion concentration uncertainties, the "weather" and "climate" targets. The "weather" goal requires an uncertainty of $<10\%$ in carbonate ion concentration, which translates to uncertainty of ± 10 µmol·kg^{-1} in C_T and A_T and ± 0.02 in pH (Newton et al. 2015). The "climate" goal is far more demanding, requiring an uncertainty of 1% in carbonate ion, which translates to an uncertainty of ± 2 µmol·kg^{-1} in C_T and A_T, and ± 0.003 in pH (Newton et al. 2015).

1.1.3 Marine Carbonate System

In the ocean, carbon dioxide exists in three different inorganic forms: as bicarbonate (HCO_3^-, ~90%), as carbonate (CO_3^{2-}, ~9%), and as aqueous carbon dioxide ($CO_2(aq)$, ~1%, Fig. 1.1). A fourth form is carbonic acid (H_2CO_3), but this is found in a much lower proportion ($<0.3\%$) than $CO_2(aq)$. The sum of carbonic acid and $CO_2(aq)$, which are chemically indistinguishable, is denoted as CO_2 (Zeebe and Wolf-Gladrow 2001), and is in thermodynamic equilibrium with the gas phase carbon dioxide $CO_2(g)$.

$$CO_2(g) + H_2O \leftrightarrows CO_2 \quad K_0 = [CO_2]/pCO_2 \qquad (1.1)$$

The CO_2 concentration is given by Henry's law with K_0 being the solubility of CO_2 in seawater. The carbonate species are then related by the following equations:

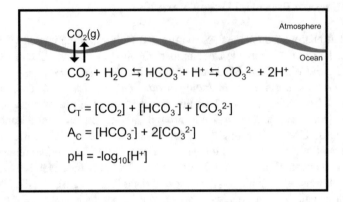

Fig. 1.1 Simplified representation of the main equations in the marine carbon system

$$CO_2 \leftrightharpoons HCO_3^- + H^+ \quad K_1 = ([H^+]\,[HCO_3^-])/[CO_2] \qquad (1.2)$$

$$HCO_3^- \leftrightharpoons CO_3^{2-} + H^+ \quad K_2 = ([H^+]\,[CO_3^{2-}])/[HCO_3^-] \qquad (1.3)$$

where K_1 and K_2 are the stoichiometric equilibrium constants (they depend on pressure, salinity, and temperature). The sum of the three dissolved forms (CO_2, HCO_3^-, and CO_3^{2-}) is called total dissolved inorganic carbon and can be denoted by several notations: C_T, DIC, TIC, ΣCO_2, TCO_2, or even "*dissic*." Here we will use C_T, and it is one of the MCS variables that can be quantitatively determined. Another is total alkalinity, which is closely related to the charge balance in seawater (the sum of the excess of proton acceptors over proton donors, Dickson, 1981). The main contributions to this come from carbonate and bicarbonate ions (carbonate alkalinity, A_C or CA, defined in Fig. 1.1), but borate ions and other minor components are also important. The concept of A_T is extensively examined in the Section 1.4.3 **Total Alkalinity**.

A useful representation of how the different components of the inorganic carbon system change in relation to each other (carbonate speciation) and their effect on seawater pH can be visualized in the so-called Bjerrum plot (Fig. 1.2). In the current ocean the speciation of the MCS follows approximately the ratio 90 HCO_3^-: 9 CO_3^{2-}: 1 ($CO_2 + H_2CO_3$), so most of the carbon is in the form of bicarbonate. The pH values where the concentration of $CO_2 + H_2CO_3$ or CO_3^{2-} equals the concentration of HCO_3^- (represented in Fig. 1.2 with vertical gray lines) correspond with the values of the first (pK_1) and second (pK_2) dissociation constants of carbonic acid, respectively.

The two equilibrium conditions (Eqs. 1.2 and 1.3), the mass balance for C_T (Fig. 1.1), and the charge balance for A_T constitute four equations with six unknown variables: $[CO_2]$, $[HCO_3^-]$, $[CO_3^{2-}]$, $[H^+]$, C_T and A_T. As a result, when two variables are known (as well as temperature and salinity, to determine the solubility of CO_2 and dissociation constants of carbonic and boric acid), the MCS is overdetermined and all other components can be calculated (also pCO_2 by the Eq. 1.1).

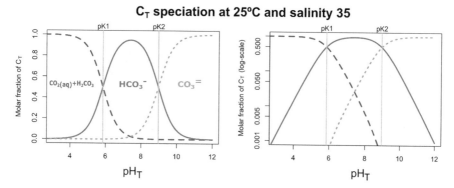

Fig. 1.2 Bjerrum plot. Molar fraction of dissolved inorganic carbon species in seawater versus pH_T in continuous (left) and logarithmic scale (right)

Currently, only pCO_2, $[H^+]$, C_T, and A_T can be measured directly with independent standards. As will be described below, other minor species contributing to A_T must also be evaluated.

1.1.4 Uncertainties in Measured and Calculated Carbonate System Variables

Generally, C_T, A_T, and pH measurements are those most typically made on samples collected at sea, with discrete measurements of pCO_2 much more infrequent for interior ocean samples. Instead, on many vessels pCO_2 is more usually measured in a continuous mode for surface waters, by pumping seawater from a depth of 2–5 meters into an automatic and integrated system that also records other surface seawater properties (temperature, salinity), along with position and meteorological variables. The global Surface Ocean Carbon Dioxide Atlas network (SOCAT, https://www.socat.info/) has recorded more than 28 million observations of sea surface pCO_2 from 1957 to 2020 (Bakker et al. 2016). The accuracy of these automatic systems is generally better than 2 ppm (2 µatm or 0.2 pascals).

For C_T and A_T, the usual accuracy in the measurement is about 2 µmol·kg^{-1}, although Certified Reference Materials produced by Dr. Andrew Dickson at the Scripps Institution of Oceanography of the University of California San Diego have shown accuracies of 0.65 µmol·kg^{-1} on average over the last two dozen batches (Bockmon and Dickson 2015). For pH the best accuracy is around 0.0055 pH units (Carter et al. 2013). The pCO_2 measurement error is close to 2 µatm for values around the atmospheric concentration, but can be up to 10 µatm at high pCO_2 of 800 µatm. With knowledge of these measurement accuracies, uncertainties in the calculated values of the other MCS variables can be robustly determined.

Currently, there are several public tools that enable the full calculation of the MCS based on thermodynamic equilibria when at least two variables are known (Table 1.1). Within these, an increasing fraction allows the option to select different

Table 1.1 software packages recommended for marine carbon system computations with uncertainty propagation errors

Software	Language	Current version	References	Link
CO2SYS	Excel-visual basic	25	Pelletier et al. (2007); Orr et al. (2018)	https://github.com/jamesorr/CO2SYS-Excel
CO2SYS	MATLAB	1.1; 2.1; 3.2.0	Van Heuven et al. (2011); Orr et al. (2018), Sharp et al. (2020)	https://github.com/jamesorr/CO2SYS-MATLAB https://github.com/jonathansharp/CO2-System-Extd
PyCO2SYS	Python	1.8.1	Humphreys et al. (2021)	https://pyco2sys.readthedocs.io/
Seacarb	R	3.3.1	Gattuso et al. (2020)	http://cran.r-project.org/package=seacarb
Mocsy	Fortran	2.0	Orr and Epitalon (2015); Orr et al. (2018)	https://github.com/jamesorr/mocsy/

dissociation constants (K_1, K_2, other key seawater components) that have been derived experimentally by different research groups. Measurement of more than two of the four carbon parameters has allowed the testing of the internal consistency of the MCS, and the appropriateness of individual dissociation constants across different environmental conditions and regions. A general consensus has formed to follow the best practices in dissociation constant selection recommended by Dickson et al. (2007): specifically the dissociation constants of Lueker et al. (2000) and the constant for sulfate dissociation of Dickson (1990). Although Lueker et al. (2000) dissociation constants are valid over a large range of salinities (19–43) and temperatures (2–35 °C), almost 40% of the ocean by volume actually has a temperature below 2 °C (Sulpis et al. 2020); for those cold temperatures, the Lueker et al. (2000) dissociation constants do not perform as well in calculations of internal consistency of the MCS (Sulpis et al. 2020). The final option for the fluoride dissociation constant, K_f, chosen from those of Pérez and Fraga (1987) or Dickson and Riley (1978), does not have much influence on the final result (Orr et al. 2018).

The packages CO2SYS (MATLAB and Excel), seacarb (R), mocsy (Fortran), and PyCO2SYS (Python) allow the user to propagate errors through the calculation scheme by setting the initial uncertainties for the input pair of carbon variables (Orr et al. 2018). In order to keep up with new developments in understanding the MCS, it is always recommended to keep your preferred software updated (Orr et al. 2015; Sharp et al. 2020; Humphreys et al. 2021). Since uncertainties should be routinely propagated (Orr et al. 2018), in this chapter only software that has uncertainty propagation implemented and that can use different parameter pairs as inputs has been included in Table 1.1. In Table 1.2, we show the mean uncertainties of calculated parameters by input pair selection when errors in the hydrographic properties and in the equilibrium constants are considered (Orr et al. 2018).

Table 1.2 Uncertainty error propagation in the marine carbon system. For each set of input pairs available, considering errors in the dissociation constants and in the main hydrographic properties (temperature and salinity) but not in nutrients. Cell with number in bold style implies assumed measured uncertainty while italic character is the computed uncertainty. Uncertainties are computed with a Monte Carlo approach based on 10^5 perturbations

Input pair	C_T ($\mu mol\ kg^{-1}$)	A_T ($\mu mol\ kg^{-1}$)	pH	pCO_2 (μatm)	CO_3^{2-} ($\mu mol\ kg^{-1}$)	$\Omega_{Aragonite}$
A_T and C_T	2	2	0.0120	14.1	2.7	0.11
A_T and pH	5.8	2	0.0055	10.3	4.6	0.13
A_T and pCO_2	5.2	2	0.0070	2	4.9	0.12
C_T and pH	2	6.2	0.0055	9.8	6	0.13
C_T and pCO_2	2	6.2	0.0074	2	5.9	0.13
pH and pCO_2	4.8	5.3	0.0055	2	8.2	0.15

As can be seen, the resulting uncertainty in the computed parameters is always larger than the analytical uncertainty of the standard measurement technique. As the improvement in analytical techniques in recent times has increased the accuracy of carbon system measurements, the uncertainty in the computations is currently dominated by the uncertainty in the dissociation constants, K_1 and K_2 (Orr et al. 2018). Comparisons between measured and computed values have shown that the selection of input pair is important: there is better agreement when a T, P-dependent parameter (pCO2 or pH) is used alongside a non T,P-dependent parameter (C_T or A_T) (Raimondi et al. 2019).

It is important to keep in mind that calculated uncertainty is dependent on the total carbon content of the system. For example, for pH: these uncertainties have been determined for both surface waters with a pH of 8.1, and for deep waters of pH 7.45 (characteristic of areas with low oxygen levels), and the uncertainties in relative terms (%) are very similar in both pH settings. Importantly, however, it must be recognized that pH is a quantity that evaluates relative changes due to its intrinsic definition on a logarithmic scale. A pH change of 0.01 thus represents a change of 2.3% of hydrogen ion concentrations at both pH $= 7.4$ and pH $= 8.1$ (and any other value), but in magnitude of the change of hydrogen ion concentrations is five times larger at 7.4 than at 8.1. The combination of Eqs.1.1 and 1.2 expressed in logarithmic form linearly relates the changes in pH to changes in $log(pCO_2)$ (also to $log[CO_2]$), meaning that uncertainties and changes in pCO_2 should therefore also be expressed in percentage terms. In fact, the pCO_2 measurement has an uncertainty that, in absolute terms, increases almost linearly with its magnitude, similar to C_T and A_T. For these, both their concentrations are far higher than the observed range in their concentrations due to bicarbonate being their main constituent (see Fig. 1.2 Bjerrum plot). However, average C_T levels are typically between 10 and 15% lower than those of A_T, almost in correspondence to the uncertainties shown in Table 1.2, where C_T values are somewhat lower than those of A_T.

This chapter is centered on open ocean carbon chemistry; readers interested in coastal environments, where the range of physicochemical variation is larger, could follow the guidelines for the measurement of carbonate chemistry of Pimenta and

Grear (2018) for example. The selection of methods included in this chapter does not pretend to be exhaustive, but they belong to the current state-of-the-art of analytical procedures. All methods require experienced operators (Mintrop et al. 2000), as can be seen in intercomparison assessments, where the percentage of uncertainty related to operator procedure is suspected to be significant (Bockmon and Dickson 2015). The methods include, where possible, quality control procedures to ensure that results are reliable and reproducible. The aim is that independent researchers, following the methods described in this chapter, should be able to produce comparable measurements of carbonate system species and pH.

1.2   Sampling Procedure: Commonalities for Marine Carbon System Parameters

While there are differences in the volume and flask type, the procedure for seawater sample collection for analysis of each of the individual parameters of the marine carbon system is very similar. Collecting discrete samples from Niskin bottles is the most widely used procedure in chemical oceanography. Firstly, with the help of a silicone tube between the Niskin output and the bottom of the flask, the flask is rinsed thoroughly with sample. Then, the flask is filled slowly and the bottle overflowed by a considerable volume. Finally, a headspace should be created within the bottle: while this can be neglected for pH samples, for C_T and A_T a 1% volume headspace is necessary (Dickson et al. 2007). The existence of larger bottle headspaces in the sample can alter the values of the carbonate variable measurements (Carter et al. 2013) so must be consistently controlled. Bubble generation should be avoided at all times during the sample collection process as this can compromise sample integrity; almost all the sampling methods explained below emphasize that exchange of CO_2 between the sample and the atmosphere should be avoided as much as possible, meaning care should be taken to avoid this during the sampling procedure. The exception is total alkalinity, which is not affected by air–sea gas exchange (Carter et al. 2014).

It is a common practice in oceanography for multiple tracers to be sampled from the same Niskin bottle. Once recovered, the water inside the Niskin begins to warm up and interact with the headspace within the bottle, which then enlarges as further water is removed. It is thus imperative to sample first for the tracers most affected by gas exchange in order of their sensitivity; for example, transient tracers (helium, chlorofluorocarbons), dissolved oxygen, carbon dioxide, in that sequence, as the time taken for a volume of water to equilibrate to atmospheric levels for CO_2 is longer than for oxygen, which is then longer than for CFCs (Hood et al. 2010). For CO_2, results are not affected as long as samples are taken not more than 10 min after the samples for other dissolved gases (Dickson et al. 2007).

Regarding sample bottle materials, glass is preferred and plastic should be avoided; the best options are high-quality borosilicate bottles for C_T and A_T, and optical glass or quartz for pH. Soda-lime glass should not be used as this contaminates MCS samples.

If discrete samples are not going to be immediately analyzed just after being collected then they should be stored in the dark and refrigerated. If samples are to be stored until a much later analysis (for example, some months) it is necessary to add mercuric chloride ($HgCl_2$) to poison the sample and preserve sample integrity: this preservative eliminates biological activity that would otherwise modify C_T concentrations. The volume of saturated solution of $HgCl_2$ to be added must not exceed 0.1% of the total volume of the sample, with the dilution effect of the poison addition needing to be accounted for in the subsequent calculation process.

1.3   Consistency and Accuracy of Analytical Techniques: The Importance of Certified Reference Materials

The use of Certified Reference Materials (CRM) is the best procedure to maximize analytical accuracy in the laboratory. They also allow for the evaluation of the quality of a laboratory, and help to ensure interlaboratory comparability and consistency. For seawater carbonate chemistry, CRMs are stable seawater samples of very well-known properties that are used to calibrate a chemical analyzer or to validate a measurement process (Riebesell et al. 2010). Their development originated in the acknowledgment that there was a great need to quantify as precisely as possible the role of the ocean in climate change and the uptake of anthropogenic CO_2 from the atmosphere (Dickson 2010a). Since their introduction, increasing research efforts investigating the ocean response to a high atmospheric CO_2 world has led to widespread use and growing demand (Bockmon and Dickson 2015).

The Dickson laboratory (Scripps Institution of Oceanography, University of California San Diego, USA) has been producing and shipping CRMs certified for C_T since 1990 (Dickson 2001), and then also for total alkalinity since 1996 (Dickson et al. 2003). The methodology to certify CRM C_T values is based on vacuum extraction of CO_2 followed by a manometric titration (Wong 1970). A similar methodology as explained in the Alkalinity section is used to certificate the total alkalinity: a two-stage, potentiometric, open-cell titration that uses coulometrically analyzed hydrochloric acid (Dickson et al. 2003). Currently, CRMs are not certified for pH, but they can be also used to assess pH analysis reproducibility (Carter et al. 2013).

In short, ocean CO_2 CRMs are made following this procedure: a large volume of filtered seawater is poisoned with $HgCl_2$ and allowed to equilibrate with atmospheric levels for a number of days while being continuously stirred. Water is then bottled in 500 mL Pyrex® reagent bottles capped with glass stoppers sealed with Apiezon-L grease (Bockmon and Dickson 2015). A reliable CRM batch must ensure homogeneity (by mixing) and stability (checked with at least 3 months of series analysis) and it is expected that CRM remains stable for at least 3 years (Bockmon and Dickson 2015).

The first intercomparison assessments identified differences between laboratories in C_T and A_T of around 20–30 $\mu mol \cdot kg^{-1}$ for analyses of the same water (Poisson et al. 1990). The subsequent widespread use of CRMs has contributed to improve the data quality and the comparability of ocean CO_2 measurements across the globe (Dickson 2010b); the latest worldwide interlaboratory comparison exercise assessing the quality of seawater CO_2 measurements (Bockmon and Dickson 2015) and QUASIMEME showed that laboratories can typically achieve C_T and A_T results that are inside the "weather" goal (± 10 µmol kg^{-1} of target) but very few are able to achieve the "climate" goal (± 2 µmol kg^{-1} of target). The same Bockmon and Dickson (2015) intercomparison also showed that global pH measurements have the least consensus. The authors also warned that the usefulness of CRM as quality control (their original function) can be limited if they are used for calibration purposes.

1.4 Methodologies for the Analytical Determination of Key Marine Carbon System Variables

1.4.1 Dissolved Inorganic Carbon

Definition
Dissolved inorganic carbon (C_T) is the sum of the concentrations of all inorganic carbon species (Eq. 1.4).

$$C_T = [HCO_3{}^-] + [CO_2] + [CO_3{}^{2-}] \tag{1.4}$$

The total inorganic carbon content in a seawater sample can be measured directly by acidifying the sample, extracting the CO_2 gas that is produced and measuring its amount (Fig. 1.3). The result is expressed in µmol·kg^{-1} of seawater.

1.4.1.1 CO$_2$ Extraction

Most C_T analysis techniques share the same procedure of CO_2 extraction by acidification. Generally, a known volume of seawater is supplied to a reaction vessel where it is acidified with acid in excess. As summarized in MCS equations (Fig. 1.1) and the Bjerrum plot (Fig. 1.2), this acidification converts all forms of C_T to gaseous CO_2 that can then be stripped from the water by purging with a CO_2-free inert gas. The liberated CO_2 gas stream is then directed to a detector for quantification, typically after passing through a condenser and/or gas dryer.

The usually recommended acid is phosphoric acid at 8.5% (Dickson and Goyet 1994) prepared from 1/10 dilution of 85% commercial orthophosphoric acid. Some

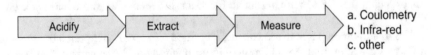

Fig. 1.3 Schematic representation of C_T analysis steps

authors also use a 1/20 dilution of the same commercial orthophosphoric acid (so 4.2%) that offers different advantages: some porous ceramic plates have been found to be affected by concentrated phosphoric acid if it is added directly to an empty stripper, releasing small amounts of CO_2 for a long time. This effect has been observed with 8.5% acid (1/10) but not with 4.2% (1/20) dilution, so we recommend the more diluted option. As an extra benefit, 1/20 dilution is much less aggressive to electromechanical components like piston pumps or valves, reducing failures and technical maintenance in the mid and long term.

Once the acid is mixed with the sample, CO_2 is usually stripped by purging it with N_2 gas flowing through the sample from a porous plate to maximize the extraction surface inside the reaction vessel. The N_2 flow carrying CO_2 and H_2O from the sample is then dried by different procedures, sometimes combined:

1. *Condenser traps.* Peltier plates are used to cool down the gas flow below the dew point. This allows the condensation of water vapor and its separation from the gas flow.
2. *Nafion pipes.* These are tubes permeable to H_2O but not to CO_2; flowing the sample gas inside the pipe while having a countercurrent dry gas flowing over the pipe's external surface allows H_2O vapor to be extracted from the sample gas.
3. *Chemical scrubbers.* Small traps of anhydrous magnesium perchlorate react with H_2O but not with CO_2. It is recommended to use magnesium perchlorate as other desiccants can also trap small amounts of CO_2 (Elia et al. 1986).

The extracted and dried CO_2 stream is then directed to a detector for quantification, of which there are many trusted techniques with the required precision.

1.4.1.2 Infrared Detection

Principle
Since CO_2 absorbs infrared radiation, infrared gas analyzers (IRGA) allow us to determine the CO_2 concentration of a gas flow by comparing the response to a reference gas of known CO_2. Analogously, as H_2O vapor also absorbs infrared radiation it is important to minimize humidity by the mechanisms previously described, and highly desirable for the analyzer to also include an H_2O detector (of which many models do) to be able to discriminate its influence. If the analyzer does not have a humidity sensor, it is necessary to completely dry the gas flow.

Since the IRGA only determines concentration, a precision mass flowmeter is essential to determine the gas flow through the cell of the analyzer at each instant of the analysis. If the flow (in mass), pressure, and temperature of the cell are known then the integration of the equation of state for ideal gases allows us to determine the moles of CO_2 extracted during the analysis.

Even in the absence of precise flow measurements, it is still possible to use this technique if the gas flow is kept constant with high accuracy. In this situation, all other variables are assumed to be constant and only the concentration values are integrated and contrasted with the analysis of reference material to establish a ratio.

¶

Technical Equipment

Currently, two commercial systems are widely known: *DIC Analyzer* from Apollo SciTech, Inc. (http://apolloscitech.com), and *AIRICA* from Marianda (http://www.marianda.com). Both systems mostly share the same scheme and even many commercial parts, as they are developments from a custom original system developed by Gernot Friederich (Friederich et al. 2002) in MBARI. Custom-made systems that use a syringe and multiport valve (Kloehn, Metrohm), a moisture trap and a IRGA analyzer (LiCOR) can also be engineered with some knowledge as they are not highly complex.

Methodological Procedure/Computation and Quality Control

A fixed volume of sample is drawn from the bottle, its temperature recorded, and sent to an acid stripper chamber. Here, it is acidified following the previously indicated CO_2 extraction procedures and bubbling inert gas strips CO_2 from the sample. The resultant gas flow containing H_2O and CO_2 is then forced through moisture traps and finally directed to the IRGA, where all xCO_2 values are recorded during the extraction time, up to a threshold value or up to a decline rate.

Depending on the sensors and features of the equipment used, the C_T can be easily calculated using the ratio of the "area" (sum of all xCO_2 values) of a reference material with known C_T to the "area" of the sample. If more sensors as pressure, temperature, and flowmeter, are available, the amount of CO_2 of each IRGA measurement interval can be calculated by the equation state for ideal gases and the C_T accounted for by summation.

Quality control is performed by comparison with known C_T reference materials.

1.4.1.3 Coulometric Titration

Principle

By definition, coulometry measures the amount of electricity (in coulombs) that is used to convert all of a chemical species to a different chemical state during an electrolysis reaction. For the determination of C_T by coulometric titration (Johnson et al. 1985, 1987, 1993, 1998), CO_2 evolved from a seawater sample is reacted in solution with ethanolamine, and the electrolysis of the byproducts of this reaction are measured by the current passed. In practice, the dried sample gas stream is directed to a coulometer analysis cell (Fig. 1.4) that takes the form of two glass chambers separated by a sintered glass frit of fine porosity. In the larger chamber, 100 ml of cathode solution is added (UIC Inc. CM300–001) [a proprietary mixture of dimethylsulfoxide (DMSO, solvent), tetraethylammonium bromide (electrolyte), ethanolamine (that absorbs CO_2), thymolphthalein (as indicator) and a small amount of water], a platinum cathode, stirrer bar, and the gas inlet tube. In the smaller sidearm chamber ~15 mL of anode solution is added (UIC Inc. CM300–002)

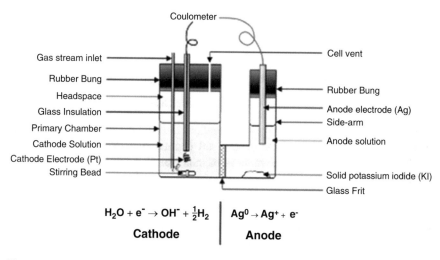

$$H_2O + e^- \rightarrow OH^- + \tfrac{1}{2}H_2 \quad \Big| \quad Ag^0 \rightarrow Ag^+ + e^-$$

Cathode   **Anode**

Fig. 1.4 Schematic of coulometer analysis cell used during CO_2 analysis following Johnson et al. (1998)

[a proprietary saturated solution of potassium iodide in water and DMSO], a small number of potassium iodide crystals and a silver anode.

During sample analysis, the resultant gas stream is bubbled into the bottom of the cathode solution where any CO_2 reacts with the dissolved ethanolamine to form hydroxyethylcarbamic acid (Eq. 1.5). The subsequent decrease in pH (increase in $[H^+]$) generated is tracked photometrically using the thymolphthalein indicator at a wavelength of 610 nm, with the cell solution's typical blue color at pH 10.5 fading to colorless at pH 9.3 as additional CO_2 reacts. The addition of CO_2 thus causes light transmittance through the solution to increase; in order to maintain transmittance constant, the coulometer passes a current between the electrodes, with OH^- ions electrogenerated at the platinum cathode by the reduction of water (Eq. 1.6), and silver ions at the silver anode (Eq. 1.7). The latter silver ions react with the iodide ions in solution in the sidearm chamber (Eq. 1.8), while the hydroxide ions react with the hydrogen ions initially generated in the main chamber (Eq. 1.9). An electric current is applied until the solution transmittance returns to its original value, indicating that the drop in pH caused by the reaction with carbon dioxide has been restored (Eq. 1.10).

$$2CO_2 + 2HO(CH_2)_2NH_2 \rightarrow 2HO(CH_2)NHCOO^- + 2H^+ \qquad (1.5)$$

$$\text{At platinum cathode}: 2H_2O + 2e^- \rightarrow H_2(g) + 2OH^- \qquad (1.6)$$

$$\text{At silver anode}: 2Ag^0(s) \rightarrow 2Ag^+ + 2e^- \qquad (1.7)$$

$$2Ag^+ + 4I^- \ (\text{saturated KI}) \rightarrow 2AgI_2^- \ (\text{at silver anode}) \qquad (1.8)$$

$$2OH^- + 2H^+ \rightarrow 2H_2O \tag{1.9}$$

Overall reaction:

$$2Ag^0(s) + 4I^- + 2CO_2 + 2HO(CH_2)_2NH_2 \rightarrow 2AgI_2^-$$
$$+ H_2(g) + 2HO(CH_2)NHCOO^- \tag{1.10}$$

The coulometer precisely tracks and integrates the total electric current passed to achieve this through coulometer counts and converted with Faraday's constant (96,489 coulombs·mol^{-1}) in the number of moles of CO_2 titrated (see calculation scheme below).

Technical Equipment/Methodological Procedure

Most coulometry-based CO_2 analysis systems are based on the SOMMA (Single Operator Multiparameter Metabolic Analyzer) developed by Kenneth Johnson (Johnson et al. 1985, 1987, 1993, 1998), which combines a sample handling and CO_2 extraction system with a coulometer. Seawater is transferred from the sample bottle to a water-jacketed glass pipette of precisely known volume and temperature through a series of pinch valves and peristaltic pumps connected by silicone tubing, avoiding contact with potential contamination from laboratory air. The sample is then transferred to a gas-tight CO_2-free glass stripper, where the sample is mixed with phosphoric acid to evolve CO_2 that is then purged from solution by a stream of oxygen-free N_2 (that is also CO_2-free, having previously been passed through a soda lime/Ascarite (II) gas scrubber). The sample gas stream is then dried by passing through a Peltier or Liebig condenser cooled to <5 °C, followed by a $Mg(ClO_4)_2$ trap. Finally, the gas is passed through a chemical scrubber (e.g., activated silica ORBO-53 tubes, Supelco Inc.) for the removal of potential interfering acidic gases such as H_2S and Cl_2 before being directed to the coulometer cell. Sample handling/ CO_2 extraction systems are available commercially, for example, from Marianda, Germany (https://www.marianda.com, Mintrop 2004), or UIC Inc., USA (https:// www.uicinc.com/carbon-analysis/) but are often bespoke made, while for coulometers UIC Inc. units continue to be extensively used, as do those from Nippon Ans Co., Japan. A comparison of systems used in global laboratories has found that accuracies of 2 μmol kg^{-1} or better are eminently achievable (Bockmon and Dickson 2015).

Computation and Quality Control

A full description of the calculation pathway is detailed in Johnson et al. (1998) and Dickson et al. (2007, SOP2) and summarized here.

Background: Ideally there should be no contribution to the coulometer signal other than its response to CO_2 derived from the seawater sample. However, as this is not always the case the background titration rate must thus be calculated by tracking the total number of coulometer counts after addition of phosphoric acid to an empty gas stripper, and running the carrier gas through the system to the coulometric cell for 10 min. If outside a specific tolerance this can be used to identify potential leaks in the system or failures in CO_2 removal scrubbers. Direct connection of the carrier gas to the coulometer, bypassing all glassware and the stripper, can be used to determine the gas blank (ideally zero).

Calibration factor: The coulometry method assumes 100% efficiency and that each faraday of electricity passed affects the conversion of one equivalent of a chemical analyte according to Eq. 1.6 (i.e., all electrons get used to create OH^- ions). However, this too is often not the case due to electronic imprecision of the instrument and the non-perfect nature of the chemicals. It is, therefore, recommended to calculate the calibration factor of the coulometer setup regularly (essentially each time a new cell is prepared with fresh solutions). This can be achieved by using a calibrated gas loop of pure CO_2 gas of known temperature and pressure being titrated or using a series of Na_2CO_3 solutions at varied concentrations. A value of counts per mole C can be calculated after correcting for any blank that can be used in the determination of sample concentrations (see below).

Sample Calculation The C_T concentration of a seawater sample is calculated as below (Eq. 1.11):

$$[C_T] = \frac{Cts^{SAMP} - (Cts^{BLK} \cdot t) - Cts^{ACID}}{C_F \cdot V \cdot \rho} \qquad (1.11)$$

where:

- $[C_T]$ is the concentration of dissolved inorganic carbon in the sample (mol kg^{-1}).
- Cts^{SAMP} is the number of coulometer counts for the seawater sample analysis (counts).
- Cts^{BG} is the number of background coulometer count rate (counts min^{-1}).
- t is the sample analysis time.
- Cts^{ACID} is the number of background coulometer counts associated with phosphoric acid addition.
- C_F is the coulometer calibration factor (counts mol·C^{-1}).
- V is the sample pipette volume at a known temperature at time of analysis (dm^3).
- ρ is the sample density at time of analysis (cm^3).

Quality Control

As the ethanolamine in solution is consumed as part of the analysis, the overall efficiency of the cell will eventually deteriorate over time to the point of necessitating fresh chemicals and a new cell to be prepared. This is typically thought to be after a cumulative total of 25–30 mg of carbon (~2–2.5 mmol) has been titrated (Johnson et al. 1998; Dickson et al. 2007). To ensure the highest quality accuracy and precision of analyses it is essential to track and monitor system performance, and several metrics can be used to do this:

- Regular background analyses to ensure stability of system and carrier gas.
- Run time of analyses, where longer analyses indicate a lowering of cell efficiency.
- Regular checking of calibration factor, for which control charts can be set up.
- Regular analysis of certified reference materials, whose outputs can be tracked on control charts.
- Regular duplicate analyses of a single sample, with the difference tracked on control charts.

1.4.2 pH

Definition

pH is a measure of the hydrogen ion concentration [H^+], i.e., the acidity of a solution. It is formally defined as the negative logarithm of the relative activity of [H^+] in aqueous solution (Buck et al. 2002) (Eq. 1.12). Operationally, as hydrogen ion activity cannot be thermodynamically measured in solutions with high ionic strength, such as seawater (Zeebe and Wolf-Gladrow 2001; Paulsen and Dickson 2020), and the [H^+] cannot be established independently without electroneutrality (Rosenberg and Klotz 2008), the activity is thus approximated as the hydrogen ion concentration in moles per kilogram (Eq. 1.13):

$$pH = - \log_{10}(a(H^+)) \tag{1.12}$$

$$pH = - \log_{10}([H^+]) \tag{1.13}$$

If the hydrogen ion concentration considered is only the concentration of the *free* species including its hydrated forms ($[H^+]_F = [H_3O^+] + [H_5O_3^+] + \ldots$), then we are defining the *free* pH scale (Eq. 1.14).

$$pH_F = - \log_{10}([H^+]_F) \tag{1.14}$$

When the hydrogen ion concentration also includes those associated with the sulfate ion ($[HSO_4^-]$) then the *total* pH scale is being considered (Eq. 1.15).

$$pH_T = - \log_{10}([H^+]_f + [HSO_4^-]) \tag{1.15}$$

Finally, if the influence of the fluoride ion ($[HF]$) is also accounted for, then pH is considered to be on the *seawater* scale (SWS) (Eq. 1.16).

$$pH_{SWS} = - \log_{10}([H^+]_f + [HSO_4^-] + [HF]) \tag{1.16}$$

There is a further pH scale, known previously as the NBS scale (now the IUPAC scale, from the International Union of Pure and Applied Chemistry) that is operationally determined from calibrated buffer standard solutions with assigned pH values (Zeebe and Wolf-Gladrow 2001). NBS comes from the US National Bureau of Standards (NBS) from where these buffer standards are derived (now known as the National Institute of Standards and Technology, NIST). This pH scale is not valid in seawater because the ionic strength of seawater ($\sim$0.7 mol·kg^{-1}) is larger than the ionic strength of NBS standard buffer solutions ($\sim$0.1 mol·kg^{-1}) (Waters and Millero 2013)—this produces electromotive potential differences that are thermodynamically not strictly well defined or standardized.

The existence of multiple pH scales in seawater creates *"unnecessary"* complications (Waters and Millero 2013) and hinders understanding of the pH parameter (Pörtner et al. 2010). This problem has its roots in that pH is defined differently in marine chemistry and physiology (Pörtner et al. 2010), two of the main sub-disciplines that undertake research on ocean acidification. Since an oceanographic point of view, we state that:

- The NBS scale is not recommended for seawater pH measurements because it is not well established operationally.
- Conceptually, the free scale is the most simple scale (Dickson 1984) and it has a powerful meaning from a biological perspective since it is the $[H^+]_F$ that affects biological processes (Riebesell et al. 2010). Even so, the uncertainties associated with the sulfate dissociation constant (K^*_{SO4}) and with the standard potential of the reaction in the ionic seawater medium precludes the use of this scale in favor of the Total scale (Dickson 1990; Zeebe and Wolf-Gladrow 2001).
- SWS has been widely used. Its difference with the Total scale is low, salinity dependent ($\sim$0.008–0.011 pH units $33 > S > 36$) and easily parameterized but is not recommended for use because it cannot be thermodynamically established.
- Use of the Total scale for measurement and reporting is the current consensus for best practice in the oceanographic community (Olsen et al. 2020). The Total scale requires buffers prepared in synthetic seawater for calibration (Hansson 1973a) and includes the effect of sulfate ion, therefore avoiding the problems with K^*_{SO4} uncertainty (Paulsen and Dickson 2020). In seawater, the amount of sulfate is large enough to create significant differences between the Total and Free scales, but it is possible to transform a pH value from one scale to another when K^*_{SO4} is known (Zeebe and Wolf-Gladrow 2001).
- Recently it has been argued that in order to fully interrogate the effects of ocean acidification the hydrogen ion concentration ($[H^+]_F$) should be used in addition to pH (Fassbender et al. 2021). This is because pH reflects relative changes in $[H^+]$ rather than absolute changes (Fassbender et al. 2021).

When comparing data from independent measurements it is very important that they share the same scales, as the use of undocumented pH scales in ocean acidification experiments makes intercomparison difficult and leads to uncertainties (Orr et al. 2009). The agreement of the community on a single pH scale, preferably the Total scale, would be advantageous from a good practice perspective.

There are mainly two techniques used in chemical oceanography today for pH analysis: (i) the potentiometric technique where an electrode sensitive to hydrogen ions is combined with a reference electrode and (ii) techniques based on spectrophotometry (see Section *"Spectrophotometric method"*).

The potentiometric determination of pH in seawater consists of pH measurements with glass/reference electrodes in closed potentiometric cells, in order to avoid the exchange of CO_2 between the sample and the atmosphere. It is an inexpensive and nondestructive technique that can provide data at a high reading rate but with the inherent caveats of closed-cell approaches, namely drift and susceptibility to noise (Dickson 1993). To solve these problems frequent calibrations based on synthetic seawater are required (Dickson 1993). However, due to their questionable quality, usage of glass electrodes is discouraged—measurements with this technique will only receive a lower quality grading (3—undefined) by networks such as GOA-ON (IOC 2019), and cannot be included in synthesis databases of hydrographic cruises such as GLODAPv2 (Olsen et al. 2016). Due to the ease of carrying out spectrophotometric measurements of seawater pH, this technique is now the standard practice for oceanography and climate change studies and will be further described below.

1.4.2.1 Spectrophotometric Method

The main principle of the original method by Clayton and Byrne (1993) is operationally described in SOP 6b by Dickson et al. (2007). The latest updates regarding dye uncertainties are also included (Yao et al. 2007; Liu et al. 2011; Carter et al. 2013; Patsavas et al. 2013).

Principle

Measuring $[H^+]_T$ spectrophotometrically using an indicator dye has become the benchmark method in the research of ocean pH since it is a *"simple, fast and precise"* technique that reports precisions of 4×10^{-4} pH units (Clayton and Byrne 1993). The development of the spectrophotometric method led to the rise and establishment of pH as a common oceanographic variable and is the reference method now commonly used across the oceanographic community (Olsen et al. 2020).

The technique is based on the distinctive absorption spectra of sulfonephthalein indicator forms. Indicator dyes of sulfonephthalein are weak acids that show different colors in the acid and the basic form, therefore absorbing light at different wavelengths depending on the pH. The use of indicator dyes in seawater is possible due to the relative constancy of seawater composition and the low concentrations of reactive tracers that keep the physicochemical characteristics of the indicator dyes relatively stable (Robert-Baldo et al. 1985). There are several sulfonephthalein indicators such as phenol red (Robert-Baldo et al. 1985), thymol blue (Byrne 1987; Zhang and Byrne 1996), cresol red (Byrne and Breland 1989; Patsavas et al. 2013), or meta-cresol purple (Clayton and Byrne 1993; Liu et al. 2011). The most appropriate sulfonephthalein indicator for full-depth water column pH profiles in the open ocean is m-cresol purple (mCP).

$$H_2I \rightleftarrows HI^+ + H^+$$

$$HI^- \rightleftarrows I^{2-} + H^+$$

$$pH = pK(HI^-) + \log_{10}\left([I^{2-}]/[HI^-]\right) \tag{1.17}$$

In this pH equation (Eq. 1.17), the first term "$pK(HI^-)$" is the acid dissociation constant for the species HI^-, and the second term "$\log_{10}([I^{2-}]/[HI^-]$" can be computed with spectrophotometric absorbance measurements. The Beer-Lambert law describes the light attenuation in aqueous solutions and specifies the measured absorbance (A_λ) as the product of the molar attenuation coefficient of the attenuating species (ε), the concentration of the attenuating species ($[c]$), and the optical path length (L) (Eq. 1.18):

$$A_\lambda = \varepsilon \cdot [c] \cdot L \tag{1.18}$$

After solving the Beer-Lambert equation for the concentration, it can be combined with Eq. 1.17 to arrive at Eq. 1.19 (the reader interested in the full development of the mathematical theory and algebraic manipulation should see Clayton and Byrne 1993):

$$pH = \log(K_1) + \log\left(\frac{\dfrac{A_1}{A_2} - \dfrac{\varepsilon_1(HI^-)}{\varepsilon_2(HI^-)[HI^-]}}{\dfrac{\varepsilon_1(I^{2-})}{\varepsilon_2(HI^-)} - \dfrac{\left(\dfrac{A_1}{A_2}\right)\varepsilon_2(I^{2-})}{\varepsilon_2(HI^-)}}\right) \tag{1.19}$$

pH is now a function of known values: the pK of the dye, the molar attenuation coefficients for the species (ε_1, ε_2), and the measured absorbances (A_1, A_2) for the wavelengths 1 and 2. Both the dissociation constant of the dye and the molar attenuation coefficients are a function of temperature, salinity, and pressure. When this dependence is included, the final equation is:

$$pH = \frac{1245.69}{T} + 3.8275 + \left(2.11 \times 10^{-3}\right)(35 - S)$$
$$+ \log\left(\frac{R - 0.0069}{2.222 - R\,0.133}\right) \tag{1.20}$$

R in Eq. 1.20 is the ratio of indicator absorbances at their maximum peak for the species I^{2-} and HI^-, corresponding to the optimal wavelengths to measure the spectrophotometric response that gives the maximum absorbance (Dickson 1993). For mCP, the wavelengths of maximum absorbance (Fig. 1.5) occur at 434 nm (for the acid species, HI^-) and 578 nm (for the base species, I^{2-}), respectively (Douglas and Byrne 2017). Therefore, (Eq. 1.21):

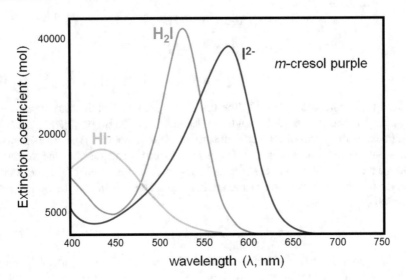

Fig. 1.5 Emission spectrum of the different species for the sulfonephthalein indicator m-cresol purple

$$R = \frac{A_{578}}{A_{434}} \tag{1.21}$$

The reproducibility or analytical precision of the spectrophotometric method with mCP indicator is 0.0004 units of pH (Clayton and Byrne 1993) but the measurement uncertainty can be an order higher due to dye impurities (Yao et al. 2007) and sample handling. Since overall uncertainty includes systematic issues (like for example, the error in assumed pK of dye) along with random contributions (for example, different sample handling between operators), the magnitude of the uncertainty is far larger than the reproducibility.

Estimates of the likely contributions to measurement uncertainty from various sources can be found in Table 1.2 of Carter et al. (2013). A summary of these are:

- Systematic error in the measured absorbances (related to spectrophotometer performance).
- Error in dye impurity correction and/or pK of dye.
- Error in the molar attenuation coefficients (ε).
- Change in pH due to dye addition (0.005).
- Poor sample handling.

Together, the overall combined uncertainty is ~0.0055 pH units (Carter et al. 2013).

Technical Equipment

- *Spectrophotometer*: A high-quality double-beam spectrophotometer is preferred over single beam versions, with good accuracy in wavelength determination and absorbance. If possible, the deuterium lamp of the spectrophotometer should be switched off and only the tungsten lamp used. As instrumental problems can directly impact pH measurements, it is essential to capture as much information as possible in cruise reports or metadata regarding spectrophotometer model and settings, but also about equipment calibration or maintenance (Álvarez et al. 2020).

- *Spectrophotometric cells*: These should be made of optical glass or quartz with a path-length of 10 cm. They are to have two ports in the upper part (so as to be outside the light beam of the spectrophotometer), capped with polytetrafluoroethylene stoppers to avoid CO_2 exchange with laboratory air. It is necessary to have as many cells as the number of samples to be collected from a specific event, plus an additional reference cell that will be inside the spectrophotometer (if it is a double-beam spectrophotometer). The use of a large path-length reduces the final concentration of the dye lessening the impact of dye uncertainty on the pH result (Takeshita et al. 2021).

- *Temperature control system*: A thermostated system is required to keep the samples at a constant temperature of 25.0 $\pm$ 0.1 °C. The system consists of two main elements: thermostatic bath and incubator. The ideal combination is to keep both the samples acclimated at constant temperature and the spectrophotometric cell chamber where the measurement is going to take place. For that, an *in-series* water circuit powered by a single thermostatic bath is needed, or the use of two thermostatic baths. Unfortunately, thermostated compartments for spectrophotometric cells are not usually available commercially, necessitating the use of a custom-made incubator connected to the thermostatic baths where pH cells are acclimated. It is important to keep the thermostated incubator and the samples inside as dry as possible. When no other option is available, samples inside spectrophotometric cells can be introduced in hermetic plastic bags and bathed directly in the thermostatic bath.

- *Micropipette:* To add the mCP dye to the sample cell, typically an adjustable repeater micropipette with dispenser syringe tips. Ideally, the tip is foil wrapped to avoid photolysis of the dye during the measurement session.

- *Indicator dye*: m-Cresol purple solution of known pH. The indicator dye should be stored so as to minimize contact with atmospheric CO_2, for example, using a collapsible container or syringe with zero headspace (Clayton and Byrne 1993). More information about the correct elaboration of the concentrated dye solution can be found in the section *Dye selection and preparation*.

Methodological Procedure

As previously explained in the section *Sampling procedure*, samples for pH are taken directly from the Niskin bottle into spectrophotometric cells. During this process bubble formation and the exchange of CO_2 between the sample and atmosphere should be avoided as much as possible. With the help of a flexible silicon tube, the samples are rinsed with the same seawater several times (4–5 times its total volume), the tube is removed carefully while seawater is still flowing and once filled sample cells are sealed with polytetrafluoroethylene caps. This is to ensure that there is no headspace or bubbles remaining inside the cells.

The amount of time required for the sample cells to reach a constant temperature of 25.0 ± 0.1 °C depends on the initial temperature of the samples and the temperature control system being used, but ~60–90 min is typically sufficient.

Once acclimated at 25 °C, the next step is to measure sample absorbances in the spectrophotometer. The sample cell is placed in the measurement compartment of the spectrophotometer as dry and clean as possible (avoid any contact with the optical windows). The first measurement is a background spectrum, performed without indicator dye at three wavelengths: 434 nm (maximum absorption wavelength for the acid form of the dye), 578 nm (maximum absorption wavelength for the basic form of the dye), and 730 nm (a non-absorbing wavelength for indicator dye that is used for quality control) (Patsavas et al. 2013).

When the first round of spectrophotometric measurements has finished, the indicator dye is added to the cell. One of the upper port caps must be removed and 75 µL of mCP (2 mM) added by the micropipette. The cap is then replaced and the sample cell shaken to homogenize the sample and dye mixture. When this has been achieved and the seawater sample is colored accordingly to its pH value, the sample cell is placed once again in the measurement compartment of the spectrophotometer and the three wavelength measurements are repeated. The optical windows must still be dry and clean.

Dye Selection and Preparation

The liquid dye solution must be prepared from a commercially purchased powder (solid state) indicator. The dye formula recommended is to dissolve 0.08 g·L^{-1} of the water-soluble mCP in a 0.7 M NaCl solution.

Ideally, the pH of the dye solution will be as close as possible to the pH of the samples to be measured. When ranges of pH are going to be measured (for example, in open ocean full-depth profiles), the central point of the expected pH range is an acceptable option. The closer the dye solution pH is to the sample pH the smaller the influence on the acid-base system (see next section, *Delta R*).

Dye impurities are one of the largest sources of uncertainties in the spectrophotometric method (Liu et al. 2011) behind only uncertainties from dissociation constants (Carter et al. 2013). With regard to dye selection, it is preferable to use purified rather than unpurified mCP (Douglas and Byrne 2017) since impurities in the dye solution affect measured absorbances (Yao et al. 2007). As purified mCP (performed via high-performance liquid chromatography (HPLC, Liu et al. 2011)) is not currently offered commercially and its availability is thus not always guaranteed (Álvarez et al. 2020), a corrective method can be applied to improve measurements made using unpurified mCP (Douglas and Byrne 2017): first, the absorbance contribution from mCP impurities for a particular lot of dye is determined, then that impurity absorption is included in the computations to determine the pH of the sample. (See Douglas and Byrne 2017, Table 1.1). This method is an inexpensive solution to improve spectrophotometric pH measurements and allows the retroactive correction of old pH datasets if the dye solution is known/preserved (Liu et al. 2011). Since impurities are unique to individual batches (Yao et al. 2007) even within the same commercial manufacturer (Douglas and Byrne 2017), the impurity absorption assessment must be performed for each lot of unpurified mCP so that the correct adjustment can be applied following sample analysis. This is essential, as the difference between corrected and uncorrected samples can be as large as 0.012–0.025 pH units (Douglas and Byrne 2017).

Delta R

The indicator dye is itself an acid-base system, meaning the addition of the water-soluble sodium salt of m-cresol purple into the sample slightly modifies the seawater pH (Clayton and Byrne 1993; Carter et al. 2013). This perturbation will be proportional to the difference between the sample acidity and the dye acidity, with the magnitude of the perturbation generally in the range of ± 0.003 pH units (Table 1.2 in Carter et al. 2013). With each different batch, it is necessary to correct the unique pH perturbation induced in the seawater sample by addition of the dye.

Using several seawater solutions in the pH range 7.4–8.2, the procedure to calculate the correction consists of repeated measurements with double addition of indicator dye. There are three measurements of the sample cell: (i) without indicator dye as a measure of the baseline, (ii) with 50 µL of dye, and (iii) with 100 µL of dye. The ratio of indicator absorbances (Eq. 1.21) for the first addition (R^{1st}) and for the second addition (R^{2nd}) are then related through the following equation:

$$\frac{\Delta R}{\Delta A} = \frac{R^{2nd} - R^{1st}}{A_{isos}^{2nd} - A_{isos}^{1st}} \tag{1.22}$$

where A_{isos} is the absorbance measured at the isosbestic wavelength for the first and the second dye addition. The isosbestic wavelength, defined as the wavelength at which the absorbance does not change during the chemical reaction (Buck et al. 2002), is 487.6 nm for mCP (Clayton and Byrne 1993) and the absorbance here is additionally used to quantify and correct any possible deviation in the amount of dye added. This has the additional advantage that the volume of dye in each addition does not need to be identical.

Since the right-hand side of Eq. 1.22 can be assumed to be a linear function ($\Delta R/\Delta A = aR^* + b$), the corrected absorbance ratio (R) that shall be used in the main pH computation equation (Eq. 1.20) is:

$$R = R^* - A_{isos}(aR^* + b) \tag{1.23}$$

Even though the impact of this correction on the final result is no larger than 0.0012 pH units when cells with a path-length of 10 cm are used (Chierici et al.

1999), it is the third greatest contribution to pH measurement uncertainty after uncertainties in the pH constants used and dye impurity deviations (Carter et al. 2013).

1.4.3 Total Alkalinity

Definition

Since it was first measured in the nineteenth century (Dickson 1992), the concept of seawater alkalinity has evolved (Dickson 1981; Zeebe and Wolf-Gladrow 2001). Experimental observations in the ocean are mainly based on alkalinity by titration; in contrast, theoretical and geological studies are focused on charge balance, following in the wake of the alkalinity concept often applied in freshwater systems (Middelburg et al. 2020). The neutral charge balance of seawater, the sum of positive minus negative charges, presents a subtle asymmetry. The charge balance of the major or conservative species is slightly positive (0.002 mol·kg^{-1} versus the 0.7 mol·kg^{-1} total ionic concentration), since ions such as bicarbonate, carbonate and borate interact with the ionic equilibrium of the water and can change their concentrations depending on the pH of the seawater. The alkalinity of seawater has been defined as the difference between the sum of major cations species minus the sum of major anions species (major ionic species >1 mg·kg^{-1} that contribute to salinity) (Skirrow 1975) which results in A_T

$$A_T = [HCO_3^-] + 2[CO_3^{2-}] + [B(OH)_4^-] \tag{1.24}$$

If minor species are also included, the definition of total alkalinity is somewhat more complex (Middelburg et al. 2020).

$$A_T = [HCO_3^-] + 2[CO_3^{2-}] + [B(OH)_4^-] + [OH^-] + [H_3SiO_4^-] + [HPO_4^-]$$
$$+ 2[HPO_4^{2-}] + 3[PO_4^{3-}] + [HS^-] + 2[S^{2-}] - [H^+]$$
$$- [HSO_4^-] - [HF] - [NH_4^+] \tag{1.25}$$

In operational terms, alkalinity is determined by potentiometric titration by adding a strong acid, usually HCl, which neutralizes weak acids. The equivalence or neutralization point is determined primarily by the major weak acid, the carbonic acid, at a pH just below 4.5, setting the reference of equivalence pH, where the Titrated Alkalinity (TA) is set to zero. In 1981, Dickson defined alkalinity (A_T) as follows: *"The total alkalinity of a natural water is thus defined as the number of moles of hydrogen ion equivalent to the excess of proton acceptors (bases formed from weak acids with a dissociation constant $K \leq 10^{-4.5}$ at zero ionic strength) over proton donors (acids with $K > 10^{-4.5}$) in one kilogram of sample."* It is conveniently stated in gravimetric units (moles per kilogram of seawater) avoiding any dependence on temperature or pressure. Dickson (1981) set the pK value of 4.5 as a reference to distinguish between hydrogen ion donor components (acids with a dissociation constant pK < 4.5) and hydrogen ion acceptors (pK ≥ 4.5) by matching the reference level to the carbonic acid equivalence point of a titration. This allows an exact definition of A_T to determine the alkalinity of a titration by knowing the dissolved constituent components as well as their dissociation constants. In the Bjerrum diagram (Fig. 1.2), the distribution of acid-base pairs for the carbonate system in seawater can be analyzed.

For the CO_2-H_2O system, at pH = 4.5 carbonic acid is by far the most dominant species and is used as a reference. Referenced to this point, we then arrive at the hydrogen ion balance, a mass balance for hydrogen ions:

$$[H^+] = [HCO_3^-] + 2[CO_3^{2-}] \tag{1.26}$$

The titration alkalinity, that is, excess of hydrogen ion acceptors over donors with respect to carbonic acid, the reference level, is then defined as follows:

$$A_T = [HCO_3^-] + 2[CO_3^{2-}] - [H^+] \tag{1.27}$$

Other acid-base systems can be incorporated into the A_T equation. This will depend on the classification of new chemical species as hydrogen ion donors or acceptors depending on the zero reference of hydrogen ions. The ions of fluoride, sulfate, borate, phosphate, and silicate as well as ammonia and hydrogen sulfide were incorporated by Dickson (1981) to finally result in:

$$A_T = [HCO_3^-] + 2[CO_3^{2-}] + [B(OH)_4^-] + [OH^-] + [HPO_4^{2-}]$$
$$+ 2[PO_4^{3-}] + [H_3SiO_4^-] + [NH_3] + [HS^-] + 2[S^{2-}]$$
$$- [H^+] - [HSO_4^-] - [HF] - [H_3PO_4] \tag{1.28}$$

The difference between this definition of A_T and the basic definition in Eq. 1.24 is that the charge balance is circumscribed to weak acids and bases such as phosphate and ammonium whose presence in seawater is usually much lower than 10 µmol·kg^{-1}, except in confined areas experiencing anthropogenic impact or extraordinary crustal, coastal or continental inputs. On the other hand, the total scale seawater pH also includes the concentration of the hydrogen sulfate ion. For convenience, here we will use the pH$_{SWS}$ scale, which also includes the given HF concentration (Hansson 1973b; Dickson 1981), defined by:

$$[H^+]_{SWS} = [H^+] + [HSO_4^-] + [HF] = [H^+](1 + S_T/K_{SO4H} + F_T/K_{FH}) \tag{1.29}$$

where S_T and F_T are the total concentration of sulfate and fluoride ions, respectively, and K_{SO4H} and K_{FH} are the acid constants of hydrogen sulfate ion and hydrofluoric acid (Hansson 1973b). Hydrogen sulfate and hydrofluoric acid represent a fixed percentage of 26 and 2%, respectively, of $[H+]_{sws}$ at pH > 3.2. For simplicity we will focus on oceanic waters where the presence of hydrogen sulfide is negligible: $[NH_3]$ and $[HS^-]$ are typically so low that they can be neglected in open ocean water. Moreover, phosphate ion concentrations are in the order of a few $\mu mol kg^{-1}$, and the presence of silicate ion is very low due to its weak dissociation constant (pK = 9.7). Eq. 1.28 can thus be simplified to:

$$A_T = \left[HCO_3^-\right] + 2\left[CO_3^{2-}\right] + \left[B(OH)_4^-\right] + \left[OH^-\right] - \left[H^+\right]_{SWS} \qquad (1.30)$$

Many processes affect alkalinity in the ocean, which we can divide into those that affect changes in the pure water itself, and processes that affect the ions dissolved in it. The variability of alkalinity in surface waters presents a very high correlation with salinity (>90%), because oceanographic processes that affect surface salinity by addition or removal of fresh water (evaporation, runoff, or ice formation/melt) also affect alkalinity. Since these processes are very intense in the surface layer of the ocean, it is possible to infer alkalinity by knowing salinity and temperature (Lee et al. 2006) with relatively low uncertainties (~0.4%). To remove this co-variability with salinity, alkalinity can be normalized with respect to salinity ($nA_T = A_T/S*35$) to infer the other biogeochemical processes that affect alkalinity, such as calcite production and dissolution (Carter et al. 2014). The variability of normalized alkalinity in the water column is driven by circulation and biogeochemical processes that induce a positive gradient toward the ocean floor where solid $CaCO_3$ is generally present. This is generically referred to as the alkalinity pump and correlates with the global thermohaline circulation (Broecker and Peng 1982). It consists of the biogeochemical consumption of alkalinity in the upper ocean layer and synthesis of $CaCO_3$ (in its aragonite or calcite isoforms) and its dissolution in the deep layer generating a net vertical downward flux of CO_2.

1.4.3.1 Titration Methodology

The classical techniques for the determination of alkalinity are based on a two-stage titration: i) addition of HCl of known volume and concentration in excess neutralizing the whole alkalinity and removing practically all CO_2 and ii) determination of excess HCl to calculate alkalinity. The determination of excess HCl is possible by direct pH measurement (Anderson and Robinson 1946) using the hydrogen ion activity coefficient (Culberson et al. 1970) or by spectrophotometric measurement using bromocresol green (Yao and Byrne 1998). The recommended standard operating procedure (SOP) designed by Dickson et al. (2007) and defined as SOP3b *"Determination of total alkalinity in seawater using an open-cell titration"* involves the determination of excess HCl through monitoring the pH as HCl is added up to pH 3. Specifically, the seawater sample is first acidified to a pH between 3.5 and 4.0 with a single aliquot of titrant. The solution is then stirred for a period of time to allow for the escape of CO_2 that has evolved, before the titration is continued until a pH of about 3.0 has been reached. The progress of the titration is monitored using a pH glass electrode/reference electrode cell, and the total alkalinity is computed from the titrant volume and e.m.f. measurements using a non-linear least-squares approach that corrects for the reactions with sulfate and fluoride ions. Specifically, SOP3b is designed for the land laboratory where it is possible to weigh the sample. The removal of CO_2 and the pH range of 3.0–3.5 makes the concentration of bicarbonate ion and all other components of Eq. 1.28 residual except for the species of Eq. 1.29. SOP3b is designed to be conducted in a thermostated (25 °C) open system with a calibrated digital thermometer readable to 0.1 °C.

$$\frac{-m_o A_T + m_{HCl} \cdot C_{HCl}}{(m_o + m_{HCl})} = [H^+] + [HSO_4^-] + [HF] \qquad (1.31)$$

Eq. 1.31 is used to estimate A_T from the second stage of titration data by means of a nonlinear least-squares procedure. The pH range of the second stage maintains very low levels of bicarbonate (<0.5 µmol kg^{-1}) and avoids the potential nonlinearity problems of Nernst's Law due to the liquid junction potential, and increased uncertainty of the hydrogen sulfate term.

Mintrop et al. (2000) show that A_T can additionally be measured by three different potentiometric techniques using a single-stage or gradual addition of HCl addition with a precision of better than 1 µmol·kg^{-1}. The first one, a closed-cell titration, was adopted as SOP3a by Dickson et al. (2007). In this, a seawater sample is placed in a closed cell of gravimetrically known volume where it is titrated with a standardized solution of HCl that has been formulated with NaCl to simulate the ionic strength of seawater, and to keep the activity coefficients stable during titration. Both the closed cell and the burette are kept thermostated at 25.0 ± 0.1 °C, with the technique assuming that the C_T remains constant except for the effect of dilution from the HCl solution. The evolution of the titration is monitored by potentiometric pH measurements with a glass electrode and its reference electrode, the gradual addition of HCl to the sample being carried out beyond the carbonic acid end point (pH ~3.2). A computer program controls the titration by recording the volume of HCl added and the e.m.f. data of the electrodes, and it is from these that the alkalinity is calculated using a least-squares procedure based on a nonlinear curve fitting approach or a modified Gran approach (Hansson and Jagner 1973; Bradshaw et al. 1981). The fit allows for several parameters to be derived, including A_T and C_T. The HCl solution is standardized by titrating weighed amounts of Na_2CO_3 dissolved in 0.7 M NaCl solutions (Mintrop et al. 2000; Dickson et al. 2007). The software assumes that nutrients such as phosphate, silicate, and ammonia are negligible and this does not affect the accuracy of alkalinity which is better than 1 µmol·kg^{-1} (Mintrop et al. 2000; Dickson et al. 2007).

A modification of SOP3a is the open-cell titration (volume approx. 120 cm^3) that the commercial VINDTA system uses (Mintrop et al. 2000). This system consists of an open water-jacketed cell, magnetic stirrer, water-jacketed pipette (nominal volume 100 mL), automated burette and pH meter, with the sample, cell, pipette, and burette cylinder maintained at 25.0 ± 0.1 °C using a circulating bath. The cell is rinsed between samples by filling and draining twice with a NaCl solution (very low alkalinity <30 µmol·kg^{-1}). Including this step lengthens the sample-to-sample to 18–20 min. Although the use of the system in the laboratory would have allowed the more accurate gravimetric determination of sample amount to be used, a calibrated pipette is used instead to allow for measurements to be easily conducted at sea. The calculation of the A_T from the titration data is performed using the same script program for curve fitting as for the closed-cell titration described above. The accuracy of the fit was between 0.2 and 0.4 $\mu mol \cdot kg^{-1}$, with the method using the CRM to calibrate the acid factor.

1.4.3.2 Method Considerations

The closed-cell method (SOP3b) requires the ionic composition of the solution to be maintained during the titration so that the equilibrium constants involved do not change within their uncertainty and the A_T can be calculated using a least-squares procedure based on a nonlinear curve fitting. For that, the HCl solution is prepared in 0.7 M NaCl (about 40 $g \cdot L^{-1}$) which implies that an unknown amount of alkalinity is incorporated due to NaCl impurities. Standardization of the HCl solution is normally performed against a well-standardized Na_2CO_3 solution. However, the inclusion of NaCl in the Na_2CO_3 standards to maintain the same ionic conditions as in seawater additionally generates small changes in the A_T. This makes the HCl standardization process somewhat more complex. Dickson et al. (2007) describe in detail the various analytical steps that must be considered.

A land-based laboratory is the ideal operational setting for A_T analysis; here, the closed-cell method is recommended for its accuracy, because the sample is perfectly quantified by weighing and working conditions are safer than at sea. During oceanographic cruises, however, it is common to deal with 100–200 samples per day and rough seas and difficult conditions, often extending over several weeks. Transporting all these samples to a laboratory on land is a major effort both in terms of the volume of material as well as ensuring methods of sample preservation until the time of analysis. Although measuring the amount of sample titrated by volume has a somewhat lower accuracy than measurement by weighing, it does have the advantage of being automated and is easy to perform onboard oceanographic vessels.

To ensure that the large number of measurements carried out under the framework of international ocean observing programs provide equally standardized data, since 1992 Andrew Dickson's lab has been supplying fully calibrated A_T and C_T standards (Dickson 2001). The intention is to have a reference independent of the quality standards applied by each laboratory. Over time it has become a general trend to use this material as a calibration standard "replacing" laboratory-made standards (usually Na_2CO_3). In fact, Mintrop et al. (2000) already practice this protocol, which has been applied by VINDTA users.

Perez and Fraga (1987) proposed a very rapid titration (<3 min) in an open flask to a final pH of 4.4, assuming that C_T loss is negligible during the titration and only changed by a small dilution; this makes it possible to determine the alkalinity titrated from a combination of the pH of the sample, the final pH and the HCl volume of the titration. The method has an advantage in that it does not require the addition of NaCl to the acid to maintain the ionic strength, and it avoids the need to add NaCl to Na_2CO_3 or borax standards. The salinity at the final pH is determined from the

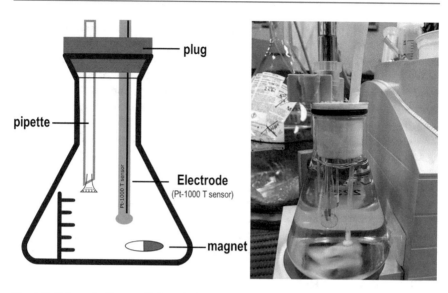

Fig. 1.6 Schematic diagram (left) and photograph (right) of the setup of the Erlenmeyer flask where the titration takes place. The pH electrode is equipped with a Pt-1000 temperature sensor

dilution with the HCl solution and along with temperature, it enables the acid constants of the carbonic and other weak acids to be determined. Standardization with CRMs additionally improves the accuracy, as small C_T losses during sample titration are replicated during CRM titrations, and it allows the method to achieve accuracy levels very similar to those published by Dickson et al. (2007) and Mintrop et al. (2000) described above. The method has been designed for use on the NBS scale, which allows it to be used in coastal waters but here we will show an update for application on the SWS scale. An automatic potentiometric titrator (DOSINO burette and Titrando 809 of Metrohm) with a downward plunger is used, avoiding the annoying accumulation of air bubbles that often interfere with the titration. An AQUATRODE glass electrode combined with a sleeve junction (Metrohm) with an internal temperature sensor allows the titration to be followed with high precision in pH measurement. The potentiometric titration is carried out with HCl (0.1 M) up to a final pH of 4.30 on the SWS scale. The instrument is calibrated using a buffer (pH 4.3) made from potassium phthalate (0.04 M) and sodium tetraborate (borax) (0.008 M) in artificial seawater (DelValls and Dickson 1998) at the final pH of the titration, although synthetic seawater can also be used. A gravimetrically calibrated Knudsen pipette (180 mL) was used to transfer the samples into an Erlenmeyer flask with a total volume of 186 mL (Fig. 1.6).

The surface area exposed to air is less than 4 cm^2 and, for an effective time of less than 2 min, generates a CO_2 loss of less than 0.5%, which allows the alkalinity to be resolved from the initial and final pH of the sample titration. The measurements are carried out in the laboratory at a stable temperature, avoiding thermostated titration cells; this makes it possible to chain one sample after another without wasting time in rinsing, with the remains of the final sample from the previous titration not affecting the next one as they have been exactly neutralized. Samples are acclimatized in the ship-lab for ~12 h to reach thermal stabilization (about lab-temperature $\pm$ 2 °C) before analysis. The titration solution of 0.1 M HCl is prepared by diluting 0.5 moles (18.231 g) of HCl (No. 61710, Titrisol Merck) with high purified MilliQ water in a 5000 $\pm$ 0.5 mL volumetric flask and measuring the temperature of the preparation to refer the molarity to the reference temperature of 20 °C. This preparation itself constitutes a high precision preparation, which can be standardized by titration against a laboratory-weighted solution of Na_2CO_3 (0.01412 moles/L). While at sea, it is convenient to use a substandard seawater, consisting of nutrient-depleted surface seawater in a large, tapped tank (~75 L). This seawater can be compared with the CRM titrations and allows estimation of the drift in A_T that occurs over the course of the day's analysis due to a small electrode drift. The values observed daily at the start of the session after the CRM analyses allow to evaluate the small drift of the substandard water and also to obtain even better results than using only the daily CRM data.

According to Skirrow (1975):

$$C_T = (A_T + D) \cdot E \qquad (1.32)$$

where D and E are the following expressions

$$D = [H^+]_{SWS} - [B(OH)_4^-] - [OH^-] = [H^+]_{SWS}$$
$$- B_T / (1 + [H^+]_{SWS}/K_B) - K'_w/[H^+]_{SWS} \qquad (1.33)$$

and

$$E = \frac{\left(1 + \frac{K'_1}{[H^+]_{SWS}} + \frac{[H^+]_{SWS}}{K'_2}\right)}{\left(1 + 2\frac{[H^+]_{SWS}}{K'_2}\right)} \qquad (1.34)$$

where $[H^+]_{SWS}$ is the total hydrogen ion concentration on the seawater scale (SWS, see Eq. 1.29) and B_T is the total borate concentration in seawater estimated from salinity (Lee et al. 2010). Regarding the equilibrium constants used: K_1' and K_2' are given by Lueker et al. (2000), K_B' by Dickson (1990), and K_W' by Dickson and Riley (1978). As only K_1' has a significant influence on the calculated A_T, an error of 0.01 in pK_1' causes an error of 0.05% in A_T. Assuming no loss of CO_2 during the quick alkalinity titration (except the dilution with the titrant solution) the following expression is obtained by equalizing the initial and final amount of C_T:

$$(A_T + D_i) * E_i = \left(A_T - \frac{N_{HCl} \cdot V_{HCl}}{W} + D_f\right) \cdot E_f \left(1 + \frac{dw_{20}V_{HCl}}{W}\right) \qquad (1.35)$$

$$A_T = \frac{N_{HCl} \cdot V_{HCl}}{W} + \frac{\left(E_i \cdot \left(D_i + \frac{N_{HCl} \cdot V_{HCl}}{W}\right) - D_f \cdot E_f \left(1 + \frac{dw_{20} V_{HCl}}{W}\right)\right)}{\left(\left(1 + \frac{dw_{20} V_{HCl}}{W}\right) \cdot E_f - E_i\right)} \qquad (1.36)$$

where i and f are the initial and final values of the titration, and V_{HCl} is the final volume of HCl solution added by the pH end point of about 4.3 that must be referred to the same temperature as N_{HCl} (20 °C). The initial pH was taken from a pH sample measured by spectrophotometric techniques. N_{HCl} is the molarity of the HCl solution (0.1 M, see above). W is the mass of the pipetted sample calculated using the volume of the pipette and the density as a function of sample salinity and temperature.

1.4.3.3 Practical Example

At sea, alkalinity measurements can be routinely performed using batches of CRM as the reference standard and oligotrophic seawater stored in a large plastic container (50 or 75 L) as a substandard. Once the 0.1 N HCl acid has been prepared (as detailed above) then calibration of the combined pH electrode and its temperature sensor is performed, for which a Knudsen pipette of ~180 mL is used (calibrated gravimetrically on land beforehand). Firstly, a number of titrations are carried out using homogeneous seawater stabilized at laboratory temperature: these are used to get the system operational, removing any bubbles that could have potentially formed in the tubing and HCl plunger and enabling the first checks of system stability and reproducibility. After these checks have been successfully completed, titrations of the substandard water are carried out (~4 repetitions, to verify its repeatability) followed by 3–6 CRM measurements (up to 3 alkalinity measurements from each CRM bottle can be achieved when the pipette is correctly filled). At this point the routine analysis of seawater samples (sampled from Niskin bottles, stored in ~600 mL borosilicate flasks) can be initiated, with up to three replicates performed for each: if the observed replication between the first two analyses is better than 0.1%, then the third replicate is not necessary. Table 1.3 gives a summary of A_T measurements performed during the OVIDE 2018 cruise (A-25 GO-SHIP repeat hydrographic section in the North Atlantic), showing analysis duration, and the number of CRM, substandard, and discrete samples performed during each alkalinity measurement session.

In each session, generally around four stations of up to 24 samples each (so 96 samples in total) were analyzed. Between each measured station (as well as at the end of the session) several substandard measurements (2–4) are performed to evaluate the minor drift in the calibration factor that typically occurs over the course of daily analysis session (8–10 uninterrupted hours). The calibration factor is then adjusted to obtain the nominal value of the corresponding CRM batch, which in this case was CRM batch 166. Table 1.3 shows the repeatability obtained with both the substandard water analyses and the CRM measurements. The repeatability of both CRMs and substandard was on average better than 0.03%. The substandard was very stable overall 10 sessions (~20 days), with a variability of less than 0.05%. These high-quality results were achieved without washing or rinsing the titration flask between samples, simply fully emptying it. The small volume left wetting the

Table 1.3 Summary of alkalinity measurements performed during the OVIDE 2018 campaign (A-25 GO-SHIP in the North Atlantic) from Figueira da Foz (Portugal) to Cape Farewell (Greenland) between 13 Jun and 14 Jul 2018. Shown are the 10 sessions of about 8–10 h duration distributed between Certified Reference Material (CRM) analyses (batch 166 with A_T 2212.56 ± 0.14 µmol-kg⁻¹ and C_T 2053.48 ± 0.30 µmol-kg⁻¹), analyses of substandard seawater consisting of natural saline and oligotrophic surface water stored in a 75 L HDPE container, and analyses of duplicate discrete samples on full bottom-to-surface profiles obtained with a CTD-rosette. The standard deviation (STD) of the differences between duplicates of each sample is shown, as well as the mean time for each analysis (min). The number of CRM analyses and their standard deviation in µmol-kg⁻¹, as well as the time taken in the overall process and the mean time for each CRM analysis are also shown. The number, mean value, and standard deviation in µmol-kg⁻¹ of previously analyzed substandard samples are also shown. The last row shows the mean values or totals for each column

Session	Analysis date (2018)	CRMs				Substandards			Samples		
		#	STD µmol-kg⁻¹	CRM analysis time (min)	Mean individual analysis time (min)	#	A_T µmol-kg⁻¹	STD µmol-kg⁻¹	#	STD µmol-kg⁻¹	Mean analysis time (min)
1	Jun-18	5	0.64	12.9	1.9	5	2348.0	±0.3	97	0.59	2.1
2	Jun-20	3	0.28	8.1	2.0	3	2346.6	±0.2	72	0.64	2.1
3	Jun-22	3	0.06	9.1	2.0	3	2345.4	±0.7	24	0.55	2.1
4	Jun-24	3	0.45	8.6	2.5	3	2345.9	±0.5	92	0.56	2.0
5	Jun-26	3	0.48	7.0	2.1	3	2347.1	±0.3	95	0.47	2.0
6	Jun-28	3	0.43	7.5	2.1	3	2347.3	±0.4	92	0.43	1.8
7	Jun-30	6	0.25	14.4	2.1	3	2346.2	±0.2	85	0.46	1.8
8	Jul-03	6	0.42	15.3	2.1	4	2347.2	±0.3	19	0.57	1.8
9	Jul-04	6	0.23	14.6	2.1	6	2347.1	±0.5	83	0.42	1.7
10	Jul-07	6	0.42	19.1	2.1	4	2347.5	±0.2	87	0.50	1.7
Total / mean		47	0.37	117	2.1	37	2346.8 ± 0.8	0.35	745	0.52	1.9

titration flask is neutralized seawater with very low levels of residual alkalinity, so does not affect the next titration. In this way, only two identical titration flasks were necessary making for easy handling and results.

1.5 Conclusions, Summary, and Future Insights

This compendium of methodologies for the analysis of marine carbonate system (MCS) species and pH can guide the next generation of chemical oceanographers dealing with marine carbon system analysis in seawater. Through these pages high-quality analytical techniques widespread in the chemical oceanography community have been explained, from the working routine inside an oceanographic vessel looking at discrete ocean samples, from the Niskin bottles to the onboard lab, but that can be easily adapted to in-land analysis when needed. They are methodologies relatively easy to reproduce and without excessive expensive technological requirements. As detailed, it is highly recommended to measure more than two carbonate system parameters on each sample whenever possible, to allow full characterization of the MCS. Even when the MCS can be solved by knowing two parameters, the resulting uncertainty in the computed parameters is always larger than the analytical uncertainty of the standard measurement technique. Therefore, when the system is overdetermined (three or more parameters measured) the internal consistency of the results increases. Reporting of results should follow best practice data standards for discrete chemical oceanographic observations since the common adoption of these protocols increases their consistency while promoting data sharing (Jiang et al. 2022).

The widespread use and installation of autonomous sensors for MCS variables like pCO_2 or pH is a current reality, while the onset of sensors for total alkalinity is the next expected stage (Seelmann et al. 2019, 2020). The progress and extension of automated sensors is indeed good news for monitoring ocean and coastal waters within the framework of climate change. This expected large volume of heterogeneous data of seawater carbonate system species and pH needs appropriate calibration and quality control in order to be valuable for validation and big data analytics. That assessment will always rely on discrete measurements performed with analytical techniques, so the subject of skillful reproducibility is quite relevant. Furthermore, routine analytical measurements ensure the detection of malfunctions and drift issues in automated carbon sensors. Appropriate access to state-of-the-art technical equipment is as important as the proper training of technical operators. Finally, data quality and the comparability of ocean CO_2 measurements should not rely exclusively on a single source of CRMs as they do today (mid-2022). International efforts must ensure that a permanent and predictable stock of CRMs is available worldwide and guaranteed with independent and ensured funds.

References

Álvarez M, Fajar NM, Carter BR et al (2020) Global Ocean spectrophotometric pH assessment: consistent inconsistencies. Environ Sci Technol 54(18):10977–10988

Anderson DH, Robinson RJ (1946) Rapid electrometric determination of alkalinity of sea water using glass electrode. Ind Eng Chem–Anal Ed 18:767–769. https://doi.org/10.1021/I560160A011

Bakker DCE, Pfeil B, Landa CS et al (2016) A multi-decade record of high-quality fCO2 data in version 3 of the Surface Ocean CO2 atlas (SOCAT). Earth Syst Sci Data 8:383–413

Bockmon EE, Dickson AG (2015) An inter-laboratory comparison assessing the quality of seawater carbon dioxide measurements. Mar Chem 171:36. https://doi.org/10.1016/j.marchem.2015.02.002

Bradshaw AL, Brewer PG, Shafer DK, Williams RT (1981) Measurements of total carbon dioxide and alkalinity by potentiometric titration in the GEOSECS program. Earth Planet Sci Lett 55: 99–115. https://doi.org/10.1016/0012-821X(81)90090-X

Broecker WS, Peng TH (1982) Tracers in the sea. Eldigio Press, New York

Buck RP, Rondinini S, Covington AK et al (2002) Measurement of pH. Definition, standards, and procedures (IUPAC recommendations 2002). Pure Appl Chem 74:2169–2200

Byrne RH (1987) Standardization of standard buffers by visible spectrometry. Anal Chem 59:1479–1481. https://doi.org/10.1021/AC00137A025

Byrne RH, Breland JA (1989) High precision multiwavelength pH determinations in seawater using cresol red. Deep Sea Res Part A Oceanogr Res Pap 36:803–810. https://doi.org/10.1016/0198-0149(89)90152-0

Carter BR, Radich JA, Doyle HL, Dickson AG (2013) An automated system for spectrophotometric seawater pH measurements. Limnol Oceanogr Methods 11:16–27. https://doi.org/10.4319/lom.2013.11.16

Carter BR, Toggweiler JR, Key RM, Sarmiento JL (2014) Processes determining the marine alkalinity and calcium carbonate saturation state distributions. Biogeosciences 11:7349–7362. https://doi.org/10.5194/bg-11-7349-2014

Chierici M, Fransson A, Anderson LG (1999) Influence of m-cresol purple indicator additions on the pH of seawater samples: correction factors evaluated from a chemical speciation model. Mar Chem 65:281–290. https://doi.org/10.1016/S0304-4203(99)00020-1

Clayton TD, Byrne RH (1993) Spectrophotometric seawater pH measurements: total hydrogen ion concentration scale calibration of m-cresol purple and at-sea results. Deep Res Part I 40:2115–2129. https://doi.org/10.1016/0967-0637(93)90048-8

Culberson C, Pytkowicz R, Hawley J (1970) Seawater alkalinity determination by the pH method. J Mar Res 28:15–21

DelValls TA, Dickson AG (1998) The pH of buffers based on 2-amino-2-hydroxymethyl-1,3-propanediol ('tris') in synthetic sea water. Deep Res Part I Oceanogr Res Pap 45:1541–1554. https://doi.org/10.1016/S0967-0637(98)00019-3

Dickson AG (2010a) The carbon dioxide system in seawater: equilibrium chemistry and measurements. In: Riebesell U, Fabry VJ, Hansson L, Gattuso J-P (eds) Guide to best practices for ocean acidification research and data reporting. Publications Office of the European Union, Luxembourg

Dickson AG (2010b) Standards for ocean measurements. Oceanography 23:34–47

Dickson AG (1984) pH scales and proton-transfer reactions in saline media such as sea water. Geochim Cosmochim Acta 48:2299–2308. https://doi.org/10.1016/0016-7037(84)90225-4

Dickson AG (1981) An exact definition of total alkalinity and a procedure for the estimation of alkalinity and total inorganic carbon from titration data. Deep Sea Res Part A, Oceanogr Res Pap 28:609–623. https://doi.org/10.1016/0198-0149(81)90121-7

Dickson AG (1990) Standard potential of the reaction: AgCl(s) + 1 2H2(g) = ag(s) + HCl(aq), and the standard acidity constant of the ion HSO4- in synthetic sea water from 273.15 to 318.15 K. J Chem Thermodyn 22:113–127. https://doi.org/10.1016/0021-9614(90)90074-Z

Dickson AG (1993) The measurement of sea water pH. Mar Chem 44:131–142. https://doi.org/10.1016/0304-4203(93)90198-W

Dickson AG (1992) The development of the alkalinity concept in marine chemistry. Mar Chem 40:49–63. https://doi.org/10.1016/0304-4203(92)90047-E

Dickson AG (2001) Reference materials for oceanic CO2 measurements. Oceanography 14:21–22

Dickson AG, Afghan JD, Anderson GC (2003) Reference materials for oceanic CO2 analysis: a method for the certification of total alkalinity. Mar Chem 80:185–197. https://doi.org/10.1016/S0304-4203(02)00133-0

Dickson AG, Goyet C (1994) Handbook of methods for the analysis of the various parameters of the carbon dioxide system in sea water. Version 2. United States: N. p., 1994. Web. https://doi.org/10.2172/10107773

Dickson AG, Riley JP (1978) The effect of analytical error on the evaluation of the components of the aquatic carbon-dioxide system. Mar Chem 6:77–85. https://doi.org/10.1016/0304-4203(78)90008-7

Dickson AG, Sabine CL, Christian JR (2007) Guide to best practices for ocean CO2 measurements. PICES Spec Publ 3:191

Douglas NK, Byrne RH (2017) Achieving accurate spectrophotometric pH measurements using unpurified meta-cresol purple. Mar Chem 190:66–72. https://doi.org/10.1016/j.marchem.2017.02.004

Elia M, McDonald T, Crisp A (1986) Errors in measurements of CO2 with the use of drying agents. Clin Chim Acta 158:237–244. https://doi.org/10.1016/0009-8981(86)90287-1

Fassbender AJ, Orr JC, Dickson AG (2021) Technical note: interpreting pH changes. Biogeosciences 18:1407–1415. https://doi.org/10.5194/bg-18-1407-2021

Friederich GE, Walz PM, Burczynski MG, Chavez FP (2002) Inorganic carbon in the central California upwelling system during the 1997–1999 El Niño-La Niña event. Prog Oceanogr 54:185–203

Gattuso J-P, Epitalon J-M, Lavigne H, Orr J (2020) seacarb: seawater carbonate chemistry. https://cran.r-project.org/package=seacarb

Hansson I (1973a) A new set of pH-scales and standard buffers for sea water. Deep Res Oceanogr Abstr 20:479–491. https://doi.org/10.1016/0011-7471(73)90101-0

Hansson I (1973b) A new set of acidity constants for carbonic acid and boric acid in sea water. Deep Res Oceanogr Abstr 20:461–478. https://doi.org/10.1016/0011-7471(73)90100-9

Hansson I, Jagner D (1973) Evaluation of the accuracy of gran plots by means of computer calculations. Application to the potentiometric titration of the total alkalinity and carbonate content in sea water. Anal Chim Acta 65:363–373. https://doi.org/10.1016/S0003-2670(01)82503-4

Hood EM, Sabine CL, Sloyan BM (2010) The GO-SHIP repeat hydrography manual: a collection of expert reports and guidelines. IOCCP Rep 14, ICPO Publ Ser 134

Humphreys M, Lewis E, Sharp J, Pierrot D (2021) PyCO2SYS v1.7: marine carbonate system calculations in python. Geosci Model Dev Discuss 15:1–45. https://doi.org/10.5194/GMD-2021-159

IOC (2019) Indicator methodology for SDG 14.3.1. IOC/EC-LI/2 Annex 6 rev. París

Jiang L-Q, Pierrot D, Wanninkhof R et al (2022) Best practice data standards for discrete chemical oceanographic observations. Front Mar Sci 8:705638. https://doi.org/10.3389/fmars.2021.705638

Johnson KM, Dickson AG, Eischeid G et al (1998) Coulometric total carbon dioxide analysis for marine studies: assessment of the quality of total inorganic carbon measurements made during the US Indian Ocean CO2 survey 1994-1996. Mar Chem 63:21–37. https://doi.org/10.1016/S0304-4203(98)00048-6

Johnson KM, King AE, Sieburth JMN (1985) Coulometric TCO2 analyses for marine studies; an introduction. Mar Chem 16:61–82. https://doi.org/10.1016/0304-4203(85)90028-3

Johnson KM, Sieburth JMN, Williams PJ, Brändström L (1987) Coulometric total carbon dioxide analysis for marine studies: automation and calibration. Mar Chem 21:117–133. https://doi.org/10.1016/0304-4203(87)90033-8

Johnson KM, Wills KD, Butler DB et al (1993) Coulometric total carbon dioxide analysis for marine studies: maximizing the performance of an automated gas extraction system and coulometric detector. Mar Chem 44:167–187. https://doi.org/10.1016/0304-4203(93)90201-X

Lee K, Kim TW, Byrne RH et al (2010) The universal ratio of boron to chlorinity for the North Pacific and North Atlantic oceans. Geochim Cosmochim Acta 74:1801–1811. https://doi.org/10.1016/j.gca.2009.12.027

Lee K, Tong LT, Millero FJ et al (2006) Global relationships of total alkalinity with salinity and temperature in surface waters of the world's oceans. Geophys Res Lett 33. https://doi.org/10.1029/2006GL027207

Liu X, Patsavas MC, Byrne RH (2011) Purification and characterization of meta-cresol purple for spectrophotometric seawater ph measurements. Environ Sci Technol 45:4862–4868. https://doi.org/10.1021/es200665d

Lueker TJ, Dickson AG, Keeling CD (2000) Ocean pCO2 calculated from dissolved inorganic carbon, alkalinity, and equations for K1 and K2: validation based on laboratory measurements of CO2 in gas and seawater at equilibrium. Mar Chem 70:105–119. https://doi.org/10.1016/S0304-4203(00)00022-0

Middelburg JJ, Soetaert K, Hagens M (2020) Ocean alkalinity, buffering and biogeochemical processes. Rev Geophys 58:e2019RG000681

Mintrop L (2004) Versatile instruments for the determination of titration alkalinity. Manual for versions 3S and 3C

Mintrop L, Pérez FF, González-Dávila M et al (2000) Alkalinity determination by potentiometry: Intercalibration using three different methods. Ciencias Mar 26(1):23–37. https://doi.org/10.7773/cm.v26i1.573

Newton JA, Feely RA, Jewett EB, et al (2015) Global Ocean acidification observing network: requirements and governance plan, Second Edition

Olsen A, Key RM, Van Heuven S et al (2016) The global ocean data analysis project version 2 (GLODAPv2)–an internally consistent data product for the world ocean. Earth Syst Sci Data 8:297–323. https://doi.org/10.5194/essd-8-297-2016

Olsen A, Lange N, Key RM et al (2020) GLODAPv2.2020–the second update of GLODAPv2. Earth Syst Sci Data Discuss 2020:1–41. https://doi.org/10.5194/essd-2020-165

Orr JC, Caldeira K, Fabry V et al (2009) Research priorities for understanding ocean acidification: summary from the second symposium on the ocean in a high-co2 world. Oceanography 22:182–189. https://doi.org/10.5670/oceanog.2009.107

Orr JC, Epitalon JM (2015) Improved routines to model the ocean carbonate system: Mocsy 2.0. Geosci Model Dev 8:485–499. https://doi.org/10.5194/gmd-8-485-2015

Orr JC, Epitalon JM, Dickson AG, Gattuso JP (2018) Routine uncertainty propagation for the marine carbon dioxide system. Mar Chem 207:84–107. https://doi.org/10.1016/j.marchem.2018.10.006

Orr JC, Epitalon JM, Gattuso JP (2015) Comparison of ten packages that compute ocean carbonate chemistry. Biogeosciences 12:1483–1510. https://doi.org/10.5194/bg-12-1483-2015

Patsavas MC, Byrne RH, Liu X (2013) Purification of meta-cresol purple and cresol red by flash chromatography: procedures for ensuring accurate spectrophotometric seawater pH measurements. Mar Chem 150:19–24. https://doi.org/10.1016/j.marchem.2013.01.004

Paulsen M-L, Dickson AG (2020) Preparation of 2-amino-2-hydroxymethyl-1,3-propanediol (TRIS) pHT buffers in synthetic seawater. Limnol Oceanogr Methods 18:504–515. https://doi.org/10.1002/LOM3.10383

Pelletier GJ, Lewis E, Wallace DWR (2007) CO2SYS.XLS: a calculator for the CO2 system in seawater for Microsoft Excel/VBA. Washingt. State Dep. Ecol. Natl. Lab. Olympia, WA/Upton, NY, USA–Open Access Libr

Perez FF, Fraga F (1987) A precise and rapid analytical procedure for alkalinity determination. Mar Chem 21:169–182. https://doi.org/10.1016/0304-4203(87)90037-5

Pérez FF, Fraga F (1987) Association constant of fluoride and hydrogen ions in seawater. Mar Chem 21:161–168. https://doi.org/10.1016/0304-4203(87)90036-3

Pimenta AR, Grear JS (2018) Guidelines for measuring changes in seawater ph and associated carbonate chemistry in coastal environments of the eastern United States

Poisson A, Culkin F, Ridout P (1990) Intercomparison of CO2 measurements. Deep Sea Res Part A Oceanogr Res Pap 37:1647–1650. https://doi.org/10.1016/0198-0149(90)90067-6

Pörtner H-O, Dickson A, Gattuso J-P (2010) Terminology and units for parameters relevant to the carbonate system. Guid to best Pract Ocean Acidif res data reporting, ed by Riebesell, Ulf, Fabry, V J, Hansson, L Gattuso, J-P Publ off Eur union, Luxemb pp 167-180

Raimondi L, Matthews JBR, Atamanchuk D et al (2019) The internal consistency of the marine carbon dioxide system for high latitude shipboard and in situ monitoring. Mar Chem 213:49–70. https://doi.org/10.1016/j.marchem.2019.03.001

Riebesell U, Fabry VJ, Hansson L (2010) Guide to best practices for ocean acidification research and data reporting. Eur Comm

Robert-Baldo GL, Morris MJ, Byrne RH (1985) Spectrophotometric determination of seawater pH using phenol red. Anal Chem 57:2564–2567. https://doi.org/10.1021/AC00290A030

Rosenberg RM, Klotz IM (2008) Chemical thermodynamics : basic concepts and methods. 563

Seelmann K, Aßmann S, Körtzinger A (2019) Characterization of a novel autonomous analyzer for seawater total alkalinity: results from laboratory and field tests. Limnol Oceanogr Methods 17: 515–532. https://doi.org/10.1002/lom3.10329

Seelmann K, Steinhoff T, Aßmann S, Körtzinger A (2020) Enhance Ocean carbon observations: successful implementation of a novel autonomous Total alkalinity analyzer on a ship of opportunity. Front Mar Sci 7. https://doi.org/10.3389/fmars.2020.571301

Sharp JD, Pierrot D, Humphreys MP, et al (2020) CO2SYSv3 for MATLAB

Skirrow G (1975) The dissolved gases-carbon dioxide. Chem Oceanogr 2

Sulpis O, Lauvset SK, Hagens M (2020) Current estimates of K_1^* and K_2^* appear inconsistent with measured CO2 system parameters in cold oceanic regions. Ocean Sci 16:847–862. https://doi.org/10.5194/os-16-847-2020

Takeshita Y, Warren JK, Liu X et al (2021) Consistency and stability of purified meta-cresol purple for spectrophotometric pH measurements in seawater. Mar Chem 236. https://doi.org/10.1016/j.marchem.2021.104018

Van Heuven S, Pierrot D, Rae JWB, et al (2011) CO2SYS v 1.1, MATLAB program developed for CO2 system calculations. ORNL/CDIAC-105b Carbon Dioxide Inf Anal Center, Oak Ridge Natl Lab US DoE, Oak Ridge, TN. doi:https://doi.org/10.1017/CBO9781107415324.004

Waters JF, Millero FJ (2013) The free proton concentration scale for seawater pH. Mar Chem 149: 8–22. https://doi.org/10.1016/J.MARCHEM.2012.11.003

Wong CS (1970) Quantitative analysis of total carbon dioxide in sea water: a new extraction method. Deep Sea Res Oceanogr Abstr 17:9–17. https://doi.org/10.1016/0011-7471(70)90084-7

Yao W, Byrne RH (1998) Simplified seawater alkalinity analysis: use of linear array spectrometers. Deep Res Part I Oceanogr Res Pap 45. https://doi.org/10.1016/S0967-0637(98)00018-1

Yao W, Liu X, Byrne RH (2007) Impurities in indicators used for spectrophotometric seawater pH measurements: assessment and remedies. Mar Chem 107. https://doi.org/10.1016/j.marchem.2007.06.012

Zeebe RE, Wolf-Gladrow D (2001) CO_2 in seawater: equilibrium, kinetics, isotopes. Chapter 1 Equilibrium. Elsevier Oceanogr Ser

Zhang H, Byrne RH (1996) Spectrophotometric pH measurements of surface seawater at in-situ conditions: absorbance and protonation behavior of thymol blue. Mar Chem 52:17–25. https://doi.org/10.1016/0304-4203(95)00076-3

Dissolved Organic Matter

2

Xosé Antón Álvarez-Salgado, Mar Nieto-Cid, and Pamela E. Rossel

Contents

Abstract

The ocean is a salty and very diluted broth of organic matter that contains about 680 Pg C. More than 97% of this organic matter is in the dissolved form. This huge amount of reduced carbon, comparable to the CO_2 accumulated in the atmosphere, results in an average concentration of 0.48 ppm (or 40.3 μmol kg^{-1}) when divided by the world ocean volume. Comparatively, salts add up to about 35,000 ppm. Low concentrations, extreme complexity of an organic matrix formed by myriads of compounds and vast interferences with the salt matrix are the main challenges associated with the accurate determination of dissolved

X. A. Álvarez-Salgado (✉)
CSIC Instituto de Investigaciones Marinas, Vigo, Spain
e-mail: xsalgado@iim.csic.es

M. Nieto-Cid
CSIC Instituto de Investigaciones Marinas, Vigo, Spain

CSIC, CNIEO, COAC, A Coruña, Spain
e-mail: mar.nietocid@ieo.csic.es

P. E. Rossel
ICBM-MPI Bridging Group for Marine Geochemistry, Institute for Chemistry and Biology of the Marine Environment (ICBM), University of Oldenburg, Oldenburg, Germany
e-mail: prossel@gfz-potsdam.de

organic matter (DOM) in the marine environment. In this chapter we aim to provide a useful overview of the procedures to effectively sample ("clean lab" protocols), process (filtration, ultrafiltration, solid-phase extraction), preserve (acidification, freezing, freeze-drying) and analyse the bulk DOM pool. Analytical methods for the characterization of its elemental (C, N, P), optical (absorption and fluorescence spectroscopy) and emerging molecular property (FT-ICR-MS, NMR) composition will be described. Out of the scope of this chapter is the determination of individual compounds (e.g. amino acids, carbohydrates) as well as the isotope characterization of DOM.

Keywords

Dissolved organic matter · Isolation techniques · Elemental (C:N:P) composition · Absorption spectroscopy · Fluorescence spectroscopy · Nuclear magnetic resonance · Mass spectrometry

---

2.1 Introduction

Organic matter in the ocean, ca. 680 Pg C (1 Pg C = 10^{15} g C), is made of a minor fraction (<3 Pg C) of living forms and a major fraction of non-living (>677 Pg C) materials (Hansell et al. 2009; Perdue and Benner 2009). Both living and non-living organic matter cover a size continuum which ranges from microbes to whales in the case of the living forms and from small organic monomers (ca. 0.0001 μm) to large particle aggregates in the case of the non-living forms (ca. 1000 μm; Fig. 2.1a; Verdugo 2012, Álvarez-Salgado and Arístegui 2015). Separation between living and

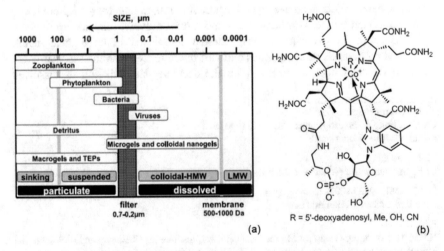

Fig. 2.1 (a) Size continuum of organic matter in the ocean. (b) Structure of B_{12} vitamin. Panel a reproduced from Álvarez-Salgado and Arístegui (2015) with permission of the Intergovernmental Oceanographic Commission of UNESCO

non-living organic matter is conceptually clear but operationally unapproachable because both are intimately linked through the microbial loop (Sherr and Sherr 1988; Azam 1998). It is also easy to conceptually separate the dissolved and particulate fractions of organic matter by stating that dissolved organic matter (DOM) is not affected by gravity acceleration and, therefore, is not prone to sedimentation, while particulate organic matter (POM) is. Operationally, POM is differentiated from DOM as the fraction of marine organic matter retained by a filter having a pore size from 0.2 μm (polycarbonate, polyethersulfone, or aluminium) to 0.7 μm (typically glass fibre) (Repeta 2015). GF/F filters (0.7 μm equivalent pore size) are preferred by biogeochemists because they are inorganic, can be pre-combusted at 450 °C to eliminate organic contamination traces and present excellent flow characteristics. However, it should be noted that about half of the bacteria ends in the DOM fraction when using these filters (Fig. 2.1a). Given the minor contribution of bacterial biomass to the organic matter pool, this interference is unimportant for the chemical analysis of DOM. However, for microbiological purposes, 0.1–0.2 μm pore-size filters are preferred.

POM, which represents 18 Pg C (Perdue and Benner 2009), is classified into a suspended (<100 μm) and a sinking (>100 μm) fraction (Fig. 2.1a). Suspended POM is commonly collected by filtration through GF/F filters of water sampled with oceanographic bottles (0.5–10 L are usually filtered) or in situ pumps, either stand-alone (100–1000 L) or ship-electricity powered (ca. 10,000 L) (McDonnell et al. 2015). Different types of sediment traps are used to sample sinking POM at fixed depths, usually from below the epipelagic or mixed layer to a few meters over the seabed (McDonnell et al. 2015). There are also Lagrangian traps that drift with the surrounding currents at a fixed depth or density level (neutrally buoyant traps; Buesseler et al. 2007). Furthermore, the "marine snow catcher" allows collecting both suspended and sinking POM (Riley et al. 2012).

DOM, which amounts 662 Pg C, is classified into "truly dissolved" (low molecular weight DOM, LMW-DOM) and colloidal (high molecular weight DOM, HMW-DOM), operationally separated by ultrafiltration through a 1 kDa membrane (Fig. 2.1a). B_{12} vitamin, with a MW size of 1.35 kDa and commonly used to test the efficiency of the 1 kDa membranes, provides an idea of what LMW- and HMW-DOM represent (Fig. 2.1b).

The focus of this chapter is on the chemical characterization of the major and less understood organic matter pool in the oceans: DOM. The DOM pool is recognized among the most complex and heterogeneous organic mixtures (Mopper et al. 2007; Dittmar and Stubbins 2014). It consists of myriads (ca. 10^4) of organic molecules (Stenson et al. 2003; Koch et al. 2005; Hertkorn et al. 2006; Riedel and Dittmar 2014) of contrasting chemical structures (e.g. monomers vs. polymers, saturated vs. unsaturated, aliphatic vs. aromatic), origins (e.g. soil vs. aquatic, terrestrial vs. marine, anthropogenic vs. natural), biogeochemical functions (e.g. carbon recycling, export or sequestration, pH buffering and metal binding capacity, UVR protection) and half-lives expanding from minutes to tens of thousands of years (Hansell 2013; Dittmar 2015; Repeta 2015).

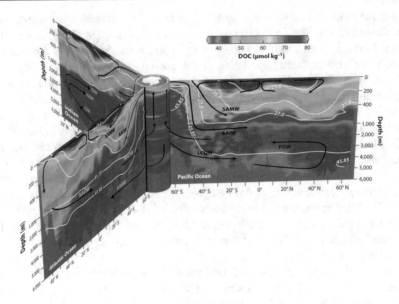

Fig. 2.2 Dissolved organic carbon (DOC) distribution in the world ocean. Isopycnals, main water masses and large-scale circulation patterns are also shown. Reproduced from Hansell et al. (2009) with permission of the Oceanography Society

Despite the huge amount of dissolved organic carbon (DOC) in the oceans (662 Pg C), when divided by the enormous world ocean volume ($1.37 \cdot 10^{18}$ m^3), it results in an average concentration of just 0.48 ppm C (or 40.3 µmol kg^{-1}). The lowest concentrations are found in the deep world ocean (>1000 m), ranging from 48 µmol kg^{-1} in the northern North Atlantic to 34 µmol kg^{-1} in the North Pacific (Fig. 2.2; Hansell et al. 2009). Conversely, higher concentrations are found in the ocean surface, particularly in subtropical gyres, where DOC values of 70–80 µmol kg^{-1} have been recorded. Concentrations can increase considerably in ocean margins, enclosed seas, continental shelves, coastal inlets and estuaries. Hence, surface DOC concentrations in ocean margins are around 100 µmol kg^{-1}, generally increasing above 300 µmol kg^{-1} in estuaries (Barrón and Duarte 2015). In the Black Sea, DOC levels are >200 µmol kg^{-1} in the surface layer and about 120 µmol kg^{-1} in subsurface anoxic water (Margolin et al. 2016). In the open Baltic Sea, DOC ranges from 260 to 480 µmol kg^{-1} in the surface layer and 200 to 330 µmol kg^{-1} in the subhalocline layer (Hoikkala et al. 2015). Conversely, other enclosed basins such as the Mediterranean Sea (Catalá et al. 2018) and the Red Sea (Calleja et al. 2018) present DOC distributions similar to the open world ocean.

Compared with these concentrations of DOC (0.4–4 ppm C), the amount of salts (35,000 ppm) is 4–5 orders of magnitude higher. Low concentrations of DOC, particularly in the deep world ocean, require clean procedures for the collection, filtration and conservation of the samples as well as during their subsequent analyses. The salt content of marine waters may affect DOM analyses in different ways, even forcing to isolate the organic matter from the salt matrix to carry out certain determinations. In this chapter we aim to provide a useful overview of the procedures to effectively sample ("clean lab" protocols), process (filtration, ultrafiltration, solid-phase extraction), preserve (acidification, freezing, freeze-drying) and analyse the bulk DOM pool. Analytical methods to characterize its elemental (carbon, nitrogen, phosphorus), optical (absorption and fluorescence spectroscopy) and emerging molecular property (FT-ICR-MS, NMR) composition will be described. Elemental, optical and molecular characterization are all key tools to disentangle the origin, fate and roles played by marine DOM in ocean biogeochemical cycles. In this chapter, we will not focus on the determination of individual organic compounds of natural (e.g. CH_4, DMSO, amino acids, carbohydrates) or anthropogenic (e.g. CFCs, PAHs, emerging contaminants) origin present in seawater.

2.2   Sample Collection and Preservation

Due to the complex nature of the marine DOM, composed by myriads of different organic compounds in very low concentrations, there are numerous potential contamination sources. Contamination sources could derive from materials used for collecting, processing and storing the samples or introduced during manipulation of volatile organic compounds in the same area where the samples are processed or measured or even by the own body of the analyst (Cauwet 1999; Mopper and Qian 2006). In this regard, materials compatible with DOM analyses, such as glass, Teflon, aluminium foil, stainless steel or high-quality polycarbonate/polypropylene (just for short-term handling), must be used; contact with organic-based supplies such as parafilm, silicone/vacuum greases, silicon tubes or organic solvents must be avoided; and powder-free vinyl gloves to prevent contamination from the fingertips grease, composed of fatty acids, must be worn (Cauwet 1999, Mopper and Qian 2006). In addition, a proper cleaning of the material is required: Pyrex glass, quartz, aluminium foil or glass/quartz fibre filters can be combusted at 450 °C for 4–12 h, while Teflon, polycarbonate/polypropylene and volumetric and delicate glass must be acid-cleaned with 1% or 0.1 M HCl and rinsed with ultrapure water. A previous wash with an oxidant solution (such as 1% commercial plain bleach) helps to eliminate persistent organic residues, followed by a throughout rinse with ultrapure water before the acid cleaning step to avoid undesired chlorine formation.

Sampling is the first step towards the characterization of marine DOM and must be accomplished according to the type of study to be conducted: bulk or molecular analysis of DOM. The latter involves the concentration of DOM isolated from seawater, which requires more processing stages, therefore potentially adding more contamination and influencing bulk DOM representativeness. While a bulk DOM analysis requires roughly 10–100 mL of sample, a molecular DOM analysis would demand from half to hundreds of litres of seawater (depending on the bulk DOC concentration), so sample collection is quite different.

Samples that are not measured right after collection need to be preserved because many organic compounds are chemically and/or biologically labile, i.e. sensitive to microbial degradation but also may be affected by photodegradation. Sample contamination/damage cannot be amended afterwards, hence the importance of these initial steps (Wurl 2009).

2.2.1 Sampling, Processing and Preservation for Bulk DOM Analyses

Seawater for bulk DOM analyses should be collected directly from the rosette rubber-free oceanographic bottles or the underwater non-toxic supply of the research vessel, avoiding the use of tubes or plastic connections, and onto acid-cleaned all-glass flasks with ground glass joints (without grease). To analyse the dissolved fraction, samples must be filtered only in all-glass/Teflon devices under positive pressure to keep the contamination to a minimum. The most common systems are (a) acid-cleaned all-glass and stainless steel syringes, with a Teflon tipped piston, coupled in-line to a Sartorius-type filter, for small volumes (Fig. 2.3a), and (b) acid-cleaned all-glass filtration systems working with low positive pressure of a high purity gas (N_2 or air), equipped with simple disc filters, for volumes up to 1 L (Fig. 2.3b). The filtrate can be collected directly in the measuring flask/cell when using these positive pressure filtration systems. On the contrary, during vacuum filtration the filtrate is collected in a Kitasato filtering flask and then transferred to the measuring flask/cell, increasing the risk of potential contamination of the sample. Therefore, vacuum filtration should be avoided. The preferred filters for DOM analyses are glass fibre filters (Whatman GF/F, nominal pore size of 0.7 μm) as they can be easily cleaned by combustion (450 °C, 4 h) and have a good rate flow (Cauwet 1999; Repeta 2015). For oceanic samples below 200 m depth, filtration procedures are regularly excluded to avoid the potential contamination. At these water depths, particulate organic carbon (POC) represents <5% and usually <1% of total organic carbon, and larger particles are not represented in the small sample volumes needed for bulk DOM measurements (Benner 2002; Mopper and Qian 2006). For the analysis of the elemental composition (C:N:P) of DOM, the filtrate can be collected in heat-sealed pre-combusted (450 °C, 12 h) glass ampoules (10–20 mL) and/or pre-combusted (450 °C, 12 h) glass vials with acid-cleaned Teflon lined caps (30 mL). Glass ampoules have been used for the analysis of DOC and total dissolved nitrogen (TDN) by high temperature combustion (see Sect. 2.3.1) although lately the use of glass vials with Teflon lined caps is getting more popular. Samples for DOC, TDN and total dissolved phosphorus (TDP) analyses by wet oxidation should be collected in glass vials. Samples for DOC analysis by wet oxidation or DOC/TDN by high temperature oxidation can be preserved by acidifying to pH <2 with a strong acid (using HCl or H_3PO_4) and refrigerated at 4 °C until analysis. They can also be preserved by freezing the vials vertically, avoiding contact between the water sample and the cap at −20 °C to −80 °C. The latter also applies to TDN and TDP by wet oxidation. Mercury chloride (Hg_2Cl_2) is not recommended to preserve the filtrate as it can affect the performance of some analytical methods (e.g. the Pt-catalyst and Cu-based halogen trap for

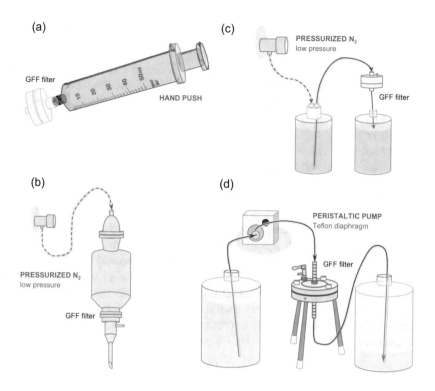

Fig. 2.3 Schemes of DOM filtration systems: (**a**) All-glass and stainless steel syringe, for small volumes; (**b**) all-glass system with N_2 positive pressure, for sample volumes up to 1 L; (**c**) all-Teflon housing filters with N_2 positive pressure for volumes up to 5 L; and (**d**) stainless-steel holder (or capsule filter system), driven with Teflon diaphragm peristaltic pump, for large volume samples (up to hundreds of litres)

DOC/TDN analyses by high-temperature oxidation or the Cu/Cd column for TDN analysis by wet oxidation). For the analyses of the optical properties of DOM (Sect. 2.3.2), the filtrate cannot be acidified or preserved with Hg_2Cl_2 since absorption and fluorescence properties of DOM are pH and Hg^+ dependent and freezing is discouraged. Therefore, samples should be maintained in the dark at 4 °C until analyses within 24–48 h of collection to prevent the alteration/spoiling of the sample.

2.2.2 Sampling Processing and Preservation for Molecular DOM Analyses

Molecular DOM analyses require large sample volumes and a time-consuming and laborious step of isolation/concentration of the samples prior to analysis. Therefore, a much lower number of samples can be processed compared with those for bulk DOM analyses. Seawater for molecular DOM analyses should be collected directly from the rosette rubber-free oceanographic bottles or the underwater non-toxic supply of the research vessel onto acid-cleaned Teflon bottles, or acid-cleaned high-quality polycarbonate/polypropylene bottles for short-term contact (for sample volumes <10 L) or acid-cleaned Teflon-coated carboys (for sample volumes >10 L). The previously described all-glass filtration systems (Fig. 2.3a, b) are not suitable for these large-volume samples, so the cleanest options are (a) an acid-cleaned Teflon filter-housing plus tubes operated with positive pressure of high purity N_2 or air for 1–5 L samples (Fig. 2.3c) or (b) a 90–142 mm $\varnothing$ stainless steel filter holder or a high-volume capsule filter driven by a Teflon diaphragm pump for very large sample volumes (Fig. 2.3d). Combusted (450 °C, 4 h) glass fibre filters (Whatman GF/F, nominal pore size of 0.7 μm) of different diameters are preferentially used with filter housings/holders, and they can also be exploited for POM analysis. Capsule filters must be previously washed with several litres of ultrapure water and checked out for contamination leaking. The same kind of bottles/carboys are used to collect the filtrate before the isolation step, while the final DOM extracts are kept in acid-cleaned Teflon bottles (ultrafiltration procedures) or pre-combusted (450 °C, 12 h) glass vials with acid-cleaned Teflon caps (solid-phase extraction) (see Sect. 2.4). The preservation of the ultrafiltration concentrates can be completed by freeze-drying the sample, obtaining a solid extract of DOM. Only the solid-phase extraction and ultrafiltration steps to concentrate and isolate DOM for molecular analyses can and should be conducted on board.

2.3   Bulk DOM Characterization

2.3.1   Elemental Analyses

Accurate knowledge of the C:N:P stoichiometry of DOM is central to understanding the role played by this pool in ocean biogeochemical cycles. The bioavailable fraction of DOM is characterized by C:N:P molar ratios around [200, 300]: [20, 25]: 1 (Fig. 2.4a; Benner 2002, Hopkinson and Vallino 2005, Lønborg and Álvarez-Salgado 2012), relatively close to the products of synthesis and early degradation of marine phytoplankton, 106: 16: 1 (Anderson 1995). Conversely, the refractory fraction of DOM, with C:N:P molar ratios around [2800, 3500]: [120, 200]: 1 (Meybeck 1982; Hopkinson and Vallino 2005; Lønborg and Álvarez-Salgado 2012) is extremely depleted in N and P, the limiting nutrients for phytoplankton and heterotrophic bacterial growth. While the bioavailable fraction of DOM plays an active role in ocean biogeochemical cycles participating in the recycling, lateral and vertical export and degradation processes of C, N and P, the refractory fraction contributes to the long-term immobilization or sequestration of these elements in the oceans (Fig. 2.2; Hansell 2013).

The contribution of the bioavailable fraction to the bulk DOM pool varies considerably with season or trophic status, from ocean surface to bottom or from the open ocean to the coast. Therefore, the C:N:P molar ratio of the bulk DOM pool could be a suitable indicator of bioavailability. In this regard, while only 20% of the

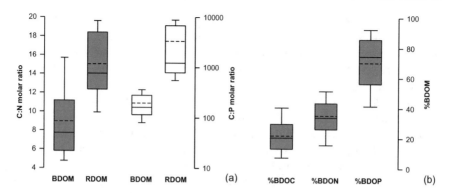

Fig. 2.4 (a) C/N and C/P molar ratios of the bioavailable (BDOM) and refractory (RDOM) fractions of DOM in the global coastal ocean. (b) Portion of the DOC, DON and DOP pools that is bioavailable in the global coastal ocean. After Lønborg and Álvarez-Salgado (2012)

bulk DOC is bioavailable in the coastal ocean, 35% and as much as 70% of the bulk DON and DOP are bioavailable, respectively (Fig. 2.4b). Furthermore, in surface waters, particularly under oligotrophic conditions, dissolved organic nitrogen (DON) and phosphorus (DOP) constitute the major reservoirs of these limiting elements (Karl and Björkman 2015; Sipler and Bronk 2015).

The analytical methods for the determination of DOC, DON and DOP are generally based on the oxidation of DOM to quantitatively produce CO_2, NO_3^- and HPO_4^{2-}, respectively (Hedges and Lee 1993; Hansen and Koroleff 1999; Statham and Williams 1999). Note that the original NH_4^+ and NO_2^- in the samples are also oxidized quantitatively to NO_3^-. For the case of DOC, the samples are decarbonated prior to determination by acidifying to pH <2 with a carbon-free strong acid (e.g. 25% H_3PO_4) and vigorous purging with carbon-free air or N_2. Therefore, the produced CO_2 after oxidation corresponds only to the initial non-volatile fraction of DOC. For the case of DON and DOP, the NO_3^- and HPO_4^{2-} produced during oxidation are added to the initial NO_3^-, NO_2^-, NH_4^+ and HPO_4^{2-} of the sample. So, total dissolved nitrogen (TDN) and phosphorus (TDP) are really measured, and DON and DOP are then obtained by subtracting the independently measured initial nutrients to TDN and TDP, respectively.

Since the 1960s DOC, TDN and TDP have been analysed by wet oxidation, commonly involving persulphate digestion, UV oxidation or a combination of both. While the CO_2 produced by DOC oxidation is usually determined by non-dispersive infrared (NDIR) absorbance, NO_3^- and HPO_4^{2-} are determined by standard colorimetric methods (Hansen and Koroleff 1999; Statham and Williams 1999; Mopper and Qian 2006; Dafner 2016). Although initially developed in the 1960s and 1970s (Sharp 1993), high-temperature oxidation (HTO) was popularized in the 1990s (Hedges and Lee 1993). In this method, DOC is oxidized to CO_2 with O_2 at high temperature (600–900 °C) in the presence or not of a catalyst, and the produced CO_2 is analysed in a NDIR detector. TDN can be analysed simultaneously by installing a chemiluminescence NO_X detector in line with the NDIR gas analyser. The works by

Suzuki et al. (1985) and Sugimura and Suzuki (1988), who reported HTO DOC and DON concentrations largely exceeding those obtained by wet oxidation methods, initiated a controversy that transformed marine organic biogeochemistry (Sharp 1993). After methodological improvements, HTO concentrations returned to the levels obtained by wet oxidation (Hedges and Lee 1993). Nevertheless, the debate contributed to focus on the previously unaccounted central role of DOC in the ocean carbon cycle (Hansell et al. 2009), to conduct broad community intercalibration exercises (Sharp et al. 2002a, b) and to create a DOC/TDN consensus reference material (CRM) program. Over the last 25 years, HTO has been the prevailing method for the determination of DOC and TDN in the oceans because oxidation efficiency and accuracy are higher and sample volume is lower than for wet oxidation methods (Cauwet 1999; Mopper and Qian 2006). Recently, an extensive collection of the DOC measurements in all ocean basins from 1994 to 2020 has been published (Hansell et al. 2021).

Seawater samples for the determination of DOC, TDN and TDP should be collected in acid-cleaned all-glass flasks, filtered through pre-combusted GF/F filters and placed in pre-combusted glass ampoules or vials (see Sect. 2.2). Samples for DOC analysis by wet oxidation or DOC/TDN by HTO can be preserved by acidifying to pH <2 with a strong acid and refrigerated until analysis or by freezing the vials. Freezing applies to TDN and TDP by wet oxidation. When the filtration step is omitted, total organic carbon (TOC), nitrogen (TN) and phosphorus (TP) will be reported.

2.3.1.1 Wet Oxidation

The most widespread wet chemical oxidation (WCO) method consists of the digestion of the sample with a strong chemical oxidant, commonly persulfate in an alkaline medium (boric acid/sodium hydroxide or sodium tetraborate) to keep the final pH of the reaction at 4–5 (McKenna and Doering 1995; Hansen and Koroleff 1999; Statham and Williams 1999; Mopper and Qian 2006; Dafner 2016) (Fig. 2.5a).

The digestion is generally reinforced by heating to 120 °C in an autoclave for at least 30 minutes. In this case, blanks, standards and samples should be placed in sealed glass ampoules for the analysis of DOC and acid-cleaned borosilicate glass or Teflon containers with Teflon screw caps for the analysis of TDN and TDP. Note that persulfate must be added in large excess, because it is partially consumed by the chloride in seawater to produce chlorine and to prevent a loss of oxidative efficiency during the course of the digestion (Mopper and Qian 2006). So, careful purification of this reagent is needed to eliminate carbon (McKenna and Doering 1995) and nitrogen (Hansen and Koroleff 1999) impurities to produce acceptable blanks. While the formed CO_2 is most frequently transported by a carrier gas, dried and determined by NDIR absorbance (Mopper and Qian 2006), NO_3^- and HPO_4^{2-} are analysed by standard colorimetric methods, either manually or by automated flow analysis (Hansen and Koroleff 1999). Since the analysis of HPO_4^{2-} involves the reduction of a phosphorus-molybdate heteropolyacid with ascorbic acid, the excess of persulfate and chlorine produced during the digestion has to be neutralized prior to

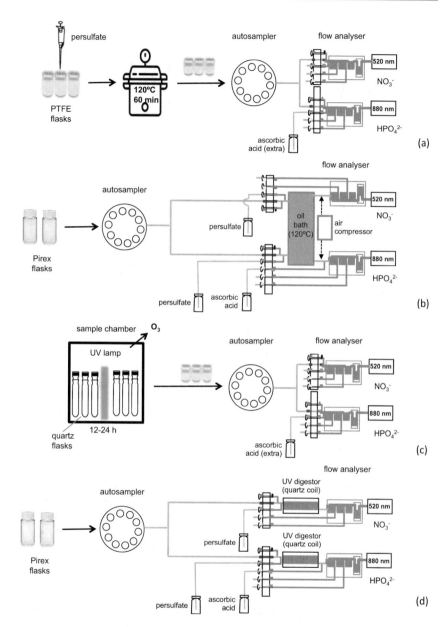

Fig. 2.5 Flow diagram of analytical procedures for (**a**) WCO manual, (**b**) WCO automated, (**c**) WCO + UVO manual and (**d**) UVO automated determination of DOC, TDN and TDP

the HPO_4^{2-} analyses by adding an extra amount of ascorbic acid (Hansen and Koroleff 1999). When determined by flow analysis, TDN and TDP can be obtained simultaneously from the same water sample. Moreover, Raimbault et al. (1999)

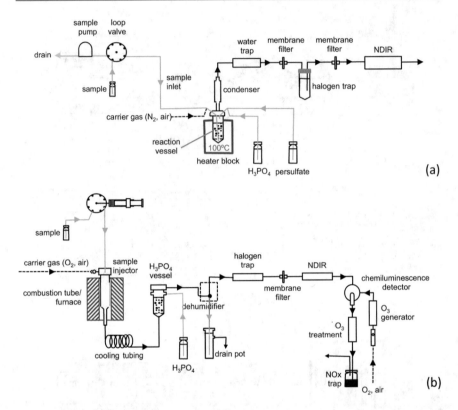

Fig. 2.6 Flow diagrams of the automated analytical determination of (**a**) DOC by WCO and (**b**) DOC and TDN by HTO. Grey lines and arrows, liquid sample and reagent flows; black lines and arrows, carrier gas and gas products

proposed a method for the colorimetric determination of the CO_2 produced by persulfate digestion using flow analysis, allowing the simultaneous determination of DOC, TDN and TDP. The method consists of pumping the digestion mixture directly from the digestion flasks into a three-channel colorimetric flow analyser. In the DOC channel, the liberated CO_2 is injected in a buffered (carbonate/bicarbonate) solution of phenolphthalein that is discoloured proportionally to the CO_2 concentration.

In the methods previously described, digestion and analytical determination are separated. However, it is possible to combine them in automated methods. Thus, McKenna and Doering (1995) optimized the determination of DOC in an automated instrument (I.O. Corporation) with a Teflon reaction vessel in a heating block. Blanks, standards and samples are injected in the reaction vessel, acidified with phosphoric acid and purged with high purity air to remove inorganic carbon. Then, persulfate is added to the reaction vessel and the digestion is conducted at 100 °C. Finally, the produced CO_2 is purged from the reaction vessel, dried and analysed in an NDIR detector (Fig. 2.6a). Likewise, Yasui-Tamura et al. (2020) proposed the automated simultaneous oxidation and colorimetric determination of TDN and TDP by installing a heating oil bath unit with a long Teflon tube and an air compressor on a two-channel flow analyser that simulates the conditions of persulfate digestion in an autoclave (Fig. 2.5b). A third channel for the colorimetric determination of CO_2 (Raimbault et al. 1999) could be added to this configuration to automatically analyse DOC, TDN and TDP by persulfate digestion from the same water sample.

DOC, TDN and TDP can also be measured by UV oxidation (UVO) (Armstrong and Tibbitts 1968), which can be easily automated by installing a quartz coil surrounding a mercury vapour lamp on a flow analyser (Fig. 2.5d). The longer the quartz coil and the higher the power of the UV lamp, the greater the exposure of the sample to UV radiation and the more efficient the photo-oxidation is. The non-stop production of oxidative free radicals in the quartz coil by the UV radiation ensures a continuous oxidative efficiency and, as no reagents are added, the blanks are much lower and reproducible than in WCO methods (Statham and Williams 1999; Mopper and Qian 2006). On the contrary, aging bulbs and inhomogeneous light exposure reduce reproducibility, and non-standardization of UV lamps reduces comparability of UVO methods (Foreman et al. 2019). To produce quantitative oxidation, an energetic lamp (>1000 W) or prolonged exposition is needed (Cauwet 1984; Statham and Williams 1999; Dafner 2016). To avoid the exposure risk to ozone in the laboratory, a low to medium pressure lamp can be used, but it could be necessary to add persulfate to the sample and/or increase the oxidation time to obtain a good recovery. Therefore, combining automated WCO and UVO allows low irradiation intensity and relatively short irradiation time, but due to the addition of persulfate, higher and less reproducible blanks are expected (Cauwet 1984). Automated methods have been developed to measure simultaneously TDN and TDP from the same water samples in a two-channel flow analyser (e.g. Dafner and Szmant 2014). Simultaneous determination of DOC, TDN and TDP would also be possible by automated WCO + UVO. The highest recoveries are obtained for DOC, which may not even need adding persulfate to the sample when using a medium pressure lamp (Statham and Williams 1999), but they are lower for some TDN and TDP compounds (Gershey et al. 1979; Dafner and Szmant 2014; Dafner 2016; Foreman et al. 2019). To obtain better TDN and TDP recoveries with lower blanks, manual methods involving exposition to energetic UV radiation for a prolonged period of time should be applied (Fig. 2.5c). In this case, acid-cleaned quartz vials with Teflon caps secured with aluminium crimp tops containing blanks, standards and samples are exposed to a UV radiation. Generally more than 1000 W for 12–24 h per batch of samples for TDN and 1–9 h per batch of samples for TDP are used (Dafner 2016). Recently, Foreman et al. (2019) revisited the UV oxidation method using UV light generated from a microwave-powered, mercury lamp of 1800 W and high-precision colorimetric analysis. After UVO for 8 h for TDN and 1.5 h for TDP per batch of samples, the recoveries are among the highest reported in the literature, and particularly a natural DOM isolate was oxidized with an efficiency >98% for both TDN and TDP.

Since the estimated errors of DON and DOP determination are $er_{DON} = \sqrt{er^2_{TDN} + er^2_{NO3} + er^2_{NO2} + er^2_{NH4}}$ and $er_{DOP} = \sqrt{er^2_{TDP} + er^2_{PO4}}$, respectively (Hansell 1993; Dafner 2016), a problem of precision remains in deep water, which is characterized by high nutrient and low DON and DOP concentrations (<10% of TDN and TDP; Hansell 1993, Foreman et al. 2019). For the case of DON, it is possible to remove NH_4^+ as NH_3 by boiling in basic medium and, subsequently, NO_2^- + NO_3^- as nitric oxide with $FeSO_4$ in acid medium. After that, DON is mineralized to NH_4^+ through the Kjeldahl method, and the produced NH_4^+ is then measured colorimetrically in a flow analyser (Doval et al. 1997). However, this procedure is so laborious and at risk of contamination increases due to the several steps included. Moreover, it is not automatable and thus not suitable for routine determination of DON (Bronk et al. 2000). For the case of DOP, it is necessary to be aware that during the colorimetric determination of HPO_4^{2-}, partial acid hydrolysis of DOP may occur, which would finally result in an equimolar underestimation of DOP. Furthermore, if sodium tetraborate is not added to persulfate, the final pH of the oxidation would be 1.5–1.8. Under these conditions, inorganic polyphosphates would be hydrolysed and/or inorganic reduced phosphorus compounds oxidized to HPO_4^{2-}, leading to an overestimation of DOP (Karl and Björkman 2015). As for the case of DON, it is also possible to separate the initial HPO_4^{2-} of the sample from DOP by applying the MAGIC procedure, which consists of the co-precipitation of HPO_4^{2-} with $Mg(OH)_2$ in alkaline solution (Karl and Tien 1992; Karl and Björkman 2015) although it is laborious and not automatable.

In summary, setting up a wet oxidation method requires careful testing of the oxidation conditions, including optimizing the amount of persulfate added to obtain maximum recoveries and lowest blanks, adjusting UV lamp intensity and exposure time and checking for bulb aging. Blanks, standards (e.g. potassium hydrogen phthalate for DOC, glycine and KNO_3 for TDN and KH_2PO_4 for TDP) and samples should follow all steps of the analytical procedures, reference seawater samples should be regularly analysed and recovery experiment with model organic compounds should be tested to check the performance of the oxidation steps. In this regard, it is well-known that certain compounds are not recovered quantitatively, such as urea for the case of TDN or adenosine monophosphate (AMP) and organic and inorganic polyphosphates for the case of UVO of TDP (Hansen and Koroleff 1999; Foreman et al. 2019). However, the abundances of these compounds in seawater can be marginal. So, an effort should be made to check recoveries with DOM concentrates of natural seawater obtained with the isolation methods described in Sect. 2.4 (e.g. Foreman et al. 2019). Standards can be prepared in UV-irradiated ultrapure water, calcined (600 °C, 1 h) NaCl in UV-irradiated ultrapure water and UV-irradiated artificial or natural low-nutrient seawater.

2.3.1.2 High-Temperature Oxidation (HTO)

Methods based on the decomposition of DOM at high temperature are classified in two types: (1) dry combustion, which consists of drying of an acidified sample and combustion of the residue in a furnace or a sealed tube (Fry et al. 1996) and (2) HTO, which involves direct aqueous injection in a high-temperature column (Hedges and Lee 1993; Mopper and Qian 2006). Dry combustion is laborious, contamination-prone and non-automatable. Conversely, HTO is the method universally used nowadays to simultaneously determine DOC and TDN in the marine environment, and it is, therefore, the focus of this section.

In HTO (Fig. 2.6b), a small volume of sample, generally from 50 to 200 µL depending on the expected concentration, is directly injected in a quartz column at high temperature. When the quartz column is filled with a catalyst, oxidation is efficient at lower combustion temperatures (680–720 °C), but preconditioning of the catalyst is necessary to obtain low and stable blanks. If catalyst is not used, blanks are lower, but combustion temperatures have to be increased (900–1000 °C) and considerable sublimation of sodium chloride, which destroys quartz parts and produces more chlorine, occurs (Cauwet 1999). The most widespread configuration nowadays involves filling the quartz column with alumina pellets impregnated with 0.5% platinum and combustion at 680 °C. Conditioning of the catalyst consists of washing the pellets with HCl (10%) and ultrapure water and subsequent drying before filling the column. Furthermore, UV-radiated ultrapure water should be repeatedly injected in the quartz column filled with the conditioned catalyst at 680 °C under a continuous flow of high purity synthetic air or oxygen, which acts as carrier and oxidant, until the blanks are low (equivalent to <5 µmol kg^{-1}) and stable (within ± 1 µmol kg^{-1}). The amount of catalyst to be added results from optimizing higher recovery and lower blank but also on minimizing overpressure during sample injection. This overpressure is created by the flash evaporation of the injected liquid water. Note that the volume occupied by 50–200 µL of water would be 217 to 868 mL at 1 atm using the ideal gas equation. Therefore, for a quartz column of 100 mL filled with 50 mL of catalyst, pressure during injection would range from 4 to 17 atm.

The products of DOC and TDN combustion, CO_2 and nitrogen monoxide (NO), are transported by the carrier gas through a strong acid solution (e.g. 25% H_3PO_4) to ensure that all the CO_2 in the carrier gas is in the form of $CO_2(g)$ rather than $CO_2(aq)$. The gas is then dried either in a permeating dryer or a Peltier cooler (cold trap) to avoid interference of H_2O during CO_2 determination and passed through a trap for the halogen gases produced during catalytic oxidation (e.g. Cu) before entering the NDIR detector for CO_2 determination. The carrier gas with the CO_2 and NO is then conducted to a chamber where the NO reacts with ozone (O_3) produced from high purity air or oxygen in an ozone generator (Fig. 2.6b). As a result of this reaction (NO + $O_3 \rightarrow NO_2^* + O_2$), excited NO_2 gas (NO_2^*) is produced, which emits light when returning to the ground state, and it recorded in a photomultiplier (Álvarez-Salgado and Miller 1998; Cauwet 1999). O_3, CO_2 and NO_x traps (e.g. activated carbon and soda lime) are placed in the outlet of the reaction chamber.

On the one side, each blank, standard or sample is injected several times, normally 3–5, and the average and standard error of the replicate measurements is reported. Since each analysis takes 3–5 min, a robust estimate of the DOC and TDN concentration can be obtained in 10–25 minutes with <1 mL of water. Furthermore, recovery experiments demonstrated that inorganic and organic nitrogen compounds are recuperated quantitatively, including urea that is not oxidized by WCO. For the case of antipyrine, with a –N–N– bond, the recovery is not quantitative (Álvarez-Salgado and Miller 1998; Bronk et al. 2000). It is not quantitative for a humic acid too (Bronk et al. 2000). On the other side, the main inconvenience of HTO methods is that seawater salts accumulate in the quartz column catalyst/filler, deteriorating the quartz and making it difficult for the carrier gas and combustion products to flow through the system, producing backpressure and the eventual breakdown of the column. Note that after a standard injection of 150 µL of a seawater sample of salinity 35, about 5 mg of salt is deposited in the catalyst/filler, i.e. 15–25 mg per sample or 1.5–2.5 g after 100 seawater samples. Furthermore, blanks have to be analysed frequently (e.g. every 5–10 samples) to check they are low and stable. The column has to be disassembled regularly to remove the salts and condition the catalyst before filling again the quartz column.

As for the case of wet oxidation, HTO methods are commonly calibrated with a mixed standard of potassium hydrogen phthalate and glycine. Furthermore, it has been demonstrated that the calibration slope does not change when the standards are prepared in ultrapure water or in low nutrient seawater (Álvarez-Salgado and Miller 1998). Consensus reference materials (CRMs) are available to daily check the performance of the instrument before starting a session of analysis (https://hansell-lab.rsmas.miami.edu/consensus-reference-material/). These CRMs include low carbon water (LCW), obtained from the Milli-Q$^{®}$ Advantage A10 System of the reference laboratory (D.A. Hansell's organic biogeochemistry lab) and a deep seawater reference (DSR) collected in the Florida Strait at 700 m depth. Mid (MSR) and surface (SSR) seawater references, collected at 150 and 5 m depth, are also available. They are all acidified with HCl at pH <2.

Finally, as for the case of DOC and TDN by HTO, it is also possible to analyse TDP from a phosphorus gas (phosphine, PH_3) obtained by reduction instead of oxidation of TDP with sodium borohydride ($NaBH_4$) at 460 °C and determination by gas chromatography (Hashimoto et al. 1987). However, the method has not became popular and does not allow for simultaneous determination of DOC, TDN and TDP from the same sample.

2.3.2 Optical Analyses

The fraction of DOM that absorbs light in the UV and visible regions of the light spectrum, generally from 240 to 750 nm, is called chromophoric DOM (CDOM; Stedmon and Nelson 2015). Main chromophores in CDOM are –N=C–C=C–N– (Schiff-base), –COOH, –COOCH$_3$, –OH, –OCH$_3$, –CH=CH–, –CH=O, –C=O, –NH$_2$, –NH–, –CH=CH–COOH, –OCH$_3$, –CH$_2$–(NH$_2$)CH–COOH (amino acid), and S-, O- or N-containing aromatic groups, although not all of them absorb at wavelengths (λ) > 240 nm (Fig. 2.7; Wozniak and Dera 2007, Mostofa et al. 2013).

Absorption spectra of natural marine waters are characterized by an exponential decay with increasing λ to undetectable levels at λ > 600 nm (Fig. 2.7). Deviations from the theoretical exponential decay curve are caused by the variable occurrence

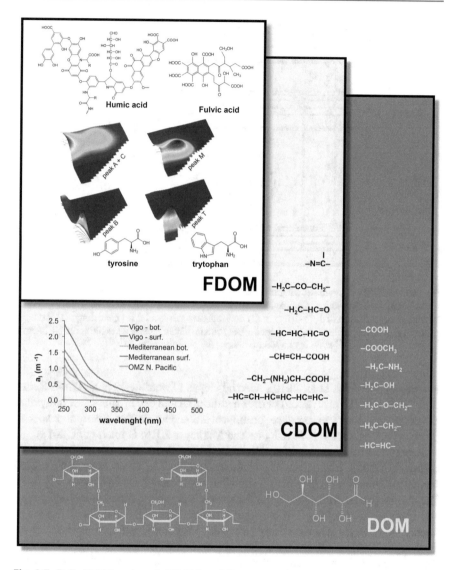

Fig. 2.7 Bulk (DOM), coloured (CDOM) and fluorescent (FDOM) pools of dissolved organic matter in the marine environment. Examples of transparent ($\lambda < 240$ nm), coloured ($\lambda > 240$ nm) and fluorescent molecules are included. Excitation emission matrices (EEMs) associated with the aromatic amino acids tryptophan and tyrosine (UV fluorophores) and to humic and fulvic acids (visible fluorophores). Absorption coefficient spectra for different ocean regions (Ría de Vigo, Mediterranean Sea and oxygen minimum zone of the Eastern North Pacific)

of multiple Gaussian-shaped absorption spectra of individual chromophores (Massicotte and Markager 2016). This variability is the reason behind the assortment of indices used as tracers for photochemical and biological processes of CDOM in the marine environment. These indices include absorption coefficients at fixed

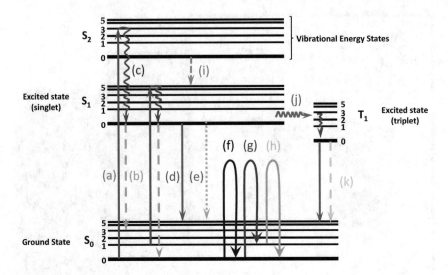

Fig. 2.8 Jablonski diagram showing transitions between electronic states relevant in fluorescence spectroscopy. (a) Excitation (absorption) 10^{-15} s; (b) emission (fluorescence) 10^{-9}–10^{-7} s; (c) vibrational relaxation 10^{-14}–10^{-11} s; (d) non-radiative relaxation; (e) quenching; (f) Rayleigh scatter; (g) Raman scatter (Stokes shifted); (h) Raman scatter (anti-Stokes-shifted); (i) internal conversion 10^{-14}–10^{-11} s; (j) intersystem crossing; (k) phosphorescence

wavelengths (Nelson et al. 2010), spectral slopes (Twardowski et al. 2004; Helms et al. 2008; Loiselle et al. 2009) or the identification and deconvolution of individual chromophores that can produce prominent shoulders in the absorption spectra (Röttgers and Koch 2012; Massicotte and Markager 2016; Grunert et al. 2018).

A portion of the coloured fraction of DOM has the capacity to emit part of the absorbed radiation in the form of blue fluorescence when returning from the excitation to the ground state (Fig. 2.8). It is called fluorescent DOM (FDOM) (Fig. 2.7). The chromophores with capacity for fluorescence are called fluorophores and generally contain aromatic rings, whose rigid structures have few vibrational degrees of freedom to relax the absorbed energy towards molecular vibration, and thus they are fluorescent (Coble 2007; Stedmon and Nelson 2015). Two types of fluorophores are the most common in natural waters: protein-like, which emit fluorescence in the UVA range ($\lambda < 400$ nm), and humic-like, which emit fluorescence in the visible range of the electromagnetic spectrum ($\lambda > 400$ nm; Coble 2007).

Protein-like fluorescence is related to the aromatic amino acids phenylalanine, tyrosine and tryptophan either free or bounded into peptides or proteins (Fig. 2.7; Coble 2007, Jørgensen et al. 2011). Although tryptophan and tyrosine both emit about 14% of the absorbed light, the former dominates the protein-like emission band since tyrosine fluorescence is negligible when coexisting with tryptophan in the same peptides because the emission energy of tyrosine is used for exciting tryptophan (Creighton 1993; Lakowicz 2006). Furthermore, phenylalanine is not always visible because it emits only 2% of the absorbed light (Lakowicz 2006) and it is highly bioavailable. These amino acids are present in all proteins, so their UV florescence could be used as a tracer of the presence and abundance of proteins in natural marine waters (Yamashita and Tanoue 2003). Humic-like fluorescence is related to the aromatic humic substances of soil (humic acids) and aquatic (fulvic acids) origin (Coble 2007). Humic substances are structurally complex by-products of organic matter decomposition mainly composed of aromatic rings with phenolic and carboxylic substituents (Fig. 2.7), characterized by bio-refractory but photodegradable nature. Therefore, identification and quantification of the main DOM fluorophores present in marine samples provide valuable insights on the terrestrial vs. marine origin and biological vs. photochemical transformations that they experience.

CDOM and FDOM are expected to be major contributors to the bulk DOM pool in estuarine waters impacted by the entry of highly coloured DOM of terrestrial origin. This influence may decrease substantially when moving to the colourless open ocean waters (Coble 2007). Contributions from 10% to 90% to the bulk DOM pool have been suggested (Twardowski et al. 2004; Coble 2007; Stedmon and Nelson 2015), but proper quantification would imply the isolation of the coloured and fluorescent fractions of DOM and the determination of their absorption, fluorescence and dissolved organic carbon (DOC) content to obtain CDOM/DOC and FDOM/DOC conversion factors that are representative for each marine ecosystem.

Although this chapter is devoted to bench CDOM and FDOM measurements of discrete seawater samples, it is noteworthy that the optical properties of DOM at specific absorption wavelengths or excitation-emission wavelength pairs can also be recorded continuously in situ with sensors installed in rosette samplers, towed vehicles, moorings or stand-alone profilers (Yamashita et al. 2015, Nelson and Gauglitz 2016, Cyr et al. 2017, Shigemitsu et al. 2020). Calibration with discrete bottle measurements is required.

2.3.2.1 Measuring CDOM

Seawater samples for the determination of CDOM absorption spectra should be collected in acid-cleaned all-glass flasks and filtered on board through 0.2–0.7 μm filters to remove cells and particles (see Sect. 2.2). Filtered samples should be preserved in glass vials/flasks, in the dark at 4 °C until determination within a few hours of collection. Long-term storage of chilled samples or freezing is not recommended.

Absorption spectra of the filtered samples are measured directly, commonly from 240 to 750 nm at 1 nm interval, in a double-beam spectrophotometer. These instruments improve the precision of the measurements by instantaneously correcting any fluctuation in the intensity of the UV and visible source lamps while recording the spectra. Methanol (10%) and acid (HCl 10%)-cleaned prismatic or cylindrical quartz cuvettes of 100 mm optical path length are generally used. Before measuring, the UV and visible lamps should be allowed to warm up for ½ hour. Then, the reference cuvette is filled with freshly produced UV-irradiated ultrapure water. Filling the measuring cell also with ultrapure water allows setting the 100% of transmittance (T). Samples have to be at room temperature and shaken

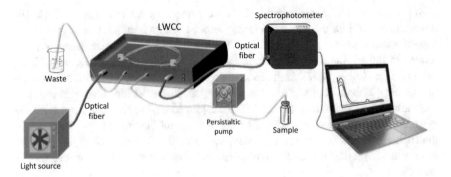

Fig. 2.9 Assembly for the determination of absorption coefficient spectra of CDOM with a LWC

before filling the measuring cuvette to avoid fogging up of the quartz walls and to eliminate bubbles. At a standard speed of 250 nm min^{-1}, collection of an absorption spectrum takes about 2 min. A sequence of measurements should begin and end with a blank of ultrapure water. Recording of additional blanks every 10 samples is recommended.

Once the absorption spectrum of the samples has been recorded, $ABS(\lambda)$, the absorption spectrum of the blank, $ABS_B(\lambda)$, is subtracted first. Then, the average absorbance from 600 to 750 nm, $ABS(600-750)$, is subtracted further to correct the residual scattering caused by micro-air bubbles or colloidal material, refractive index differences between the sample and the reference or light attenuation that is not related to organic matter (Green and Blough 1994). This λ-independent correction instead of the λ-dependent correction of scattering proposed by Bricaud et al. (1981) is preferred because the low differences between sample and baseline absorbance at long wavelengths did no produce measurable differences between both corrections at short wavelengths. Finally, to obtain the absorption coefficient spectrum, $a(\lambda)$ in m^{-1}, the blank and scattering-corrected absorption spectrum is divided by the cell path length (l) (generally 0.1 m) and multiplied by 2.303 to convert from decadic ($ABS = -\log_{10}T$) to Naeperian ($ABS = -\ln T$) scale:

$$a(\lambda) = 2.303 \cdot \frac{ABS(\lambda) - ABS_B(\lambda) - ABS(600 - 700)}{l} \qquad (2.1)$$

The limit of detection of present-day double-beam spectrophotometers is around 0.001 absorbance units (AU) or 0.02 m^{-1} for a 100 mm cell path length. Therefore, any measurement below 0.06 m^{-1}, i.e. three times the detection limit, should not be reliable. Such a low value of $a(\lambda)$ is common in surface open ocean waters at visible wavelengths. In these cases, the analysis of the CDOM spectra should be restricted to the range where $a(\lambda) > 0.06$ m^{-1}.

Alternatively, a long path length liquid waveguide capillary cell (LWCC) coupled with spectrophotometric detection can be used (Fig. 2.9) (Gimbert and Worsfold 2007). Capillary cells of path length ranging from 0.5 to 5 meters are commercially available. The recommended path length for the determination of the absorption coefficient spectrum of an open ocean sample is 1 m. With this path length, the limit of detection decreases to $0.002 \, m^{-1}$, allowing reliable measurements of the absorption coefficient at visible wavelengths, and, at the same time, maximum absorption, usually $<1.5 \, m^{-1}$ at 240 nm for surface open ocean samples, produces an acceptable maximum absorbance of <0.65 UA. The LWCC are usually made of a solid Teflon AF tubing (Type I) or a fused silica capillary tubing with an external Teflon AF coating (Type II) (Floge et al. 2009). Since Teflon AF is a low refractive index polymer, these cells are designed to minimize absorbance biases due to changes in the refractive index between samples of different salinity and the waveguide, which is very convenient when measuring seawater samples, although small offsets and noisy signals can be produced (Lefering et al. 2017). The LCCW is illuminated axially using fibre-optic cables attached to commercially available deuterium-halogen light source and a miniature spectrometer. Finally, a peristaltic pump is needed to force the ultrapure water and the samples through the LCCW at a recommended rate of $1 \, mL \, min^{-1}$. In this regard, stable pumping of the sample through the capillary cell was found to improve measurement precision over measurements made with the sample kept stationary (Lefering et al. 2017). As for the case of a convectional spectrophotometer, the deuterium-halogen light sources have to warm up ½ h. Furthermore, the LCCW should be lubricated daily for 5 min with a surfactant solution of Triton X-10 to facilitate capillary flow. Apart from the tenfold increase in sensitivity, the LCCW has a very small volume (about 125 μL for a 1 m cell), which means that 1 mL of sample is enough to clean the circuits and the cell and to perform repeated measurements of the sample in just 1 minute. The main disadvantage of this configuration is that it works as a single-beam spectrophotometer, and, therefore, variations in the intensity of the source lamps or the efficiency of the detector when recording the absorption spectra will not be corrected. This inconvenience is partly overcome by using the average of the repeated records of each sample. Furthermore, the quality of both the light sources and the detector is crucial to keep the detection limit at 0.001 AU. Finally, if not certified by the provider, the effective path length of the LWCC has to be obtained experimentally.

2.3.2.2 Processing CDOM Measurements

CDOM indices derived from absorption spectra can be classified into quantitative and qualitative (Fig. 2.10). Quantitative indices provide information about the concentration of different CDOM pools and consist of absorption coefficients at specific wavelengths. $a(254)$, $a(325)$ and $a(350)$ are among the most commonly reported in marine biogeochemical studies (Fig. 2.10a). $a(254)$ frequently exhibits a positive linear relationship with the concentration of DOC in coastal waters influenced by continental runoff (Spencer et al. 2009, 2012; Chen et al. 2011) as well as in open ocean waters (Lønborg and Álvarez-Salgado 2014; Catalá et al. 2018). This is because the conjugated carbon double bonds that absorb at 254 nm are abundant forms of carbon in the DOM pool (Fig. 2.7). Furthermore, $a(254)$ is not directly influenced by the incident solar UV radiation (Del Vecchio and Blough 2002) given that very few solar photons of wavelength < 295 nm reach the Earth's surface (Fichot and Benner 2011). Conversely, $a(325)$ represents a more aromatic

Fig. 2.10 (a) Absorption coefficient spectra of CDOM samples collected in the North Pacific during the Malaspina 2010 circumnavigation. The spectra present two shoulders centred at 305 nm (UV chromophore, NO_3^-) and 415 nm (visible chromophore). (b) The original spectrum is decomposed into an exponential decay and c two Gaussian-type chromophores. Wavelengths of common absorption coefficients and spectral slopes are also indicated

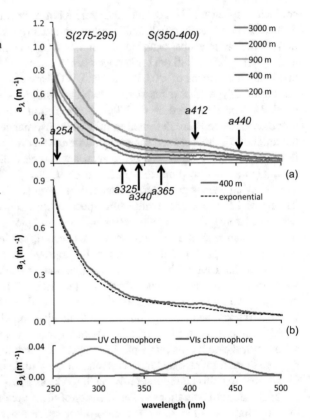

Table 2.1 Common fluorophores identified in natural marine samples by peak picking of EEMs. Adapted from Coble (2007) and Stedmon and Nelson (2015)

Type	Label	λ_{Ex} (nm)	λ_{Em} (nm)	Nature
UV	B	270–280	300–310	Tyrosine-like, protein-like
UV	T	270–280	340–360	Tryptophan-like, protein-like
UV	N	280	370	Unknown
Visible	A	230–260	380–460	Terrestrial humic-like
Visible	M	290–320	370–420	Marine humic-like
Visible	C	320–360	420–480	Terrestrial humic-like

moiety of CDOM produced during the microbial degradation of organic matter that responds to the absorption of solar UVA radiation (Nelson et al. 2004). $a(325)$ exhibits a direct relationship with apparent oxygen utilization (AOU) and can be used as a tracer for microbial production and/or photodegradation processes (Nelson et al. 2010; Iuculano et al. 2019). It should also be noticed that a common humic-like fluorophore of marine origin has its maximum excitation wavelength at around 325 mm (see Table 2.1). $a(350)$ represents a larger and more complex aromatic moiety than $a(325)$, and it is also around the maximum excitation wavelength of another common humic-like fluorophore of terrestrial origin (Table 2.1). In ocean colour studies, in which accurate retrievals of $a(\lambda)$ are needed to improve satellite estimates of chlorophyll, absorption coefficients are also determined at visible wavelengths, with $a(412)$ and $a(443)$ being the most common (e.g. Siegel et al. 2002; Mannino et al. 2014). $a(443)$ corresponds approximately to the midpoint of the absorption band of photosynthetic pigments (Siegel et al. 2002).

Qualitative indices provide information about the origin (terrestrial vs. marine), chemical structure (mean molecular weight, aromaticity) and biogeochemical processes (microbial synthesis vs. degradation, photohumification vs. photodegradation) experienced by the CDOM pool. Among the most common indices proposed in the literature are $SUVA_{254}$, $a(254/365)$, s $(300–600)$, $s(275–295)$, $s(350–400)$ and S_R (Fig. 2.10a). The carbon-specific UV absorbance ($SUVA_{254}$) is the ratio between decadic $a(254)$ and DOC, i.e. $a(254)/$ $(2.303 \cdot DOC)$, with DOC expressed in ppm, and it is positively correlated with aromaticity (Weishaar et al. 2003). The dimensionless absorption coefficient ratio at 254 nm/365 nm, $a(254/365)$, is a proxy for the average molecular weight (MW) of CDOM with high values of $a(254/365)$ indicating a lower average MW (De Haan and De Boer 1987; Engelhaupt et al. 2003; Berggren et al. 2010). The spectral slope (in nm^{-1}) over a wide spectral range, e.g. 300–600 nm, also exhibits a negative relationship with the average MW of CDOM (Twardowski et al. 2004; Stedmon and Nelson 2015). The slope s can be obtained by adjusting the CDOM spectrum to the exponential decay function (Fig. 2.10a):

$$a(\lambda) = a(\lambda_0) \cdot e^{-s \cdot (\lambda - \lambda_0)} \qquad (2.2)$$

The value of s depends on the spectral range used in the fit, 300 to 600 nm in our case, and on the fitting model, a linear regression of the Ln-transformed $a(\lambda)$ or, preferably, a non-linear fitting (Twardowski et al. 2004, Stedmon and Nelson 2015). In this regard, spectral slopes over narrower wavelength ranges have been suggested. Helms et al. (2008) proposed the spectral slope calculated over the wavelength range 275 nm to 295 nm, $s(275–295)$, and the ratio (S_R) of $s(275–295)$ and the spectral slope over 350 nm to 400 nm, $s(350–400)$, $S_R = s(275–295) / s(350–400)$ as proxies to the MW and origin of CDOM. Again, steeper slopes are inversely related to the MW and the microbial degradation of DOM and directly related to photodegradation of CDOM. Twardowski et al. (2004) also tried other fitting functions, obtaining that the hyperbolic model was statistically the most useful descriptor of CDOM absorption spectra, but the exponential decay model is still the most popular in the literature.

In opposition to the estimation of spectral slopes over wider or narrower wavelength ranges, Loiselle et al. (2009) proposed to obtain a continuous slope curve, s (λ). To estimate $s(\lambda)$ at a given wavelength, λ_0, $a(\lambda)$ from λ_0 to $\lambda_0 + 49$ nm is adjusted to Eq. 2.2 using a non-linear fitting. This promising approach has been applied to lake ecosystems, but its use has not been generalized yet, particularly in marine sciences. $s(\lambda)$ allows a quick identification of the prominent shoulders produced by individual chromophores in $a(\lambda)$ (Fig. 2.10b, c). Assuming that the absorption coefficient spectra of the chromophores that distort the exponential decay of CDOM with increasing wavelength (Eq. 2.2) have a Gaussian shape, the complete equation to be fitted would be (Massicotte and Markager 2016):

$$a(\lambda) = a(\lambda_0) \cdot e^{-s \cdot (\lambda - \lambda_0)} + \sum\nolimits_{i=0}^{n} \frac{\varphi_i}{\sigma_i \cdot \sqrt{2\pi}} \cdot e^{-\frac{(\lambda - \mu_i)^2}{2 \cdot \sigma_i^2}} + \epsilon \qquad (2.3)$$

where φ_i (m^{-1}) is the height of the Gaussian peak, μ_i (nm) is the position of peak maximum, σ_i (nm) is the width of the peak, and ϵ is the residual after fitting of the full model. Following Massicotte and Markager (2016), $a(\lambda)$ is fitted first to the exponential decay function (first term on the right of Eq. 2.3), and the residuals of this fitting are then fitted to the Gaussian function (second term on the right of Eq. 2.3). The Gaussian decomposition procedure is generally applied to $\lambda < 500$ nm, because $a(\lambda)$ at longer wavelengths is often close to the detection limit. The optimal number of Gaussian peaks is obtained by minimizing the Bayesian information criterion (BIC) score, which penalizes the number of parameters of the model to avoid overfitting. Massicotte and Markager (2016) developed a R package named "cdom" with the tools to perform these calculations that is available at the CRAN website (https://cran.r-project.org).

More recently, Omanović et al. (2019) have published "ASFit", a very useful tool for CDOM absorption analysis that includes the calculation of all the parameters described in this section. The compiled version and the user manual of "ASFit" are available at https://sites.google.com/site/daromasoft/home/asfit. The source code can be downloaded from https://github.com/darioomanovic/ASFit.

2.3.2.3 Measuring FDOM

As for the case of CDOM, seawater samples for the determination of FDOM should be collected in acid-cleaned all-glass flasks, immediately filtered through 0.2–0.7 μm filters and preserved in glass vials/flasks in the dark at 4 °C until determination within a few hours of collection (Sect. 2.2). Samples have to be acclimatized to room temperature (20 °C), and then, in an acid-cleaned 10 mm path length prismatic cuvette with 4 quartz walls, determine its FDOM in a spectrofluorometer. Given the low natural FDOM of marine samples, particularly in open ocean waters, a high range instrument with a high signal-to-noise ratio (>500:1 minimum, >5000:1 optimal) should be used.

FDOM can be determined at specific excitation/emission wavelength pairs ($\lambda_{Ex}/\lambda_{Em}$), in the form of emission, excitation or synchronous spectra or as excitation-emission matrices (EEMs) (Coble 2007; Stedmon and Nelson 2015). While emission spectra are recorded at a fixed λ_{Ex} and excitation spectra at a fixed λ_{Em}, fluorescence in synchronous spectra varies in both wavelengths and is recorded at $\lambda_{Em} = \lambda_{Ex} + \Delta\lambda$, with $\Delta\lambda$ being a constant λ difference. Finally, EEMs are obtained by concatenating excitation, emission or synchronous spectra, generally covering λ_{Ex} from 240 to 500 nm and λ_{Em} from 300 to 600 nm (Fig. 2.11; Stedmon and Nelson 2015). Determination of FDOM at the predetermined $\lambda_{Ex}/\lambda_{Em}$ of the most common protein- and humic-like fluorophores (Table 2.1; Hudson et al. 2007, Coble 2007) is

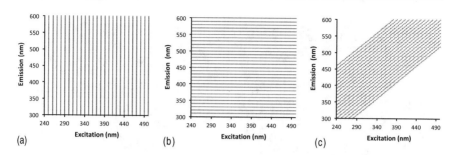

Fig. 2.11 (**a**) Excitation, (**b**) emission and (**c**) synchronic scans and EEMs

very quick; in less than 1 minute, replicate measurements can be made at several λ_{Ex}/λ_{Em}. However, following this procedure it is not possible to notice the natural variability observed in the position of the λ_{Ex}/λ_{Em} maxima, and the occurrence of new fluorophores would not be perceived. In this regard, it is well-known that the position of fluorescent amino acids in proteins determines the intensity and λ_{Ex}/λ_{Em} of the protein-like fluorescence (Lakowicz 2006). These inconveniences are overcome by measuring EEMs, which takes more time, generally 15–30 min, and requires an elaborated post-processing of the raw data.

EEMs are commonly recorded as a concatenation of emission scans from λ_{Em} 300 to 600 nm, at 2 nm increments, starting at λ_{Ex} 240 nm and ending at 500 nm, recording an emission scan at 10 nm step increments. Excitation and emission bandwidths are generally 5 mm (Fig. 2.11). EEMs have to be first corrected for Rayleigh and Raman scattering and inner filter effects. Rayleigh and Raman scattering occur when a molecule is illuminated by a photon with insufficient energy to completely excite the molecule (Fig. 2.8; Larsson et al. 2007). Rayleigh scattering takes place when the light emitted has the same energy than the exciting light, i.e. when $\lambda_{Em} = \lambda_{Ex}$, which appears as a prominent diagonal band in all EEMs. A second Rayleigh scattering band appears at $\lambda_{Em} = 2 \cdot \lambda_{Ex}$. In Raman scattering, the emitted light has different energy than the exciting light, and this energy shift is a constant value for each molecule that depends on its vibrational energy levels in the ground state. In an aqueous solution, the only molecule in sufficient concentration to produce Raman scattering in fluorescence measurements is water. Since the molecule of water relaxes to a lower energy vibrational state than the original, i.e. $\lambda_{Em} > \lambda_{Ex}$, the Raman scattering of water appears as a slightly curved band at longer λ_{Ex} than the Rayleigh scattering band. As for the case of the Rayleigh band, a second Raman band appears at $2 \cdot \lambda_{Ex}$ in the EEMs.

The areas of the EEMs affected by the Rayleigh scattering bands, from $\lambda_{Em} -$ 20 nm to $\lambda_{Em} +$ 20 nm, are not usable and have to be masked, and the Raman scattering band is corrected by subtracting the EEM of UV-irradiated freshly produced ultrapure water to the EEM of the sample. After these corrections, the usable part of the EEM restricts to the diagonal area in between the two Rayleigh scattering bands (Fig. 2.12). Therefore, although not very popular, a concatenation of synchronous scans would be a more efficient and somewhat quicker way to record the EEM

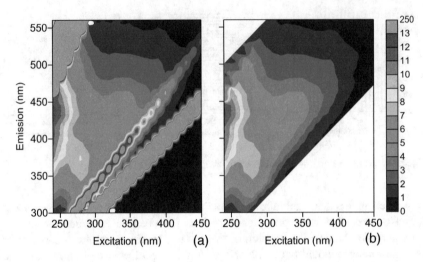

Fig. 2.12 (**a**) Untreated EEM, (**b**) treated EEM. Colour scale in fluorescence units (FU)

of a natural water sample (Fig. 2.11c). The first scan would start at λ_{Ex} 240 nm and λ_{Em} 260 nm ($=\lambda_{Em}$ + 20 nm) and be followed with increments of 1 nm in both λ_{Exc} and λ_{Em} up to λ_{Ex} 500 nm and λ_{Em} 520 nm. Subsequent scans would be recorded starting at λ_{Em} 240 nm and λ_{Em} 260 nm, 280 nm and so on up to the last scan that would start at λ_{Ex} 240 nm and λ_{Em} 460 nm ($=\lambda_{Ex}$ + 220 nm) and end at λ_{Exc} 500 nm and λ_{Em} 720 nm (Nieto-Cid et al. 2005).

Inner filter effect (IFE) is produced by the absorption of the excitation and emission light by the chromophores in the sample, resulting in a reduction of the recorded fluorescence intensity. The per cent reduction depends on the absorption coefficient spectrum of the sample, $a(\lambda)$. The IFE is corrected as follows (Ohno 2002):

$$FDOM_{corr} = FDOM_{obs} \cdot 10^{\frac{a(\lambda)_{Ex}+a(\lambda)_{Em}}{2 \cdot 2.303} \cdot l} \qquad (2.4)$$

where $FDOM_{obs}$ and $FDOM_{cor}$ are the observed and corrected fluorescence of the sample, $a(\lambda)_{Ex}$ and $a(\lambda)_{Em}$ are the absorption coefficients (in m^{-1}) at the excitation and emission wavelengths, and l is the path length of the fluorescence cuvette (0.01 m). Given the exponential decay of $a(\lambda)$ in natural water samples, the IFE is more significant in the UV range (Fig. 2.10). Since in open ocean samples $a(\lambda)$ is usually <1.5 m^{-1}, the IFE would be <1.5% and it is not corrected. Usually, the IFE correction is applied in coastal samples with $a(\lambda) > 10$ m^{-1}, i.e. when the IFE represents >10% of the fluorescence intensity (Stedmon and Bro 2008). Raman scattering and IFE corrections need also to be applied to the determination of FDOM at specific $\lambda_{Ex}/\lambda_{Em}$ pairs and to emission, excitation or synchronous spectra.

Next step is standardization of the measurements, which is a tricky question in fluorescence spectrometry. Three procedures have been commonly applied in natural water samples: quinine sulphate units (QSU), Raman units (RU) and normalized fluorescence units (NFlU). Standardization in QSU consists of preparing a calibration curve with quinine sulphate in 0.05 M sulphuric acid (in $\mu g\ L^{-1}$) and measuring fluorescence at $\lambda_{Ex}/\lambda_{Em}$: 350 nm/450 nm to obtain a conversion factor from instrument fluorescence units (FU) into QSU. It should be noticed that different forms of quinine sulphate are commercially available in such a way that when using quinine sulphate monohydrate (formula weight, 440.51 g mol^{-1}), the conversion factor is 12.5% higher than when using quinine hemisulphate monohydrate (formula weight, 391.47 g mol^{-1}). Standardization in RU consists of acquiring an emission scan of freshly produced ultrapure water at λ_{Ex} 350 nm and calculating the area between λ_{Em} 381 and 426 nm following the trapezoidal rule of integration (Lawaetz and Stedmon 2009) and its baseline correction. This area, which integrates the signal due to the Raman scattering of water, is then used to convert instrument FU into RU. Standardization in RU has become more popular because it does not require the tedious preparation of standard solutions, which is particularly convenient when measuring on board. Note that when normalizing to QSU or RU, the same conversion factors apply to any $\lambda_{Ex}/\lambda_{Em}$ in the scan or EEM. On the contrary, when standardizing in NFlU, specific conversion factors are obtained for each $\lambda_{Ex}/\lambda_{Em}$ pair. While quinine sulphate is used to standardize the $\lambda_{Ex}/\lambda_{Em}$ pairs of the visible humic-like fluorophores, the aromatic amino acid tryptophan has been proposed to standardize the UV protein-like $\lambda_{Ex}/\lambda_{Em}$ pairs (Nieto-Cid et al. 2005). Furthermore, commercially available (Strana) prismatic blocks of methacrylate injected with P-terphenyl and tetraphenyl butadiene can be used to check the performance of the spectrofluorometer in the protein- and humic-like regions of the EEM, respectively. P-Terphenyl fluoresces between 310 nm and 600 nm (max. at 338 nm) when exciting at 295 nm and tetraphenyl butadiene between 365 and 600 nm (max. at 422 nm) when excited at 348 nm (Catalá et al. 2015).

Standardization in NFlU has the advantage that measurements are fully comparable among instruments. This procedure can be applied when measuring at selected $\lambda_{Ex}/\lambda_{Em}$ pairs (Table 2.1). On the contrary, when standardizing in QSU and RU, comparability among instruments is secured only if instrument-specific excitation and emission factors to correct for sensitivity variation with λ are applied. The manufacturer usually supplies these correction factors, but they might change with the use of the instrument or after replacing gratings, detectors or other components.

The drEEM toolbox (http://dreem.openfluor.org/; Murphy et al. 2013) for MATLAB is currently available to process raw EEMs to obtain fully corrected (Rayleigh and Raman scattering, IFE) and standardized EEMs ready for post-processing.

2.3.2.4 Processing of FDOM Measurements

When $\lambda_{Ex}/\lambda_{Em}$ pair measurements are conducted at the predetermined peak maxima of the most common protein- and humic-like fluorophores (Table 2.1), once corrected for Raman scattering and IFE and standardized to QSU, RU or NFlU, no further processing is required. The protein-like peaks (B, T) allow tracing the variability of a labile FDOM pool with low sensitivity to UV radiation and prone to microbial utilization. Conversely, the humic-like peaks (A, C, M) represent a refractory DOM pool that is sensitive to UV radiation and is produced during microbial degradation processes, exhibiting a strong positive correlation with AOU (Nieto-Cid et al. 2006; Yamashita and Tanoue 2008; Jørgensen et al. 2011; Catalá et al. 2015). Although this relationship could also be caused by transformations of terrestrial humic-like materials in the open ocean (Andrew et al. 2013), culture experiments have also demonstrated that these materials are produced in situ in the oceans (Jørgensen et al. 2014). Humic-like peak M associates with humic substances of marine origin (Coble 2007; Stedmon and Nelson 2015) although it has also been related to eukaryote respiration (Romera-Castillo et al. 2011). Conversely, peaks A and C associate with humic substances of terrestrial origin although peak C has also been related to prokaryote respiration. Since λ_{Ex} of peak A is <295 nm, this humic-like peak is not so sensitive to photodegradation as peaks M and C, with λ_{Ex} in the UV-A band. Since λ_{Em} of peak C (>420 nm) is longer than of peak M (<420 nm), the former represents a more complex aromatic moiety because the energy difference between ground and excited states decreases as the size of the aromatic compound increases, resulting in longer λ_{Em} (Stedmon and Nelson 2015). Therefore, simple and quick FDOM measurements at predetermined λ_{Ex} / λ_{Em} pairs provide useful information about the origin and photochemical/ microbial transformations experienced by the different protein and humic-like FDOM pools.

Useful indices can be obtained by combining peak measurements, as the labile to refractory FDOM ratio, (B + T)/(A + M + C); the terrestrial humic-like photodegradation ratio, A/C; or the marine to terrestrial humic-like ratio, M/C (Kowalczuk et al. 2013; Martínez-Pérez et al. 2019). Fluorescence and absorption measurements can also be combined to produce fluorescence to absorption ratios as M/$a(325)$ and C/$a(340)$, which are proxies to the fluorescence quantum yield of these humic-like moieties, i.e. to the portion of the absorbed light that can be emitted as fluorescence. The higher M/$a(325)$ and C/$a(340)$, the more humificated or less photodegraded are the humic-like moieties (De Haan 1993; Lønborg et al. 2010). Finally, C-specific fluorescence ratios as T/DOC and M/DOC can be used as proxies to the contribution of the protein- and humic-like FDOM to the DOC pool (Nieto-Cid et al. 2006).

Furthermore, there is an assortment of additional indicators such as the fluorescence (FI), humification (HIX) and biological (BIX) indices (Gabor et al. 2014). FI is calculated as the ratio of the emission at 470 nm and 520 nm when excited at 370 nm, and it is a proxy for the microbial vs. terrestrial origin of FDOM (Cory and McKnight 2005). HIX is calculated dividing the integrated emission scan from 435 to 480 nm by the integrated emission scan from 300 to 345 nm when both excited at 254 nm, and it is a proxy of the degree of humification (Zsolnay et al. 1999). BIX is determined as the ratio of the emission at 380 nm and the maximum intensity between 420 and 435 nm when excited at 310 nm and points to the organic matter recently produced from biological activity (Huguet et al. 2009). It should be noted that while FDOM peaks and the FI are obtained at predetermined $\lambda_{Ex}/\lambda_{Em}$ pairs, emission scans have to be obtained to calculate HIX and BIX.

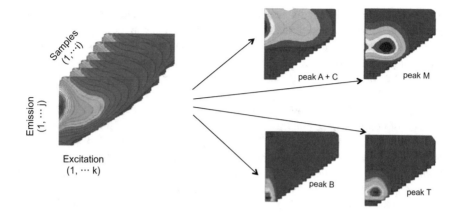

Fig. 2.13 PARAFAC applied to the EEM dataset of the MALASPINA 2010 circumnavigation arranged in a three-way structure (λ_{Ex}, λ_{Em} and sample) and decomposed into four PARAFAC components, two humic-like (peaks A + C and M) and two protein-like (peaks B and T). See Table 2.1. After Murphy et al. (2013)

Finally, processing the information contained in a large EEM dataset requires the application of multivariate data analysis tools. Principal component analysis, Laplacian operator and Nelder-Mead optimization algorithm or self-organizing maps have been proposed to explore patterns in EEM datasets, as well as to discriminate and locate potential peaks (Boehme et al. 2004; Butturini and Ejarque 2013; Ejarque-González and Buturini 2014). However, the most widespread approach is parallel factor analysis (PARAFAC) (Stedmon and Bro 2008). PARAFAC decomposes the combined fluorescence signal of all the fluorophores present in an EEM dataset into components that correspond to individual fluorophores (Stedmon and Bro 2008; Murphy et al. 2013). PARAFAC decomposes a three-dimensional array of data (sample, λ_{Ex} and λ_{Em}) (Fig. 2.13) into a set of trilinear terms (a, b and c) fitting to Eq. 2.5 by minimizing the sum of squares of the residuals (e):

$$x_{ijk} = \sum_{f=1}^{F} a_{if} \cdot b_{jf} \cdot c_{kf} + e_{ijk} \tag{2.5}$$

where $i = 1, \ldots, I$ represents the sample; $j = 1, \ldots, J$ represents λ_{Em}; $k = 1, \ldots, K$ represents λ_{Ex}; x_{ijk} represents the measured fluorescence intensity of sample i at $\lambda_{Em} = j$ and $\lambda_{Ex} = k$; e_{ijk} represents the residual of the fitting; $f = 1, \ldots, F$ corresponds to a PARAFAC component; and a_{if}, b_{jf} and c_{kf} are the outputs of the model. a_{if} represents the concentration of component f in sample i; b_{jf} and c_{kf} are the normalized emission and excitation intensities of component f at $\lambda_{Em} = j$ and $\lambda_{Ex} = k$ in such a way that multiplying vectors b and c produces the EEM of component f. Therefore, PARAFAC relies on that (1) two components cannot have the same intensity or EEM; (2) the normalized excitation spectra of a component is invariant across emission wavelengths and vice versa; (3) fluorescence intensity of each component correlates linearly with concentration (Lambert-Beer law); and (4) total fluorescence intensity is due to the linear combination of the components so that quenching, interactions between fluorophores or changes in the electronic environment of the fluorophores are negligible (Stedmon and Bro 2008, Murphy et al. 2013).

The success of PARAFAC modelling relies on the specification of the right number of components by the user, which is confirmed using split-half validation and random initialization. The drEEM toolbox (http://dreem.openfluor.org/; Murphy et al. 2013) for MATLAB is currently available to run PARAFAC modelling. Furthermore, a repository of published fluorescence EEM is also available online (https://openfluor.lablicate.com/; Murphy et al. 2014) allowing comparison with the components obtained from a new EEM dataset based on the combined Tucker's congruence coefficient of the excitation and emission spectra.

2.4   Fractionation and Isolation

Instruments used nowadays for molecular characterization of DOM (Sect. 2.5) still have constraints about sample purification (low-salt, low-ash...), to avoid interferences and get enough concentration for a good analytical signal. Unlike freshwater studies, high vacuum-low temperature rotary evaporation or direct freeze-drying of the samples is not an option in marine waters, where >35 kg of salts per gram of DOC are present. Hence there is the need of isolation and concentration steps before analysis, which is yet an overwhelming mission due to this huge amount of salts, the low DOM concentrations and the enormous number of different organic compounds in marine samples. The perfect isolation method should (a) recover all DOM, (b) be unbiased (minimize the chemical or physical alteration of the sample), (c) handle large volumes of water in short times and (c) remove all inorganic salts (Mopper et al. 2007). However, none of the actual methodologies fulfil all these requirements. Currently, there exist three main isolation techniques, each employing different extraction principles based in physical or chemical properties: (a) tangential-flow ultrafiltration (based on molecular size), (b) adsorption onto a solid phase (based on hydrophobicity) and (c) reverse osmosis coupled with electrodialysis (based on polarity). These methods differ in the percentage of DOM recovered, the purity of the isolated DOM, the chemical or size fraction isolated, the rate of recovery, equipment and operating costs and the time extent of the process (Dittmar and Stubbins 2014). So far, no method has been able to recover 100% of DOM.

2.4.1   Ultrafiltration

Although small molecules are the major fraction of organic matter in the ocean, in the 1990s, with the advances on the HTO methodologies (Sect. 2.3.1.2), it was exposed that a larger than expected part of the DOC pool was colloidal (Koike et al. 1990; Wells and Goldberg 1991). About 75% of the DOC in the ocean passes through a tangential-flow ultrafiltration (UF) membrane with a cut-off of 1000 Da and a pore size of about 1 nm (Benner et al. 1992). This fraction is designated as permeate, low molecular weight (LMW) DOM or truly dissolved (Fig. 2.1a). The fraction retained in the UF membrane constitutes the DOM concentrate, and it is labelled as retentate, high molecular weight (HMW) DOM or colloidal and ranges from 10% to 40% of the DOM pool (higher at the surface; Table 2 in Mopper et al. 2007). As the UF membrane is highly permeable to water and inorganic solutes, the HMW fraction is concentrated and contains a much smaller fraction of inorganic salts than the LMW fraction. The residual salts in the HMW DOM are partially washed away by dia-filtration against ultrapure water, adding an extra loss of organic carbon (Perdue and Benner 2009). Despite these low recoveries, in general the HMW DOM samples isolated using UF have bulk properties that are compositionally representative of the bulk DOM pool (Benner et al. 1997). Particularly, UF procedures isolate C and N compounds equally, in such a way that the C:N ratios of the HMW extracts ranged within the normal values for the bulk DOM regardless of the low recoveries (Benner et al. 1992). Knowledge of the chemical composition of marine organic matter varies with molecular size: a larger fraction of macromolecular organic matter has been characterized at the molecular level (60–80% of carbon in marine plankton and 30–40% in sinking particles) and a smaller percentage of 15–20% of the carbon in suspended POM. However, only 8–14% of carbon in the HMW DOM and 2–5% of carbon in the LMW DOM are currently known (Perdue and Benner 2009).

UF systems to isolate HMW DOM are usually equipped with membrane coils to process large-volume samples (10^2 L; Fig. 2.14). Hundreds of litres of filtered seawater (DOM drum) are recirculated through the membrane coil applying a low pressure of 5 psi. The LMW fraction is tangentially permeated across the coil and eliminated from the system (LMW drum), while the higher fraction is concentrated until a final volume of 2 L. Low pressures are essential to minimize the loss of HMW DOM, which slows down the process taking about 2–3 days to ultrafiltrate 200 L of seawater. Different membranes were tested during an intercomparison exercise (Buesseler 1996), but in spite of the choice of membrane, the UF results depend more directly on physical than on chemical properties (Perdue and Benner 2009), as long as the contamination is reducing by using housings, tubing, pumps and containers made of Teflon or similar.

More compact systems, such as stirring UF cells (Fig. 2.15), equipped with simple disc membranes (cut-off of 1000 Da) are used to concentrate small-volume samples or to size-fractionate the sample in a more effective way (Martínez-Pérez et al. 2017a). This system is driven by a high pressure (55 psi, using high purity N_2) to force the LMW DOM to tangentially cross the membrane by means of a continuous and vigorous stirring. The efficiency of these UF systems is checked using a solution of vitamin B12 (1355 Da; Sigma, 50 mg L^{-1}; Fig. 2.1b). Regarding this standardization with vitamin B12, previous works reported recovery efficiencies of about 80% for the large-volume systems (Guo and Santschi 1996) and > 90% for the

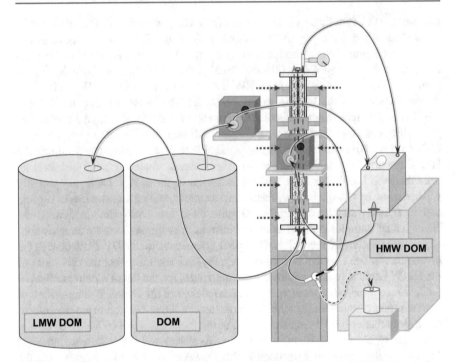

Fig. 2.14 Tangential UF system equipped with a membrane coil (cut-off of 1000 Da) to isolate DOM from large-volume samples (10^2 L)

ultrafiltration cell (Martínez-Pérez et al. 2017a). This difference is due to the general decrease of the retention rate with increasing concentration factors (sample volume/ retentate volume), i.e. higher concentration factors involve a larger loss of the HMW DOM retained by the UF membrane, and also an increase of the UF processing time (Guo and Santschi 1996). Consequently, the percentage of the high and low molecular fractions can only be evaluated among samples isolated with similar concentration factor (Martínez-Pérez et al. 2017a). UF performed with UF cells involved lower coefficient factors, which increases the percentage of HMW DOM recovered compared to systems equipped with UF coil membranes (Martínez-Pérez et al. 2017a).

2.4.2 Solid-Phase Extraction

Isolation by solid-phase extraction (SPE) is based on the selective adsorption of the DOM on a solid-phase extractant such as polystyrene-divinylbenzene polymers (SDVB) or bonded silica (C_{18}). These two types of sorbents present contrasting extraction mechanisms affecting the type of isolated molecules: C_{18} mainly retains nonpolar compounds, while SDVB polymers target a wider range of compounds,

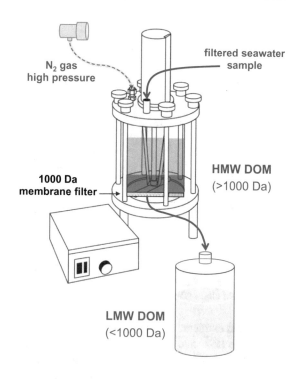

Fig. 2.15 Stirring UF cell fitted with a disc membrane (cut-off of 1000 Da) to size-fractionate more effectively the sample using small-volume samples (2 L)

N$_2$ gas high pressure

filtered seawater sample

HMW DOM (>1000 Da)

1000 Da membrane filter

LMW DOM (<1000 Da)

from nonpolar to polar (Ferrer and Barceló 1999). Among the SDVB sorbents, the most exploited on marine samples are the XAD and PPL resins (Table 1 in Waska et al. 2015). Classically, XAD polymers were used since the late 1970s (Aiken et al. 1979) yielding recoveries from 5% to 63% (averaging $34 \pm 16\%$; Perdue and Benner 2009), but involving a laborious process, including several steps of pre-cleaning and the sequential combination of different XAD resins. Recently, the PPL sorbents have been postulated as the best option for the SPE of marine DOM as they confirmed to retain highly polar to nonpolar molecules from large volumes of water (Dittmar et al. 2008). The use of these commercially pre-packed cartridges not only simplified the handling and improved the reproducibility but also increased the recoveries up to 40–75% (Green et al. 2014). In the open ocean, higher recoveries correspond to deep samples, where the DOM is more hydrophobic. Additionally, PPL SPE DOM is completely salt- and ash-free and yet more representative of the total bulk than C$_{18}$ SPE extracts (Dittmar et al. 2008). However, PPL extraction shows a preferential recovery of C over N, favouring DOM isolates with higher C:N ratios than the bulk DOM (Green et al. 2014), but still capturing the short-scale variation of this SPE DON (Osterholz et al. 2021).

SPE methodologies involve six main steps that can be more or less elaborated according to the type of sorbent. In particular, for PPL polymer, a scheme of the procedure is shown in Fig. 2.16. (1) The cartridge with the resin is activated/conditioned in advance. (2) The filtered sample is acidified to pH 2 beforehand.

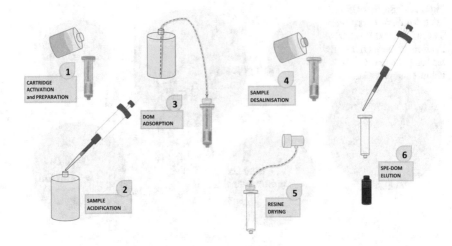

Fig. 2.16 Steps for solid-phase extraction using PPL resins: (1) cartridge activation and preparation, (2) sample acidification, (3) DOM absorption, (4) sample desalinization, (5) resin drying and (6) SPE-DOM elution/desorption

(3) DOM is extracted on board by passing the filtrate through the sorbent by gravity, cutting the flow rate down to 10 mL min^{-1}, as a compromise between ensuring sufficient interaction between sorbent and the organic molecules and a relatively expeditious process. (4) After extraction, cartridges are rinsed with acidified ultrapure water (pH 2) to remove remaining salts. (5) Next, the sorbents must be dried by flushing with inert gas as N_2 or Ar and (6) eluted with methanol to extract the SPE DOM. Methanol extracts can be stored in amber vials at -20 °C for long periods of time. Although complete processing including elution as soon as possible after sampling is preferable, if conditions do not allow handling of solvents on-site, the cartridges might be stored at 4 °C or -20 °C after desalting and prior to elution, to continue the extraction process when arriving at the laboratory.

PPL cartridges of 1 g of sorbent can extract the DOM of 5 L without a cost in efficiency. Larger samples can also be handled using bigger cartridges, for example, about 2500 L of seawater were processed using a custom-made cartridge of 500 g of PPL sorbent (Green et al. 2014). At a rate flow of 10 mL min^{-1}, it will take less than 9 h to pass a 5 L sample using a 1 g of resin. Taking into account that several samples can be processed at the same time and considering that this is a relative low effort/budget methodology, the PPL SPE isolation approach can be selected to perform oceanographic studies, as hundreds of samples can be achieved in a month-long cruise (Green et al. 2014). Additional cartridges for SPE have been evaluated (e.g. Li et al. 2017). Although these cartridges have lower recoveries compared to the PPL, they have the potential to target different compounds of interest.

2.4.3 Coupled Reverse Osmosis/Electrodialysis

The coupled reverse osmosis/electrodialysis (RO/ED) combines a water-selective membrane (reverse osmosis) which retains organic and inorganic solutes from water (concentration process) and a piled arrangement of anion and cation exchange membranes (electrodialysis). During RO/ED small charged solutes are transferred from the sample into a less saline solution, thus desalting the sample. The RO/ED was projected to be the ultimate marine DOM concentration methodology as it was proven to recover approximately the 70–75% of DOM from oceanic waters (Green et al. 2014). Nonetheless, the need of expensive and specialized equipment, together with the laborious handling, reduced the application of RO/ED in marine samples to essentially one research group (Green et al. 2014). Regarding the representativeness of the isolates, the C:N ratios and the optical properties of the RO/ED samples are within the range of the source water C:N values (Koprivnjak et al. 2009). Nevertheless, the final RO/ED isolates contain an elevated proportion of inorganic forms of boron and silicon, so they are unsuitable for certain analysis (Green et al. 2014).

The RO/ED experimental procedure to isolate DOM samples comprises three distinct operational stages. Firstly, the conductivity of the seawater samples is reduced using electrodialysis, which causes the removal of most of the ionic solutes (70%) and also a proportionate decrease in the osmotic pressure of the sample. During this phase, with an appropriate low pressure, water can be eliminated at an acceptable speed by reverse osmosis. Subsequently, both reverse osmosis and electrodialysis are used together to remove most of the water in the sample until reaching the final selected volume while maintaining a low conductivity. Finally, electrodialysis is used to lower the conductivity of the sample to a final low value. Although 99.997% of conductivity is removed, only roughly the 25% of the isolated sample is DOM, the remainder being residual water and sea salts. It takes 68 h to process 200 L of seawater using this approach (Perdue and Benner 2009).

2.5 Molecular Characterization

As mentioned before, the determination of specific compounds of the DOM pool, such as amino acids or monosaccharaides, is out of the scope of this chapter. Although the determination of specific compounds has generated significant advances in the knowledge of DOM composition and cycling, it is important to highlight that the fraction of DOM characterized by these specific molecular analyses comprise <10% of oceanic DOC (Benner 2002; Repeta 2015). Conversely, the most auspicious molecular techniques discussed here, mass spectrometry and nuclear magnetic resonance spectroscopy, although do not allow for the full identification of new compounds, aim to determine a broader fraction of the marine DOM, targeting other features of the DOM molecular composition. Characteristics as the distribution of main C, N and P functional groups or the major classes of compounds that contribute to the labile and refractory fractions of DOM, or changes in aromaticity, offer a more complete perspective, triggering the advances on the knowledge of DOM composition, which has reached 60–70% of the oceanic DOM (Repeta 2015).

2.5.1 Ultrahigh-Resolution Mass Spectrometry

The molecules present in marine DOM are highly diluted in a salt matrix, and because salts interfere with mass spectrometric analysis, organic species need to be isolated and concentrated to enable DOM molecular characterization (see Sect. 2.4).

Currently, ultrahigh-resolution mass spectrometers used for the analysis of natural organic matter include ion mobility quadrupole time-of-flight (IM Q-TOF-MS; Lu et al. 2018), Orbitrap (Hawkes et al. 2016; Phungsai et al. 2016; Simon et al. 2018; Pan et al. 2020) and Fourier transform ion cyclotron mass spectrometers (FT-ICR-MS; Stenson et al. 2003, Koch et al. 2005, Hertkorn et al. 2008, Zark and Dittmar 2018). Among these different analytical techniques, FT-ICR-MS was first reported in 1974 (Comisarov and Marshall 1974) and is currently one of the most recognized techniques to effectively characterize complex organic mixtures on a molecular level (Steen et al. 2020). The high resolution provided by FT-ICR-MS makes this instrumentation the most advanced mass spectrometer able to resolve complex organic mixtures providing elemental formulas over a wide range of masses. Because of the potential of this instrumentation in improving our understanding of marine DOM molecular composition, this section is focused on FT-ICR-MS analysis.

2.5.1.1 Ionization Techniques for FT-ICR-MS Analysis

All mass spectrometers need the molecules to be in the gas phase and charged (i.e. ionized) in order to be analysed. Currently, there are several commercially available ionization sources that can be coupled to FT-ICR-MS. The choice of ionization technique will influence the obtained mass spectra. Thus, depending on the sample nature, one can select the most appropriate technique to target the organic compounds of interest.

Electrospray Inoization (ESI) Among the ionization techniques, electrospray ionization was the first one to make the analysis of natural organic mixtures using FT-ICR-MS possible (McIntyre et al. 1997), and it is currently the most widely used for the analysis of DOM (Schmidt et al. 2011; Lechtenfeld et al. 2014; Seidel et al. 2014; Hansman et al. 2015; Medeiros et al. 2015; Osterholz et al. 2016; Martínez-Pérez et al. 2017b; Gomez-Saez et al. 2017; Rossel et al. 2017, 2020; Waska et al. 2019). The ESI technique allows to directly infuse a solution with the dissolved organic molecules to the FT-ICR-MS. Moreover, ESI is a "soft" ionization technique since the ionization takes place without fragmenting the molecules. Thus, the combination of ESI with FT-ICR-MS allows the simultaneous detection of intact molecules present in the DOM pool across a large mass range and without the need

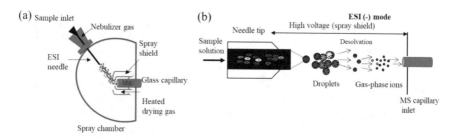

Fig. 2.17 (**a**) Representation of an electrospray ionization (ESI) source and (**b**) the ion formation

of additional steps to reduce DOM complexity prior to analyses (Dittmar and Stubbins 2014; Novotny et al. 2014).

The ESI technique is more efficient for polar compounds and typically causes protonation ([M + H]$^+$) and deprotonation ([M-H]$^-$) of molecules (basic and acidic, respectively; Henriksen et al. 2005, Marshall and Rodgers 2008, Konermann et al. 2013). The ions are formed by applying high voltage to a very fine tip where the molecules in the solvent are introduced by droplets in the spray. The initially large droplets are further reduced in size by evaporation, with the help of nitrogen gas, forcing the charges in the molecules to get closer until they repeal each other. This process results in the division of the droplets into smaller ones, a process that is repeated several times until the solvent is completely evaporated remaining only single, charged molecules (a process known as desolvation; Fig. 2.17). The positive ionization mode favours the detection of nitrogen-containing compounds (such as pyridinic homologues, some primary amines), while negative ESI favours the detection of species that are easily deprotonated like organic acids, phenols, S-containing compounds and "neutral nitrogen" species (Koch et al. 2007; Marshall and Rodgers 2008; D'Andrilli et al. 2010; Mapolelo et al. 2011). In both positive and negative modes, adducts, such as those formed with K$^+$, Li$^+$, NH$_4^+$ and Cl$^-$, can occur, respectively (Koch et al. 2007). The presence of these adducts can further complicate the spectra, and thus, they should be carefully evaluated.

Atmospheric pressure photoionization (APPI) and **atmospheric pressure chemical ionization (APCI)** are also commonly used to study complex organic mixtures (Purcell et al. 2006; Hertkorn et al. 2008; Hockaday et al. 2009; D'andrilli et al. 2010; Ventura et al. 2020). These techniques ionize a wider range of compound classes, and thus their mass spectra are usually more intensively populated by masses when compared to ESI (Marshall and Rodgers 2008). Moreover, they can produce, in addition to protonated molecules, radical ions (M$^{+\cdot}$; Kondyli and Schrader 2019), which require higher resolution to properly be identified. These ionization techniques also ionize low polarity and nonpolar compounds (Hanold et al. 2004; Purcell et al. 2006; Marshall and Rodgers 2008), thus extending the analysis of compounds not efficiently ionized by ESI. One big advantage is that they are less affected by salts, potential contaminants and ion suppression from the sample matrix

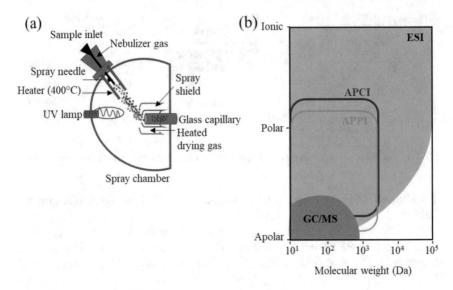

Fig. 2.18 (a) Representation of an atmospheric pressure photoionization (APPI) source and (b) comparison of the range of compounds ionized by the different techniques (after Hoffmann and Stroobant 2007). In the case of APCI, the UV lamp is replaced by a corona discharge needle in panel **a**

compared to ESI (Hanold et al. 2004; Hertkorn et al. 2008; Hockaday et al. 2009; Ventura et al. 2020).

In APPI, the sample molecules are mixed with a solvent that depending on the type used can increase the abundance of ions detected. Before entering the ionization chamber at atmospheric pressure, the solvent is evaporated with the assistance of a heated nebulizer (Fig. 2.18). Then the mixture of molecules and solvent is exposed to a krypton lamp that emits photons resulting in the ionization of the molecules (photoionization) that are measured by the mass spectrometer. Because not all analytes are ionized directly by this mechanism, dopant molecules are frequently used to increase the ionization efficiency. In this assisted ionization, the dopant (e.g. toluene, acetone) is photoionizable and acts as intermediate to ionize the sample molecules (Hoffmann and Stroobant 2007). Similar to APPI, APCI also uses atmospheric pressure, but it reaches the ionization using gas-phase ion molecule reaction with the solvent, which works as a reactant ion. The molecules in solution are introduced from a direct inlet probe into a pneumatic nebulizer where the mobile phase is transformed into a thin fog by the help of high-speed nitrogen beam. With the assistance of heat, the mobile phase is evaporated and the compounds in the gas flow are carried along a corona discharge electrode where ionization takes place. The APCI technique can produce positive or negative ions by different mechanisms. For the positive ionization mode, proton transfer or adduct of reactant gas ion can occur, while in the negative mode, proton abstraction or adduct formation has been described (Hoffmann and Stroobant 2007).

Additional Ionization Techniques Other more recent ionization techniques used for the analysis of organic mixtures include matrix-assisted laser desorption/ionization (MALDI) and laser desorption ionization (LDI) (Cao et al. 2015; Blackburn et al. 2017; Solihat et al. 2019). However, contrary to the abovementioned ionizations, which require solutions for the analyses, with MALDI and LDI techniques the sample is analysed in solid phase, so the lasers can ablate and ionize the compounds in the surface of the sample. As DOM is already in solution, these ionization techniques are not used but can be applied for standards like Suwannee River fulvic acid, for which LDI has shown to provide molecular information that is complementary to ESI (Blackburn et al. 2017). Another recent technique that utilizes sample placed in a surface is the paper spray ionization (PSI). For PSI, paper tips prepared with chromatographic paper are used as surface for the sample solution. A solvent-spray voltage is applied to the paper tip, which is placed in front of the mass spectrometer inlet. Although this ionization technique provides similar spectra to those obtained by ESI, it has the advantage of being less affected by salts, and it allows to work with less material (Kim et al. 2018). Another potential application of these ionization techniques is the analysis of DOM obtained by UF, which after desalting and freeze-drying can also be analysed in solid-phase state.

2.5.1.2 FT-ICR-MS Analysis

Once the sample molecules are ionized in the ionization source (see Sect. 2.5.1.1 for ionization techniques), they are transferred with the help of ion optics to the mass analyser (Fig. 2.19a). In the FT-ICR-MS, the mass to charge ratio (m/z) of the ions injected will be determined based on their cyclotron frequency in a magnetic field. Because ion trajectories in a magnetic field are curved, the ions are "trapped" in a circular trajectory, which is the principle of the ion cyclotron or penning trap (Hoffmann and Stroobant 2007). The charge of the ions will allow them to respond to the magnetic field running through the centre of the ion cyclotron cell (Fig. 2.19a).

The ions are rotating in tiny orbits, and the speed is related to their mass. Emitted radio frequency pulses by the electrodes in the excitation plates will allow the

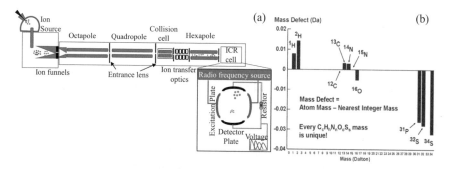

Fig. 2.19 (a) Schematic diagram of an FT-ICR-MS with its ICR cell inside a high magnetic field and (b) atomic mass defects for common chemical elements (reproduced from Marshall and Rodgers (2008) with permission from the National Academy of Science of the USA)

separation of the ions by weight. For the range of frequency emitted, each ion will respond to only one, a particular ion frequency that is a function of its mass. The ion will absorb the radio frequency energy and with it will increase its orbit around the magnetic field. After the radio frequency is removed, the ions rotating at their larger orbits will rotate in phase as "packets" of ions extending to a maximum radius approaching the detector plates. This will result in a stream of negatively charged electrons that are equal in charge to the packet of ions traveling to the detector's electrode. In this process the electrons will hunt the ions and a resistor will measure the voltage (the indirect measurement of the circling ions in the ICR cell) (Marshall et al. 1998; Hoffmann and Stroobant 2007). The instrument takes this information, and later the computer performs the Fourier transform of the data that provide the mass spectra illustrating *m/z* of the ions (molecules) in the analysed sample. Accordingly, the intensity of each mass peak in the mass spectrum is the abundance of ions in the mass analyser, which is considered to be related to the abundance of molecules in the sample solution. However, as indicated above, different ionization techniques play a key role in the ions detected by the mass spectrometer, and therefore one single ionization technique does not allow to see all the ions (molecules) present in the sample in one analysis.

Currently, with a FT-ICR-MS it is possible to determine the mass of a molecule with an accuracy superior to 0.1 mDa (for comparison, the mass of an electron is 0.5 mDa; Dittmar and Stubbins 2014). In order to calculate the molecular formula for each of the exact masses detected in the mass spectra, it is necessary to use the mass defect. Each isotope of every element has a distinct mass defect, which is the slight deviation of its mass from the integer mass given by the number of protons and neutrons (by definition ^{12}C is zero). Thus, if the mass of a molecule is measured with high accuracy, the combination of the exact masses of the elements to get that exact measured mass is unique (Fig. 2.19b; Marshall and Rodgers 2008). Accordingly, just based on the exact measured masses, it is possible to estimate molecular formulas of individual molecules present and, thus, to get an insight of DOM molecular diversity. Based on the FT-ICR-MS analysis of marine DOM, this pool is composed of molecules within the small size range of 250 to 550 Da (Fig. 2.20). It should be noted that the thousands of molecular formulas identified in the mass spectra are expected to be represented by a multitude of different isomers (Zark et al. 2017), i.e. the same molecular formula can be represented by compounds with different arrangements of atoms.

2.5.1.3 Applications and Visualizations to Assess DOM Complexity

Several methods are used to exploit the data provided by FT-ICR-MS measurements. These include different visualizations that will reduce the data dimension for easier evaluation and interpretation (e.g. the **van Krevelen diagram**, **Kendrick mass defect analysis**). Additionally, several indices have been described to highlight certain compound groups and molecular properties in the DOM (e.g. the **double bond equivalent** and the **aromaticity index**).

The **van Krevelen diagram** (VKD) is one of the most common visualizations to display the molecular formulas derived from FT-ICR-MS analysis. The VKD was

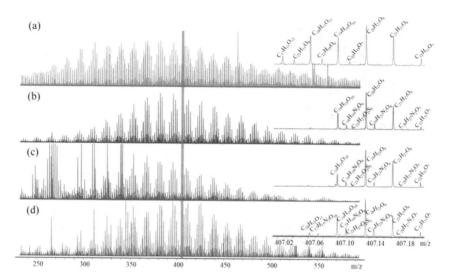

Fig. 2.20 Mass spectra from FT-ICR-MS measurements in ESI (negative mode) of organic mixtures from land to the ocean floor. Presenting the whole mass spectra and one exemplary nominal mass (407 Da, highlighted with the grey bar) of (**a**) Suwannee River fulvic acid reference material, (**b**) deep ocean DOM from 700 m water depth (Pacific Ocean), (**c**) hydrothermal fluid (250 °C) from 800 m water depth (Mid-Atlantic ridge) and (**d**) surface sediment porewater from 2400 m water depth (Arctic Ocean)

initially used to assess the evolution of oil and coal samples (van Krevelen 1950; Hatcher et al. 1989; Curiale and Gibling 1994). Later, Kim et al. (2003) underlined the advantage of using this visualization to facilitate the interpretation of complex mass spectral data obtained by FT-ICR-MS. In the VKD it is possible to represent major biogeochemical compounds (because known molecular classes use a specific location in the diagram), and they can also provide a hint of the diagenetic history of the compounds (due to gain or loss of C, H, O, N atoms) (Kim et al. 2003). In the VKD, the molecular formulas are presented according to their elemental ratios, usually hydrogen over carbon (H/C) and oxygen over carbon (O/C) (Fig. 2.21a).

Using these ratios, different molecular categories have been defined in order to make the evaluation of data easier. However, it should be considered that the assignments of these molecular categories only indicate that the molecular formula detected has the same space in the VKD as a known compound, although its molecular structure or functional groups may be different. Accordingly, it is common to report the name of the molecular category-like, e.g. peptide-like, to indicate that the molecular formula, which contains nitrogen, uses the same compositional space as peptides, but due to the fact that FT-ICR-MS does not provide structural information, it is not possible to call this molecular formula peptide. The definition of these molecular categories should consider several molecular constraints (elemental ratios, aromaticity and heteroatom content) and must be clearly reported.

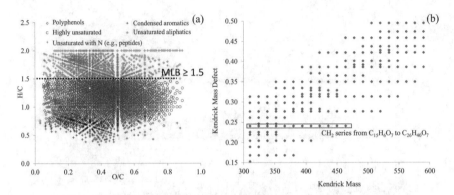

Fig. 2.21 (**a**) van Krevelen diagram (VKD) with commonly described molecular categories and the threshold of the molecular lability index (MLB) and (**b**) Kendrick mass defect analysis highlighting a CH_2 homologous series with the same Kendrick mass defect (0.240) and Z^* (-12)

Kendrick mass defect analysis has been widely used for the analysis of ultrahigh-resolution mass spectral data and allows to separate the mass peaks from the mass spectra that belong to same homologous family. Consequently, compounds that differ in composition by specific structural units (e.g. CH_2, COOH, CH_2O, etc.) can be displayed in a plot and those homologously related will follow a diagonal or horizontal line (Fig. 2.21b) (Hughey et al. 2001; Kim et al. 2003). Homologous series differing in CH_2 can be obtained with the following equations:

$$Kendrick\ mass = IUPAC\ mass\ measured \times 14/14.01565 \qquad (2.6)$$

$$Kendrick\ mass\ defect = (nominal\ mass - Kendrick\ mass) \qquad (2.7)$$

where nominal mass is the mass rounded to the nearest integer. Based on these equations, the members of an alkylation series will differ by exactly 14 Da and will have the same Kendrick mass defect. Consequently, it is possible to assign with high confidence the molecular formulas for the members of a homologous series in the lower mass range and then extend it to higher masses (Marshall and Rodgers 2008). Note that, as mentioned above, these equations can be extended to other structural units or reoccurring mass differences (Stenson et al. 2003; Merder et al. 2020).

In order to "isolate" the same alkyl chain homologue, it is necessary to calculate Z^* (Hsu et al. 1992) as follows:

$$Z^* = (modulus\,(nominal\,mass/14)\text{-}14) \qquad (2.8)$$

where the modulus is the remainder of the division of the nominal mass by 14, a residue that will be the same for all compounds that are part of the same series. Therefore, compounds with the same Kendrick mass defect and Z^* are part of the same homologous series (Fig. 2.21b).

The **double bond equivalent** (DBE) is a well-established tool in mass spectrometric analysis, and it represents the number of double bonds and rings in the molecule (McLafferty and Turecek 1993). The DBE is calculated based on the following equation:

$$DBE = 1 + 0.5(2C - H + N + P) \qquad (2.9)$$

where C, H, N and P represent the number of carbon, hydrogen, nitrogen and phosphorus atoms in each molecular formula. Based on this equation, higher DBE values will be obtained if the number of hydrogen atoms decreases due to increase in the unsaturation of the molecule (Koch and Dittmar 2006). However, the DBE does not consider the C=O double bonds, which are commonly found in DOM. Thus, to account for the presence of heteroatoms, like O, a different index was proposed.

The **aromaticity index** (AI) was introduced by Koch and Dittmar (2006) to provide a useful criterion for the identification of aromatic structures present in natural organic matter analysed by ultrahigh-resolution mass spectrometry. These aromatic compounds can derive from burning biomass and fossil fuels as well as petrogenic sources, and, thus, they can play a major role in the carbon cycle (Kramer et al. 2004; Dittmar and Koch 2006; Wagner et al. 2018, 2019; Rossel et al. 2017). The AI_{mod} indicates the density of C=C double bonds in the molecule and can be calculated for each molecular formula according to the following equation:

$$AImod = \frac{1 + C - 0.5O - S - 0.5(N + P + H)}{C - 0.5O - N - S - P} \qquad (2.10)$$

where O and S are the number of atoms of oxygen and sulphur in each molecular formula. For the AI_{mod}, the threshold of >0.5 and ≥ 0.67 are considered unambiguous indicators of the occurrence of aromatic and condensed aromatic compounds, respectively (Koch and Dittmar 2006, 2016). Note that in some cases the AI_{mod} can give values below 0 if the number of oxygen atoms exceeds the number of π bonds in the molecule. In this situation the AI_{mod} is defined as zero because it is unnecessary to estimate C=C double bonds for a compound without carbon atoms (Koch and Dittmar 2006).

Additional indices: the **nominal oxidation state of carbon** (NOSC) was proposed to determine the degradation potential of compounds present in natural organic matter. The idea behind it was that organic compounds with higher NOSC were easier to degrade under anoxic conditions by microorganisms (LaRowe and Van Cappellen 2011). The NOSC index is calculated as follows:

$$NOSC = 4 - \frac{(4C + H - 3N - 2O - 2S)}{C} \qquad (2.11)$$

assuming that the elements are in their initial oxidation state (H + 1, O = −2, N = −3 and S = −2) (Riedel et al. 2012).

The **molecular lability index** (MLB) was proposed by D'Andrilli et al. (2015) to assess the presence of labile compounds in natural organic matter. For this index, the authors proposed that organic compounds with an H/C ≥ 1.50 represent more labile chemical species, while those below this threshold correspond to more recalcitrant or less labile compounds. This threshold is based on the occurrence of known chemical classes in the VKD such as aliphatics and aromatics above and below the threshold, respectively (Fig. 2.21a). Based on this index, it has been reported that glacial ecosystems contain more labile organic matter than marine and freshwater environments (D'Andrilli et al. 2015).

The **Marine DOM diversity indices** were reported by Menges et al. (2017). Using different diversity measurements such as richness of molecular formulas, abundance-based diversity and functional molecular diversity, the authors showed temporal and spatial variability in DOM molecular diversity. The script to calculate functional diversity, which is used to determine the reactivity of a molecule, has been recently published (https://github.com/andreamentges/functional-diversity-index).

2.5.1.4 Data Processing

Ultrahigh-resolution via FT-ICR-MS provides large amount of data for single samples. Over the years several approaches have been reported to facilitate molecular formula assignments (Hughey et al. 2001; Kujawinski and Behn 2006; Koch et al. 2007; Tolić et al. 2017). Recently, tools of open access have been reported to facilitate the assignment of the molecular formulas (according to previously published criteria) but also to align the samples that need to be compared, therefore obtaining a sample set in a single file (Savory et al. 2011; Tolić et al. 2017; Leefmann et al. 2019; Merder et al. 2020). This advance in automatization for data analysis allows now to compare several samples at the same time (big datasets).

For big datasets, statistical analyses are essential to get insights of compositional differences among samples. Differences in composition are given by the presence or absence of compounds (i.e. above or below the detection limit) or based on the relative abundance of ions (molecules) detected. Accordingly, it is possible to use the mass peak intensity, which is frequently normalized to the total intensities of mass peaks in each sample, to represent relative abundance or presence/absence of molecular formulas. To work with presence and absence of molecular formulas, methods based on Jaccard dissimilarity are used, while for the comparison of relative abundances, Bray-Curtis dissimilarities are used.

Commonly used statistical methods for the evaluation of compositional changes in a dataset that can be performed directly with the relative abundances are principal component analysis (PCA), principal coordinate analysis (PCOA), non-metric multidimensional scaling (NMDS) and cluster analysis (Fig. 2.22). Among these analyses, PCOA and NMDS are more appropriate for the evaluation of data from FT-ICR-MS, because they are less influenced by zero entrance in the dataset (i.e. zero has less weight in the analysis, which is suitable for data where zero just represents below detection limit and not necessarily absence).

It is also possible to extend the statistical analyses to get an idea on how the changes in composition between samples are related to the environmental factors or how key parameters evaluated under experimental conditions modify the molecular composition of the DOM. This type of evaluations can be performed by redundancy analysis (RDA), PCOA and partial least square analysis (PLS) (e.g. Osterholz et al. 2015, 2016; Seidel et al. 2014; Rossel et al. 2017, 2020; Catalá et al. 2020). The VKD can also be extended to a 3D plot if information from the statistical analysis is

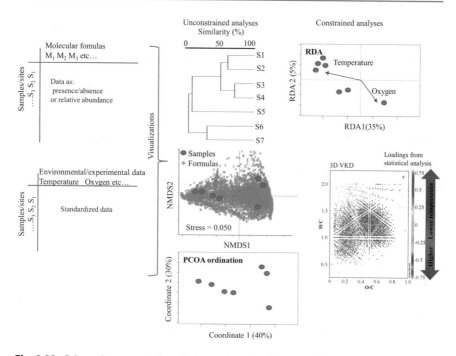

Fig. 2.22 Schematic representation of changes in molecular composition by cluster, NMDS and PCOA (unconstrained analyses) and the relation between the compositional variability and the environmental factors (constrained analysis). Blue circles represent the samples, and the more distant to each other they are, the more dissimilar their molecular composition. In the constrained analysis (e.g. RDA), environmental factors are represented by the red arrows. The relation between the environmental factors, as example temperature, and the sample compositions is displayed in a 3D VKD using the loadings of the first RDA axis

included. For instance, VKD could display in the z-axis the information of the intensities, masses or heteroatom content of those molecular formulas that correlate to certain environmental factors (3D in VKD in Fig. 2.22).

2.5.1.5 Limitations

Despite the big amount of data provided by FT-ICR-MS analysis, this method, like any other, has its limitations. The information provided by the FT-ICR-MS, represented by the signal intensity, depends on many factors and not just in the abundance of molecules in solution. As indicated previously, the ionization technique used and the efficiency at which the molecules in the sample are ionized and, thus, are "visible" to the mass analyser are crucial. Consequently, the obtained data are a semiquantitative measurement of the relative abundance of molecules under the specific conditions used in the instrument in combination with the methods applied during sample preparation.

In order to select the most appropriate method for DOM analysis, it is necessary to consider the nature of the DOM material and the research questions that need to be addressed. It is also highly recommended to monitor instrumental variability using reference samples during the analysis (i.e. reference material analysed between samples to evaluate the stability of the instrumental signal). Additionally, procedural blanks (i.e. ultra-clean water treated, extracted and analysed as the samples) and replicate analysis are also essential to corroborate the obtained data. Moreover, it is necessary to consider that statistical analysis always provides a result that is given by a correlation, but it is the scientists who need to interrogate the outcome of the analysis. In the future, validation of both FT-ICR-MS data and statistical analysis may be improved by comparing the obtained information with data publicly available. Making information publicly available in data depositories will allow us to find, access, integrate and reuse data (Wilkinson et al. 2016), promoting new discoveries and with the potential to deliver new knowledge about the DOM molecular composition.

2.5.2 Nuclear Magnetic Resonance Spectroscopy

Nuclear magnetic resonance (NMR) spectroscopy is the most powerful tool for the molecular level characterization of DOM in environmental samples through homo- and heteronuclear correlation experiments (Simpson et al. 2011). NMR experiments can be performed on the material in solid, gel, liquid and gas phases, although for marine DOM samples only solid- and liquid-state (also denoted as solution-state) studies have been conducted (Mopper et al. 2007). Atomic nuclei with a non-zero nuclear spin number, i.e. with an odd number of protons or neutrons in its nucleus (e.g. ^1H, ^{13}C, ^{15}N, ^{31}P), possess magnetic moment and behave like tiny bar magnets. In the absence of an external magnetic field, these magnetic moments in a sample orient in random directions such that the resultant net magnetization is zero. When a strong magnetic field is applied, these nuclei either align or oppose in the direction of the applied magnetic field and occupy the nuclear energy levels according to the Boltzmann population distribution. Application of a radio frequency pulse for a few microseconds flips the equilibrium bulk magnetization into transverse plane, and this precessing nuclear transverse magnetization induces an oscillating electric field in the coil near to it known as NMR signal or free induction decay (FID). Fourier transformation of the time domain signal (FID) results in an NMR spectrum with several signals in frequency units (δ in ppm). The effective magnetic field experienced by each nucleus varies depending upon the electron density surrounding the nucleus, which results in the shielding (shifts towards lower ppm) and deshielding (shifts towards higher ppm) of the resonances. The signals in the NMR spectrum appear as J-multiplets due to scalar coupling interactions with adjacent nuclei. NMR determinations do not change the chemical nature of the sample, so it is a non-destructive analysis (Mopper et al. 2007).

The two main nuclei studied by NMR are protons (^1H) and carbons (^{13}C). With its high gyromagnetic ratio and 99.98% natural isotopic abundance, proton is the highly sensitive nuclei in NMR, while carbon is relatively insensitive due to the low (1.1%) natural abundance of NMR active ^{13}C and low gyromagnetic ratio. The NMR

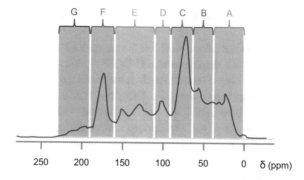

Fig. 2.23 ^{13}C NMR spectrum with molecular group structures assigned to the various spectral regions. A, paraffinic carbons from lipids and biopolymers (δ 0–45); B, methoxyl, mainly from lignin, and amino groups (δ 45–60); C, carbohydrate carbons (δ 60–90); D, carbohydrate anomeric and H-substituted aromatic carbons (δ 90–120); E, C- and O-substituted aromatic carbons, mainly from lignin and tannins (δ 120–160); F, carboxylic and aliphatic amide carbons from degraded lignin and fatty acids (δ 160–190); G, aldehyde and ketone carbons (δ 190–220) (after Mopper et al. 2007 and Dria et al. 2002)

analysis of dissolved organic nitrogen (DON) is very limited due to the low natural isotopic abundance (0.37%) of ^{15}N and its low gyromagnetic ratio. Another attractive and sensitive probe for the NMR structural characterization of marine DOM is the ^{31}P due to its 100% natural isotopic abundance and relatively higher gyromagnetic ratio. However its relative elemental abundance in marine DOM is low (Sect. 2.3.1); still it would be the sensitive nuclei in NMR compared to ^{13}C and [15]N. Pioneering works on NMR spectroscopy of oceanic DOM revealed difficulties to interpret these spectra due to the vast diversity of major functional groups present in the samples, which generate complex spectra with broad and overlapping peaks. Advances in NMR techniques and the increase in the number of studies allowed establishing a correspondence between NMR chemical shift ranges and molecular structures, by matching the unknown spectral data to published ^{1}H, ^{13}C, ^{15}N and ^{31}P NMR databases. For environmental DOM samples, there are several published works detailing chemical shift ranges for the most common chemical structures (Knicker and Lüdemann 1995; Kögel-Knabner 1997; Dria et al. 2002). In these studies, the selected spectral regions are assigned to the key structural groups, which are integrated to estimate the relative abundance of these molecular classes in marine DOM. Figure 2.23 illustrates a characteristic ^{13}C NMR spectrum for marine DOM, displaying the correspondences with the main functional groups found in the sample. Proton and carbon NMR spectra are complementary, each showing the majority of molecular structures that can be identified by NMR and that most of them can be assigned to carbohydrates (Repeta 2015).

2.5.2.1 Solid- Vs. Liquid-State NMR Spectroscopy of Marine DOM Samples

NMR is a very versatile technique that can be performed on solids or solutions. Regarding solid-state NMR, marine extracted DOM samples must be isolated from its salt matrix (including bound paramagnetic trace metals) and processed as a solid (usually by freeze-drying) before analysis. Therefore, it is suitable for HMW DOM isolates obtained by UF (see Sect. 2.4). For liquid-state NMR, the sample must be dried and then re-dissolved in an appropriate deuterated solvent. The solvent of choice is often determined by the solubility of the sample. As, per definition, DOM is soluble, it is likely suited to liquid-state NMR studies, so deuterated solvents can be used directly on the salt-free SPE DOM dry elutes or the freeze-dry HMW DOM isolates without additional sample preparation. Considering that modern liquid-state NMR generally produces higher-resolution spectra than the solid-state NMR and that there is a variety of one-dimensional (1D) and two-dimensional (2D) approaches that can be applied, i.e. involving one or two nuclei, respectively, liquid-state NMR seems the ideal choice for the study of marine DOM. However, both methodologies have their own advantages and drawbacks. Solid-state NMR methods are useful when the material is not soluble in any of the solvents and not amenable for crystallization. In solid-state NMR, the sample is placed inside the receiver coils of the NMR probe as a solid (most concentrated), which is very appropriate as NMR spectroscopy is an inherently insensitive technique; therefore, it is usually necessary to maximize the amount of sample analysed. In liquid NMR, the intensity of the detected signal is dependent on the amount of DOM that can be dissolved into approximately 600 μL of solvent in a 5 mm NMR tube. However, dissolving too much DOM usually results in organic matter aggregation, which can lower the signal-to-noise ratio and spectral resolution. Solvent effects that may alter chemical shifts are avoided in solid-state NMR. Nevertheless, there are also disadvantages to solid-state NMR. One is sample handling. Perfect packing of the small rotors with solid material is a crucial and very tedious process for obtaining stable magic angle spinning. Often rotors crash, cause overheating and damage the probe. Another shortcoming is the longer delay time (due to spin-lattice relaxation processes), which causes longer acquisition times (Mopper et al. 2007), although using the cross-polarization technique, which involves the polarization transfer from protons to heteronuclei (^{13}C or ^{15}N), short recovery delays are required. The main disadvantage of solid-state NMR techniques is the very poor spectral resolution due to strong anisotropic interactions such as direct dipole-dipole interactions and chemical shift anisotropy. Also, the fact that the instrumentation for solids NMR is not standard requires special NMR probes, high power decoupling and high signal generation power levels. As long as DOM is soluble (as it should be), liquid NMR is the best approach for the structural analyses of DOM. Of course, if one has both liquid- and solid-state NMR capabilities available, it is advisable to try both to obtain highly complementary structural information of the DOM. To date, the ^1H nucleus has been the most widely used probe for liquid-state NMR studies of DOM, whereas the other nuclei (^{13}C, ^{15}N and ^{31}P) are often analysed by solid-state NMR (Simpson and Simpson 2009).

2.5.2.2 One-Dimensional Solid- and Liquid-State NMR Spectroscopy

In the 1990s and early 2000s, several key works based on 1D NMR were published to present innovative insights regarding the molecular composition and structure of marine DOM (Table 2.2). In order to apply these NMR procedures to DOM, these pioneering studies used tangential UF (see Sect. 2.4) to isolate DOM, obtaining, in general, compositional changes with water depth. Thus, in upper oceanic waters, complex polysaccharides are the most important components of HMW DOM (about 50%), as demonstrated by solid-state ^{13}C NMR spectra (Benner et al. 1992) and liquid-state ^{1}H NMR spectra (Aluwihare et al. 1997), while in deeper layers of the ocean, carbohydrates are minor, being mostly replaced by aliphatic structures. Both ^{13}C and ^{1}H NMR spectra reveal that marine DOM is mostly aliphatic, attributed to its predominantly autochthonous source, and that aromatic functional groups are not significant contributors to the identified chemical structures (Benner et al. 1992). Aluwihare et al. (1997) showed that biosynthetically derived acetylated carbohydrate polymers, associated with many amino sugars, with chemical shifts in the range of approximately 1–2 ppm, contribute significantly to the DOM in ocean water. Overall, HMW DOM is composed of two key components: a polysaccharide fraction referred to as acylated polysaccharide (APS; Aluwihare et al. 1997) or heteropolysaccharide (HPS; Hertkorn et al. 2006) and a carboxylic acid-alkyl carbon-rich fraction referred to as carboxyl-rich alicyclic molecules (CRAM; Hertkorn et al. 2006).

Regarding the studies of DOM by ^{15}N NMR, McCarthy et al. (1997) established that the majority of organic nitrogen of marine HMW DOM from surface and deep waters of the North Pacific Ocean is composed of amide structures (mostly as microbial peptidoglycan; McCarthy et al. 1998). This first oceanic DOM ^{15}N NMR spectrum displayed one broad peak in the region of the spectrum that is conventionally assigned to amide N. Accordingly, the authors suggested that the majority of HMW DON exists as non-hydrolysable amide biopolymers, perhaps associated with chitin or peptidoglycans from bacteria, i.e. nitrogenous substances that are resistant to bacterial degradation, while the hydrolysable amino acids only have a marginal contribution to their DOM samples (McCarthy et al. 1997). A small contribution of nitrogen heterocyclic structures (such as indole- or pyrrole-like) was also observed, revealing a higher impact in the deep-sea samples, which is

Table 2.2 Major molecular groups found in marine HMW DOM (20–40% of the bulk DOM) by means of 1D NMR spectroscopy analyses

Type	Compounds		Reference
^{13}C	Polysaccharides	25–50%	Benner et al. (1992)
^{1}H	Carbohydrates	60–80%	Aluwihare et al. (1997)
	Acetates	5–7%	
^{15}N	Amides	65–86%	McCarthy et al. (1997)
	Nitrogen heterocyclics	6–26%	
^{31}P	Phosphorus esters	75%	Clark et al. (1998)
	Phosphonates	25%	

compatible with the fact that these structures are among the most stable organic N forms and they are found in petroleum, coals and marine sediments (McCarthy et al. 1997). A subsequent study located two distinct pools of organic nitrogen in the ocean, one composed by N-acetyl amino polysaccharides (which would include peptidoglycan) and another comprise of amide N (Aluwihare et al. 2005). In the same way as ^{15}N, only a few studies applying ^{31}P solid-state NMR have been reported for marine DOM. Clark et al. (1998) examined HMW DOM from the North Pacific Ocean by solid-state ^{31}P NMR spectroscopy, obtaining signals associated with phosphate esters and a minor peak for phosphonate structures. The strong chemical shift anisotropy for P compounds induced a dominance of the spinning sidebands, causing the broadening of the NMR peaks. ^{1}H and ^{13}C NMR spectra from RO/ED marine DOM isolates (see Sect. 2.4) were acquired for the first time in 2009, observing a divergence from previous studies based on lower-yield concentration methodologies (Koprivnjak et al. 2009). Remarkably, 1D NMR analysis of RO/ED samples comprised a lower proportion of carbohydrates and a higher proportion of alkyl carbon (probably CRAMs) compared to 1D NMR studies of HMW DOM extracts.

2.5.2.3 Two-Dimensional Liquid-State NMR Studies

Very few works have applied 2D NMR to molecular structure research of DOM. The cause behind the lack of these studies is likely the relatively large amount of DOM material needed for these analyses (10–100 mg, ideally >50 mg), which is time-consuming and expensive to isolate (Green et al. 2014). The major advantage of 2D NMR over 1D NMR is the ability to distinguish between the overlapping signals existing in large molecules to obtain more detailed structural information by deciphering spectral dispersion and connectivity data. There are numerous types of 2D liquid-state NMR spectroscopy carefully summarized in the "top ten" (prioritizing maximum volume of information in the shortest extent of time) NMR experiments for the study of natural organic matter in solution (Table 1 of Simpson et al. 2011). Among all these experiments, COrrelation SpectroscopY (COSY) provides connectivity information between neighbouring protons that are J-coupled to each other, whereas TOtal Correlation SpectroscopY (TOCSY) provides correlations between all the protons of an unbroken spin network. Figure 2.24 displays clear examples of COSY (panel **a**) and TOCSY (panel **b**) and the differences between them. On the other hand, during a heteronuclear single quantum correlation (HSQC) experiment, magnetization is transferred from a highly abundant nucleus, ^{1}H, to the inherently less sensitive and less abundant nuclei (e.g. ^{13}C or ^{15}N) that is directly attached to the proton. This transfer of magnetization results in spectral cross-peaks that intersect the chemical shifts of ^{1}H and the ^{13}C (or ^{15}N) nuclei to which it is bonded, providing information on ^{1}H–^{13}C (or ^{15}N) one-bond correlations. Heteronuclear multiple bond correlation (HMBC) relies on the long-range carbon (or nitrogen)-proton J-couplings and is used to identify the carbon (or nitrogen) and protons that are two, three, four and rarely five bonds apart from each other. Figure 2.24 also shows the bases and results from HSQC (panel **c**) and HMBC (panel **d**) experiments in a very comprehensible way.

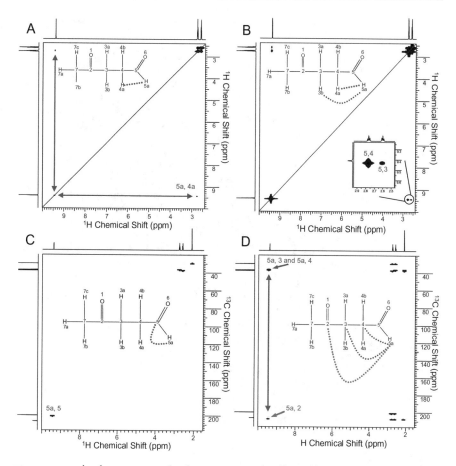

Fig. 2.24 (a) ^{1}H–^{1}H COSY, (b) ^{1}H–^{1}H TOCSY, (c) ^{1}H–^{13}C HSQC and (d) ^{1}H–^{13}C. HMBC spectra of 4-oxopentanal. For clarity, only key assignments have been given as an example. Reproduced from Simpson and Simpson (2009) with permission of John Willey and Sons

Hertkorn et al. (2006, 2013) were the first achieving 2D NMR spectroscopy on marine DOM isolated by UF (Hertkorn et al. 2006) and SPE (Hertkorn et al. 2013) at several depths in the Pacific and Atlantic oceans, respectively. These studies were able to identify some of the major components present in oceanic DOM. Figure 2.25 shows an HSQC spectrum from Hertkorn et al. (2006) where key structural components as heteropolysaccharides, CRAMs, aliphatic compounds and peptides are identified. While previous studies based on 1D liquid- and solid-state NMR have debated the presence of some of these structures (e.g. Benner et al. 1992; Aluwihare et al. 1997), these 2D NMR studies demonstrate its existence and also estimate its contribution through the water column. Comparing surface and deep waters, heteropolysaccharides accounted for 60.1% and 27.1% of DOM, respectively, while CRAMs constituted 22.9% and 50.9%, respectively. Additionally, peptides

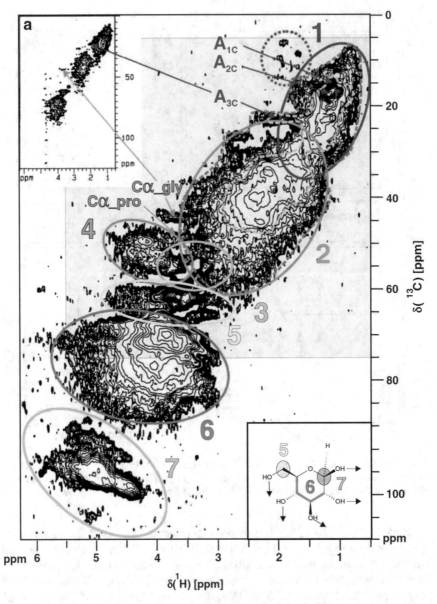

Fig. 2.25 2D ¹H–¹³C HSQC NMR spectrum of surface oceanic HMW DOM. The main assignments can be summarized in seven groups of major constituents as 1 = methyl bound to carbon and sulphur (dotted circle), and in the lower left corner, branched purely aliphatic CH pairs and polymethylene, 2 = methylene and methine cross-peaks without direct bonds to heteroatoms, 3 = low intensity cross-peaks from methoxyl, 4 = cross-peaks representative mainly of α-CH in proteins and vicinal dicarboxylic acids, 5 = carbohydrate methylene cross-peaks, 6 = carbohydrate methine cross-peaks and 7 = anomeric units in carbohydrates. A_{1C} = (poly)alanine-CH$_3$; A_{2C} = methylated carbohydrates; A_{3C} = N-acetyl carbohydrate. Reproduced from Hertkorn et al. 2006 with permission from Elsevier Science

comprised 23.6% and 20.7% of DOM in surface and deep waters, respectively (Hertkorn et al. 2006). In some cases, CRAMs account for more than 50% of DOM, potentially making them the most abundant molecule type on Earth.

Acknowledgements We wish to express our gratitude to Dr. Sahithya Phani Babu Vemulapalli and Prof. Thorsten Dittmar for their critical reading of the chapter and their useful suggestions. XAA-S and MN-C acknowledge financial support from the Spanish "Plan Estatal" project e-IMPACT (grant number PID2019-109084RB-C22). PER acknowledges financial support from the European Research Council (ERC) under the European Union's Horizon 2020 Research and Innovation Programme (ERC Synergy Grant "Deep Purple" under grant agreement No. 856416). PER also thanks the funding from Schlumberger oil service and ERC project ABYSS (Grant N° 294757 to Antje Boetius) for their support during the 10 years of work on FT-ICR-MS at the laboratory of Thorsten Dittmar. MJ Pazó helped with the preparation of some figures.

References

Aiken GR, Thurman EM, Malcolm RL, Walton HF (1979) Comparison of XAD macroporous resins for the concentration of fulvic acid from aqueous solution. Anal Chem 51:1799–1803
Aluwihare LI, Repeta DJ, Chen RF (1997) A major biopolymeric component to dissolved organic carbon in surface sea water. Nature 387:166–169
Aluwihare LI, Repeta DJ, Pantoja S, Johnson CG (2005) Two chemically distinct pools of organic nitrogen accumulate in the ocean. Science 308:1007–1010
Álvarez-Salgado X, Arístegui J (2015) Organic matter dynamics in the canary current. In: Valdés L, Déniz-González D (eds) Oceanographic and biological features in the canary current large marine ecosystem, IOC tech. Ser. No. 115, IOC-UNESCO, Paris, pp 151–160
Álvarez-Salgado XA, Miller AEJ (1998) Simultaneous determination of dissolved organic carbon and total dissolved nitrogen in seawater by high temperature catalytic oxidation: conditions for accurate shipboard measurements. Mar Chem 62:325–333. https://doi.org/10.1016/S0304-4203(98)00037-1
Anderson LA (1995) On the hydrogen and oxygen content of marine phytoplankton. Deep-Sea Res I Oceanogr Res Pap 42:1675–1680. https://doi.org/10.1016/0967-0637(95)00072-E
Andrew AA, Del Vecchio R, Subramaniam RA, Blough NV (2013) Chromophoric dissolved organic matter (CDOM) in the equatorial Atlantic Ocean: optical properties and their relation to CDOM structure and source. Mar Chem 148:33–43
Armstrong FAJ, Tibbitts S (1968) Photo-chemical combustion of organic matter in seawater for nitrogen, phosphorus and carbon determination. J Mar Biol Assoc UK 48:143–152
Azam F (1998) Microbial control of oceanic carbon flux: the plot thickens. Science 280:694–696. https://doi.org/10.1126/science.280.5364.694
Barrón C, Duarte CM (2015) Dissolved organic carbon pools and export from the coastal ocean. Global Biogeochem Cycles 29:1725–1738. https://doi.org/10.1002/2014GB005056
Benner R (2002) Chemical composition and reactivity. In: Hansell DA, Carlson CA (eds) Biogeochemistry of marine dissolved organic matter. Academic Press, pp 59–85
Benner R, Biddanda B, Black B, McCarthy M (1997) Abundance, size distribution, and stable carbon and nitrogen isotopic compositions of marine organic matter isolated by tangential-flow ultrafiltration. Mar Chem 57:243–263
Benner R, Pakulski JD, McCarthy M, Hedges JI, Hatcher PG (1992) Bulk chemical characteristics of dissolved organic—matter in the ocean. Science 255:1561–1564
Berggren M, Laudon H, Haei M, Ström L, Jansson M (2010) Efficient aquatic bacterial metabolism of dissolved low-molecular-weight compounds from terrestrial sources. ISME J 4:408–416. https://doi.org/10.1038/ismej.2009.120

Blackburn JWT, Kew W, Graham MC, Uhrin D (2017) Laser desorption/ionization coupled to FTICR mass spectrometry for studies of natural organic matter. Anal Chem 89:4382–4386. https://doi.org/10.1021/acs.analchem.6b04817

Boehme J, Coble P, Conmy R, Stovall-Leonard A (2004) Examining CDOM fluorescence variability using principal component analysis: seasonal and regional modeling of three-dimensional fluorescence in the Gulf of Mexico. Mar Chem 89:3–14

Bricaud A, Morel A, Prieur L (1981) Absorption by dissolved organic matter of the sea (yellow substance) in the UV and visible domains. Limnol Oceanogr 26:43–53

Bronk DA, Lomas MW, Glibert PM, Schukert KJ, Sanderson MP (2000) Total dissolved nitrogen analysis: comparisons between the persulfate, UV and high temperature oxidation methods. Mar Chem 69:163–178. https://doi.org/10.1016/S0304-4203(99)00103-6

Buesseler K (1996) Introduction to "use of cross-flow filtration (CFF) for the isolation of marine colloids". Mar Chem 55:vii–viii

Buesseler KO, Antia AN, Chen M, Fowler SW, Gardner WD, Gustafsson O, Harada K, Michaels AF, van der Loeff M, Sarin M (2007) An assessment of the use of sediment traps for estimating upper ocean particle fluxes. J Mar Res 65:345–416

Butturini A, Ejarque E (2013) Technical note: dissolved organic matter fluorescence—a finite mixture approach to deconvolve excitation-emission matrices. Biogeosciences 10:5875–5887

Calleja ML, Ansari MI, Røstad A, Silva L, Kaartvedt S, Irigoien X, Morán XAG (2018) The mesopelagic scattering layer: a hotspot for heterotrophic prokaryotes in the Red Sea twilight zone. Front Mar Sci 5:259. https://doi.org/10.3389/fmars.2018.00259

Cao D, Huang H, Hu M, Cui L, Geng F, Rao Z et al (2015) Comprehensive characterization of natural organic matter by MALDI- and ESI-Fourier transform ion cyclotron resonance mass spectrometry. Anal Chim Acta 866:48–58. https://doi.org/10.1016/j.aca.2015.01.051

Catalá TS, Martínez-Pérez AM, Nieto-Cid M, Álvarez M, Otero J, Emelianov M, Reche I, Arístegui J, Álvarez-Salgado XA (2018) Dissolved organic matter (DOM) in the open Mediterranean Sea. I. Basin wide distribution and drivers of chromophoric DOM. Prog Oceanogr 165: 35–51. https://doi.org/10.1016/j.pocean.2018.05.002

Catalá TS, Reche I, Fuentes-Lema A, Romera-Castillo C, Nieto-Cid M, Ortega-Retuerta E, Calvo E, Álvarez M, Marrasé C, Stedmon CA, Álvarez-Salgado XA (2015) Turnover time of fluorescent dissolved organic matter in the dark global ocean. Nat Commun 6:5986. https://doi.org/10.1038/ncomms6986

Catalá TS, Rossel PE, Álvarez-Gómez F, Tebben J, Figueroa FL, Dittmar T (2020) Antioxidant activity and phenolic content of marine dissolved organic matter and their relation to molecular composition. Front Mar Sci 7:984. https://doi.org/10.3389/fmars.2020.603447

Cauwet G (1984) Automatic determination of dissolved organic carbon in seawater in the sub-ppm range. Mar Chem 14:297–306

Cauwet G (1999) Determination of dissolved organic carbon (DOC) and nitrogen (DON) by high temperature combustion. In: Grashoff K, Kremling K, Ehrhardt M (eds) Methods of seawater analysis, 3rd edn. Wiley-VCH, Weinheim, pp 407–420

Chen H, Zheng B, Song Y, Qin Y (2011) Correlation between molecular absorption spectral slope ratios and fluorescence humification indices in characterizing CDOM. Aquat Sci 73:103–112. https://doi.org/10.1007/s00027-010-0164-5

Clark LL, Ingall ED, Benner R (1998) Marine phosphorus is selectively remineralized. Nature 393: 426

Coble PG (2007) Marine optical biogeochemistry: the chemistry of ocean color. Chem Rev 107: 402–418. https://doi.org/10.1021/cr050350

Comisarov MB, Marshall AG (1974) Fourier transform ion cyclotron resonance spectroscopy. Chem Phys Lett 25:282–283. https://doi.org/10.1016/0009-2614(74)89137-2

Cory RM, McKnight DM (2005) Fluorescence spectroscopy reveals ubiquitous presence of oxidized and reduced quinines in dissolved organic matter. Environ Sci Technol 39:5142–8149

Creighton TE (1993) Proteins, structure and molecular properties, 2nd edn. Freeman, New York

Curiale JA, Gibling MR (1994) Productivity control on oil shale formation-Mae Sot Basin, Thailand. Org Geochem 21:67–89. https://doi.org/10.1016/0146-6380(94)90088-4

Cyr F, Tedetti M, Besson F, Beguery L, Doglioli AM, Petrenko AA, Goutx (2017) A new glider-compatible optical sensor for dissolved organic matter measurements: test case from the NW Mediterranean sea. Front Mar Sci 4:89. https://doi.org/10.3389/fmars.2017.00089

D'andrilli J, Cooper WT, Foreman CM, Marshall AG (2015) An ultrahigh-resolution mass spectrometry index to estimate natural organic matter lability. Rapid Commun Mass Spectrom 29: 2385–2401. https://doi.org/10.1002/rcm.7400

D'Andrilli J, Dittmar T, Koch BP, Purcell JM, Marshall AG, Cooper WT (2010) Comprehensive characterization of marine dissolved organic matter by Fourier transform ion cyclotron resonance mass spectrometry with electrospray and atmospheric pressure photoionization. Rapid Commun Mass Spectrom 24:643–650. https://doi.org/10.1002/rcm.4421

Dafner EV (2016) An assessment of analytical performance of dissolved organic nitrogen and dissolved organic phosphorus analyses in marine environments: a review. Int J Environ Anal Chem 96:1188–1212. https://doi.org/10.1080/03067319.2016.1246662

Dafner EV, Szmant AM (2014) A modified segmented continuous flow analysis method for simultaneous determination of total dissolved nitrogen and phosphorus in marine environments. Limnol Oceanogr Methods 12. https://doi.org/10.4319/lom.2014.12.577

De Haan H (1993) Solar UV-light penetration and photodegradation of humic substances in peaty Lake water. Limnol Oceanogr 38:1072–1076

De Haan H, De Boer T (1987) Applicability of light absorbance and fluorescence as measures of concentration and molecular size of dissolved organic carbon in humic Lake Tjeukemeer. Water Res 21:731–734. https://doi.org/10.1016/0043-1354(87)90086-8

Del Vecchio R, Blough NV (2002) Photobleaching of chromophoric dissolved organic matter in natural waters: kinetics and modeling. Mar Chem 78:231–253. https://doi.org/10.1016/S0304-4203(02)00036-1

Dittmar T (2015) Reasons behind the long-term stability of dissolved organic matter. In: Hansell DA, Carlson CA (eds) Biogeochemistry of marine dissolved organic matter, 2nd edn. Academic Press, Boston, pp 369–385

Dittmar T, Koch B, Hertkorn N, Kattner G (2008) A simple and efficient method for the solid-phase extraction of dissolved organic matter (SPE-DOM) from seawater. Limnol Oceanogr Methods 6:230–235

Dittmar T, Koch BP (2006) Thermogenic organic matter dissolve in the abyssal ocean. Mar Chem 102:208–217. https://doi.org/10.1016/j.marchem.2006.04.003

Dittmar T, Stubbins A (2014) Dissolved organic matter in aquatic systems. In: Holland HD, Turekian KK (eds) Treatise on geochemistry. Elsevier, Amsterdam, pp 125–156. https://doi.org/10.1016/B978-0-08-095975-7.01010-X

Doval MD, Fraga F, Perez FF (1997) Determination of dissolved organic nitrogen in seawater using Kjeldahl digestion after inorganic nitrogen removal. Oceanol Acta 20:713–720

Dria KJ, Sachleben JR, Hatcher PGJ (2002) Solid-state carbon-13 nuclear magnetic resonance of humic acids at high magnetic field strengths. J Environ Quality 31:393–401

Ejarque-González E, Buturini A (2014) Self-organising maps and correlation analysis as a tool to explore patterns in excitation-emission matrix data sets and to discriminate dissolved organic matter fluorescence components. PLoS One 9:e99618. https://doi.org/10.1371/journal.pone.0099618

Engelhaupt E, Bianchi TS, Wetzel RG et al (2003) Photochemical transformations and bacterial utilization of high-molecular-weight dissolved organic carbon in a southern Louisiana tidal stream (Bayou Trepagnier). Biogeochemistry 62:39–58. https://doi.org/10.1023/A:1021176531598

Ferrer I, Barceló D (1999) Validation of new solid-phase extraction materials for the selective enrichment of organic contaminants from environmental samples. Trends Anal Chem 18:180–192

Fichot CG, Benner R (2011) A novel method to estimate DOC concentrations from CDOM absorption coefficients in coastal waters. Geophys Res Lett 38:L03610. https://doi.org/10.1029/2010GL046152

Floge SA, Hardy KR, Boss E, Wells ML (2009) Analytical intercomparison between type I and type II longpathlength liquid core waveguides for the measurement of chromophoric dissolved organic matter. Limnol Oceanogr Methods 7:260–268. https://doi.org/10.4319/lom.2009.7.260

Foreman RK, Björkman KM, Carlson CA, Opalk K, Karl DM (2019) Improved ultraviolet photo-oxidation system yields estimates for deep-sea dissolved organic nitrogen and phosphorus. Limnol Oceanogr Methods 17:277–291. https://doi.org/10.1002/lom3.10312

Fry B, Peltzer ET, Hopkinson CS Jr, Nolin A, Redmond L (1996) Analysis of marine DOC using a dry combustion method. Mar Chem 54:191–201

Gabor R, Baker A, McKnight D, Miller M (2014) Fluorescence indices and their interpretation. In: Coble P, Lead J, Baker A, Reynolds D, Spencer R (eds) Aquatic organic matter fluorescence, Cambridge Environmental Chemistry Series. Cambridge University Press, Cambridge, pp 303–338. https://doi.org/10.1017/CBO9781139045452.015

Gershey RM, Mckinnon MD, Williams PJLEB, Moore RM (1979) Comparison of three oxidation methods used for the analysis of dissolved organic carbon in seawater. Mar Chem 7:289–306

Gimbert LG, Worsfold PJ (2007) Environmental applications of liquid-waveguide-capillary cells coupled with spectroscopic detection. TrAC Trends Anal Chem 26:914–930

Gomez-Saez G, Pohlabeln AM, Stubbins A, Marsay CM, Dittmar T (2017) Photochemical alteration of dissolved organic sulfur from sulfidic porewater. Environ Sci Technol 51:14144–14154. https://doi.org/10.1021/acs.est.7b03713

Green NW, Perdue EM, Aiken GR, Butler KD, Chen H, Dittmar T, Niggemann J, Stubbins A (2014) An intercomparison of three methods for the large-scale isolation of oceanic dissolved organic matter. Mar Chem 161:14–19. https://doi.org/10.1016/j.marchem.2014.01.012

Green SA, Blough NV (1994) Optical absorption and fluorescence properties of chromophoric dissolved organic matter in natural waters. Limnol Oceanogr 39:1903–1916

Grunert BK, Mouw CB, Ciochetto AB (2018) Characterizing CDOM spectral variability across diverse regions and spectral ranges. Global Biogeochem Cycles 32:57–77. https://doi.org/10.1002/2017GB005756

Guo L, Santschi PH (1996) A critical evaluation of the cross-flow ultrafiltration technique for sampling colloidal organic carbon in seawater. Mar Chem 55:113–127

Hanold KA, Fischer SM, Cornia PH, Miller CE, Syage JA (2004) Atmospheric pressure photoionization. 1. General properties for LC/MS. Anal Chem 76:2842–2851. https://doi.org/10.1021/ac035442i

Hansell D, Carlson C, Repeta D, Schlitzer R (2009) Dissolved organic matter in the ocean: a controversy stimulates new insights. Oceanography 22:202–211

Hansell DA (1993) Results and observations from the measurement of DOC and DON in seawater using a high-temperature catalytic oxidation technique. In: Hedges JI, Lee C (eds) Measurements of dissolved organic carbon and nitrogen in natural waters. Mar chem, vol 41, pp 195–202

Hansell DA (2013) Recalcitrant dissolved organic carbon fractions. Annu Rev Mar Sci 5:421–425. https://doi.org/10.1146/annurev-marine-120710-100757

Hansell, DA, Carlson CA, Amon RMW, Álvarez-Salgado XA, Yamashita Y, Romera-Castillo C, Bif MB (2021) Compilation of dissolved organic matter (DOM) data obtained from the global ocean surveys from 1994 to 2020 (NCEI Accession 0227166). NOAA National Centers for Environmental Information. Dataset. DOI: https://doi.org/10.25921/s4f4-ye35. Accessed [date]

Hansen HP, Koroleff F (1999) Determination of nutrients. In: Grasshoff K, Kremling K, Ehrhardt M (eds) Methods of seawater analysis, 3rd revised and extended edition. Wiley, Weinheim, pp 149–228. https://doi.org/10.1002/9783527613984.ch10

Hansman RL, Dittmar T, Herndl GJ (2015) Conservation of dissolved organic matter molecular composition during mixing of the deep water masses of the Northeast Atlantic Ocean. Mar Chem 177:288–297. https://doi.org/10.1016/j.marchem.2015.06.001

Hashimoto S, Fujiwara K, Fuwa K (1987) Determination of phosphorus in natural water using hydride generation and gas chromatography. Limnol Oceanogr 32:729–735

Hatcher PG, Lerch HE III, Bates AL, Verheyen TV (1989) Solid-state ^{13}C nuclear magnetic resonance studies of coalified gymnosperm xylem tissue from Australian brown coals. Org Geochem 14:145–155. https://doi.org/10.1016/0146-6380(89)90068-5

Hawkes JA, Dittmar T, Patriarca C, Tranvik LJ, Bergquist J (2016) Evaluation of the Orbitrap mass spectrometer for the molecular fingerprint analysis of natural dissolved organic matter (DOM). Anal Chem 88:7698–7704. https://doi.org/10.1021/acs.analchem.6b01624

Hedges JI, Lee C (eds) (1993) Measurements of dissolved organic carbon and nitrogen in natural waters. Proceedings NSF/NOAA/DOE workshop, Seattle, WA, USA, 15–19 July, 1991. Mar Chem 41:1–3

Helms JR, Stubbins A, Ritchie JD, Minor EC, Kieber DJ, Mopper K (2008) Absorption spectral slopes and slope ratios as indicators of molecular weight, source, and photobleaching of chromophoric dissolved organic matter. Limnol Oceanogr 53:955–969. https://doi.org/10.4319/lo.2008.53.3.0955

Henriksen T, Juhler RK, Svensmark B, Cech NB (2005) The relative influences of acidity and polarity on responsiveness of small organic molecules to analysis with negative ion electrospray ionization mass spectrometry (ESI-MS). J Am Soc Mass Spectrom 16:446–455. https://doi.org/10.1016/j.jasms.2004.11.021

Hertkorn N, Benner R, Frommberger M, Schmitt-Kopplin P, Witt M, Kaiser K, Kettrup A, Hedges JI (2006) Characterization of a major refractory component of marine dissolved organic matter. Geochim Cosmochim Acta 70:2990–3010. https://doi.org/10.1016/j.gca.2006.03.021

Hertkorn N, Frommberger M, Witt M, Koch BP, Schmitt-Kopplin P, Perdue EM (2008) Natural organic matter and the event horizon of mass spectrometry. Anal Chem 80:8908–8919. https://doi.org/10.1021/ac800464g

Hertkorn N, Harir M, Koch BP, Michalke B, Schmitt-Kopplin P (2013) High field NMR spectroscopy and FTICR mass spectrometry: powerful discovery tools for the molecular level characterization of marine dissolved organic matter. Biogeosciences 10:1583–1624

Hockaday WC, Purcell JM, Marshall AG, Baldock JA, Hatcher PG (2009) Electrospray and photoionization mass spectrometry for the characterization of organic matter in natural waters: a qualitative assessment. Limnol Oceanogr Methods 7:81–95. https://doi.org/10.4319/lom.2009.7.81

Hoffmann E, Stroobant V (2007) Mass spectrometry: principles and application, 3rd edn. John Wiley and Sons LTD

Hoikkala L, Kortelainen P, Soinne H, Kuosa H (2015) Dissolved organic matter in the Baltic Sea. J Mar Sys 142:47–61. https://doi.org/10.1016/j.jmarsys.2014.10.005

Hopkinson CS, Vallino JJ (2005) Efficient export of carbon to the deep ocean through dissolved organic matter. Nature 433:142–145

Hsu CS, Qia K, Chen YC (1992) An Innovative approach to data analysis in hydrocarbon characterization by on-line liquid chromatography-mass spectrometry. Anal Chim Acta 264:79–89. https://doi.org/10.1016/0003-2670(92)85299-L

Hudson N, Baker A, Reynolds D (2007) Fluorescence analysis of dissolved organic matter in natural, waste and polluted waters—a review. River Res Appl 23:631–649

Hughey CA, Hendrickson CL, Rodgers RP, Marshall AG, Kendrick QQ (2001) Mass defect spectrum: a compact visual analysis for ultrahigh-resolution broadband mass spectra. Anal Chem 73:4676–4681. https://doi.org/10.1021/ac010560w

Huguet A, Vacher L, Relexans S, Saubusse S, Froidefond JM, Parlanti E (2009) Properties of fluorescent dissolved organic matter in the Gironde estuary. Org Geochem 40:706–719. https://doi.org/10.1016/j.orggeochem.2009.03.002

Iuculano F, Álvarez-Salgado XA, Otero J, Catalá TS, Sobrino C, Duarte C, Agustí S (2019) Patterns and drivers of UV absorbing chromophoric dissolved organic matter in the euphotic layer of the open ocean. Front Mar Sci 6:320

Jørgensen L, Stedmon CA, Granskog MA, Middelboe M (2014) Tracing the long-term microbial production of recalcitrant fluorescent dissolved organic matter in seawater. Geophys Res Lett 41:2481–2488

Jørgensen L, Stedmon CA, Kragh T, Markager S, Middelboe M, Søndergaard M (2011) Global trends in the fluorescence characteristics and distribution of marine dissolved organic matter. Mar Chem 126:139–148. https://doi.org/10.1016/j.marchem.2011.05.002

Karl DM, Björkman KM (2015) Dynamics of dissolved organic phosphorus. In: Hansell DA, Carlson CA (eds) Biogeochemistry of marine dissolved organic matter. Academic Press, Burlington, pp 233–334

Karl DM, Tien G (1992) MAGIC: a sensitive and precise method for measuring dissolved phosphorus in aquatic environments. Limnol Oceanogr 37:105–116

Kim D, Lee J, Kim B, Kim S (2018) Optimization and application of paper-based spray ionization mass spectrometry for analysis of natural organic matter. Anal Chem 90:12027–12034. https://doi.org/10.1021/acs.analchem.8b02668

Kim S, Kamer RW, Hatcher PG (2003) Graphical method for analysis of ultrahigh-resolution broadband mass spectra of natural organic matter, the van Krevelen diagram. Anal Chem 75: 5336–5344. https://doi.org/10.1021/ac034415p

Knicker H, Lüdemann HD (1995) N-15 and C-13 CPMAS and solution NMR studies of N-15 enriched plant material during 600 days of microbial degradation. Org Geochem 23:329–341

Koch BP, Dittmar T (2006) From mass to structure? An aromaticity index for high-resolution mass data of natural organic matter. Rapid Commun Mass Spectrom 20:926–932. https://doi.org/10.1002/rcm.2386

Koch BP, Dittmar T (2016) Erratum of: from mass to structure: an Aromaticity index for high-resolution mass data of natural organic matter. Rapid Commun Mass Spectrom 30:250. https://doi.org/10.1002/rcm.7433

Koch BP, Dittmar T, Witt M, Kattner G (2007) Fundamentals of molecular formula assignment to ultrahigh resolution mass data of natural organic matter. Anal Chem 79:1758–1763. https://doi.org/10.1021/ac061949s

Koch BP, Witt M, Engbrodt R, Dittmar T, Kattner G (2005) Molecular formulae of marine and terrigenous dissolved organic matter detected by electrospray ionization Fourier transform ion cyclotron resonance mass spectrometry. Geochim Cosmochim Acta 69:3299–3308. https://doi.org/10.1016/j.gca.2005.02.027

Kögel-Knabner I (1997) 13C and ^{15}N NMR spectroscopy as a tool in soil organic matter studies. Geoderma 80:243–270

Koike I, Hara S, Terauchi K, Kogure K (1990) Role of sub—micrometre particles in the ocean. Nature 345:242–243

Kondyli A, Schrader W (2019) Evaluation of the combination of different atmospheric pressure ionization sources for the analysis of extremely complex mixtures. Rapid Commun Mass Spectrom 34:e8676. https://doi.org/10.1002/rcm.8676

Konermann L, Ahadi E, Rodriguez AD, Vahidi S (2013) Unraveling the mechanism of electrospray ionization. Anal Chem 85:2–9. https://doi.org/10.1021/ac302789c

Koprivnjak JF, Pfromm PH, Ingall E, Vetter TA, Schmitt-Kopplin P, Hertkorn N et al (2009) Chemical and spectroscopic characterization of marine dissolved organic matter isolated using coupled reverse osmosis–electrodialysis. Geochim Cosmochim Acta 73:4215–4231

Kowalczuk P, Tilstone GH, Zabłocka M, Röttgers R, Thomas R (2013) Composition of dissolved organic matter along an Atlantic meridional transect from fluorescence spectroscopy and parallel factor analysis. Mar Chem 157:170–184. https://doi.org/10.1016/j.marchem.2013.10.004

Kramer RW, Kujawinski EB, Hatcher PG (2004) Identification of black carbon derived structures in a volcanic ash soil humic acid by Fourier transform ion cyclotron resonance mass spectrometry. Environ Sci Technol 38:3387–3395. https://doi.org/10.1021/es030124m

Kujawinski EB, Behn MD (2006) Automated analysis of electrospray ionization Fourier transform ion cyclotron resonance mass spectra of natural organic matter. Anal Chem 78:4363–4373. https://doi.org/10.1021/ac0600306

Lakowicz JR (2006) Principles of fluorescence spectroscopy. Springer

LaRowe DE, Van Cappellen P (2011) Degradation of natural organic matter: a thermodynamic analysis. Geochem Cosmochim Acta 75:2030–2042. https://doi.org/10.1016/j.gca.2011.01.020

Larsson T, Wedborg M, Turner D (2007) Correction of inner-filter effect in fluorescence excitation-emission matrix spectrometry using Raman scatter. Anal Chim Acta 583:357–363. https://doi.org/10.1016/j.aca.2006.09.067

Lawaetz AJ, Stedmon CA (2009) Fluorescence intensity calibration using the Raman scatter peak of water. Appl Spectrosc 63:936–994

Lechtenfeld OJ, Kattner G, Flerus R, McCallister SL, Schmitt-Kopplin P, Koch BP (2014) Molecular transformation and degradation of refractory dissolved organic matter in the Atlantic and Southern Ocean. Geochim Cosmochim Acta 126:321–337. https://doi.org/10.1016/j.gca.2013.11.009

Leefmann T, Frickenhausee S, Koch BP (2019) UltraMassExplorer: a browser-based application for the evaluation of high-resolution mass spectrometric data. Rapid Commun Mass Spectrom 33:193–202. https://doi.org/10.1002/rcm.8315

Lefering I, Röttgers R, Utschig C, McKee D (2017) Uncertainty budgets for liquid waveguide CDOM absorption measurements. Appl Optics 56:6357–6366. https://doi.org/10.1364/AO.56.006357

Li Y, Harir M, Uhl J, Kanawati B, Lucio M, Smirnov KS, Koch BP, Schmitt-Kopplin P, Hertkorn N (2017) How representative are dissolved organic matter (DOM) extracts? A comprehensive study of sorbent selectivity for DOM isolation. Water Res 116:316–323. https://doi.org/10.1016/j.watres.2017.03.038

Loiselle SA, Bracchini L, Dattilo AM, Ricci M, Tognazzi A, Cózar A, Rossi C (2009) The optical characterization of chromophoric dissolved organic matter using wavelength distribution of absorption spectral slopes. Limnol Oceanogr 54. https://doi.org/10.4319/lo.2009.54.2.0590

Lønborg C, Álvarez-Salgado XA (2012) Recycling versus export of bioavailable dissolved organic matter in the coastal ocean and efficiency of the continental shelf pump. Global Biogeochem Cycles 26:GB3018. https://doi.org/10.1029/2012GB004353

Lønborg C, Álvarez-Salgado XA (2014) Tracing dissolved organic matter cycling in the eastern boundary of the temperate North Atlantic using absorption and fluorescence spectroscopy. Deep Sea Res - Part I 85:35–46. https://doi.org/10.1016/j.dsr.2013.11.002

Lønborg C, Álvarez-Salgado XA, Richardson K, Martínez-García S, Teira E (2010) Assessing the microbial bioavailability and degradation rate constants of dissolved organic matter by fluorescence spectroscopy in the coastal upwelling system of the Ría de Vigo. Mar Chem 119:121–129. https://doi.org/10.1016/j.marchem.2010.02.001

Lu K, Gardner WS, Liu Z (2018) Molecular structure characterization of riverine and coastal dissolved organic matter with ion mobility quadrupole time-of-flight LCMS (IM Q-TOF LCMS). Environ Sci Technol 52:7182–7191. https://doi.org/10.1021/acs.est.8b00999

Mannino A, Novak MG, Hooker SB, Hyde K, Aurin D (2014) Algorithm development and validation of CDOM properties for estuarine and continental shelf waters along the northeastern U.S. coast. Remote Sens Environ 152:576–602. https://doi.org/10.1016/j.rse.2014.06.027

Mapolelo MM, Rodgers RP, Blakney GT, Yen AT, Asomaning S, Marshall AG (2011) Characterization of naphthenic acids in crude oils and naphthenates by electrospray ionization FT-ICR mass spectrometry. Int J Mass Spectrom 300:149–157. https://doi.org/10.1016/j.ijms.2010.06.005

Margolin AR, Gerringa LJA, Hansell DA, Rijkenberg MJA (2016) Net removal of dissolved organic carbon in the anoxic waters of the Black Sea. Mar Chem 183:13–24. https://doi.org/10.1016/j.marchem.2016.05.003

Marshall AG, Hendrickson CL, Jackson GS (1998) Fourier transform ion cyclotron resonance mass spectrometry: a primer. Mass Spectrom Rev 17:1–35. https://doi.org/10.1002/(SICI)1098-2787 (1998)17:1<1::AID-MAS1>3.0.CO;2-K

Marshall AG, Rodgers RP (2008) Petroleomics: chemistry of the underworld. PNAS 105:18090–18095. https://doi.org/10.1073/pnas.0805069105

Martínez-Pérez AM, Álvarez-Salgado XA, Arístegui J, Nieto-Cid M (2017a) Deep-ocean dissolved organic matter reactivity along the Mediterranean Sea: does size matter? Sci Rep 7:5687. https://doi.org/10.1038/s41598-017-05941-6

Martínez-Pérez AM, Catalá TS, Nieto-Cid M, Otero J, Álvarez M, Emelianov M, Reche I, Álvarez-Salgado XA, Arístegui J (2019) Dissolved organic matter (DOM) in the open Mediterranean Sea II: basin–wide distribution and drivers of fluorescent DOM. Progr Oceanogr 170:93–106

Martínez-Pérez AM, Osterholz H, Nieto-Cid M, Álvarez M, Dittmar T, Álvarez-Salgado XA (2017b) Molecular composition of dissolved organic matter in the Mediterranean Sea. Limnol Oceanogr 62:2699–2712. https://doi.org/10.1002/lno.10600

Massicotte P, Markager S (2016) Using a gaussian decomposition approach to model absorption spectra of chromophoric dissolved organic matter. Mar Chem 180:24–32

McCarthy M, Pratum T, Hedges J, Benner R (1997) Chemical composition of dissolved organic nitrogen in the ocean. Nature 390:150–154

McCarthy MD, Hedges JI, Benner R (1998) Major bacterial contribution to marine dissolved organic nitrogen. Science 281:231–234

McDonnell AMP, Lam PJ, Lamborg CH, Buesseler KO, Sanders R, Riley JS, Marsay C, Smith HEK, Sargent EC, Lampitt RS, Bishop James KB (2015) The oceanographic toolbox for the collection of sinking and suspended marine particles. Prog Oceanogr 133:17–31. https://doi.org/10.1016/j.pocean.2015.01.007

McIntyre C, Batts BD, Jardine DR (1997) Electrospray mass spectrometry of groundwater organic acids. J Mass Spectrom 32:328–330. https://doi.org/10.1002/(SICI)1096-9888(199703)32:3<328::AID-JMS480>3.0.CO;2-M

McKenna JH, Doering PH (1995) Measurement of dissolved organic carbon by wet chemical oxidation with persulfate: influence of chloride concentration and reagent volume. Mar Chem 48:109–114

McLafferty FW, Turecek F (1993) Interpretation of mass spectra, 4rd edn. University Science Books, Mill Valley, CA

Medeiros PM, Seidel M, Ward ND, Carpenter EJ, Gomes HR, Niggemann J et al (2015) Fate of the Amazon River dissolved organic matter in the tropical Atlantic Ocean. Glob Biogeochem Cycles 29:677–690. https://doi.org/10.1002/2015GB005115

Menges A, Feenders C, Seibt M, Blasius B, Dittmar T (2017) Functional molecular diversity of marine dissolved organic matter is reduced during degradation. Front Mar Sci 4:194. https://doi.org/10.3389/fmars.2017.00194

Merder J, Freund JA, Feudel U, Hansen CT, Hawkes JA, Jacob B et al (2020) ICBM-Ocean: processing ultrahigh-resolution mass spectrometry data of complex molecular mixtures. Anal Chem 92:6863–6838. https://doi.org/10.1021/acs.analchem.9b05659

Meybeck M (1982) Carbon, nitrogen and phosphorus transport by world rivers. Am J Sci 282:401–450. https://doi.org/10.2475/ajs.282.4.401

Mopper K, Qian J (2006) Water analysis: organic carbon determinations. In: Meyers RA, Miller MP (eds) Encyclopedia of analytical chemistry. https://doi.org/10.1002/9780470027318.a0884

Mopper K, Stubbins A, Ritchie JD, Bialk HM, Hatcher PG (2007) Advanced instrumental approaches for characterization of marine dissolved organic matter: extraction techniques, mass spectrometry, and nuclear magnetic resonance spectroscopy. Chem Rev 107:419–442. https://doi.org/10.1021/cr050359b

Mostofa KMG et al (2013) Colored and Chromophoric dissolved organic matter in natural waters. In: Mostofa K, Yoshioka T, Mottaleb A, Vione D (eds) Photobiogeochemistry of organic matter. Environmental science and engineering (environmental engineering). Springer, Berlin, Heidelberg. https://doi.org/10.1007/978-3-642-32223-5_5

Murphy KR, Stedmon CA, Graeber D, Bro R (2013) Fluorescence spectroscopy and multi–way techniques. PARAFAC. Anal Methods 5:6557–6566. https://doi.org/10.1039/c3ay41160e

Murphy KR, Stedmon CA, Wenig P, Bro R (2014) OpenFluor- an online spectral library of auto-fluorescence by organic compounds in the environment. Anal Methods 6:658–661. https://doi.org/10.1039/c3ay41935e

Nelson NB, Carlson CA, Steinberg DK (2004) Production of chromophoric dissolved organic matter by Sargasso Sea microbes. Mar Chem 89:273–287. https://doi.org/10.1016/j.marchem.2004.02.017

Nelson NB, Gauglitz JM (2016) Optical signatures of dissolved organic matter transformation in the global Ocean. Front Mar Sci 2. https://doi.org/10.3389/fmars.2015.00118

Nelson NB, Siegel DA, Carlson CA, Swan CM (2010) Tracing global biogeochemical cycles and meridional overtuning circulation using chromophoric dissolved organic matter. Geophys Res Lett 37:L03610. https://doi.org/10.1029/2009GL042325

Nieto-Cid M, Álvarez-Salgado XA, Gago J, Pérez FF (2005) DOM fluorescence, a tracer for biogeochemical processes in a coastal upwelling system (NW Iberian Peninsula). Mar Ecol Prog Ser 297:33–50

Nieto-Cid M, Álvarez-Salgado XA, Pérez FF (2006) Microbial and photochemical reactivity of fluorescent dissolved organic matter in a coastal upwelling system. Limnol Oceanogr 51:1391–1400. https://doi.org/10.4319/lo.2006.51.3.1391

Novotny NR, Capley EN, Stenson AC (2014) Fact or artifact: the representativeness of ESI-MS for complex natural organic mixtures. J Mass Spectrom 49:316–326. https://doi.org/10.1002/jms.3345

Ohno T (2002) Fluorescence inner-filtering correction for determining the humification index of dissolved organic matter. Environ Sci Technol 36:742–746. https://doi.org/10.1021/es0155276

Omanović D, Santinelli C, Marcinek S, Gonnelli M (2019) ASFit - an all-inclusive tool for analysis of UV-vis spectra of colored dissolved organic matter (CDOM). Comput Geosci 133:104334

Osterholz H, Kilgour DPA, Storey DS, Lavik G, Ferdelman TG, Niggemann J, Dittmar T (2021) Accumulation of DOC in the South Pacific subtropical gyre from a molecular perspective. Mar Chem 231. https://doi.org/10.1016/j.marchem.2021.103955

Osterholz H, Kirchman DL, Niggemann J, Dittmar T (2016) Environmental drivers of dissolved organic matter molecular composition in the Delaware estuary. Front Earth Sci 4:95. https://doi.org/10.3389/feart.2016.00095

Osterholz H, Niggemann J, Giebel H-A, Simon M, Dittmar T (2015) Inefficient microbial production of refractory dissolved organic matter in the ocean. Nat Commun 6:7422. https://doi.org/10.1038/ncomms8422

Pan Q, Zhuo X, He C, Zhang Y, Shi Q (2020) Validation and evaluation of high-resolution orbitrap mass spectrometry on molecular characterization of dissolved organic matter. ACS Omega 5:5372–5379. https://doi.org/10.1021/acsomega.9b04411

Perdue E, Benner R (2009) Marine organic matter. In: Xing B, Senesi N, Huang PM (eds) Biophysico-chemical processes involving natural nonliving organic matter in environmental systems. John Wiley and Sons, Hoboken, NJ, pp 407–449

Phungsai P, Kurisu F, Kasuga I, Furumai H (2016) Molecular characterization of low molecular weight dissolved organic matter in water reclamation processes using Orbitrap mass spectrometry. Water Res 100:526–536. https://doi.org/10.1016/j.watres.2016.05.047

Purcell JM, Hendrickson CL, Rodgers RP, Marshall AG (2006) Atmospheric pressure photoionization Fourier transform ion cyclotron resonance mass spectrometry for complex mixture analysis. Anal Chem 78:5906–5912. https://doi.org/10.1021/ac060754h

Raimbault P, Pouvesle W, Diaz F, Garcia N, Sempere R (1999) Wet-oxidation and automated colorimetry for simultaneous determination of organic carbon, nitrogen and phosphorus dissolved in seawater. Mar Chem 66:161–169

Repeta DJ (2015) Chemical characterization and cycling of dissolved organic matter. In: Hansell DA, Carlson CA (eds) Biogeochemistry of marine dissolved organic matter, 2nd edn. Academic Press, pp 21–64

Riedel T, Bieste H, Dittmar T (2012) Molecular fractionation of dissolved organic matter with metal salts. Environ Sci Technol 46:4419–4426. https://doi.org/10.1021/es203901u

Riedel T, Dittmar T (2014) A method detection limit for the analysis of natural organic matter via Fourier transform ion cyclotron resonance mass spectrometry. Anal Chem 86:8376–8382. https://doi.org/10.1021/ac501946m

Riley JS, Sanders R, Marsay C, Le Moigne FAC, Achterberg EP, Poulton AJ (2012) The relative contribution of fast and slow sinking particles to ocean carbon export. Global Biogeochem Cycles 26:GB1026. https://doi.org/10.1029/2011GB004085

Romera-Castillo C, Sarmento H, Álvarez-Salgado XA, Gasol JM, Marrasé C (2011) Net production and consumption of fluorescent colored dissolved organic matter by natural bacterial assemblages growing on marine phytoplankton exudates. Appl Environ Microbiol 77:7490–7498. https://doi.org/10.1128/AEM.00200-11

Rossel PE, Bienhold C, Hehemann L, Dittmar T, Boetius A (2020) Molecular composition of dissolved organic matter in sediment porewater of the Arctic deep-sea observatory HAUSGATEN (Fram Strait). Front Mar Sci 7:428. https://doi.org/10.3389/fmars.2020.00428

Rossel PE, Stubbins A, Rebling T, Koschinsky A, Hawkes JA, Dittmar T (2017) Thermally altered marine dissolved organic matter in hydrothermal fluids. Org Geochem 110:73–86. https://doi.org/10.1016/j.orggeochem.2017.05.003

Röttgers R, Koch BP (2012) Spectroscopic detection of a ubiquitous dissolved pigment degradation product in subsurface waters of the global ocean. Biogeosciences 9:2585–2596

Savory JJ, Kaiser NK, McKenna AM, Xian F, Blakney GT, Rodgers RP et al (2011) Parts-per-billion Fourier transform ion cyclotron resonance mass measurement accuracy with a "walking" calibration equation. Anal Chem 83:1732–1736. https://doi.org/10.1021/ac102943z

Schmidt F, Koch BP, Elvert M, Schmidt G, Witt M, Hinrichs K-U (2011) Diagenetic transformation of dissolved organic nitrogen compounds under contrasting sedimentary redox conditions in the black sea. Environ Sci Technol 45:5223–5229. https://doi.org/10.1021/es2003414

Seidel M, Beck M, Riedel T, Waska H, Suryaputra IGNA, Schnetger B et al (2014) Biogeochemistry of dissolved organic matter in an anoxic intertidal creek bank. Geochim Cosmochim Acta 140:418–434. https://doi.org/10.1016/j.gca.2014.05.038

Sharp JH (1993) The dissolved organic carbon controversy: an update. Oceanography 6:45–50. https://doi.org/10.5670/oceanog.1993.13

Sharp JH, Carlson CA, Peltzer ET, Castle-Ward DM, Savidge KB, Rinker KR (2002a) Final dissolved organic carbon broad community intercalibration and preliminary use of DOC reference materials. Mar Chem 77:239–253

Sharp JH, Rinker KR, Savidge KB, Abell J, Benaim JY, Bronk D, Burdige DJ, Cauwet G, Chen WH, Doval MD, Hansell DA, Hopkinson C, Kattner G, Kaumeyer N, McGlathery KJ, Merriam J, Morley N, Nagel K, Ogawa H, Pollard C, Pujo-Pay M, Raimbault P, Sambrotto R, Seitzinger S, Spyres G, Tirendi F, Walsh TW, Wong CS (2002b) A preliminary methods comparison for measurement of dissolved organic nitrogen in seawater. Mar Chem 78:171–184. https://doi.org/10.1016/S0304-4203(02)00020-8

Sherr E, Sherr B (1988) Role of microbes in pelagic food webs: a revised concept. Limnol Oceanogr 33:1225–1227. https://doi.org/10.4319/lo.1988.33.5.1225

Shigemitsu M, Uchida H, Yokokawa T, Arulananthan K, Murata A (2020) Determining the distribution of fluorescent organic matter in the Indian Ocean using in situ fluorometry. Front Microbiol 11:3163. https://doi.org/10.3389/fmicb.2020.589262

Siegel DA, Maritorena S, Nelson NB (2002) Global distribution and dynamics of colored dissolved and detrital organic materials. J Geophys Res 107:3228. https://doi.org/10.1029/2001JC000965

Simon C, Roth V-N, Dittmar T, Gleixner G (2018) Molecular signals of heterogeneous terrestrial environments identified in dissolved organic matter: a comparative analysis of orbitrap and ion cyclotron resonance mass spectrometers. Front Earth Sci 6:138. https://doi.org/10.3389/feart.2018.00138

Simpson AJ, McNally DJ, Simpson MJ (2011) NMR spectroscopy in environmental research: from molecular interactions to global processes. Prog Nucl Magn Reson Spectrosc 58:97–175. https://doi.org/10.1016/j.pnmrs.2010.09.001

Simpson AJ, Simpson MJ (2009) Nuclear magnetic resonance analysis of natural organic matter. In: Senesi N, Xing B, Huang PM (eds) Biophysico-chemical processes involving natural nonliving organic matter in environmental systems. John Wiley & Sons, Hoboken, New Jersey, pp 589–650

Sipler RE, Bronk DA (2015) Dynamics of marine dissolved organic nitrogen. In: Hansell DA, Carlson CA (eds) Biogeochemistry of marine dissolved organic matter. Academic Press, Cambridge, MA, pp 127–232. https://doi.org/10.1016/B978-0-12-405940-5.00004-2

Solihat NN, Acter T, Kim D, Plante AF, Kim S (2019) Analyzing solid-phase natural organic matter using laser desorption ionization ultrahigh resolution mass spectrometry. Anal Chem 91:951–957. https://doi.org/10.1021/acs.analchem.8b04032

Spencer RGM, Butler KD, Aiken GR (2012) Dissolved organic carbon and chromophoric dissolved organic matter properties of rivers in the USA. J Geophys Res 117:G03001. https://doi.org/10.1029/2011JG001928

Spencer RGM, Stubbins A, Hernes PJ, Baker A, Mopper K, Aufdenkampe AK et al (2009) Photochemical degradation of dissolved organic matter and dissolved lignin phenols from the Congo River. J Geophys Res 114:G03010. https://doi.org/10.1029/2009JG000968

Statham PJ, Williams PJB (1999) The automated determination of dissolved organic carbon by ultraviolet photooxidation. In: Grasshoff K, Kremling K, Ehrhardt M (eds) Methods of seawater analysis. https://doi.org/10.1002/9783527613984.ch16

Stedmon CA, Bro R (2008) Characterizing dissolved organic matter fluorescence with parallel factor analysis: a tutorial. Limnol Oceanogr Methods 6:572–579. https://doi.org/10.4319/lom.2008.6.572

Stedmon CA, Nelson NB (2015) The optical properties of DOM in the ocean. In: Hansell DA, Carlson CA (eds) Biogeochemistry of marine dissolved organic matter. Elsevier, pp 481–508. https://doi.org/10.1016/B978-0-12-405940-5.00010-8

Steen AD, Kusch S, Abdulla HA, Caki N, Coffinet S, Dittmar T et al (2020) Analytical and computational advances, opportunities, and challenges in marine organic biogeochemistry in an era of "omics". Front Mar Sci 7:718. https://doi.org/10.3389/fmars.2020.00718

Stenson AC, Marshall AG, Cooper WT (2003) Exact masses and chemical formulas of individual Suwannee River fulvic acids from ultrahigh resolution electrospray ionization Fourier transform ion cyclotron resonance mass spectra. Anal Chem 75:1275–1284. https://doi.org/10.1021/ac026106p

Sugimura Y, Suzuki Y (1988) A high-temperature catalytic oxidation method for the determination of non-volatile dissolved organic carbon in seawater by direct injection of a liquid sample. Mar Chem 24:105–131

Suzuki Y, Sugimura Y, Itoh T (1985) A catalytic oxidation method for the determination of total nitrogen dissolved in seawater. Mar Chem 16:83–97

Tolić N, Liu Y, Liyu A, Shen Y, Tfaily MM, Kujawinski EB, Longnecker K et al (2017) Formularity: software for automated formula assignment of natural and other organic matter from ultrahigh-resolution mass spectra. Anal Chem 89:12659–12665. https://doi.org/10.1021/acs.analchem.7b03318

Twardowski MS, Boss E, Sullivan JM, Donaghay PL (2004) Modeling the spectral shape of absorption by chromophoric dissolved organic matter. Mar Chem 89:69–88

van Krevelen (1950) Graphical-statistical method for the study of structure and reaction processes of coal. Fuel 29:269–284

Ventura GT, Rossel PE, Simoneit BRT, Dittmar T (2020) Fourier transform ion cyclotron resonance mass spectrometric analysis of NSO-compounds generated in hydrothermal altered sediments from the Escanaba Trough, northeastern Pacific Ocean. Org Geochem 149:104085. https://doi.org/10.1016/j.orggeochem.2020.104085

Verdugo P (2012) Marine Microgels. Annu Rev Mar Sci 4:375–400

Wagner S, Bandes J, Spencer RGM, Ma K, Rosengard SZ, Moura JMS, Stubbins A (2019) Isotopic composition of oceanic dissolved black carbon reveals non-riverine source. Nat Commun 10: 5064. https://doi.org/10.1038/s41467-019-13111-7

Wagner S, Jaffé R, Stubbins A (2018) Dissolved black carbon in aquatic ecosystems. Limnol Oceanogr Lett 3:168–185. https://doi.org/10.1002/lol2.10076

Waska H, Brumsack H-J, Massmann G, Koschinsky A, Schnetger B, Simon H, Dittmar (2019) Inorganic and organic iron and copper species of the subterranean estuary: origins and fate. Geochim Cosmochim Acta 259:211–232. https://doi.org/10.1016/j.gca.2019.06.004

Waska H, Koschinsky A, Ruiz-Chancho MJ, Dittmar T (2015) Investigating the potential of solid-phase extraction and Fourier-transform ion cyclotron resonance mass spectrometry (FT-ICR-MS) for the isolation and identification of dissolved metal–organic complexes from natural waters. Mar Chem 173:78–92

Weishaar JL, Aiken GR, Bergamaschi BA, Fram MS, Fugii R, Mopper K (2003) Evaluation of specific ultraviolet absorbance as an indicator of the chemical composition and reactivity of dissolved organic carbon. Environ Sci Technol 37:4702–4708

Wells ML, Goldberg ED (1991) Occurrence of small colloids in sea water. Nature 353:342–344

Wilkinson M, Dumontier M, Aalbersberg IJ, Appleton G, Axton M, Baak A et al (2016) The FAIR guiding principles for scientific data management and stewardship. Sci Data 3:160018. https://doi.org/10.1038/sdata.2016.18

Wozniak B, Dera J (2007) Light absorption in seawater. Atmospheric and oceanographic sciences library 33. Springer, New York

Wurl O (2009) Practical guidelines for the analysis of seawater. CRC Press, Boca Raton

Yamashita Y, Lub C-J, Ogawa H, Nishioka J, Obata H, Saito H (2015) Application of an in situ fluorometer to determine the distribution of fluorescent organic matter in the open ocean. Mar Chem 177:298–305. https://doi.org/10.1016/j.marchem.2015.06.025

Yamashita Y, Tanoue E (2003) Chemical characterization of protein-like fluorophores in DOM in relation to aromatic amino acids. Mar Chem 82:255–271

Yamashita Y, Tanoue E (2008) Production of bio-refractory fluorescent dissolved organic matter in the ocean interior. Nat Geosci 1:579–582

Yasui-Tamura S, Hashihama F, Ogawa H, Nishimura T, Kanda J (2020) Automated simultaneous determination of total dissolved nitrogen and phosphorus in seawater by persulfate oxidation method. Talanta Open 2:100016. https://doi.org/10.1016/j.talo.2020.100016

Zark M, Christoffers J, Dittmar T (2017) Molecular properties of deep-sea dissolved organic matter are predictable by the central limit theorem: evidence from tandem FT-ICR-MS. Mar Chem 191: 9–15. https://doi.org/10.1016/j.marchem.2017.02.005

Zark M, Dittmar T (2018) Universal molecular structures in natural dissolved organic matter. Nat Commun 9:3178. https://doi.org/10.1038/s41467-018-05665-9

Zsolnay A, Baigar E, Jimenez M, Steinweg B, Saccomandi F (1999) Differentiating with fluorescence spectroscopy the sources of dissolved organic matter in soils subjected to drying. Chemosphere 38:45–50

Trace Metals

3

Rob Middag, Rebecca Zitoun, and Tim Conway

Contents

Abstract

In this chapter an overview of sampling and analytical techniques for the marine trace metals (and their stable isotope ratios) is given, focusing largely on the six bio-essential transition metals (iron, manganese, copper, nickel, zinc and cobalt). The aim of this chapter is to introduce the reader to the breadth of techniques and methods currently available to study the biogeochemical cycles of trace metals and their isotopes in the ocean. We note that we do not cover all existing and historical techniques as some are no longer used, some remain immature for trace metal studies, and some are just emerging or are still being developed. A more detailed focus on the methods used by the authors is also provided. We anticipate the continuing development and refinement of methods; as with any expanding and developing scientific field, novel strategies and techniques continuously come and go. For further background reading on marine trace metal distribution

R. Middag (✉) · R. Zitoun
Department of Ocean Systems, Royal Netherlands Institute for Sea Research (NIOZ), Texel, Netherlands
e-mail: rob.middag@nioz.nl

T. Conway
College of Marine Science, University of South Florida, St Petersburg, FL, USA

© The Author(s), under exclusive license to Springer Nature Switzerland AG 2023
J. Blasco, A. Tovar-Sánchez (eds.), *Marine Analytical Chemistry*,
https://doi.org/10.1007/978-3-031-14486-8_3

and key biogeochemical processes in the ocean, the reader is referred throughout the chapter to appropriate overviews, articles and textbooks available online, including the freely available GEOTRACES electronic atlas and data products, as well as the GEOTRACES 'cookbook'.

Keywords

Marine trace metal biogeochemistry · Marine trace metal sampling · Marine trace metal sample handling · Marine trace metal analysis · Marine metal stable isotope analysis · Marine trace metal speciation analysis

3.1 Introduction

3.1.1 Trace Metals in the Ocean

Seawater comprises all naturally occurring chemical elements including metals (Bruland et al. 2014). Metals are often defined to include the 40 transition metal elements (periodic table: d-block; 31 natural and 9 artificial metal elements), the rare earth elements (lanthanides and actinides, both f-block), the s-block elements except hydrogen (H) and helium (He) (e.g. lithium (Li), magnesium (Mg)) and some elements from the p-block (aluminium (Al), gallium (Ga), indium (In), tin (Sn), thallium (Th), lead (Pb) and bismuth (Bi)). Most metals occur at trace concentrations in the ocean, hence the common terminology 'trace metal'. Six of the transition metals are essential in small quantities for all living organisms both on land and in the sea, needed for maintaining metabolic processes such as carbon uptake, photosynthesis and nitrogen fixation (de Baar et al. 2018; Goldhaber 2003; Sunda 2012; Vraspir and Butler 2009). Listed in order of their typical abundance in the living cells of marine unicellular biota, the bio-essential metals are iron (Fe), zinc (Zn), manganese (Mn), copper (Cu), nickel (Ni) and cobalt (Co) (Bruland et al. 1991; de Baar et al. 2018; Morel et al. 2014). Some biota also appear to utilize cadmium (Cd) under some conditions to substitute for Zn and Co in enzymes such as carbonic anhydrase (Lane and Morel 2000; Lane et al. 2005; Price and Morel 1990). Other trace metals have no known biological function (e.g. Al; (Adams and Chapman 2007) or are entirely toxic (e.g. mercury (Hg) and arsenic (As)) (Hunter 2008; Tercier Waeber et al. 2012). However, it should be recognized that all 'essential' elements can be toxic at high concentrations (Shah 2021; Tercier Waeber et al. 2012), and there can be a narrow optimum concentration range, as, for example, seen in the ocean for Cu (e.g. Bruland et al. 1991). In the ocean, the naturally low concentrations of trace metals mean that most are beneficial or quite harmless for biota, but elevated concentrations of some metals due to, for example, anthropogenic environmental pollution can result in localized negative effects (Shah 2021; Tercier Waeber et al. 2012). For instance, the transition metals chromium (Cr), Pb, Ni, Cu, Zn, silver (Ag) and Cd are common industrial by-products and can enter the aquatic system via untreated waste waters, accumulate in sediments and become toxic for benthic

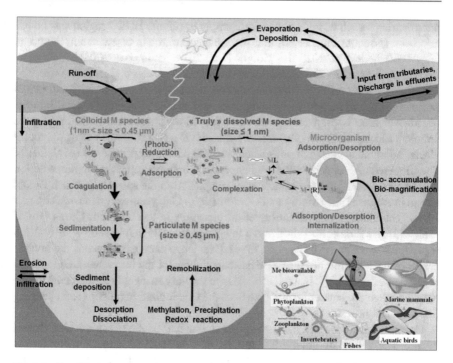

Fig. 3.1 Simplified schematic of trace metal (M) cycles in the marine environment. The conceptual representation includes metal sources, sinks and internal biogeochemical and physical processes that drive metal cycling and fate (Tercier Waeber et al. 2012). Note that the size definition of particulate, dissolved and colloidal trace metals used in the schematic is one of many operationally defined definitions, dependent on the filter sizes used. M^+, free metal ions; ML, organic metal complexes (M complexed with various organic ligands (L)); MY, inorganic metal complexes

organisms (Shah 2021). Fortunately, in many nations most waste waters are nowadays treated to remove pollutants, but in other nations the purification treatments of waste waters have yet to begin.

The growth of phytoplankton, the organisms which form the base of the marine food web, depends on the availability of both sunlight and nutrients in the upper water column. Nutrients include carbon and the 'macro'-nutrients such as nitrogen, phosphorus and silicate but also include so called 'micro'-nutrients, i.e. the six transition metals mentioned above, of which Fe (Fig. 3.1) is arguably the most important (Bruland et al. 2014). A scarcity of one or several of these trace metals in surface waters can limit overall primary productivity and change the community structure and functioning of marine ecosystems. Micronutrient limitation is especially acute in some remote ocean environments, commonly referred to as High Nutrient, Low Chlorophyll (HNLC) regions, where upwelling of deep water has provided ample macronutrients for phytoplankton, but a paucity of micronutrients such as Fe means that macronutrients cannot be fully utilized, and as a result growth is lower than one may expect (Moore et al. 2013). However, the role of micronutrients such as Fe is not limited simply to remote HNLC regions, as Fe has also been shown to limit or co-limit phytoplankton growth elsewhere in the ocean (e.g. Achterberg et al. 2013; Le Moigne et al. 2014; Mills et al. 2004; Moore et al. 2009; Moore et al. 2006; Nielsdóttir et al. 2009; Ryan-Keogh et al. 2013).

The importance of Fe for influencing marine ecosystem productivity and global carbon cycling, in particular, has long been recognized. Varying Fe input into the ocean over time is even thought to have played an important role in driving the dramatic and regular sawtooth shifts in atmospheric carbon dioxide during the recent glacial-interglacial cycles of our planet (Martin et al. 1990; Yamamoto et al. 2019). The first reliable experimental evidence for Fe limitation was by Fe addition experiments (bioassays) in surface water samples from the sub-Arctic North Pacific Ocean in August 1987 (Martin and Gordon 1988). One year later, in austral summer 1988, Fe limitation was demonstrated in bioassays in the Southern Ocean (Buma et al. 1991; de Baar et al. 1990), and a few years later the first in situ Fe fertilization experiment was done (Martin et al. 1994). Ever since these experiments, the investigation of the bio-limiting role of Fe has become one of the major research topics in ocean plankton ecology. Nevertheless, due to the very stringent requirements to rule out inadvertent Fe contamination (see Sect. 3.2), Fe limitation remains a challenging line of plankton research.

Ocean productivity and the biogeochemical cycles of the ocean are also shaped by the availability of other trace metals, co-limitation by two or more factors (e.g. light and Fe) and/or variability in nutrient requirements between species and environmental conditions (e.g. Arrigo 2005; Buma et al. 1991; de Baar et al. 1990; Morel et al. 2014; Saito et al. 2008). For instance, Fe together with other bio-essential trace metals (Mn, Co, Ni, Cu, Zn) can influence the taxonomic composition of key ecological communities with wide reaching influences on overall marine primary productivity (Boyd et al. 2017; Hutchins and Boyd 2016; Moore et al. 2013; Twining and Baines 2013).

Besides being essential micronutrients, trace metals and other minor elements can serve as useful tracers of human activity (e.g. radioactive tracers) or of the physical, geological and chemical processes that shape biogeochemical cycles in the oceanic water column and in the sediment on the sea floor below (Anderson 2020). Notably, isotope ratios of trace metals can provide information of internal oceanic processes (biological, scavenging and redox cycling) and external sources and sinks of metals to/from the ocean (Conway et al. 2021). Briefly, biochemical and geochemical reactions can lead to small, but measurable, mass-dependent fractionation of the isotope ratio of a certain element (e.g. Fe; Dauphas et al. 2017). Each fractionation process can lead to distinctly different isotope ratio signatures that can be used to 'fingerprint' processes, sources and sinks of certain elements in the ocean interior. Accordingly, the use of isotope ratios of trace metals such as Fe, Zn, Cd, Ni, Cu and Cr has exploded in the last decade as powerful tracers of oceanic trace metal cycling, fluxes and transport (Conway et al. 2021). In addition to modern water column measurements, trace metals or trace metal isotope ratios preserved in geologic archives also provide valuable information about the past ocean. For a recent comprehensive summary of the state of the field for using trace metals and their isotopes as proxies for past ocean productivity and other processes, see Horner et al. (2021).

3.1.2 Trace Metal Concentrations and Distributions

Concentrations of dissolved trace metals in seawater range from sub fmol kg^{-1} at the lower end to 10 μmol kg^{-1} at the upper end (μ (micro) $= 10^{-6}$, n (nano) $= 10^{-9}$, p (pico) $= 10^{-12}$, and f (femto) $= 10^{-15}$), where the concentration of 10 μmol kg^{-1} is an arbitrary upper value typically used to define the separation of trace elements from the major elements in seawater (Bruland et al. 2014; de Baar et al. 2018). Concentrations in seawater are often reported in units of mol kg^{-1} (e.g. nmol kg^{-1}), or mol L^{-1} (e.g. nmol L^{-1} which is the same as nM), where mol kg^{-1} units are generally recommended. Reporting in mol per mass units is recommended because mass, unlike volume, is unaffected by temperature- and/or pressure-derived changes in the density of a sample that can occur during or after collection. Volume can be affected by reasonably large changes in pressure and/or temperature between sample collection depth and later laboratory analysis; as such, the reported values in mol L^{-1} may differ (slightly) between the environmental conditions at which the sample was collected and those of analysis. Nevertheless, when analyses (see Sect. 3.5) are based on volume rather than mass, it can be argued the results should be reported in the units they were obtained in (e.g. L^{-1} in case of volume-based measurements) with the required information for any subsequent conversion (e.g. to nmol kg^{-1}) in the metadata (see Sect. 3.6) as a variety of different units is possible and use often depends on traditions within (sub-)disciplines.

Seawater contains a complex 'soup' of truly dissolved metal ion species, organic-complexed metals and small particles, the sum of which, for ease, is commonly referred to as 'dissolved'. Dissolved metals are thus typically operationally defined as any metal species that has passed through a filter, usually 0.4 or 0.2 μm in pore size (Fig. 3.1), with the latter becoming the common standard in recent decades (Cutter et al. 2017). Smaller filtration sizes and techniques (i.e. ultra-filtration; see Sect. 3.4.2 for more information) are sometimes also employed to look at concentrations of different fractions of the 'dissolved' pool of trace metals in seawater (e.g. Fitzsimmons et al. 2015b; Homoky et al. 2021), for example, colloids (~0.02–0.4 μm in size; see Fig. 3.1).

The distributions of the different trace metals in the ocean are each controlled by a different combination of biological, chemical and physical processes that lead both to different spatial and temporal patterns for each metal and to distinct relationships between the distributions of different trace and major elements (e.g. Bruland et al. 2014; de Baar et al. 2018). Ultimately, trace metals are added to the ocean from external boundary 'sources' such as rivers, margin and deep-sea sediments, wind-blown dust from continents (especially deserts), anthropogenic pollution and from venting from submarine hydrothermal activity (Anderson et al. 2014; de Baar et al. 2018). Similarly, trace metals are removed from the ocean by so-called sinks, which also occur at ocean boundaries and thus are often at or near the same locations as sources (e.g. marine sediments, hydrothermal vent deposits) (Bruland et al. 2014; de Baar et al. 2018). Within the ocean, active uptake by biota and passive particle scavenging act as internal sinks of trace metals. Trace metals associated with particulate matter can also be released back into the dissolved phase via bacteria-induced degradation, grazing or abiotic dissolution mechanisms, either in shallow surface waters or deeper in the water column or by respiration processes in the underlying sediment (e.g. Anderson et al. 2014). As such, particles can also act as internal ocean sources for trace metals. Generally, for most trace metals, their 'final' sink from the ocean is permanent burial in marine sediments (e.g. Bruland et al. 2014; de Baar et al. 2018).

The combination of specific sources, sinks and internal cycling that is unique to each trace metal then merges with the general physical ocean circulation, resulting in characteristic oceanographic dissolved distributions of trace metals that reflect their intrinsic biogeochemical behaviour and chemistry (Aparicio-González et al. 2012; Bruland et al. 2014; de Baar et al. 2018). Historically, dissolved trace metals have been broadly grouped into several definitions: (1) conservative, trace metals with a relatively narrow concentration range that varies in concert with salinity; (2) nutrient type, trace metals that are taken up by phytoplankton in the surface ocean and are regenerated with depth, leading to depleted surface concentrations and elevated deep concentrations; (3) scavenged, trace metals with strong particle interactions that are removed from the ocean; or (4) hybrid, trace metals which do not fit into a single distribution type (Aparicio-González et al. 2012; Bruland et al. 2014). In fact, as more becomes known, most dissolved trace metals exhibit aspects of some or all of these distributions.

3.1.3 Pioneering Marine Trace Metal Biogeochemistry

The overall very low concentrations of (dissolved and particulate) trace metals in the ocean meant that insight into their distributions and roles in the ocean was hindered until clean sampling and analytical techniques started to become available in the 1970s (see Sect. 3.2, Box 3.1) (Bruland and Lohan 2003; Protti 2001). More recently, in 2006, the international GEOTRACES programme that aims to identify the processes and quantify the fluxes that control the distributions of key trace elements and isotopes in the ocean has made major leaps in assessing the concentration, distribution and biogeochemical cycling of Fe and other trace metals in the global ocean (Anderson 2020; Henderson et al. 2018). A key focus point of GEOTRACES is the inter-comparability of methods and data quality between the ~36 countries involved (GEOTRACES-Group 2006), making sure that sampling efforts and measurements by different nations and laboratories give comparable results that can be combined into data products (Mawji et al. 2015). Prior to the inception of the GEOTRACES programme in the early 2000s, data was often not inter-comparable, owing to the difficulty of trace metal sampling without contamination and analysis using instruments with suitable enough resolution to quantify low trace metal concentrations (Johnson et al. 2007). Indeed, such challenges had meant that before the initiation of the GEOTRACES field programme, dissolved Fe had been measured at only 25 locations worldwide down to a depth of 2000 m depth (Anderson et al. 2014; Anderson 2020; GEOTRACES-Group 2006). Currently, hundreds of full-depth profiles of various trace metals, including Fe, can be found in the GEOTRACES inventory database, and this number is quickly expanding (Schlitzer et al. 2018; Anderson 2020). For example, GEOTRACES has now facilitated ocean transects of Fe in all ocean basins, illuminating both the external sources/sinks and internal cycling of Fe and building on previous understanding (e.g. Abadie et al. 2017; Conway and John 2014b; Ellwood et al. 2018; Fitzsimmons et al. 2015b; Klunder et al. 2012; Klunder et al. 2011; Moffett and German 2020; Nishioka et al. 2020; Rijkenberg et al. 2014; Saito et al. 2013; Tagliabue et al. 2017; Tonnard et al. 2020). However, despite this explosive increase in oceanic trace metal data and a number of decades of high-quality insights, scientists are still working to understand the details of the processes and factors that drive the distributions of many trace metals in the ocean, as well as the interactions between micronutrients and marine microbes that drive marine productivity and marine carbon sequestration. Consequently, it is still very challenging for scientists to make accurate predictions about the role of trace metals in influencing climate change in the past, present and future ocean.

Box 3.1: Technical Advances and Trace Metal Clean Techniques
The importance of trace metal clean techniques together with technical advances in analytical methods and instrumentation for trace metal chemistry is very nicely illustrated by the 'lure of gold story' of the mid-nineteenth century (Pilson 2012). The text below retells the story that was published by Pilson (2012).

Gold gained much attention in 1872 when it was announced that the waters of the English Channel contained 65 mg of gold in each ton of water (Sonstadt 1872). At the time, the value of gold in one ton of water was only about 6.5 cents, much lower than today's value. However, despite this, as well as a later assessment by Svante Arrhenius (~1900) that reported only 6 mg per ton in the English Channel (the equivalent of 0.6 cents per ton (Jensen et al. 2020; Riley et al. 1965)), societal interest in extracting gold in seawater persisted. The vastness of the ocean opened up the possibility of immense wealth even at the lowest estimates of 6 mg of gold per ton of water. Indeed, it was thought that if extraction of gold from the ocean was possible, such an endeavour would generate enough wealth to make every living person on Earth a millionaire twice over. Thus, in the first four decades of the twentieth century, patent after patent was issued at patent and trademark offices around the world for gold extraction methods and techniques from seawater. Fritz Haber, a Nobel prize laureate, was one of the many researchers that jumped onto the bandwagon of developing oceanic gold extraction methods. His interest was sparked by the

(continued)

Box 3.1 (continued)
promise to help pay off the German World War I debt. However, with his newly developed procedure, Haber found gold concentrations of only 0.004 mg per ton seawater (20 picomoles per kg). Back then, this was only worth about 0.004 cents per ton. This finding crushed any hopes and prospects of ever extracting gold from the sea economically. The Dow Chemical Company further illustrated the uneconomic nature during the 1950s when processing nearshore seawater to extract magnesium and bromine and also investigating the extraction of gold as a side project. The company processed 15 tons of seawater, but extracted only 0.09 mg of gold with an estimated value of 0.01 cent. This value stood in stark contrast to the exorbitant cost of ~ $50,000 that the Dow Chemical had spent on the extraction process. A more recent study quantified a concentration of about 50–200 femtomoles gold per kg seawater, a level two orders of magnitude less than was accepted decades earlier (Falkner and Edmond 1990). In fact, Falkner and Edmond (1990) deduced that the relatively 'large' amount of gold extracted by Dow Chemical was probably caused by 'unclean' reagents and containers that provided the bulk of the gold collected. Using modern values, we can now estimate that the entire world's ocean contains 'only' 14,000 tons of gold, five times less than the world gold holdings in central banks in early 2021 (69,400 tons) as published by the World Gold Council.

3.1.4 Future Challenges in Marine Trace Metal Biogeochemistry

Currently, the global marine Fe cycle, and those of other trace metals, is undergoing major changes because of ocean acidification, stratification, warming, deoxygenation, anthropogenic pollution and land use change, amongst other factors (Hutchins and Boyd 2016; Tagliabue et al. 2017). These changes raise questions about how future change will affect marine ecosystems, marine primary productivity and carbon uptake by the ocean, underlining the importance of studying the biogeochemical cycle of bio-essential metals such as Fe in the past and present ocean. Here, global ocean biogeochemical models come into play, since they are important tools that aid in our understanding of the impacts of future change and test hypotheses regarding biogeochemical processes (Tagliabue et al. 2016). The current generation of models do a reasonable job when it comes to macronutrients such as phosphate, but as of yet, they do a much poorer job of reproducing oceanic micronutrient distributions and as such vary widely in their predictions, especially for Fe (Tagliabue et al. 2016). The poorer predictive power of models for Fe in the ocean than for the macronutrients results largely from (i) the complex biogeochemistry of Fe, (ii) the short residence time of Fe and (iii) the insufficiently constrained sinks and sources (Tagliabue et al. 2016; Tagliabue et al. 2017). This uncertainty in the marine Fe cycle, given its known importance for global carbon cycling, raises large challenges for climate and earth system modelers and thus creates hurdles for civil society and stakeholders to evaluate and implement appropriate climate change actions and policy. The distributions of trace metals such as Ni, Zn and Cd are perhaps easier to incorporate into models because their behaviour is more predictable when considering the ocean from a three-dimensional perspective in which both ocean circulation and the biological pump play an important role (e.g. Middag et al. 2020; Tagliabue et al. 2016; Vance et al. 2017; Weber et al. 2018). However, the interaction of these other bio-essential trace metals and their effect on marine ecosystem functionality and biogeochemical fate are only just emerging, which undoubtedly will reveal currently unknown roles or feedback loops in the future. With more interdisciplinary science, more data on spatial and temporal variability and more ocean observations using new technologies and methods, trace metal scientists will most probably gain more insight into the fate and behaviour of trace metals and their isotopes in the ocean. Such knowledge will be essential for providing a holistic understanding of the processes that are essential to achieve a safe, sustainable, clean, healthy and predictable ocean, as is the goal of the UN Decade of Ocean Science for Sustainable Development.

3.2   Trace Metal Clean Procedures

In the early 1970s, technical advances in analytical chemistry and instrumentation with high sensitivities and low detection limits (Bruland et al. 2014; Protti 2001) paved the way for marine chemists to improve their understanding of concentrations, distributions, speciation (i.e. the chemical form of an element) and associated biogeochemical behaviours of trace metals in the marine environment (Bruland and Lohan 2003). However, the extremely low concentrations of many trace metals in natural seawater – commonly in the nanomolar range and often lower than the detection limits of common analytical techniques – together with the omnipresence of such metals in the terrestrial environment (Bruland and Lohan 2003; EPA 1996; Protti 2001; Richter 2003; Sander et al. 2009) complicated contamination-free analysis of trace metal samples both on ships and in the laboratory. Consequently, along with the development of powerful analytical techniques came the recognition and appreciation of the importance of using rigorous trace metal clean techniques in all stages of sampling and laboratory work (sample collection, storage, handling, treatment and analysis) to avoid the inadvertent introduction of contamination (Bruland et al. 1979, 2003; Cutter et al. 2017; EPA 1996; Sander et al. 2009).

The concept of trace metal clean laboratories for environmental analyses (Patterson et al. 1976) was pioneered by Patterson after he recognized that measurements of environmental Pb concentrations were often too high because of the often inadvertent introduction of Pb to the samples during sample collection, handling and storage (Patterson 1965). This realization resulted in Patterson's development of rigorous trace metal clean protocols for the elemental analyses of environmental samples in the early 1970s (EPA 1996; Patterson et al. 1976). This introduction of trace metal clean protocols led to the first oceanographically reliable and consistent trace metal data in the late 1970s with the first dissolved Cd and Zn data published between 1976 and 1978, respectively (Boyle et al. 1976; Bruland et al. 1978; Martin et al. 1976). Concurrently, the Geochemical Ocean Sections Studies (GEOSECS) programme (1972–1978) was conducting global surveys designed to investigate the three-dimensional distributions of chemical, isotopic and radiochemical tracers in the ocean (Chester 1990) (see Box 3.2). However, the GEOSECS sampling system did not permit uncontaminated samples to be collected for certain easily contaminated trace metals such as Zn, Fe and Pb (Bruland and Franks 1983; Pilson 2012). Thus, the first reliable vertical profiles of dissolved Fe were only published in 1981 (Landing and Bruland 1981). By the 1980s, thanks to analytical advances in trace metal chemistry and the use of trace metal clean techniques, two notions were adopted: (i) seawater concentrations of many trace elements are a factor of 10–1000 lower than previously believed (e.g. compare Brewer 1975 and Bruland and Franks 1983), and (ii) trace metals have well-defined distributions in the world's ocean that relate to physical, chemical and biological features of the water column (Chester 1990; Pilson 2012), i.e. trace metal distributions are 'oceanographically consistent'. For instance, in the 1920s, Fe in seawater was said to be 1–25 μmol L^{-1} followed by values of 20 nmol L^{-1} –5.3 μmol L^{-1}in the 1930s and values of 60 nmol L^{-1} –0.1 μmol L^{-1} in the 1950s (de Baar 1994 and references therein), whereas today, values of <2 nmol L^{-1} in surface waters are the consensus (e.g. Anderson 2020; Schlitzer et al. 2018 and references therein).

> **Box 3.2: Evolution of Trace Metal Clean Procedures**
> Preventing water samples from becoming contaminated during sampling and analytical processing constitutes one of the greatest difficulties encountered in marine trace metal analysis. Therefore, it is imperative that extreme care is taken to avoid contamination when collecting and analysing samples for trace metals. Before the 1970s, marine chemists did not commonly follow rigorous trace metal clean procedures, did not have clean materials and equipment, and did not pay enough attention to sample handling in order to avoid the inadvertent introduction of contamination (Chester 1990). However, even with the gradual introduction of trace metal clean protocols from the late 1970s onwards and with care being taken during sample collection and analysis, samples taken before the GEOTRACES programme, i.e. before 2006, were scarce and sometimes still contaminated and inconsistent with oceanographic features.
>
> For example, the GEOSECS programme was the first major ocean programme to generate geochemical ocean sections in all three major ocean basins in the years 1972–1978. While focusing on the ocean carbon cycle, samples were also collected for several other tracers, that is, various isotopes and trace elements. At the time, the GEOSECS vertical oceanic profiles of Fe

(continued)

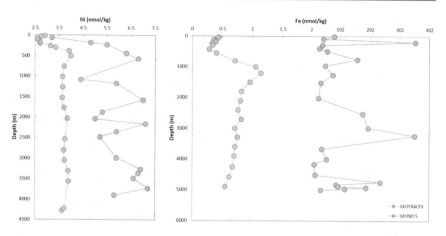

Fig. 3.2 Comparison of Ni and Fe profiles collected and measured during GEOSECS in the 1970s and GEOTRACES in 2011 and 2017. Displayed GEOSECS and GEOTRACES stations were located in the same ocean basins (Ni and Fe stations are 118 km and 137 km apart, respectively). These data show how endeavours such as the GEOTRACES programme added much to the ability of marine chemists to collect reliable and oceanographically consistent trace metal data. GEOSECS data for Ni (Station 3; North Atlantic) and Fe (Sargasso Sea) are from Sclater et al. (1976) and Brewer et al. (1972), respectively. GEOTRACES data for Ni are from GA02 station 5 (North Atlantic; Middag et al. 2020) and data for Fe are from GA03 station 8 (North Atlantic; Schlitzer et al. 2018; data originates from various investigators). Please note the change in scale for the x-axis (Fe concentration) for the Fe profiles

---

Box 3.2 (continued)

and Ni (orange profiles below) were deemed to be major breakthroughs and were considered to be some of the very first reliable vertical profile datasets (Fig. 3.2), with values much lower than previously thought (e.g. Bruland 1983). However, later work has shown the profiles for these elements overestimated concentrations and were too variable (see below), likely due to contamination and the limits of techniques available.

However, GEOSECS was the role model for GEOTRACES that started its ocean sections campaign between 2007 and 2008 with GEOTRACES expeditions that were part of the International Polar Year. The GEOTRACES vertical profiles shown here (blue profiles; Fig. 3.2) were collected in 2011 and 2017, some 35 years after the also shown GEOSECS data. These improved profiles (lower, more accurate concentrations, which are also oceanographically consistent) illustrate the significant advances made in both the collection of samples and measurement of trace metal concentrations since GEOSECS.

While apparently 'clean' to the regular laboratory scientist, the exceptionally low levels of trace metals in seawater mean that many common laboratory procedures or equipment (e.g. glassware) have the potential to contaminate samples beyond the point where the 'true' trace metal concentrations can be established. There are numerous pathways by which samples may become contaminated, and potential sources include metallic or metal-containing labware or gloves (e.g. talc gloves that contain high levels of Zn), containers, sampling equipment (improperly cleaned and stored), chemicals, reagents, reagent water and atmospheric inputs (EPA 1997). Additionally, human contact (hair, dead skin, exhalation) and dust particles in a laboratory can be a significant source of trace metal contamination to samples (EPA 1997). In the following sections, we describe the procedures required to make ultraclean trace metal measurements in the most pristine marine environments in order to produce reliable, accurate and reproducible trace metal data. The section below, however, only gives a general overview of trace metal clean concepts and principles, and various laboratories have developed slightly different procedures depending on their research objective and traditions. For more details on trace metal clean techniques, the reader is also referred to the GEOTRACES Cookbook (Cutter et al. 2017).

3.2.1 Trace Metal Clean Environment

A clean laboratory atmosphere, in which the contact of the sample with particles in the air and other surfaces is minimized, is one of the key features of trace metal clean practices. Such a clean atmosphere is ideally provided by a clean laboratory, which consists of one or more rooms that are kept under positive pressure by drawing air through a series of filters that effectively remove particles, including ultrafine particles (<0.5 μm), from the atmosphere (Sander et al. 2009). To be classed as a trace metal clean working space, the working space must comply with the class 100 standard by the Federal Standard 209 or the equivalent ISO 5 class standard by ISO-14644-1 (Fig. 3.3a) (Cutter et al. 2017; Goldberg 1996). Cleanrooms or working spaces that fall under these two standards implement high-efficiency particulate air (HEPA) filter systems to obtain a permissible density of less than 3500 (<0.5 μm) dust particles per m^3 of air (Tovar-Sánchez 2012). Ambient outdoor air in a typical urban area contains 35,000,000 particles (> 0.5 μm) per m^3 of air, and ordinary activities performed by people generate millions of particles every minute (Goldberg 1996; Nehme 2020). For instance, the rate of particle emission during normal human speech ranges from 1 to 50 particles per second, which equates to 60,000–3,000,000 particles per m^3 (Asadi et al. 2019). If full clean room facilities are not available or needed, a clean working atmosphere should be provided via ISO class 5 or class 100 laminar flow benches, or a non-metal glove box fed by particle-free air or nitrogen (EPA 1997). Care must be taken to avoid metallic components such as screws to mount screens of laminar flow benches, as they will cause contamination (Sander et al. 2009).

Onboard ships, when samples are collected and prepared for storage or onboard analysis, full clean room facilities are not usually available, even though a shipboard environment is more prone to cause contamination (Fe ship, metallic structure, Zn anodes, paint, engine exhaust, waste water, etc.) than a land-based facility (Gillain et al. 1982; Sander et al. 2009; Tovar-Sánchez 2012). When permanent clean room

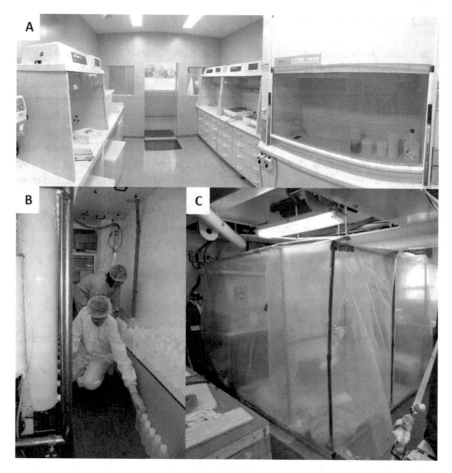

Fig. 3.3 Photographs of examples of trace metal clean facilities. Photo of a land-based ISO-6 clean room laboratory with ISO-5 laminar flow hoods (**a**) (photograph credit: Tim Conway, University of Florida, USA), the NIOZ trace metal clean container (**b**) (photograph credit: Loes Gerringa, NIOZ, Netherlands) and a trace metal clean working 'bubble' on board a research vessel (**c**) (photograph credit: Gert van Dijken, Stanford University, USA)

facilities are not available on a ship, converted shipping containers commonly act as designated mobile clean rooms on research expeditions. Such a mobile laboratory is fitted with a HEPA air filtration system and laminar flow benches to comply with clean room standards (Fig. 3.3b). If such a mobile laboratory is unavailable, a standard shipboard laboratory can be converted into a temporary clean room, commonly referred to as a 'bubble'. A bubble consists of a polyvinylchloride (PVC) plumbing tubing structure (or similar material) covered by plastic film that is also used for lining the walls and benches (Fig. 3.3c). To accord with clean room regulations, clean air is usually provided via HEPA filtered air that also keeps the working environment inside the bubble over-pressured (Sander et al. 2009).

3.2.2   Trace Metal Clean Practices

Apart from the atmosphere, the second most likely source of contamination for samples comes from the human investigator (see Box 3.3). To minimize and avoid this source of contamination, a strict trace metal clean working procedure must be followed during all phases of sampling and laboratory work. It is recommended that protective clothing is worn in all laboratory operations since humans are the main contamination risk in cleanrooms, particularly through the shedding of particles from personal clothing, exacerbated by movement (Goldberg 1996). A study with test subjects that wore cotton tracksuits vs cleanroom uniforms in an ISO 5 cleanroom showed that test subjects wearing tracksuits shed on average 34,955,780 particles (>0.5 μm) per minute while walking, while test subjects in full cleanroom attire shed on average only 106,328 particles (>0.5 μm) per minute while walking (Cleanroom-Technology 2011). The clean room uniform commonly comprises a clean room coverall, disposable plastic gloves (powder-free), a hair net, dedicated plastic shoes or plastic foot covers and eye protection (e.g. EPA 1997; EPA 1996; Sander et al. 2009; Tovar-Sánchez 2012). Often, two pairs of gloves are used by the human analyst, one for ultraclean handling and one for clean (i.e. dirtier) handling. If it is even suspected that gloves have become contaminated, work must be halted, the contaminated gloves removed and a new pair of clean gloves put on (EPA 1997; EPA 1996). In addition, all surfaces that equipment, samples, reagents and standards come into contact with are potential sources of contamination, and thus all equipment and work surfaces should be wiped with a lint-free cloth prior to use, or at least on a regular basis, to remove dust. All apparatuses and laboratory equipment used for trace metal work must be non-metallic, and glass materials and coloured plastics should be avoided (EPA 1997; EPA 1996; Tovar-Sánchez 2012). When not being used, laboratory equipment should be covered with clean plastic wrap, stored in a clean bench or plastic box or bagged in clean polyethylene bags (colourless zip-type bags are recommended) (EPA 1997; Tovar-Sánchez 2012).

Box 3.3: Trace Metal Clean Practices
The key requirement for reliable and contamination-free trace metal data is compliance with trace metal clean practices throughout the entire process, from equipment preparation all the way through to sample collection and eventual analysis (Sander et al. 2009). Two of the most important factors in avoiding and minimizing sample contamination are (1) an awareness of potential sources of contamination including the position of the investigator's arms and hands relative to the airflow, open samples and reagents and the flow and direction of the investigators breathing and (2) strict attention to work being done (EPA 1996; EPA 1997). Therefore, it is imperative that trace metal clean procedures are carried out by well-trained and experienced personnel (EPA 1997; Tovar-Sánchez 2012).

3.2.3   Trace Metal Clean Sample Bottles

Appropriate container material for sample storage is also key in trace metal chemistry, notably when considering the general long contact times (days, weeks, months or years) between seawater sample and container wall. Two opposing aspects are important here: (1) sample contamination by different kinds of container materials and (2) trace metal losses by surface adsorption to the container wall (Gillain et al. 1982). Both processes are dependent on the surface to volume ratio and the sample bottle material (Jensen et al. 2020). Generally, samples for dissolved trace metal analysis are acidified for storage to avoid 'wall adsorption' of metals and thus undermeasurement of the 'true' concentration of a metal of interest in the sample (Cutter et al. 2017; EPA 1997). Both fluoropolymers (specifically fluorinated ethylene propylene (FEP), perfluoroalkoxy alkane (PFA) or polytetrafluorethylene (PTFE)) and low-density polyethylene (LDPE) bottles are used for acidified sample storage owing to their low intrinsic trace metal composition and low levels of metal adsorption (Cutter et al. 2017; Noble et al. 2020). High-density polyethylene (HPDE) bottles have been shown to contain organometallic trialkyl aluminium (Al) compounds and are thus deemed unsuitable for dissolved Al analysis Cutter et al. 2017, but can be used for most other metals. Generally, LDPE bottles are recommended for sample storage of samples reserved for dissolved and particulate trace metal analysis as well as for speciation and isotope analysis. Fluoropolymer bottles are chemically and thermally more resistant compared to LDPE bottles and are also deemed 'cleaner' due to their commonly lower metal blanks (e.g. Gasparon 1998; Noble et al. 2020 and references therein). However, the high cost and environmental impact of fluoropolymer may be a limiting factor for the use of this material.

3.2.4   Trace Metal Cleaning Procedures for Sample Bottles

Trace metal analysis results can easily become inaccurate if sample bottles are contaminated. To remove potential sources of contamination in sample bottles, all sample bottles should be thoroughly cleaned, both to remove dust and any metals that could exchange with the sample during storage. This cleaning goes beyond usual cleaning of labware and often involves soaking the equipment in soap or acidic solutions in order to remove organics and/or leach metals from the plastic itself (Apte et al. 2002; Cutter et al. 2017; Sander et al. 2009). The procedure used to prepare bottles for seawater samples is typically different between each laboratory and individually assessed for suitability. Differences arise because groups implement methods based on historical experience or differences in intended sampling objective or metal of interest (Apte et al. 2002). Today, however, most laboratories use similar methods which have been standardized by advice from the international GEOTRACES programme and the accumulated experience of the community since the 1970s. For example, the minimum effective cleaning procedure recommended by GEOTRACES for analysing sub-nanomolar levels of most trace metals in seawater involves soaking of the bottle in alkaline detergent to remove organic residues (grease and fat), soaking in diluted hydrochloric or nitric acid to mobilize and desorb solid phase and/or adsorbed contaminants from the bottle wall, followed by exhaustive rinsing with ultra-high purity water (UHPW) (Cutter et al. 2017). The cleaned bottles are then double bagged using at least two (resealable) polyethylene bags and stored until use either empty or filled with dilute high purity acid (Apte et al. 2002; Cutter et al. 2017; Sander et al. 2009); however, it should be noted that bottles cleaned in such a way might not be useable for some speciation studies (see Sect. 3.5.3). Bottles should be handled at later stages using clean gloves, and the final 'clean' steps should be carried out in a dedicated clean working space. Obviously, all work involving acids should be carried out safely in well-vented areas with the correct personal safety precautions (Apte et al. 2002).

3.2.5   Trace Metal Clean Reagents

Systematic contamination of samples may often also be caused by using chemical reagents or water of insufficient purity during processing and analysis (Bowie and Lohan 2009; Sander et al. 2009). This type of contamination is usually indicated by the systematic measurement of unexpectedly high metal concentrations or by high procedural 'blanks', which in the latter case is the amount of metal involuntarily added to a low-metal or ultrapure water sample during processing and analysis (using the same analytical steps as for the actual samples; see Sect. 3.6 for more information on procedural blanks; Bowie and Lohan 2009; Sander et al. 2009). Thus, all chemicals and reagents used for the analysis of trace metals must be of high purity, typically denoted as 'ultrapure' grade, which are relatively expensive (or for some reagents simply not available). As such, many laboratories utilize chemical or physical procedures to reduce trace metal impurities by removing metals from lower-grade reagents (Bowie and Lohan 2009). Reagent cleaning methods are common practice in many trace metal laboratories, for example, using clean sub-boiling distillation methods to obtain purified acids or reagents or isopiestic distillation for purifying ammonia (e.g. Sander et al. 2009 and references therein). Dilutions of reagents for trace metal methods or rinsing of clean equipment must be carried out using ultra-high purity water (UHPW), produced from deionized water by commercially available filtration systems, and defined with a resistivity of >18.18 MΩ·cm. Similar UHPW may also be prepared by sub-boiling distillation of deionized water. To verify the purity of reagents, reagent blanks and/or process blanks should be determined regularly (see Sect. 3.6). Furthermore, to avoid contamination of clean reagents, reagent preparation, handling and manipulation must be performed under rigorous trace metal clean conditions (see Sect. 3.2.1 and 3.2.2), and reagents must be stored and dispensed into acid-cleaned fluoropolymer or LDPE bottles (see Sect. 3.2.3) (EPA 1997). Equipment such as pipette tips or measuring cylinders that may be used for dispensing reagents must also be checked and/or acid cleaned or acid rinsed prior to use to prevent metal contamination of the samples.

3.3   Trace Metal Clean Sample Collection

Marine trace metal chemists still depend on the collection of water samples for trace metal analysis because, unlike for some physical and other chemical oceanographic parameters, instruments to make in situ measurements for trace metals are either not yet mature or not readily available (Capodaglio et al. 2001; Grand et al. 2019). The sections below will focus on the collection of samples from both the shallow and the deep oceanic water column for trace metal analysis. We note that aerosol dust sampling and sediment-water interface sampling procedures for trace metal analysis also exist; however, these methods fall outside the scope of this chapter and are not discussed here.

The three main difficulties a marine trace metal chemist is confronted with at sea during sample collection are (Gillain et al. 1982):

1. *Representative samples* – obtaining a sample that accurately represents, both in time and space, the conditions of the water chemistry of the system targeted for the study is challenging. This concern is of primary importance for a relevant description of the system of interest (Wilde and Radtke 1998). Where the trace metal or parameter of interest is expected to vary dramatically in space or time (e.g. coastal settings), high spatial- and/or temporal-resolution sampling is needed to account for this variability. For metals with conservative (i.e. invariant, or salinity related) distributions, fewer sampling points may be needed. Obtaining representative samples also includes a need to store or process (e.g. filtration) samples accordingly to avoid artefacts from storage (e.g. bottle adsorption or contamination levels). For samples that are sensitive to rapid chemical alterations (e.g. Fe^{2+} oxidation), special time-sensitive precautions may need to be deployed in the field.
2. *Minimizing contamination* – the main sources of contamination in the field are the sampling platform, the personnel and the sampling device. For open ocean environments, where large research vessels are required, the vessel must be considered as the main source of contamination in surface waters. Generally, research ships, owing to their Fe structure and other metal features such as propellers, Zn anodes, paint and engine exhausts, act as large sources of trace metals to the immediate environment (Gillain et al. 1982; Tovar-Sánchez 2012). Thus, care must be taken when collecting surface samples or when deploying sampling devices, for example, by considering the ships' draught and wind as well as current direction relative to sources of contamination (Gillain et al. 1982; Tovar-Sánchez 2012). Where possible, such as in coastal studies, surface samples may be collected from small non-metallic boats.
3. *Sampler choice* – reliable devices and techniques to minimize and eliminate contamination during surface water and deepwater sampling have been developed by the community over several decades (Tovar-Sánchez 2012) and are now standardized and intercompared by programmes such as GEOTRACES (e.g. Cutter and Bruland 2012; Middag et al. 2015a). The volume of sample needed, the sampling depth and the element/s of interest are key factors when

choosing a sampling device. Although the choice of appropriate samplers for collecting trace metal samples is fundamental, adequate cleaning treatments (i.e. flushing the sampler with low trace metal seawater multiple times before use, and conditioning of the device), verifying cleanliness and correct handling of the devices are also vital (Capodaglio et al. 2001).

Commonly used trace metal clean samplers and analytical techniques are detailed in the following sections. The description of samplers is supposed to guide the reader through available and mature low-cost and high-priced sampling devices that are currently in use for trace metal sample collection by the scientific community. At present, the trace metal clean rosette (see Sect. 3.3.1.1) is the workhorse of the trace metal community and the backbone for the collection of large datasets within the GEOTRACES programme (see De Baar et al. (2008) for a brief history of GEOTRACES sampling systems). This chapter also introduces low-cost devices that provide lower-resolution solutions, facilitating relatively low-cost trace metal observations and allowing the filling of current gaps in data coverage in environments not accessible with a rosette system. The low-cost systems can produce equally high-quality data if trace metal clean procedures are followed (see Sect. 3.2), but as for all systems, rigorous intercomparison of results and cross-validation of protocols are recommended (Cutter et al. 2017).

3.3.1 Dissolved Trace Metal Sampling

3.3.1.1 Depth Profile Sampling

When sampling below the surface ($\sim > 10$ m), the choice of the sampling device is made based on the study objective, analytical requirements, characteristics of the system and the available capabilities. General approaches used for deepwater and vertical profile sampling are akin to those used by surface seawater sampling activities (see Sect. 3.3.1.2) and commonly include pump set-ups and/or discrete bottle sampling. Care should be taken not to touch the bottom with the deployed water sampler, as disturbed sediments and associated metals could contaminate the sample or damage the sampler. The deepest sampling depth is commonly 5–10 m above the bottom. Further, it has to be noted that instrument deployments during strong currents can result in an error (underestimation) of the deployment depth (Turk 2001), if not checked by pressure sensors attached to the instrument itself. Additionally, if the deployment platform is drifting, care has to be taken that the bottom depth remains deep enough to avoid running the instrument aground.

Discrete Bottle Sampler Systems

Bottle samplers allow marine chemists to obtain discrete samples from specific water depths, both in shallow and deep waters (e.g. Cunliffe and Wurl 2014; Cutter et al. 2017; Van Dorn 1956). This section covers the use of bottle samplers in deep waters (for use in surface waters, see Sect. 3.3.1.2). Bottle samplers can be obtained in different sizes and generally consist of a cylinder or 'bottle' with stoppers at each end

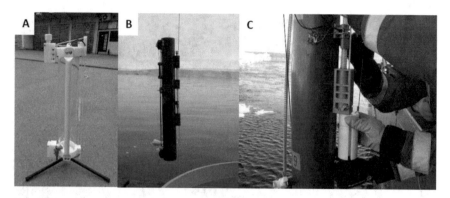

Fig. 3.4 Photographs of a custom-built acrylic bottle sampler (**a**) (photograph credit: Dario Omanović, Rudjer Boskovic Institute, Croatia) and a commercially available GO-FLO sampling bottle attached to a Kevlar cable (**b**), or with messenger (**c**) (photograph credit: Gert van Dijken, Stanford University, USA)

(Cunliffe and Wurl 2014) that can be closed at a desired depth, either manually using a messenger or electronically. The electronic version is preferred in deep waters. To minimize contamination, trace metal clean bottle samplers are often made of transparent acrylic or fluoropolymer-lined opaque PVC, and their interior is totally free from metal parts (Fig. 3.4; see Box 3.4; Cunliffe and Wurl 2014). The earliest version of a (non-trace metal clean) bottle sampler is commonly referred to as Van Dorn sampler, but since the establishment of the GEOTRACES programme, external spring bottles such as internally fluoropolymer-coated ~10–12 L Niskin-X and GO-FLO bottles are most commonly used (Fig. 3.4), obtained from General Oceanics (see Cutter and Bruland (2012) for more details), or Ocean Test Equipment (slightly different samplers). Different laboratories have also constructed custom-designed bottle samplers based on requirements, with an example being the Royal Netherlands Institute for Sea Research (NIOZ) Titan system which makes use of custom-designed 'Pristine' bottles (see Rijkenberg et al. (2015)) (Fig. 3.5d).

Bottle samplers can be individually or serially attached directly to a non-metallic cable (e.g. Kevlar) to allow contamination-free sampling or mounted on a carousel (Fig. 3.5), which is often referred to colloquially as a CTD rosette (because of sensors measuring conductivity, temperature and depth), or just a rosette, to allow marine chemists to obtain discrete water samples from various depths (Bruland et al. 1979; Cunliffe and Wurl 2014; Van Dorn 1956). The latter is commonly used for depth profiling during GEOTRACES expeditions. However, regular rosette systems, like regular bottle samplers, are too contaminating to collect pristine water samples for trace metal analysis. Modifications are thus required. Such modifications typically take the form of coating the regular rosette frame with epoxy or replacing it with titanium and removing any sacrificial metal anodes. Such 'trace metal clean' rosettes are then loaded with trace metal bottle samplers. One of the first such trace metal clean rosette-based systems for collecting trace metal samples was designed by

Fig. 3.5 Photographs of various trace metal clean rosette samplers. Photographs of the Japanese Niskin-X sampler system (**a**) (photograph credit: Taejin Kim, Pukyong National University, South Korea), a commercially available CTD rosette system from General Oceanics with GO-FLO bottles (**b**) (photograph credit: Antonio Tovar Sánchez, ICMAN (CSIC), Spain), a modified small CTD system (**c**) (photograph credit: Antonio Tovar Sánchez, ICMAN (CSIC), Spain) and the custom-built NIOZ sampling system (**d, e**) (with its lightproof samplers rather than the original PVDF samplers; Rijkenberg et al. 2015) that can be transferred to a custom-designed clean laboratory container for processing (photograph credit: Loes Gerringa, NIOZ, Netherlands)

the Trace Metal/Plankton Group at Moss Landing Marine Laboratories, first used by Murray et al. (1992) and described by Sanderson et al. (1995). Today, 'trace metal clean' rosettes are commercially available for ocean sampling (e.g. Seabird Scientific or General Oceanics) and are widely used for clean water sampling. Currently, a number of different varieties of trace metal multiple bottler samplers in various size ranges are in use, such as the GO-FLO trace metal rosette used by the US GEOTRACES programme (Cutter and Bruland 2012), the NIOZ Titan system

(De Baar et al. 2008) and the Japanese Niskin-X sampler system (Obata et al. 2017) (Fig. 3.5).

Similar to the individual bottle deployment, a rosette arrangement allows samples to be taken at different water depths with a vertical resolution of ~5 m (Strady et al. 2008). One of the major advantages of a rosette sampler, however, is the possibility for simultaneous collection of multiple samples at one depth. This simultaneous collection is especially useful when large volumes of water are needed for experimental work such as culturing/incubation studies or when studying elements and isotopes with inherently low seawater concentrations (e.g. radium (Ra)). Further advantages of a rosette system are: (i) faster deployment compared to the deployment of multiple bottle samplers on a cable, (ii) higher-resolution sampling capability since more bottles can be used during a single deployment and (iii) higher reliability in relation to messengers (see below) that sometimes fail to trip bottles at the desired depth that is often estimated rather than measured, in contrast to a rosette system (Sanderson et al. 1995).

When deploying individual bottles on a Kevlar cable, a (plastic covered) weight should be attached to the bottom of the cable, several meters below the last sampler, to keep the cable taut. Prior to deployment, the bottles should be attached to the cable and armed (closure system ready for use). Each sampler, except the one closest to the bottom, will be equipped with a plastic-coated metal weight or 'weighted messenger' attached to the Kevlar cable via a lanyard. When the bottles are at their desired depth, the first messenger will be dropped down the cable by the investigator which closes the first sampler by tripping the spring-loaded valve (closure system) (Cunliffe and Wurl 2014). This mechanism also causes the next messenger to drop, closing the subsequent samplers in rapid succession. Enough time has to be allowed for the messengers to trip each sampler—which can take up to 1 h in 6000 m water depth – before winching the cable to the surface (Measures et al. 2008). By touching the cable with one hand, it is possible to feel a strong 'thump' on the cable as each messenger triggers the subsequent sampler. Upon recovery, plastic gloves can be placed over the spigots of bottles before the bottles are transferred to a clean room for sample collection and filtration via the sampling valves/spigots (Cutter and Bruland 2012). It is critical to note and record if there are any leaks from the samplers or any open samplers upon retrieval, since leakages and open bottles may affect the integrity of the sample and/or may result in contamination. When using bottle samplers to collect shallow waters (<100 m), it is also critical to be aware of, and to avoid, sources of surface contamination, for example, the wake or plume of trace metals associated with a research ship. Thus, discrete bottle sampler systems (either individual or on a rosette) are usually not used to sample water shallower than 10 m for trace metals.

Some of the individual bottle samplers and rosette systems can be deployed with bottle samplers in the open position, while others can be deployed with bottle samplers in the closed position – since they open themselves automatically at a fixed depth (usually ~10 m) to avoid contamination of the sampler by the surface microlayer (SML) which is particularly rich in trace metals (Caroli et al. 2001). During descent, the open bottle samplers are flushed. Commonly, individual samplers on a cable are closed prior to ascent, while rosette systems close samplers during the ascent. Rosette systems either close at pre-programmed depths (using a pressure sensor) or are triggered electronically via the conducting cable at the desired depth (Fitzsimmons and Boyle 2012; Measures et al. 2008). The bottles can either be closed on the fly (usually at winch speeds of 0.3 m/s), so the bottles are always moving into clean water that has not been in contact with the rosette frame (to avoid possible contamination of the water via the frame), or after $1-2$ min after reaching the desired depth to allow the temperature and salinity readings of the CTD sensor to equilibrate. The latter is commonly done for titanium systems that pose minimum risk of contamination due to the absence of sacrificial anodes and other contaminating metal components.

As with regular rosette systems, trace metal rosettes are also commonly equipped with various sensors (e.g. for oxygen, fluorometer and transmissometer) including conductivity, temperature and pressure (CTD) sensors providing real-time readouts during deployment if deployed via a conducting hydro wire. After recovery, the rosette is secured, plastic covers are often immediately placed on top of the bottle samplers, and plastic gloves are placed over the spigots. Typically, bottle samplers are then removed individually from the rosette frame and carried into a dedicated clean laboratory/bubble, where they are secured to a purpose-built rack for sub-sampling (Cutter and Bruland 2012). An alternative approach is to use a custom-built option like the NIOZ Titan system (Fig. 3.5), where the bottle samplers remain on a custom-built titanium frame, and the whole frame is transferred to a custom-designed clean laboratory container for processing, without needing to remove individual bottles (Rijkenberg et al. 2015). The NIOZ Titan system has been proven to be effective and clean (Middag et al. 2015a) and has some other advantages over commercial systems. For example, the Titan system was motivated by problems with GO-FLO bottles, specifically that their closure system is notoriously fickle in cold waters (Measures et al. 2008). To address this issue, the Titan system houses 24 polypropylene (lightproof) or PVDF samplers (23 L), so-called pristine samplers, with butterfly valves that close the bottles hydraulically. A drawback of this system is its size and weight, limiting deployment from smaller ships and requiring a strong winch and cable.

Box 3.4 Cleaning Bottle Samplers

There is some discussion about whether cleaning of water samplers mounted on a CTD rosette (i.e. GO-FLO and trace metal Niskin bottles) is needed or desirable before and between system deployments. If these bottles are cleaned, acid concentrations should be kept low (0.1 M HCl is recommended in the GEOTRACES 'Cookbook'; Cutter et al. 2017), and no acid should contact the outside of the bottle, the nylon components in particular.

Other discrete, bottle-based sampling systems that can be used for depth profiles are the MITESS (moored in situ trace element serial sampler; Bell et al. 2002) or an autonomous underwater vehicle (AUV) sampler. Although the former was designed for moorings (see below), the system can also be deployed on a hydro wire to collect vertical trace metal profiles in a so-called VANE mode (Fitzsimmons and Boyle 2012). In the VANE mode, the MITESS is loaded into a weathervane-type PVC and polycarbonate structure that orients the sampler upstream of the potentially contaminating hydro wire (Fitzsimmons and Boyle 2012). The system then autonomously opens and closes a pre-cleaned sampling bottle at a desired depth. Trace metal samplers based on AUV systems are currently being developed and tested for trace metal clean sampling of mid-waters and deep waters in areas that are not easily accessible with research vessels, that is, areas near or under ice shelves, sea ice, and icebergs.

Pumping System on CTD Rosette

A so-called pump CTD system enables water sampling with higher volume (effectively unlimited volume) and higher resolution (vertical resolution of 1 m) relative to the typical bottle sampler rosette (Strady et al. 2008). This configuration allows for the detection of small vertical structures in trace metal distribution across interfaces (e.g. the nutricline or the redoxcline) or the halocline (Strady et al. 2008). The CTD pump system combines a rosette and a pump system, i.e. a peristaltic pump, and was developed in 2001 in collaboration between IOW (Institut für die Ostseeforschung, Warnemünde) and MPI (Max Planck Institute for Microbiology, Bremen). The system consists of a submersible CTD rosette with fluoropolymer-coated Niskin bottles, an acoustic Doppler current profiler (ADCP), a pump probe and a digital flow meter for the water stream. The flow rate of the system at ~300 m can be up to 2.9 L min^{-1} (Strady et al. 2008), and the water is pumped directly through a nylon hose to a clean laboratory on board ship for sub-sampling. However, the use of a pump typically limits the application of the system to a depth down to around 350 m (Strady et al. 2008).

Moored in Situ Serial Samplers

Moored in situ trace metal samplers have been developed for time series sampling to resolve temporal and seasonal variabilities in trace metal concentrations in various marine environments (Bell et al. 2002). Such in situ samplers collect and preserve samples for later laboratory analysis. Moored in situ samplers can be very useful (Bell et al. 2002), especially for established monthly time series stations such as at the Southern Ocean Time Series (SOTS; Trull et al. 2010), Bermuda Atlantic Time Series Study (BATS; Michaels and Knap 1996) and the Hawaiian Ocean Time Series (HOT; Karl and Lukas 1996). Currently, there are various moored samplers available that are suitable for trace metal work, with the most notable examples being MITESS (Bell et al. 2002) and ACE (autonomous clean environmental sampler; Fig. 3.6c, d; van der Merwe et al. 2019). Other systems such as PRISM (portable remote in situ metal; Mueller et al. 2018) and ANEMONE (advanced natural environmental monitoring equipment; Okamura et al. 2013) are also available, and others will likely be developed. Both the MITESS and the ACE samplers are self-powered and can be deployed for 6–12 months at various depths on standard deep-

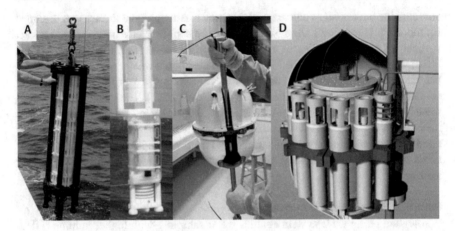

Fig. 3.6 Photographs of the MITESS (**a**) (photograph credit: Edward Boyle, Massachusetts Insitute of Technology, USA), the MITESS module in the VANE configuration (**b**) (photograph credit: Jessica Fitzsimmons, Texas A & M University, USA) and the ACE sampler (**c**) and module (**d**) (photograph credits: Pier van der Merwe, University of Tasmania, Australia)

sea moorings. A comparative advantage of the moored samplers compared to the commonly used rosette sampler is that the deployment itself requires no trace metal expertise, since the entire sampler is prepared in a clean room and no additional handling is necessary (Bell et al. 2002; van der Merwe et al. 2019).

The MITESS collects unfiltered 500 mL samples at any depth by opening and closing a sample bottle lid at a predefined depth. The time-controlled bottles are originally filled with high purity dilute acid that is replaced by denser seawater during sampling via passive density-driven flow (Bell et al. 2002). These samples are preserved over the deployment time of several months at pH 2.5 by the diffusion of high purity acid out of a diffusion chamber inside of the bottle. The sampler itself is made entirely out of ultra-high molecular weight polyethylene (UHMW) and can hold up to 12 bottles that are individually controlled by independent modules so that the failure of a single unit does not affect the entire set of the time series samples (Bell et al. 2002). The MITESS is programmed by wireless communication, and the electronic board retains a record of the timing of bottle opening and closure (Bell et al. 2002). A comparison of Fe data of GO-FLO and MITESS on the GEOTRACES IC2 expedition in 2009 did not show any differences between the two sampling systems, indicating that these samplers can be used interchangeably to collect trace metal samples, either on moorings or for discrete samples in the 'VANE' mode (Fitzsimmons and Boyle 2012). Potential issues with MITESS, however, are that the seawater is not filtered prior to being acidified (meaning that some portion of ocean particulate material will be dissolved during deployment) and that biofouling may occur, since the sample intake is not physically removed from the sampler body.

In contrast to MITESS, the ACE sampler collects filtered (0.2 μm, polyethersulfone filter membrane) samples into 65 mL fluoropolymer containers.

The time-controlled sampler works by drawing seawater through up to 12 individual intake tubes via acid-washed filters into the UHP filled sample bottles using individually programmable micro-peristaltic pumps (density displacement mechanisms; van der Merwe et al. 2019). The intake tubes are maintained in an upstream position relative to the device to minimize contamination during sampling (van der Merwe et al. 2019). A key advantage of this system is that samples are filtered to remove particles, and intake pots are made of PFA which together with their small surface area reduces biofouling and thus potential sampling artefacts and contamination (van der Merwe et al. 2019). However, a disadvantage of the ACE system is the fact that the filtered samples can only be acidified back in the laboratory after recovery of the sampler which might result in low-biased results due to wall absorption.

ROV-Based Discrete Samplers

Remotely operated vehicles (ROVs) with manipulator arms that operate on spatial resolutions on the single cm scale are a useful tool for sampling high trace metal environments such as under ice, near sediments or within fluids from hydrothermal vents and seeps along oceanic spreading centres, subduction zones and subsurface volcanoes. Multiple samplers such as the isobaric gas-tight sampler (IGT; Seewald et al. 2002), the titanium syringe sampler (Majors sampler; Von Damm et al. 1985) and the Kiel pumping system sampler (KIPS; Garbe-Schönberg 2006) are commonly in use to collect trace metal samples at a depth of up to 4000 m and can be

Fig. 3.7 Photographs of KIPS (**a**), Majors (**b**) and IGT (**c**) samplers attached on the manipulator arm of the ROV MARUM QUEST (photograph credit: MARUM – Centre for Marine Environmental Sciences, University of Bremen, Germany). Photo (**d**) shows the sample collection system of the KIPS device (photograph credit: Dieter Garbe-Schönberg, Christian-Albrechts-University Kiel, Germany)

easily attached to a ROV manipulator arm (Fig. 3.7). These samplers are made of titanium or other inert materials, are acid and temperature resistant to withstand the hot and acidic conditions of hydrothermal working areas, are fully remotely controlled and are filled through a titanium nozzle or snorkel which can be directly inserted into the vent orifice or other localized trace metal sources. The Majors sampler can collect samples of up to 750 mL, but is non-gas-tight and designed to release pressure during ROV recovery. The IGT sampler can collect samples of 150 mL and is gas-tight up to 450 bar to prevent degassing of the sample during ascent, thereby avoiding precipitation and chemical alterations of the prevalent trace metals (Seewald et al. 2002). This sampler can thus be used to characterize major, trace, semi-volatile and volatile components (Seewald et al. 2002). The KIPS device was specifically designed for trace metal clean work (Garbe-Schönberg 2006). This device consists of a manipulator operated titanium nozzle with PFA (perfluoroalkoxy) tubing that leads to the PFA sampling flasks that are non-gas-tight. The latest version has up to seven PFA sampling flasks that are remotely controlled by motor-driven open-close valves (Garbe-Schönberg, pers. commun.). The PFA flasks have a volume of 750 mL, and in situ filtration and/or in situ fixation units can be added in-line. Large volume sample bags up to 10 L have also been successfully filled using the KIPS system. Various other sensors and probes (i.e. temperature, pH, oxygen) can be mounted on either of the samplers to record in situ parameters at the point of sampling, which can be transmitted directly to the ROV control room in real time. While such ROV-based samplers have proven their use in high trace metal environments, they are not commonly utilized in open ocean situations, due to the elevated level of background contamination compared to traditional bottle samplers.

3.3.1.2 Surface Sampling

There are three general approaches that can be used for surface water sampling (0–10 m): (1) pumping water to the surface from the depth of interest; (2) sampling by bottles lowered to an appropriate depth by line, sampling device or pole and then closed manually, automatically (pressure triggered) or by a signal from the surface; and (3) adsorbing the metals or compounds of interest on an appropriate material lowered to the desired depth (Capodaglio et al. 2001). Additionally, for sampling the microlayer at the sea surface, special approaches have been developed (see below). For trace metal clean sampling of surface waters that are not easily accessible, such as areas near ice shelves, sea ice and icebergs, drone sampling systems are currently being developed, spearheaded by the University of Tasmania, Australia.

Discrete Bottle Samplers

As described in Sect. 3.3.1.1, bottle samplers can be individually or serially attached to a Kevlar cable to allow marine chemists to obtain discrete water samples from various depth intervals, including near the surface if this can be done without contamination (Bruland et al. 1979; Cunliffe and Wurl 2014; Matamoros 2012; Van Dorn 1956). Individual bottle samplers can be deployed in two configurations, i.e. horizontal (type alpha) or vertical (type beta), depending on the study objectives

(Cunliffe and Wurl 2014; Matamoros 2012). Type alpha samplers are ideal for sampling at the thermocline, narrow stratification layers or just above the bottom sediment (Cunliffe and Wurl 2014; Matamoros 2012). For more information on the bottle samplers, deployment and recovery, the reader is referred to Sect. 3.3.1.1.

Continuous Flow Samplers

Pumping systems, i.e. peristaltic pumps and diaphragm pumps (preferably an all-fluoropolymer inert type pump), with an extended inlet tube are frequently used by marine chemists to allow continuous and high-volume trace metal sampling (bulk sampling) of the near-surface water column (1–10 m) (Fig. 3.8) (Cunliffe and Wurl 2014; Tovar-Sánchez 2012). Prior to sample collection, it is recommended to condition the pre-cleaned tubing by pre-rinsing before collecting the unfiltered or filtered (in-line filtration) sample into an acid-cleaned sample container (Cunliffe and Wurl 2014), preferably in a clean space. To avoid contamination during sampling, the tubing should be extended ~3–4 m away from the sampling platform by attaching it to a plastic telescope bar or by deploying it via a boom or crane, which is usually the case for towed sampling devices as described next.

There are several versions of towed sampling devices (often referred to colloquially as 'tow-fish' or 'towed fish') that are deployed by marine trace metal chemists from a moving ship (Bowie and Lohan 2009; Cunliffe and Wurl 2014; Cutter et al. 2017; McDonnell et al. 2015; Tovar-Sánchez 2012). The simplest type consists of a subsurface torpedo-shaped heavy vehicle (Fig. 3.8). The water intake PTFE tube is attached to the nose of the towed fish, oriented into the oncoming water and connected to a PTFE diaphragm pump or a large peristaltic pump on board which supplies the sample water directly into the shipboard clean space (Cunliffe and Wurl 2014; Cutter et al. 2017; McDonnell et al. 2015; Vink et al. 2000). For underway surface sampling, commonly at relatively low speeds, the system is deployed from a boom or crane outside the bow wake of the ship to avoid sample contamination (Cunliffe and Wurl 2014; Cutter et al. 2017; McDonnell et al. 2015). Faster speeds are possible with this system if there is little or no swell and the towed fish remains

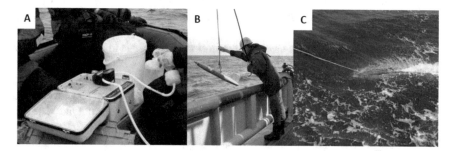

Fig. 3.8 Photographs of a peristaltic pump sampling system deployed via a rubber boat (**a**) (photo credit: Antonio Tovar Sánchez, ICMAN (CSIC), Spain) and a towed-fish sampling system deployed via a crane for continuous flow sampling of near-surface seawater (**b, c**) (photograph credit: Loes Gerringa, NIOZ, Netherlands)

outside of any breaking bow waves Cutter et al. 2017. Various sensors can be attached to the fish to provide accurate depth and temperature data. It is important to note, however, that most pumps are often not self-priming and may not be able to lift water to a height greater than 10 m (Cunliffe and Wurl 2014).

Passive Samplers

Passive sampling techniques are based on the diffusion of a metal of interest from the seawater onto a collecting medium (the passive sampler), owing to Dickian molecular diffusion and a greater binding affinity of the metal of interest with the passive sampler relative to seawater (Knutsson 2013). The metals will concentrate on the passive sampler until a steady-state concentration gradient from seawater to the passive sampler is reached (Knutsson 2013; Zhang and Davison 1995). Passive sampling devices can be deployed for long periods of time (often days or months before saturation is reached) to provide long-term, time-weighted averages of the concentration of a metal in the water column or to accumulate sufficient concentration of a metal for analysis (Allan et al. 2008; Zhang and Davison 1995). Consequently, the use of passive samplers is beneficial in investigations where concentrations of metals are low and/or fluctuate widely (Allan et al. 2008). Passive samplers can provide a more representative picture of overall trace metal concentrations in a system of interest compared to active sampling techniques that commonly just sample one point in time (Allan et al. 2008; Davison and Zhang 1994; Zhang and Davison 1995). However, it is important to note that trace metal data from passive samplers do not equate to trace metal data from active samplers, since passive samplers exclusively sample the labile fraction of metals in situ, that is, the metal fraction that can easily diffuse through, and be adsorbed by, the passive sampler. This feature excludes various phases of the dissolved pool – for example, metals that are strongly bound to organic ligands – and thus passive sampler metal measurements are generally lower than dissolved metal concentrations. Therefore, passive samplers provide information on the supposedly 'bioavailable' metal fraction, i.e. the metal fraction that can be taken up by marine organisms, and consequently, passive samplers offer more toxicologically relevant data relative to active samplers (Allan et al. 2008) (see Box 3.5). While passive samples can be deployed in deeper waters on moorings, they are more commonly used in shallow waters in coastal areas.

The main passive samplers used for monitoring trace metals in marine waters are the diffusive gradient in thin film (DGT) device and the Chemcatcher (Fig. 3.9; Schintu et al. 2014). DGTs were developed by Zhang and Davison (1995) and consist of a small piston-like plastic device containing a Chelex 100 layer as a receiving phase overlaid with a well-defined diffusion layer of polyacrylamide hydrogel protected by a filter membrane (Fig. 3.9) (Allan et al. 2008; Schintu et al. 2014). The Chemcatcher comprises a fluoropolymer sampler body that retains a chelating disk as a receiving phase overlaid with a cellulose acetate diffusion-limiting membrane (Allan et al. 2008; Schintu et al. 2014). A comparison study of the DGT device and the Chemcatcher demonstrated that the two sampling devices provided similar information and were able to integrate concentrations reliably

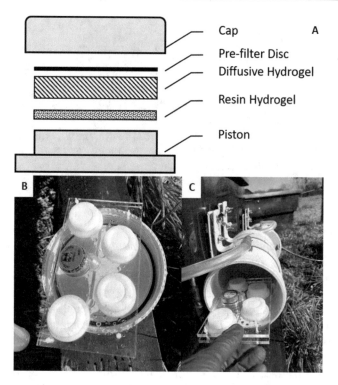

Fig. 3.9 Cross section of a functional DGT assembly (Figure courtesy of Billie Benedict, University of Otago, New Zealand) (**a**), and photograph of a diffusive gradient in thin film (DGT) passive sampler with DGTs and temperature data loggers mounted in acrylic plate holders for easy deployment and retrieval (**b, c**) (photograph credit: Amir Mohammadi, University of Waikato, New Zealand)

during their deployment period in surface waters (Allan et al. 2008). For field deployment, DGTs and Chemcatchers should be fixed between Perspex plates attached to a rope and a buoy to hold the devices in place (stationary) (Fig. 3.9). The time of deployment and retrieval must be recorded by the investigator to the nearest minute for later calculation of metal concentrations (Knutsson 2013). After recovery, samplers must be rinsed with UPHW water, preferably under clean conditions, and placed into two plastic bags for their transport in ice boxes back to the land-based laboratory. Gloves must be worn at all times when handling the passive sampling devices.

Box 3.5: Limitations of Passive Sampling Devices
Environmental factors can affect passive sampling including biofouling, presence of dissolved organic carbon (DOC), water turbulence and changes in

(continued)

Box 3.5 (continued)

temperature and salinity (Schintu et al. 2014). All these factors can alter the uptake rates of metals using the passive samplers and can thus create bias in field evaluations (Schintu et al. 2014). In particular, most of the environmental variables that can influence passive sampling measurements in the field are not fully accounted for in the laboratory-based calibration studies (i.e. quantification of the metal diffusion coefficient (D_e) through the diffusion layer of the sampler), which introduces uncertainty in the time-weighted averages of the metal concentration estimates.

Sea Surface Microlayer (SML) Sampler

Sampling devices that are commonly used to sample the sea surface microlayer (SML), defined as the top 1–1000 μm of the surface ocean, for trace metals include plate-, tube-, and screen-samplers (Fig. 3.10) (Cunliffe and Wurl 2014; Tovar-Sánchez et al. 2014; Tovar-Sánchez et al. 2020). These devices are usually deployed from shore or from a non-metallic boat. Two sampling materials are typically chosen for the samplers owing to their characteristic hydrophobicity, namely, borosilicate glass and plexiglass (plate- and tube-sampler). In the field, after preconditioning of the sampler (i.e. dipping it into surface water), the sampler is dipped into the ocean until most of the surface area is submerged and then withdrawn through the SML at a slow rate while wearing polyethylene gloves (Ebling and Landing 2015). After recovery, the sampler is held over a receiving bottle for the sample to drip off (Ebling and Landing 2015). The process is repeated until the desired volume of sample is acquired. To reduce contamination issues of the sample during sample handling (e.g. exposure to airborne particles), rotating glass drum samplers are

Fig. 3.10 Photographs of a sea surface microlayer (SML) plate sampler (left) (photo credit: Antonio Tovar Sánchez, ICMAN (CSIC), Spain), and a SML drum sampler prototype (right) (photograph credit: Dario Omanović, Rudjer Boskovic Institute, Croatia)

gaining more and more attention (Fig. 3.10) (Cunliffe and Wurl 2014). The drum sampler can be towed over the water surface to sample the SML via capillary force, and the sample can then be collected into pre-cleaned containers (Cunliffe and Wurl 2014).

Pole Sampler

One of the simplest methods to collect surface water samples is the manual collection of the water sample into pre-cleaned containers by submerging the sample bottle either directly from a non-metallic small boat or by using a non-metallic telescoping 'pole' (Fig. 3.11) (Tovar-Sánchez 2012). The bottle can be attached to the bottom of the pole with non-metallic clamps or secured on the pole via a plastic frame (Bowie and Lohan 2009; Turk 2001). The pole sampler can be deployed from shore or a non-metallic boat or even from a larger research vessel if conditions permit. The sampler should be deployed into the direction of the current to avoid sampling water that has been in contact with the sampling platform. Prior to collecting the sample, the sampling container should be conditioned (two or three times) with seawater below the SML. The pole is then submerged with the open bottle upside down, and at the desired depth, the system is turned to fill the bottle with ambient seawater. It is recommended that the investigator closes the bottle below the surface (wearing gloves) to avoid contamination from the SML (Cunliffe and Wurl 2014), which is particularly rich in trace metals (Capodaglio et al. 2001). To allow the collection of water samples at a specific depth, the bottle can be plugged with a non-contaminating silicone stopper attached to a line that the investigator pulls when the bottle is at the designated depth – the depth can be marked on the pole (Turk 2001). After recovery of the system, the sample bottles are immediately double bagged in polyethylene bags and processed in a clean room environment as soon as possible. When sampling for dissolved species, water is usually filtered as quickly as possible and then acidified with ultrapure reagents either shipboard or back on land (see Sect. 3.4.1).

Fig. 3.11 Photographs of a pole sampler for collecting near-surface water samples directly from the coast or from an inflatable rubber boat (photograph credit: Dario Omanović, Rudjer Boskovic Institute, Croatia)

3.3.2 Particulate Trace Metal Sampling

Oceanic particles are an important, yet perhaps less quantified, part of the oceanic trace metal inventory. However, with advances in particle collection and analysis coming in recent years linked to large-scale field programmes such as GEOTRACES, this trace metal fraction is gaining more and more attention in the scientific community (Fig. 3.1) (e.g. McDonnell et al. 2015). Sampling for particulate metals can be done using ship-board filtration from trace metal bottle samplers, if their volume is sufficient to collect enough particles to measure the element of interest (see Sect. 3.3.2.1) (McDonnell et al. 2015; Planquette and Sherrell 2012) or by larger volume in situ filtration systems (see Sect. 3.3.2.2). GEOTRACES inter-calibration efforts have shown that there is no systematic difference between particulate trace metals collected by direct bottle filtration and by in situ filtration, suggesting that these sampling strategies can be used interchangeably (Planquette and Sherrell 2012).

3.3.2.1 Bottle Sampler Collection

Once bottle samplers as used for dissolved metal sampling are back on deck, particles can be collected from the samplers, directly by pressurising the samplers to allow filtration (in-line filtration) or by filtering sub-samples from secondary containers after sub-sampling (off-line filtration) (Fig. 3.12). There are advantages of using bottle samplers for particle collection, namely, that the particulate metals collected can be related directly to dissolved metals measured from the exact same

Fig. 3.12 Photograph of a custom-built off-line particulate trace metal set-up (photograph credit: Mathijs van Manen, NIOZ, Netherlands). For the off-line particulate trace metal set-up, water samples were collected from the bottle samplers right after recovery into secondary containers to decrease between-cast turnaround time (Cutter et al. 2017)

depth and that multiple sample depths can easily be collected from a single rosette cast. The disadvantages are that only relatively small volumes can be filtered (10s of L) and thus there may be insufficient particulate material to measure some low-level trace metals of interest and that there is the possibility of particle loss by settling in the sampler prior to filtration. The latter requires that the investigator mixes samplers regularly and limits filtration time to 1–2 h (Cutter et al. 2017; McDonnell et al. 2015; Planquette and Sherrell 2012).

3.3.2.2 In Situ Filtration

In contrast to bottle sampler filtration, in situ filtration techniques allow the collection of very large volume (e.g. ~500 L; Twining et al. 2015b) size-fractionated samples of marine particulate matter from a single depth in the water column (Cutter et al. 2017), although multiple samplers can be deployed in sequence on a non-metallic cable to obtain a depth profile of particles in one single cast (McDonnell et al. 2015). Several titanium and stainless steel in situ systems are currently in use, including the ship-powered multiple unit large volume in situ filtration system (MULVFS, deployable to 1000 m depth; Bishop et al. 2012; Bishop et al. 1985), the battery-powered in situ McLane Research Laboratories Large Volume Water Transfer System (WTS-LV; referred to as 'McLane pumps'; Fig. 3.13; deployable to 5500 m depth in water temperatures from 0 to 50 °C; Morrison et al. 2000) and the Challenger Oceanic Stand-Alone Pump System (SAPS; deployable to 6000 m depth; Fig. 3.13). Generally, multiple filters can be used for size fractionation in the samplers, and various filter types are available

Fig. 3.13 Photographs of a McLane in situ pump (left) (photo credit: Alex Fox, Science Writer) and a Challenger Oceanic Stand-Alone Pump System (SAPS) (right) (photo credit: Maeve Lohan, University of Southampton, United Kingdom)

depending on the metal and the particle size of interest (McDonnell et al. 2015). Depending on the system used, the target depth, the filter used and the prevalent particle concentration, large volumes of seawater can be filtered per cast with a pump speed of 1–50 L min^{-1} (Bishop et al. 2012; McDonnell et al. 2015). Conventionally, in situ pump systems are programmed to sample for several hours, but this obviously depends on the research objective and sample region.

Underway and towed sampling systems can also be used for the collection of particulate trace metals in surface waters via systems such as the towed fish using in-line filtration (Hales and Takahashi 2002; McDonnell et al. 2015). While these systems improve spatial and temporal resolutions in the upper water column as well as minimize the amount of ship time dedicated to sampling, particles might disintegrate or flocculate during collection (due to turbulent fluid environments in the tubing from the underway samplers to the ship) which inhibits quantitative assessments of size distributions (McDonnell et al. 2015).

3.4 Trace Metal Clean Sample Handling and Storage

Trace metal clean sample protocols have to be applied during all stages of sample handling and storage (see Sect. 3.2.2), especially during sample manipulation, e.g. acidification. Samples should be processed as quickly as possible after recovery of the sampling device to minimize loss of trace metals by absorption on samplers and/or bottles and avoid chemical alteration and/or speciation changes. The following sections illustrate required sample processing and handling steps at (near) ambient conditions.

3.4.1 Dissolved Trace Metal Samples

Large volume samples for operationally defined dissolved metals should be filtered through 0.2 μm cartridge filters such as Pall AcroPak capsules (Cutter et al. 2017). Investigators will typically choose a specific filter type and protocol tailored to their element of interest, choosing filters which have historically been shown to have low contamination. Different laboratories also follow different pre-cleaning protocols and use different filter brands, but a typical process is that the filters are cleaned with mild HCl, rinsed with UHPW and stored in UHPW before use (Cutter et al. 2017). Care must be taken to match the filter material and type with an appropriate cleaning method – for example, some filter types such as cellulose acetate filters should not undergo cleaning procedures besides rinsing with UHPW or sample media since they degrade under acidic conditions. Other materials may also not tolerate harsh acid cleaning.

Filtration of large volumes of water from bottle samplers through 0.2 μm cartridge filters is most efficient under positive pressure (filtered nitrogen (N_2) or compressed air) or vacuum (max. 0.5 bar) (Fig. 3.14a). However, care should be taken to avoid excessive pressure to prevent the risk of exploding bottles and/or the

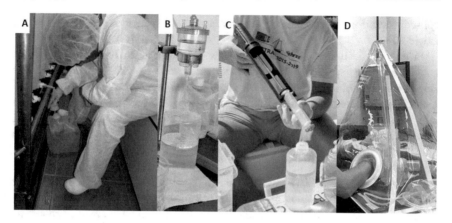

Fig. 3.14 Photographs of various filter set-ups for dissolved trace metals: in-line bottle filtration through Sartobran cartridge filters directly from the sampler (**a**) (photograph credit: Micha Rijkenberg, NIOZ, Netherlands), off-line filtration using an acid-cleaned vacuum filtration unit (**b**) (photograph credit: Dario Omanović, Rudjer Boskovic Institute, Croatia), off-line filtration through acid-cleaned plastic syringes (**c**) (photograph credit: Dario Omanović, Rudjer Boskovic Institute, Croatia) and syringe filtration within a 'glove box' under an inert gas atmosphere (**d**) (photograph credit: Andrea Koschinsky, Jacobs University Bremen, Germany)

rupture or lysis of algal cells retained by the filter which may release intracellular metals into the sample (Apte et al. 2002; Cutter et al. 2017). Gravity filtration is not recommended for large volume samples over 0.2 μm filters owing to the slow flow rate which can lead to absorption or chemical alterations of the sample (Fig. 3.14) (Cutter et al. 2017). New filter capsules should be flushed, e.g. with ~0.5 L sample seawater prior to use and with ~0.2 L sample seawater in between different samples. One filter can be used for multiple depth profiles, preferably working from the surface to the deep, or filters can be dedicated to certain depth intervals, i.e. surface and deep ocean (Cutter et al. 2017). However, reusing of filters should be done with extreme care, especially if gradients are expected in the study region, for example, when going from particulate-rich samples around hydrothermal vents or near-sediment to surface samples. Filtration of small sample volumes can also be done using acid-cleaned plastic syringes with pre-cleaned filters, but this is typically too time-consuming for large samples (Fig. 3.14c). When sampling waters which are anoxic or from low-oxygen environments, once bottle samplers are brought to the surface, samples should be processed within a 'glove box' under an inert gas atmosphere (Fig. 3.14d). This approach ensures that the integrity of the sample is maintained, i.e. minimization of the ratio of oxidation and precipitation reactions which may change the phase (dissolved to particulate) and/or speciation of trace metals of interest (e.g. US-Geological-Survey 2006). However, it is important to note that a headspace of an inert gas such as N_2 has been shown to facilitate outgassing of CO_2 which can lead to changes in pH of the sample with potential consequences for dissolved metal concentrations and speciation (Fitzsimmons and Boyle 2012).

Prior to collection of a filtered seawater sample into an acid-cleaned sample collection bottle (see Sect. 3.2.4), it is recommended to condition and rinse the empty sample bottles (including the cap) at least three times with the filtered seawater sample, each of which is discarded to waste, before finally filling the bottle with the sample. Sample bottles for dissolved trace metal or isotope analysis should be filled to the bottle shoulder to ensure that bottles are filled to the same amount and thus acidified to a similar acid concentration later on (Cutter et al. 2017). Ideally, acidification of the sample to below pH 2 should be carried out as soon after filtration as possible (Cutter et al. 2017), in order to avoid wall adsorption that can take a long time to resolubilize (Jensen et al. 2020). Sometimes, however, shipboard acidification is not practical. In this case, filtered samples that are stored unacidified should be left for an appropriate time after acidification and before processing (typically several months), in order to resolubilize metals which have adsorbed to the container walls. The preferential method of sample acidification is to add a volume of concentrated ultraclean HCl to achieve a final concentration of either 0.012 or 0.024 M HCl in the sample, depending on the element of interest and the preference of the research group (Cutter et al. 2017). Use of HNO_3 for acidification is typically avoided because it complicates commercial transport of these samples (Cutter et al. 2017). Following acidification, sample bottles should be tightly closed and double bagged in resealable polyethylene bags for storage (preferably at room temperature and in the dark) until analysis. Labels should be put both on the sample bottle and the bag, so that samples can be kept organized.

For dissolved metal speciation (organic ligand) samples, filtered samples (0.2 μm) should be stored in acid-clean bottles, kept at natural pH (without acidification) and either stored in the fridge (+4 °C) or frozen (−4 °C or −20 °C) until voltammetric analysis in the home laboratory or measured 'fresh' directly on-board ship (Bruland et al. 2000; Buck et al. 2012; Padan et al. 2020; Sander et al. 2005). In all cases, speciation samples must be stored in the dark in order to prevent photodegradation of the prevalent ligands, and it should be verified that the pre-cleaning procedure does not result in leaching of acid into the sample (i.e. a gentle acid cleaning procedure should be used). The most appropriate storage procedure of metal speciation samples, which avoids changes in speciation parameters pending analyses, is still a topic of discussion in the marine chemistry community (e.g. Buck et al. 2012).

3.4.2 Size-Fractionated Dissolved Trace Metal Samples

Current understanding of the cycling of metals is largely based on observations of the dissolved metal fraction, which is usually operationally defined as everything that passes through a filter with 200 nanometre pores (0.2 μm; see Fig. 3.1). However, such a sharp boundary does not reflect the continuum in which metals are actually present in seawater, which ranges from truly dissolved molecules (<0.02 um soluble fraction) to nanoparticles (<100 nm) via colloidal size

(<200 or <400 nm) and even larger-size particles (Santschi 2018). Ultimately, all size cut-offs are arbitrary operational definitions, and at some point, the question whether something is a very small particle, or a relatively large molecule becomes a philosophical question. Perhaps the more interesting scientific consideration is, however, at what point molecular Brownian motion becomes dominant over gravitational settling (Honeyman and Santschi 1989; Wells and Goldberg 1992) or, in other words, whether a substance behaves like a particle or a dissolved substance. Such behaviour is of course a function of both particle size and density, as well as other factors such as temperature (e.g. Farley and Morel 1986) that are beyond the scope of the discussion in this chapter, but should occur somewhere around the transition from the colloidal to the particulate size class.

Metal size fractionation studies generally focus on the difference in size classes of particles and colloids before and after filtration, i.e. trace metal levels of particles in solution (the filtrate) and retained on the filter (the retentate) (Bergquist et al. 2007; Fitzsimmons and Boyle 2014b; Fitzsimmons et al. 2015a; Fitzsimmons et al. 2015b; Ussher et al. 2010). The most common practice used to classify trace metal concentrations beyond the dissolved and particulate fraction is referred to as ultrafiltration. Ultrafiltration (UF) is a pressure-driven filtration process that separates particulate matter from truly colloidal and soluble compounds using an ultrafine membrane media. For this, a 0.2 μm filtered seawater sample – with commonly used capsule filters that are not suitable for studying the material retained on the filter – undergoes another so-called UF step, often using either cross-flow filtration (CFF) or Anopore filter membranes. With this step, the soluble metal fraction can be obtained in the filtrate, and the difference between the dissolved (0.2 μm fraction) and soluble (UF fraction) gives the calculated colloidal metal fraction. The Anopore filter membranes have a pore size of 0.02 μm, whereas CFF filters are often defined by the cut-off size of molecules they let pass through, e.g. 10 kDa in case of a Millipore Pellicon XL (PLCGC) filter (Fitzsimmons and Boyle 2014a; Jensen et al. 2020), complicating direct comparison of size fractions between the two filtration techniques. There are advantages and disadvantages of both filtration techniques— for example, CFF is quicker for filtering larger volumes of seawater, but it is a more complex technique and requires more expensive equipment, more cleaning and more training. In contrast, Anopore filter units are cheaper and simpler to user, but much slower for filtering larger volumes. However, Anopore may be the better choice if only small volumes of sample (e.g. <150 mL) are required for analysis of the parameter of interest. For a more comprehensive discussion of the pros and cons of each filtration technique, when deciding which to use, we refer the reader to Fitzsimmons and Boyle (2014a).

3.4.3 Particulate Trace Metal Samples

Akin to the definition for dissolved metals, the 'particulate' trace metal phases are also operationally defined, based on particle size (see Sect. 3.3.2; Box 3.6). To collect particulate trace metals from bottle samplers, the use of pre-cleaned

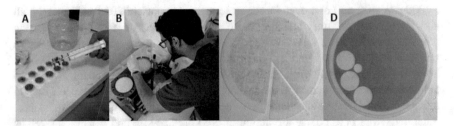

Fig. 3.15 Photographs of a stacked syringe filter system (**a**) (photograph credit: Dario Omanović, Rudjer Boskovic Institute, Croatia), and an in situ pump filter holder disassembled inside of a HEPA filtered clean air bubble (**b**) (photograph credit: Alex Fox, Science Writer, picturing Vinicius Amaral, University of California, USA). The filter holder shown in (**b**) can contain several filters for in situ size fractionation. Picture (**c**) shows a 51 μm polyester mesh filter and (**d**) shows a 0.8 μm polyethersulfone (PES) filter. Filters (**c,d**) can be sub-sampled for multi-investigator use (photograph credit: Daniel Ohnemus, Skidaway Institute of Oceanography, USA)

polycarbonate or fluoropolymer filters holders with polyethersulfone (PSE) or mixed cellulose ester filters of diameters between 25 and 47 mm is recommended (Cutter et al. 2017) (Fig. 3.15). Depending on the type of analysis, it is usually advised to use the smallest filter diameter to maximise the particle loading per filter area and thus ensure a sufficient sample to filter blank ratio. The implemented filter pore size varies depending on the research objective, but 0.45 μm is currently the GEOTRACES standard for particulate trace metal sampling. If continuous size fractionation of the particulate trace metal pool is of interest, it is also possible to use different filter pore sizes with a single syringe in a stagged configuration (Fig. 3.15a).

For the actual filtration, the acid-cleaned filter holders with pre-cleaned filters should be connected to a pressurized bottle, container or pump outlet using acid-cleaned tubing (Fig. 3.14a). Trapped air in the filter holders should be cleared by unscrewing the holder to allow a small volume of water to flow around the filter, before the sample water can pass through the filter. The volume of water passing the filter must be recorded for later quantification (i.e. by collecting the water in a secondary container or measuring cylinder) and can vary drastically depending on the area of interest and associated particle loading in the water column. If the filter clogs, filtration should be stopped, and the filtrate volume should be recorded. To avoid filter rupture (pressure built up), the filtering rate should not exceed about one drop per second (Cutter et al. 2017). As stated by Cutter et al. (2017), filtration times >2 h should be avoided to prevent speciation changes and particles settling within the bottle. It is also important to seal and tighten the filter holder appropriately (i.e. avoid miscalculation of filtrate volume) and to ensure that the filter lies flat for successful filtration (i.e. trapping all particles on the filter). Leaking membrane filter holders should be identified and recorded since they can be a major source of contamination. Further, each filter holder should be marked with a unique number, so that samples can be kept organized.

To collect particulate trace metals from in situ pumps, various filter sizes (both membrane diameter and pore size) and plastic filter types can be used depending on the research objective and the preferred filter digestion method (McDonnell et al. 2015). GEOTRACES intercalibration expeditions have shown that cleaned polyethersulfone (PES) filters are a good choice for trace metal work (Bishop et al. 2012). Akin to the filtration step for the bottle samplers, multiple filter plates with different pore sizes can be paired for in-line size fractionation work (Fig. 3.15) with particles typically defined as suspended, slowly sinking or fast sinking (Riley et al. 2012) or relatively large- and small-size fractions based on used filter sizes (Lam et al. 2015). Most filter holders also contain a baffle system, i.e. a prefilter (plastic film or grid cover), sitting on top of the first filter to reduce turbulence, distribute particles evenly across the filter and minimize particle loss during pump retrieval (Bishop et al. 2012). An advantage of the in situ sampler is the possibility to distribute filter sub-samples to multiple investigators owing to the large filter holder size of many commercially available in situ pumps (~142 mm for standard McLane pumps and MULVFS) (Fig. 3.15c, d) (McDonnell et al. 2015).

> **Box 3.6: Ultrafiltration for Colloids and Particulates**
>
> It is also possible to study particulate trace metals collected on filters during ultrafiltration (UF; see Sect. 3.4.2). Commonly, particulate trace metals are studied on filters with pore sizes of 0.2 μm, 0.4 μm or larger; however, smaller fractions can be studied even though this process is more time-consuming and thus less practical. The longer processing time is problematic for maintaining the integrity of most trace metals (i.e. precipitation, chemical alteration, adsorption), especially for Fe given its tendency for wall absorption when samples are not acidified on time scales of hours (Fitzsimmons and Boyle 2012). Consequently, when relatively large ultrafiltered volumes are needed for analysis of the 'soluble' phase, preference is given to filters with a fast flow rate to avoid artefacts and alterations occurring during filtration (Jensen et al. 2020). However, none of those commonly used 'fast' filters allow assessment of retained material. For instance, cross-flow filtration (CFF) filters are not designed to capture the colloids that partly end up on the filter with the majority in the retentate. By contrast, Anopore filters may capture the colloid fraction, but are made electrochemically by the anodic oxidation of aluminium and contain particulate Fe inclusions; the latter makes these filters unsuitable for common chemical leaching and/or digestion techniques as well as microscopy techniques aimed at identifying and quantifying colloidal Fe (Fitzsimmons and Boyle 2014a). Thus, investigators interested in the colloidal fraction of trace metals need to implement UF with filters that are free of metals (low filter blank) such as polycarbonate (PC) filters or polyethersulfone (PES) filters (Cullen and Sherrell 1999). However, previous testing of these two filters resulted in slow flow rates, and thus these filters were deemed unsuitable for UF studies of trace metals (Fitzsimmons and Boyle 2014a). Other colloid separation techniques such as reverse osmosis-electrodialysis

(continued)

Box 3.6 (continued)
(Koprivnjak et al. 2009) or flow field-flow fractionation techniques (Santschi 2018) do exist but have, to the best of our knowledge, so far, not been used to study colloidal trace metals in a contamination-free manner.

When filtration is complete, either on board or using an in situ pump (see Sects. 3.3.2.1 and 3.3.2.2), residual water in the filter holder has to be removed using syringes or vacuum to reduce the residual sea salt matrix for analytical simplicity (Cutter et al. 2017). Inside a laminar flow bench, filter holders should be disassembled, and filters should be removed using plastic acid-cleaned forceps before storing them individually in a clean labelled petri dish, tube or similar, at $-20\ °C$ freezer to physically stabilize the sample. For larger filters, the primary filters may be cut up using a ceramic acid-leached blade scalpel to provide sub-samples. Photo documentation of the filter before and after sub-sampling can be of use to document filter heterogeneity Cutter et al. 2017. Once filters are removed, the filter holders should be rinsed with acid and UHPW before next use.

Prior to analysis of the particulate trace metals, the filters and/or the material on them needs to be digested (Sherrell and Boyle 1992). Various full or partial digestion techniques and protocols are available depending on the metal of interest, the filter used and whether the filter should stay intact or not. After digestion the digest can be (re-)diluted with a specific matrix solution prior to analysis (Sherrell and Boyle 1992). Rather than a full digestion of the particulate metal pool, researchers can also carry out partial digestions or 'leaches' with leachates of various strength to characterize a specific portion of the particulate trace metal pool, e.g. labile, refractory, bound to carbonates, bound to organic matter, etc. (Berger et al. 2008; Tessier et al. 1979). Sequential leaching techniques can also be used to characterize multiple fractions of the particulate trace metal pool. Although more time-consuming and costly, sequential leaching methods provide more detailed information about the origin and fate of trace metals in the study area (Tessier et al. 1979). Overall, many protocols exist for digestion and leaching methods that are not further detailed here but can be found elsewhere (e.g. Ohnemus et al. 2014; Rauschenberg and Twining 2015; Twining et al. 2015a).

3.5   Sample Processing and Analytical Techniques

Once a clean sample of seawater is collected, filtered, acidified (if appropriate for the element and analytical technique of interest) and stored, the next challenge is to analyse the sample for trace metal concentrations (see Sect. 3.5.1), isotopic composition (see Sect. 3.5.2) or speciation (see Sect. 3.5.3). This chapter has already discussed the challenges of collecting and processing contamination-free samples, and such procedures must be maintained throughout analysis to generate accurate results. However, contamination-free analysis is not trivial since most analytical methods aim to pre-concentrate the trace metals into a smaller size sample for analysis, and such an approach usually involves the use of multiple reagents, equipment and steps. Further, the challenges of pre-concentration and/or measuring the very low concentrations of trace metals and isotope ratios in seawater accurately are compounded by the sea-salt matrix, which contains very high concentrations of major ions such as Na^+, Ca^{2+} and Cl^- (at typical salinity these ions are present at 35 g kg^{-1}), all of which can interfere with the signal of interest.

As mentioned in Sect. 3.2, the advent of trace metal clean sampling and handling techniques in combination with advances in modern analytical chemistry and instrumentation was critical to obtain a first-order understanding of the concentration of trace metals in seawater, later followed by insights in both their speciation (see Sect. 3.5.3) and isotopic composition (see Sect. 3.5.2). Notably the availability of the graphite furnace as the sample introduction system to an atomic absorption spectrometer (GF-AAS) in the mid-1970s was pivotal, but this was superseded by flow injection techniques (FIA) which pre-concentrated the metal(s) of interest onto a column (Sohrin and Bruland 2011; Worsfold et al. 2014) and which have since largely been supplanted by the (even) more sensitive and powerful high-resolution (HR) inductively coupled plasma mass spectrometers (ICP-MS) (Sohrin and Bruland 2011). While flow injection methods are often still used shipboard, ICP-MS is the current standard for determination of trace metals and their isotopes (TEIs) in shore-based laboratories as it provides high sensitivity, high accuracy, low limits of detection (LOD), linear response to the analyte(s) over a wide dynamic range and high sample throughput and is a powerful tool for the simultaneous determination of multiple elements.

Here we describe ICP-MS techniques for trace metal analysis in Sect. 3.5.1.1, shipboard flow injection techniques (FIA) for trace metal analysis in Sect. 3.5.1.2., systems for in situ trace metal analysis in Sect. 3.5.1.3, multi-collector ICP-MS techniques for trace metal isotope analysis in Sect. 3.5.2 and voltammetry techniques for trace metal speciation in Sect. 3.5.3. Throughout, for further reading, we mainly refer the reader to synthesis review articles or textbook chapters as a starting point, rather than attempting to cite all of the available research on the topic.

3.5.1 Trace Metal Concentration Measurement Techniques

3.5.1.1 ICP-MS Techniques

Several varieties of ICP-MS exist (Olesik 2014), including high-resolution sector field (HR-SF) and quadrupole instruments that differ in how they separate analytes from each other and how they resolve interferences, which is beyond the scope of this chapter. A HR-SF-ICP-MS is typically required in laboratories analysing seawater samples, principally in order to resolve interferences from the argon (Ar) carrier gas typically used by ICP-MS instruments (Sohrin and Bruland 2011; Wuttig et al. 2019). In an ICP-MS, a sample is vaporized in the sample introduction system, and its elements are atomized and then ionized in an Ar plasma. The resulting ions enter the vacuum inside the instrument through two interface cones, i.e. the sampler cone and the skimmer cone, which focus and guide the ions into the mass spectrometer. The mass analyser separates ions according to their distinct mass to charge ratios (m/z) via magnetic and electrostatic fields before they reach and are measured at the detector. Variations in the magnetic and electrostatic fields allow detection of different ions based on their m/z ratio, where charge is usually +1 (or +2) in the plasma.

However, different elements can have ions with the same m/z ratio, and this must be carefully addressed. For example, Fe and Ni have isotopes at mass 58 (57.9332744 and 57.9353429 amu, respectively), and thus any signal measured at m/z 58 will have contributions from both $^{58}Fe^+$ and $^{58}Ni^+$. Such interferences are described as 'isobaric' and can be avoided by measuring another isotope of the element of interest or by doing a subtraction correction by measuring another isotope of the element without interference and applying a natural abundance ratio. Typically, isobaric interference peaks are too close together in m/z to be separated by HR-SF-ICP-MS, unlike many 'polyatomic' interferences (see below). A second potential isobaric interference type comes from doubly charged ions (e.g. $^{116}Sn^{++}$ also has a m/z of 58), but these are typically only formed at low levels in the plasma and as such only become a problem if that element is present at high concentrations in the sample.

The second potential type of interference on ICP-MS is known as a 'polyatomic' interference and arises from polyatomic molecules that are formed by the combination of two (or more) molecules in the plasma. For example, when measuring Fe, the most abundant isotope is ^{56}Fe, which does not have isobaric interferences. However, in the case of $^{56}Fe^+$, several polyatomic interferences cause problems, especially $^{40}Ar^{16}O^+$ from the carrier gas and $^{40}Ca^{16}O^+$ that can come from the seawater matrix. Another particularly problematic polyatomic interference arising from the seawater matrix is MoO^+ which interferes with Cd measurements. Such interferences can cause issues for low-resolution instruments such as quadrupole ICP-MS, which may not be able to resolve the interference from the peak of interest. HR-SF-ICP-MS addresses many polyatomic interferences by using a higher 'resolution', which allows peaks that are close to each other to be separated (resolved). To obtain a higher resolution, narrower 'slits' are used that further constrict the ion beam and thus allow better separation of ions with very similar m/z. This separation achieves the goal of avoiding interferences, but comes at the expense of sensitivity (signal size) as less ions reach the detector. Alternatively, modern quadrupole ICP-MS instruments can be used, utilising 'reaction cells' as an alternative method to minimize polyatomic interferences for seawater applications (Jackson et al. 2018). In such instruments the m/z of the interference (or the target analyte) is changed in the reaction cell via a chemical reaction, allowing subsequent separation. Here we have focused on the more widespread use of HR-SF-ICP-MS.

Avoiding or correcting for interferences can be challenging, depending on the sample matrix and the relative concentrations of interferences and analytes of interest. Thus, the operation of a HR-SF-ICP-MS or any other types of ICP-MS for the determination of trace metals requires substantial expertise that varies with the application and objective of the research and the element of interest. Further discussion on this is beyond the scope of this chapter. In the next sections, we focus on matrix removal (salt matrix) and pre-concentration applications. These applications are required before most analytical techniques can be implemented, including ICP-MS. This section is then followed by some more detail on ICP-MS analysis techniques commonly used within the trace metal community.

Matrix Removal and Pre-Concentration Prior to ICP-MS Analysis

From a limit of detection perspective, the concentrations of most marine trace metals should be measurable directly using a HR-SF-ICP-MS without pre-concentration. However, the salt matrix prevents such direct injection of marine samples into the instrument as the high concentration of salts (Na^+, Ca^{2+}, Cl^-) leads to interferences and clogging of the cones and introduction system, resulting in substantial reduction and variations in signal sensitivity as well as inaccuracy. Thus, most ICP-MS seawater trace metal concentration techniques (as well as FIA; Section 3.5.1.2) and isotope ratio techniques (Section 3.5.2.2) involve a matrix removal step prior to analysis.

A matrix removal step typically also has the added benefit of constituting a significant pre-concentration step as analytes of interest are concentrated at the same time as the analyte is isolated from the (interfering) major ions in the matrix. Pre-concentration methods include co-precipitation with Mg, solvent extraction and/or solid-phase extraction (SPE) with chelating resins (Sohrin and Bruland 2011), where the latter is now the most commonly used technique in trace metal chemistry (Wuttig et al. 2019).

Typically, for SPE, a chelating ligand is immobilized by covalent bonds on a stationary resin phase packed in a column or sometimes added to the resin as beads (Lee et al. 2011b). A seawater sample solution of which the pH is adjusted to the right range, depending on the chelating resin and analyte of interest, is subsequently passed over the resin column where the analyte of interest forms a chelate with the resin's binding sites and is thus retained. Major matrix ions such as the sea salts (Ca, Na, K etc.) are usually not retained by the resin at certain pH, and hence the pH of the sample solution plays a key role during matrix removal methods using SPE. The retention efficiency (recovery) of the analyte of interest on the resin is also dependent on the seawater sample pH and thus should be observed carefully and adjusted if necessary using appropriate reagents (buffers or acids) (Sohrin and Bruland 2011).

After retention, the analyte of interest is eluted from the resin with an appropriate acidic eluent. Typically, this is done with a much smaller volume than the original seawater sample, leading to a substantial pre-concentration factor (Fig. 3.16). A range of chelating resins exist and have been used historically by trace metal chemists, but notably Nobias PA-1 has become increasingly popular (Sohrin and Bruland 2011). This chelating resin is also used in the column material of the commercially available SeaFAST system (Elemental Scientific (ESI)), an automated pre-concentration system for undiluted seawater (Lagerström et al. 2013; Wuttig et al. 2019). This system is currently considered 'state of the art' and in use in many trace metal clean laboratories around the world, including the Royal Netherlands Institute for Sea Research (NIOZ; the Netherlands). The following section details

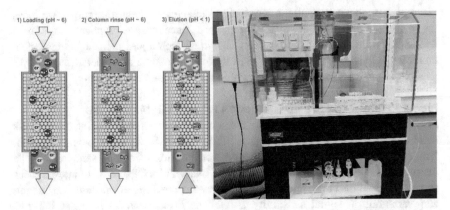

Fig. 3.16 Illustration of a solid-phase extraction (SPE) of metals on a Nobias Chelate-PA1 column in three steps (right) (figure courtesy of Elemental Scientific (ESI)), and a photograph of a SeaFAST (left) (photograph credit: Patrick Laan, NIOZ, Netherlands). SPE Extraction: (1) Loading: in this step seawater at the right pH passes over the column, and analyte(s) of interest (M^+) are retained, whereas the matrix (salts) mostly passes through. (2) Rinsing: in this step the remaining seawater and most salts are rinsed from the column. (3) Elution: the analyte(s) of interest are eluted from the column with elution acid that can subsequently be analysed by ICP-MS

some of the typical procedures, analysis steps and experiences gained by the NIOZ laboratory team using the SeaFAST system, as an example of a commonly used pre-concentration and salt removal method.

SeaFAST: Automated Extraction of Metals from Seawater

The SeaFAST system consists of an autosampler, a syringe pump module and valves that pre-concentrate metal(s) from acidified seawater samples onto the resin, and then either elute the analyte(s) directly into an ICP-MS system when using the in-line configuration (see Box 3.7) or collect the eluent in small vials for later analysis during the off-line mode (Lagerström et al. 2013). The latter mode is used in the NIOZ laboratory (Gerringa et al. 2020) and other trace metal clean facilities (e.g. Rapp et al. 2017; Wuttig et al. 2019) as it saves on instrument runtime and allows for better multi-resolution analysis on the ICP-MS (see Box 3.7). After off-line extraction (i.e. extracting the metals from the seawater matrix with the SeaFAST system), a whole batch of collected samples can be run right after each other using the ICP-MS with minimum idle time of the plasma, especially when using a high-throughput sample introduction system (e.g. the double loop MicroFAST MC (Elemental Scientific (ESI)) as in use in the NIOZ laboratory). Lagerström et al. (2013) describe the SeaFAST system in more details, but general steps regarding sample loading, pre-concentration, matrix removal and metal elution steps are briefly described below (Fig. 3.16).

Box 3.7: SeaFAST in-Line Versus off-Line Configuration
When using the SeaFAST in the in-line mode, the ICP-MS is running and idle, while the SeaFAST is still in the pre-concentration step, resulting in higher (and wasted) analysis costs. Additionally, the analytes are detected as an elution curve, meaning that the signal builds up to a maximum and decreases again in the shape of a peak. An ICP-MS generally does not measure analytes simultaneously but has to cycle through the analytes of interest which takes time especially if switching between resolution (low or high resolution to resolve interferences; see Sect. 3.5.1.1) is required, limiting the number of analytes that can be measured in an elution curve (i.e. after a given amount of time, the peak has passed so there is only a limited number of elements that can be measured in that time). Nevertheless, the in-line mode can be useful for method development of new analytes of interest, for example, testing when an analyte is eluted off the column with a given eluent. When using the SeaFAST off-line, the collected eluent is homogeneous in concentration, enabling analysis of as many analytes as the eluent volume allows, where obviously a greater elution volume leads to lower sensitivity as the analyte concentration is diluted.

During operation of the SeaFAST system, an autosampler probe moves into the sample and fills a sample loop (typically 10 mL) using an integrated vacuum pump. The sample loop is 'overfilled', i.e. the loop gets rinsed by the first ~2 mL of sample which goes to waste (i.e. 12 mL is taken up of which 10 ml stays in the loop). Subsequently, by syringe pump action, the 10 mL sample is pushed from the loop, buffered to the appropriate pH using an ammonium acetate buffer (see Box 3.8) and immediately passed over the Nobias PA-1 pre-concentration resin column. Prior to mixing, the buffer solution is passed over a 'clean-up' column with the same chelating resin as the pre-concentration column to minimize any trace metal contribution from the buffer solution. Depending on the needed pre-concentration factor, sequential 10 mL aliquots can be pre-concentrated over the column, where at NIOZ, 2×10 mL (20 mL) is commonly pre-concentrated for low-metal open ocean samples. After sample loading, UHPW is pushed over the pre-concentration column to remove the residual salt matrix remaining in the column. The ensuing elution of the chelated metals on the resin then occurs in the reverse direction (compared to sample loading) with elution by ultra-pure 1.5 M nitric acid into a clean sample vial using pressurized N_2 gas as a carrier. The volume (and strength) of elution acid can be varied depending on required ICP-MS analysis time and/or the desired pre-concentration factor. At NIOZ, a volume of 350 µL is typically used, resulting in a pre-concentration factor of ~57× (20 mL sample preconcentrated into 0.35 mL). A choice of sample vials can be used, including LDPE, PFA, polyvinylidene difluoride (PVDF) or polypropylene (PP), as long as they are rigorously cleaned before initial use and between different samples (see Sect. 3.2.3).

Box 3.8: Solid-Phase Extraction (SPE) and pH

At NIOZ, a pH of 5.8 ± 0.2 (obtained with a more dilute buffer than recommended by the manufacturer) has been determined optimal for the suite of routinely measured metals using the SeaFAST (Middag et al. 2015a), whereas a slightly higher pH and a more concentrated buffer are often used in other laboratories (Wuttig et al. 2019). There is a trade-off between pH and recovery, notably for Mn and Fe where a somewhat lower pH leads to better recovery for Fe, but below a pH of 5.5, recovery for Mn becomes non-quantitative (Middag et al. 2015a). Thus, it is important to check the pH of the seawater that passed the pre-concentration column regularly (easily done at the waste outlet of the SeaFAST) and monitor recovery in every extraction run. However, different laboratories report slightly different optimum pH, implying that the recovery of metals might vary between different set-ups depending on local laboratory conditions and practices.

The SeaFAST columns are cleaned after elution with 1.5 M nitric acid to eliminate carry-over (memory effects) of subsequent samples. Nevertheless, some carry-over can still occur – for example, Rapp et al. (2017) reported a carry-over of 0.5–1.3% for Fe and Ni, and Wuttig et al. (2019) also observed carry-over of <1% for some elements of interest. The carry-over effect can be minimized by pre-concentrating samples from presumably low to high initial trace metal concentrations or by processing of low-metal seawater and/or dummy samples (i.e. where the eluent is not collected but deposited in the waste of the system) in between samples.

At NIOZ, prior to pre-concentration using the SeaFAST system, an aliquot (30 mL) of each filtered and acidified seawater sample is pipetted into an acid-cleaned FEP bottle, followed by addition of hydrogen peroxide (see Box 3.9) and an internal standard (see Box 3.10). In addition to major ions, seawater contains organic material, i.e. ligands, that can chelate metals (see Sect. 3.5.3 on speciation). Some of these organic ligands are destroyed by acidification after sampling, but others can persist in the sample, interfering with the extraction of metals from seawater via the chelating resins. In particular, Co and Cu chelated to dissolved organic ligands can pass through the pre-concentration column without binding to the resin (see Box 3.9; Fig. 3.17) (e.g. Biller and Bruland 2012; Lagerström et al. 2013; Middag et al. 2015a; Rapp et al. 2017; Wuttig et al. 2019). Using UV digestion after hydrogen peroxide addition (see Box 3.9) of samples prior to extraction destroys the organic ligands that chelate trace metals in solution which would otherwise outcompete the resin's functional groups (see Box 3.9).

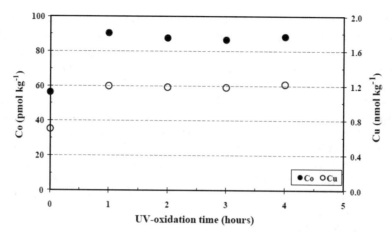

Fig. 3.17 Concentrations of Cu and Cd in acidified natural seawater, determined as a function of UV digestion time (Biller and Bruland 2012)

Box 3.9: UV Digestion

Rapp et al. (2017) and Wuttig et al. (2019) assessed the influence of quartz and FEP vessels on UV digestion efficiency and contamination. No difference between a FEP bottle and a quartz cuvette was observed with regard to the efficiency of UV digestion using either vessel material. However, contamination from quartz vessels was observed for Pb, Ti, Fe and Zn, whereas an increase in Ti was observed in the PTFE bottle during UV digestions (Rapp et al. 2017; Wuttig et al. 2019). These tests suggest that FEP bottles are suitable for UV digestion with the added advantage of being suitable for rigorous cleaning protocols due to their inert behaviour when subjected to hot acids, thus decreasing contamination from the digestion vessel. Moreover, FEP bottles can be placed directly into the SeaFAST autosampler after the digestion step, minimizing further sample handling.

At NIOZ, samples inside FEP bottles are irradiated in a custom-built UV box containing 4 TUV 15 W/G15 T8 fluorescent tubes for 4 h (Fig. 3.17) after addition of clean hydrogen peroxide (final concentration ~ 30 μM). The addition of hydrogen peroxide leads to the formation of reactive radicals during the UV irradiation, assisting in the breakdown of organic ligands. Using this procedure, an increase in concentration of 17% and 15–50% for Cu and Co, respectively, has been reported after UV digestion of samples (Wuttig et al. 2019). After irradiation, the samples are usually left for at least another 4 h to cool to room temperature and to let any leftover radicals react prior to the SeaFAST pre-concentration step.

A typical batch of samples processed at NIOZ, besides actual samples, consists of calibration standards (standard additions), reference samples and blanks (acidified

UHPW). Natural seawater containing low concentrations of metals (e.g. North Atlantic surface water) is used as the matrix for standard additions, but if that is not available, low-metal seawater can be made by passing it over a chelating resin column. A calibration line is made by adding increasing amounts of in-house prepared multi-element stock standard solution (from high purity commercial standard solutions) with natural isotopic abundances to known volumes of the low-metal seawater (typically 30 mL). The highest added concentration depends on expected concentrations in the samples where the multi-element stock standard should be designed so that the highest standard addition is approximately 120% of the highest expected sample concentration. Usually, a calibration is prepared in duplicate where one set is extracted preceding the samples and one set extracted after the samples, in order to be able to account for any drift (changes in instrument sensitivity) or changes in recovery of the chelating resin over time.

An alternative approach for calibration is the use of the isotope dilution technique, which has been successfully used for seawater samples with SeaFAST extraction in the past (e.g. Lagerström et al. 2013; Rapp et al. 2017). This approach involves the addition of a known volume of a 'spike' made by dissolution of a highly purified single isotope of an element. For example, for Fe, where natural Fe abundance is 92% for ^{56}Fe and just 2% ^{57}Fe, a spike may be prepared that is 99% ^{57}Fe. This way, by adding a known amount of ^{57}Fe spike to a known amount of acidified seawater, prior to SeaFAST pre-concentration, and then measuring the ratio of ^{56}Fe/^{57}Fe in the pre-concentrated sample via ICP-MS, the original amount of natural sample Fe can be calculated (and then converted to a concentration using the volume or mass of the original sample). The benefit of the isotope dilution method is that ^{56}Fe/^{57}Fe ratios are not affected by recovery efficiency (i.e. efficiency of resin in retaining the element of interest), and so the sample concentration of Fe obtained is unaffected by incomplete recovery on the SeaFAST, variable recovery of different samples or changes in intensity of the signal on the ICP-MS during the analytical run. However, isotope dilution is not possible for monoisotopic elements, such as Mn and Co, and thus multiple elemental analysis requires a combination of isotope dilution and standard additions (if those elements are of interest). It also takes time and money to purchase and prepare spike solutions. Further, spike addition should be broadly matched to the expected concentration of the metal of interest (with some flexibility), and so it can be complicated to prepare a multi-element spike that is appropriate for a range of seawater samples (e.g. when Cd can vary from 0.00003 to 1 nmol kg^{-1} in a North Atlantic seawater depth profile, while Fe only varies from 0.1 to 0.6 nmol kg^{-1}; Schlitzer et al. 2018).

It is also worth noting that the methods used for trace metal isotopic analysis in Sect. 3.5.2 also often make use of isotope dilution techniques on larger volumes of seawater to generate concentration datasets. Ultimately, the choice between SeaFAST standard addition and SeaFAST isotope dilution (or other methods such as flow injection analysis) often depends on the capacity, capability and experience with one or the other method in a given laboratory, as both calibration methods have been shown to produce accurate, comparable and publishable results—and both have led to datasets included in GEOTRACES data products (see Sect. 3.6).

At NIOZ, besides calibration standards, a suite of in-house reference samples is processed in every SeaFAST extraction run (see also Sect. 3.5.1.4). These reference samples are usually (acidified) sub-samples of a large sample stored in 20 L batches in large acid-cleaned containers and are used to track the consistency within and between extraction runs. At NIOZ, North Atlantic deep water, North Atlantic surface water and column-cleaned low trace metal seawater (natural seawater passed over a large Nobias PA1 column prior to acidification) are typically used as in-house reference samples as they cover a wide range of trace metal concentrations. Triplicate samples of the in-house reference samples are extracted with the start and end calibration of each run and at least once in between the actual samples. In selected extraction runs, community consensus reference samples (CCRS) from the GEOTRACES programme (see https://www.geotraces.org/standards-and-reference-materials/) are also run to verify the in-house reference samples to these CCRS. The CCRS were collected by trained experts at various global locations into large volume seawater containers and then sub-sampled into smaller LDPE bottles for distribution within the trace metal community. For example, the SAFe samples were collected from the SAFe station in the North Pacific and were made available to investigators by Ken Bruland at UC Santa Cruz (Johnson et al. 2007). The CCRS have been traditionally used (and accuracy and precision reported) by investigators measuring trace metal concentrations. However, most of these samples are now exhausted at the source, with laboratories using up what stocks exist. Thus, given the scarcity of the CCRS, these are usually measured only once or twice at NIOZ for all extraction runs for a specific expedition or project.

Besides the calibrations and reference samples, sample 'blanks' are also processed in each extraction run to quantify any contamination introduced via reagents, sample handling and instrumental processing. At NIOZ, acidified UHPW is used as such a 'blank'. The metal measured in the blank is then subtracted from the measured sample concentration to account for any added contamination. It is important to verify the cleanliness and performance of the SeaFAST system prior to extracting 'real' samples for a project by running and analysing blanks and in-house reference samples to ensure satisfactory performance of the SeaFAST, i.e. good recovery and low blanks (ideally at least a factor 2 lower than the lowest observed concentrations). Analytical considerations regarding measurement precision, accuracy, blanks and data evaluation are further detailed in Sects. 3.5.1.4 and 3.6.

Box 3.10: Internal Standard Addition for ICP-MS

At NIOZ, the internal standard contains a mixture of In and Lu owing to both their low (sub pmol/L) concentrations in natural seawater and their resin recoveries of >98% using Nobias chelating PA1 resin (Middag et al. 2015a). Internal standards are used to allow the pre-concentration factor of the SeaFAST method to be determined without the need to measure the weight

(continued)

Box 3.10 (continued)

of the sample before and after the extraction step. The pre-concentration factor is accounted for since the ratios between the internal standard and the trace metal of interest do not change after addition of the standard spike to the sample and remain the same in the sample and in the eluent when the recoveries are quantitative. Therefore, any change in the volume of the sample or the eluent (by, for example, evaporation) becomes irrelevant since the ratio of the element of interest to the internal standard will stay unchanged. During ICP-MS data analysis, the signal of the trace metals of interest is normalized to the In and Lu signal which accounts for the pre-concentration factor as well as any variations in sensitivity, i.e. drift.

ICP-MS Analysis Following Extraction

When a full set of samples with calibration lines, reference samples and blanks has been extracted with the SeaFAST, the extracts can be measured on the ICP-MS. It is also worth mentioning that prior to the SeaFAST being available, trace metals were concentrated for ICP-MS using resin beads, Mg co-precipitation or in-house-built chelating resin column systems (e.g. Biller and Bruland 2012; Lee et al. 2011a; Sohrin and Bruland 2011; Wu 2007). In all cases, the methods end up with a small volume of acid that contains the pre-concentrated trace metals of interest ready for analysis by ICP-MS.

After tuning of the relevant ICP-MS instrument (at NIOZ, samples are analysed using a Thermo Fisher Element HR SF-ICP-MS; Fig. 3.18) and verifying the background signals of the analytes of interest are low and stable, the analytical run can start. Especially on instruments that are also used for other samples, the background concentrations of contamination-sensitive elements like Fe, Zn or Pb may gradually decrease with time till a stable background level is reached. It should be noted that tuning the instrument for optimal performance (e.g. signal size, signal stability, peak shape, oxide interferences, resolution, etc.) requires specialized expertise and is beyond the scope of this chapter.

At NIOZ, first an eluent standard addition calibration curve (standard additions to eluent acid) is measured to assess recovery, together with some elemental standards to calculate the polyatomic interference of MoO^+ on Cd, if Cd is an analyte of interest (see Biller and Bruland (2012) for more details), followed by the SeaFAST extracts. After every 12 SeaFAST extracts, a drift standard is measured followed by two separate sub-samples of the eluent acid (the acid taken from the reagent bottle, without going through the SeaFAST). The drift standard is a standard from the eluent standard addition calibration curve (~2/3 of the maximum concentration in this case) that is used to correct for any element specific drift that is different from the internal standard element (and thus cannot be accounted for by the internal standard). Such a drift occurs, for example, if the change in sensitivity during an analytical run for an analyte of interest is different from the change in sensitivity for the used internal

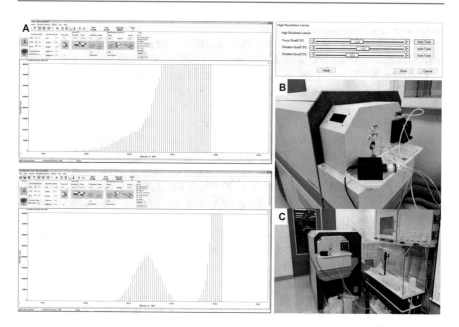

Fig. 3.18 Thermo Fisher Element HR-SF-ICP-MS set-up. Scans at the mass of ^{56}Fe in medium resolution on the NIOZ HR-SF-ICP-MS with (panel **a**, lower graph) and without (panel **a**, upper graph) tuning to resolve ^{56}Fe from ^{40}Ar^{16}O (**a**), Thermo Fisher Element HR-SF-ICP-MS (**b**) and MicroFAST sample introduction system (**c**) (Figure and photograph credit: Patrick Laan, NIOZ, Netherlands)

standards (In and Lu). Changes in sensitivity are very common when running SeaFAST extracts, especially in long analytical runs, due to build-up of traces of salt still present in such samples on the ICP-MS cones. Element-dependent drift, however, is quite rare in our experience, but should be monitored and corrected for if needed. The two separate sub-samples of the elution acid are necessary to check (and correct for) any changes in metal background concentration originating from the sample introduction system or the ICP-MS itself during the analytical run. The first sample solely serves the purpose of reducing any carry-over to the second elution acid sample, which is actually used for the background correction of the seawater samples. This background correction, especially for very low open ocean seawater concentrations, can significantly influence the results, and thus care should be taken in applying the right background correction. Overall, measured values are always corrected for the background (metals in the elution acid and derived from the ICP-MS and sample introduction system) and subsequently normalized to the internal standard. Thereafter, a blank (metals added during the SeaFAST extraction and sample handling) correction is applied, and last, if needed, an element-dependent drift correction or interference correction (e.g. for Cd) is applied.

3.5.1.2 Flow Injection Analysis

Flow injection techniques were the standard for measuring trace metal concentrations such as Fe prior to the development of ICP-MS techniques (Sohrin and Bruland 2011) and remained popular into the GEOTRACES programme. At the present day, while matrix removal followed by ICP-MS analysis is now the most used method (that also has the added benefit of multiple element detection), there is one very significant drawback: it cannot be used at sea. Therefore, there is a valuable use for flow injection analysis techniques (FIA) on board ship (or when a more expensive ICP-MS is not available back on shore). FIA systems can be used for shipboard determinations of Al, Mn, Fe, Cu, Zn and Cd (e.g. Middag et al. 2015b; Rijkenberg et al. 2018; van Hulten et al. 2017). A FIA set-up only allows one trace metal to be measured at a time, but the system is quite compact, can easily be used at sea in a shipboard (mobile) clean laboratory or 'bubble' and is often automated with electronic valves and an autosampler that allows semi-autonomous operation of the system (Worsfold et al. 2013). Shipboard metal determination is a highly valuable technique used to guide at sea sampling efforts (e.g. if looking for high-concentration features such as hydrothermal plumes), to check for inadvertent contamination of bottle samplers, to measure short-lived species such as dissolved Fe^{2+} or to inform shipboard experiments such as Fe addition bioassays (e.g. Bowie et al. 2002; Cutter and Bruland 2012; de Baar et al. 1990). More recently, sequential injection analysis has been used for determination of individual marine trace metal concentrations (Grand et al. 2016), but this method that uses minimal reagents and is amenable to autonomous deployment is not yet mature and is still being developed.

In a FIA system, the analyte of interest is separated from the interfering bulk seawater matrix by being pumped through a chelating resin column (see Sect. 3.5.1.1 for more details on chelating resins). After this separation and subsequent elution, a reaction between the analyte of interest and the reagent leads to the formation of a complex or a reaction that can be detected using a specific detector (Figs. 3.19 and 3.20), e.g. chemiluminescence can be detected using a photomultiplier or a formed coloured analyte-reagent complex can be detected using a spectrometer (Bowie et al. 2004). The loading time in a FIA system is standardized and automated, which should ensure that the same amount of sample is loaded onto the column every time. Reagents are continuously pumped through the FIA system using a peristaltic pump, where automated valves either pass buffered seawater sample (denoted load), deionized water (DI, often UHPW; denoted rinse) or elution acid (denoted elute) over the chelating resin column.

Here, we describe an example using a chemiluminescence reaction for the analysis of Fe concentrations in seawater (Figs. 3.19 and 3.20), using SPE. During the loading of the sample of interest, prevalent metals including Fe are retained by the resin column and thereby separated from the seawater matrix. The loading pH is crucial, since Fe binds at a much lower pH compared to Mn, allowing either metal to be measured by changing the loading pH (Klunder et al. 2011; Middag et al. 2011). Acidified samples are usually buffered in-line, and the loading pH can be changed by using a stronger or different buffer solution, where the resulting loading pH should be regularly checked. Samples for Fe should be acidified to a pH < 2 prior to

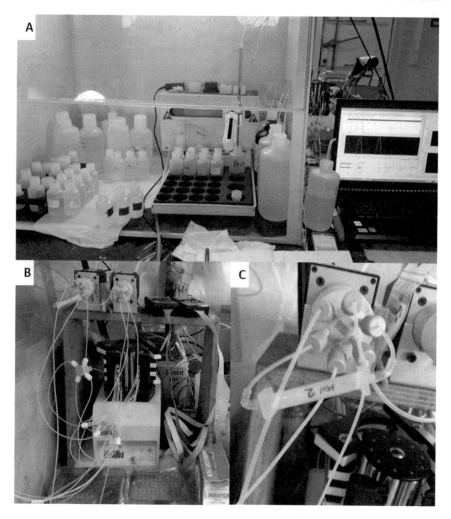

Fig. 3.19 Photo of a shipboard flow injection analysis (FIA) set-up including an autosampler, a computer and a water bath (**a**), a peristaltic pump and two automated valves (**b**) and a resin column on a valve (**c**) (photograph credits: Patrick Laan, NIOZ, Netherlands)

analysis (Johnson et al. 2007) and left to equilibrate at least 12 hrs (Klunder et al. 2011) or heated to reduce this time (Lohan et al. 2006). Depending on the used method, a reducing (e.g. sodium sulfate) or oxidizing reagent (e.g. hydrogen peroxide) should be added to ensure Fe is present only as either Fe^{2+} or Fe^{3+} (Bowie et al. 2004; Johnson et al. 2007), or FIA can also be used to measure Fe^{2+} specifically if the redox speciation of Fe is the research aim (e.g. Bowie et al. 2002).

During the rinsing step, the seawater matrix is washed from the column and goes to waste (i.e. a dedicated waste line). At the same time, the reagent stream (in our example, luminol and peroxide as reactants and ammonium hydroxide to maintain

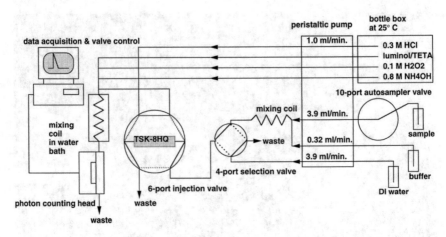

Fig. 3.20 Schematic overview of a flow injection system (FIA) for iron in seawater (de Jong et al. 1998)

the required reaction pH) flows continuously via the detector providing a baseline signal, i.e. the reaction between luminol and peroxide already produces chemiluminescence. Luminol (5-amino-2,3-dihydro-1,4-phthalazinedione) is a compound that emits light upon oxidation with reactive oxygen species, a reaction catalysed by Fe (Borman et al. 2009). After loading and rinsing, the retained sample Fe is eluted from the column and is injected into the reagent stream that is kept at optimum temperature in a water bath, where the Fe catalyses the reaction between luminol and peroxide, resulting in increased chemiluminescence. The increase in chemiluminescence scales to the amount of Fe after pre-concentration. With this system, one first observes a dip in the baseline as first the UHPW in the column (result of the rinsing step) is injected, followed by a peak due to the catalysing effect of the Fe (Figs. 3.19 and 3.20).

Ideally, the peak is sharp and the signal returns back to baseline right away, but FIA systems are known to be 'temperamental'. Thus, it can take significant tuning to get good peaks and sufficient sensitivity. Moreover, the chemiluminescence-based FIA method is prone to drift in sensitivity, for example, due to changes in temperature. In particular, a sample shows a higher signal at a higher temperature, and thus drift standards have to be run regularly to correct for temperature effects (Bowie et al. 2003; Bowie et al. 2004; Floor et al. 2015; Worsfold et al. 2019c). The calibration is done using standard additions, where care should be taken to measure samples in the calibrated and linear range of the chemiluminescence response (Borman et al. 2009). Moreover, despite the automated and standardized loading time, it has been shown that variations of up to 5% in the loaded mass of seawater samples occur. Combined with variation in sensitivity during a run and the uncertainty associated with the calibration (estimation of the sensitivity), this leads to an overall uncertainty in the order of 10–15% for the chemiluminescence-based FIA system for Fe (Floor et al. 2015). Thus, proper consideration should always be given to the analytical uncertainty associated with measurements of trace metals (see Sects. 3.5.1.4 and 3.6) to avoid overinterpretation of the data (Worsfold, et al. 2019).

3.5.1.3 In Situ Metal Analysis Systems

The sampling and analysis approaches detailed thus far largely deal with discrete samples that are representative for a given moment in time. However, to capture the spatial and temporal variabilities of trace metals at sub-nanomolar concentrations in dynamic (open) ocean settings, ideally in situ sensors or analysers that capture data continuously are implemented (Grand et al. 2019). Progress is being made in this area, and a combination of wet chemical analysers with electrochemical sensors may become available in the coming years or decades, but currently developed systems lack the accuracy and precision needed for oceanic trace metal concentrations and speciation. Readers interested in the state of the art and remaining challenges are referred to Grand et al. (2019) for an in-depth overview.

3.5.1.4 Data Quality Control for Trace Metal Concentration Measurements

Here, we briefly describe analytical considerations that are needed to produce consistent high-quality trace metal concentration data, including 'blanks', detection limits and uncertainty (accuracy and reproducibility) of analytical techniques. For a more in-depth discussion of good practices to obtain and report good quality data, the reader is referred to Worsfold et al. (2019) and Wurl (2009).

Obtaining low and consistent blanks, commonly referred to as 'procedural blanks', is key for both validating the method of choice and for ensuring contamination-free trace metal analysis during the implemented procedure (Wurl 2009). The blank is defined as the amount of analyte that is added (inadvertently) during the overall procedure, and differentiation can be made between various components of the procedure such as a sampling blank (contamination during taking of the sample), storage blank (contamination between sample collection and analysis) and/or analysis blank (contamination from the analytical procedure, including sample handling). The focus usually lies on the analysis blank associated with the use of reagents, i.e. systematic contamination, as this can be quantified. Nevertheless, levels of contamination during sampling and storage should also be assessed as detailed in the previous sections (Sects. 3.2, 3.3 and 3.4). The analytical limit of detection (LOD), which is a criterion for the performance of an analytical method and/or technique, indicates the lower limit at which a technique can differentiate between a signal due to background noise and a signal derived from the target analyte. Depending on the method, this can be closely related to the instrumental LOD (i.e. the limit at which the used instrument can differentiate between an analyte signal and background noise), but often the details of the analytical procedure are responsible for changes in the LOD (e.g. variation in extraction efficiency with SPE columns, or variations in pipetted sample volumes) and hence the analytical LOD should be determined throughout the whole procedure, not just from the instrumental LOD. The LOD is inherently linked to the blank value and the variations therein, and so the LOD is commonly defined as three times the standard deviation of the analytical blank (Wurl 2009). To be able to calculate a standard deviation and subsequently the LOD, the blank should be measured at least three times; however, a better approach is to measure the blank more often throughout an analytical run and also use the blank values from discrete runs over a longer time period to calculate the LOD in order to account for any intra- and inter-run variability. For trustworthy trace metal data, the metal concentration of interest of the target sample must be higher than the LOD (Wurl 2009). Common values for blanks and LODs of the six bio-essential metals from NIOZ for the SeaFAST technique can be found in Table 3.1 (Seyitmuhammedov 2021), showing a similar range as those generated by other approaches (e.g. Biller and Bruland 2012; Rapp et al. 2017).

Expressing uncertainty (error) on any dataset is vital. This uncertainty is typically expressed via accuracy (do we have the right number?) and precision (how reproducible is the number?). For seawater trace metal concentration data, accuracy of data can be compared in two ways: (1) measuring trace metal concentrations in a reference material (RM) and comparing these values to published or consensus values and/or (2) comparing measurements made by multiple laboratories in the same natural samples or on samples collected from the same location (see Sect. 3.6). Such activities are essential steps to confirm that the obtained results have acceptable accuracy relative to the variability being studied Worsfold et al. 2019; Wurl 2009). Certified reference materials (CRM) for ICP-MS can either be commercially available certified RMs and/or obtained from bodies such as the US National Institute for Standards and Technology (NIST). For good practice, RMs should have the same matrix as the sample of interest, their concentrations should be in the same range as those of the target samples, and obtained concentrations should agree with published concentrations (within the uncertainty range) of the RM (Worsfold et al. 2019; Wurl 2009). Because of the challenges of collecting and storing open-ocean seawater, there is no commercial CRM available for open ocean trace metal concentrations, and trace metal investigators thus often use in-house reference samples (e.g. a large volume homogenized seawater sample), or aliquots of reference seawater that GEOTRACES has made available upon request (e.g. Cutter et al. 2017; Johnson et al. 2007; see https://www.geotraces.org/standards-and-reference-materials/) for which the average accepted values are known as 'community consensus values'. Analysts will typically make (and report) regular measurements of these reference samples as RM alongside samples in order to assess the accuracy of the technique (see Sect. 3.5.1.1). When the procedure returns inaccurate reference sample values, action is needed by the investigator to identify and eliminate sources for the inconsistency of the data (Wurl 2009).

Evaluating the reproducibility (precision) of trace metal concentration data is also essential and should be done via replicate measurements of samples, blanks and RMs (Worsfold et al. 2019). Additionally, it is recommended to evaluate precision at concentrations appropriate to those of the samples being studied—for instance, the precision of low-concentration samples may be much poorer than high-concentration samples. Commonly, duplicate or triplicate measurements are used to assess precision. However, it should be noted that triplicate measurements of a single sample are not necessarily representative of the overall precision, if sampling

Table 3.1 Typical blank values, detection limits (LOD) and an example of a CCRS (Geotraces Surface Pacific; GSP, consensus values reported 2020; https://www.geotraces.org/standards-and-reference-materials/) results for various dissolved trace metals (DFe, DCu, DZn, DMn, DNi, DCo and DCd) of using a SeaFAST pre-concentration step with subsequent HR-SF-ICP-MS analysis at NIOZ. The procedural blank was UHPW acidified to the same pH as seawater samples. Please note that the LOD ($3 \times$ SD) may appear different from three times the reported SD due to rounding

	HR-SF-ICP-MS						
	DMn pmol L⁻¹	**DFe** nmol L⁻¹	**DCo** pmol L⁻¹	**DNi** nmol L⁻¹	**DCu** pmol L⁻¹	**DZn** nmol L⁻¹	**DCd** Pmol L⁻¹
Blank	2	0.03	2	0.03	7	0.03	3
SD	1	0.01	1	0.04	7	0.02	3
n	24	24	24	24	24	24	24
LOD	4	0.02	3	0.1	20	0.07	8
	DMn nmol/L	**DFe** nmol/L	**DCo** pmol/L	**DNi** nmol/L	**DCu** nmol/L	**DZn** nmol/L	**DCd** pmol/L
GSP 239	0.79	0.15	7	2.56	0.60	0.04	3
±SD	0.01	0.01	1	0.03	0.01	0.02	2
n	14	14	14	14	14	14	14
Consensus value	0.78 ± 0.04	0.16 ± 0.05	-	2.6 ± 0.1	0.57 ± 0.06	0.03 ± 0.06	2 ± 2

also contributes to uncertainty. For this it is best to collect replicate samples in successive sampling efforts at the same depth and location, which is done during some expeditions (e.g. sampling at the same depth twice on overlapping casts by the US GEOTRACES programme), but this is not currently routine practice for all sampling endeavours, usually due to the associated extra time and cost. For RMs and blanks, replicate measurements performed in a single day enable statements on the repeatability, while replicate measurements over longer time periods can be used to determine laboratory reproducibility, accounting for drifts in detector response or laboratory conditions, i.e. temperature (Worsfold et al. 2019). For blanks, the average values are commonly subtracted from the sample values obtained in the same run. In all cases, it is recommended that data is reported as means including standard deviations (Worsfold et al. 2019). Often samples are measured only once, as trace metal analysis is time-consuming and costly. As a compromise, the standard deviation (SD) for samples can be calculated as the square root of the sum of the internal and external SD. The internal SD is the SD of the ICP-MS measurement (instrumental precision), and the external SD is the SD of the specific analyte of an (in house) RM measured regularly throughout an analytical run (Gerringa et al. 2021a), where it is recommended to use RMs at different concentration levels as the precision is likely lower at lower concentrations. Good precision (% SD) for trace metal concentrations assessed by replicate analyses of RM or seawater by SeaFAST, ICP-MS, FIA and isotope dilution from isotopic techniques, from a range of laboratories, has been assessed as ± 0–5%. Reported precision is often better for less contamination-prone metals such as Cd and also better at higher trace metal concentrations (e.g. Conway et al. 2013; Jensen et al. 2020; Jensen et al. 2019; Lagerström et al. 2013; Middag et al. 2015a; Minami et al. 2015; Rapp et al. 2017; Rijkenberg et al. 2014; Wuttig et al. 2019), but may range from ± 2 to 20% for FIA depending on the metal (e.g. Resing et al. 2015; Sedwick et al. 2015; Wyatt et al. 2014).

However, while reporting the precision (% SD) of good quality data is key, it is perhaps more realistic to estimate and report the overall or so-called 'combined' uncertainty of the obtained data (Worsfold et al. 2019). Such combined uncertainty considers the contributions from all the uncertainty contributions during sample analysis, including analyte uncertainty, blank uncertainty, pipette uncertainty for all volumes pipetted (sample and reagents), slope uncertainty of the calibration line, uncertainty of metal concentrations in reagents and volume uncertainties (Worsfold et al. 2019). Such accounting for the combined uncertainty may lead to larger relative uncertainties obtained for both ICP-MS and FIA techniques (e.g. 10–30%; Clough et al. 2015; Rapp et al. 2017; with more uncertainty at the lowest concentrations), but can also be of similar size to internal precision for ICP-MS (see Sect. 3.5.2.3). Larger uncertainty from combined uncertainty highlights that reporting only instrumental precision on data can underestimate the overall uncertainty of the obtained data, depending on the technique. For more information on combined uncertainty including example applications, the reader is referred to Worsfold et al. (2019) and references therein. Perhaps the best way to assess data precision and accuracy is to compare data from the same samples or same location produced by different laboratory groups, analytical techniques and/or sampling systems. Indeed, within the GEOTRACES programme, besides analysis and reporting of RMs, there has been a strong emphasis on intercalibration using crossover stations (see Sect. 3.6) to assess the comparability of datasets obtained via different sampling and/or analytical techniques.

3.5.2 Trace Metal (Fe, Ni, Cu, Zn, Cd) Isotope Ratio Measurement Techniques

3.5.2.1 Background

With the onset of improved clean collection and handling techniques and the development of HR multi-collector ICP-MS (MC-ICP-MS) (e.g. Douthitt 2008) has come the ability to measure the stable isotopic ratios of transition metals even at the low concentrations of these metals found in seawater (pmol L^{-1}-nmol L^{-1}). The promise of such measurements is that isotopic ratios may offer more insight into marine trace metal cycling than concentrations alone. The first isotopic measurements for Cd, Cu and Zn in seawater were published in 2006 (Bermin et al. 2006; Lacan et al. 2006), followed by Fe in 2007 (De Jong et al. 2007). Despite these efforts, by the official beginning of the GEOTRACES field programme in 2010, data remained sparse, with only about 10–50 data points having been published for each element (Conway et al. 2021). Since 2010, however, stimulated largely by GEOTRACES, the field of transition metal isotopes in seawater has undergone a figurative explosion, adding Cr and Ni isotopes to the toolbox and with data coverage increasing dramatically (Conway et al. 2021). The other bio-essential trace metals discussed in this chapter, Co and Mn, are monoisotopic for stable isotopes, and thus ratios cannot be determined. The first ocean 'sections' of dissolved trace metal isotope ratios were published for the Atlantic in 2014 (e.g. Zn, Conway and John 2014a); Fe, Conway and John 2014b)), and now a number of other such sections are published and available from several groups in the GEOTRACES Intermediate Data Products (Schlitzer et al. 2018). Indeed, sufficient data now exists for each isotope system to provide comprehensive insights into the biogeochemical cycling of each trace metal. This insight includes tracing Fe and Zn sediment sources to the ocean (Homoky et al. 2016), balancing the oceanic Zn and Cu budgets (Little et al. 2014; Little et al. 2016), understanding redox cycling of Fe and Ni (Rolison et al. 2018; Vance et al. 2016) and showing how scavenging, biological uptake and water mass mixing influence Ni, Zn and Cd (e.g. Abouchami et al. 2011; Archer et al. 2020; John and Conway 2014; Ripperger et al. 2007).

Here, we briefly describe the chemical and analytical challenges for making measurements of transition metal isotope ratios in seawater and the methods developed by different groups to overcome these—including the typical methods used at the University of South Florida (USF) for measuring $\delta^{56}Fe$, $\delta^{66}Zn$ and $\delta^{114}Cd$ in seawater. For further reading on the distributions of the trace metal isotope ratios throughout the ocean, as well as the current state of understanding of the processes which influence these distributions, we refer the reader to several recent synthesis review articles (e.g. Anderson 2020; Horner et al. 2021 and references therein).

3.5.2.2 Chemical Processing for Trace Metal Isotope Analysis

There are two critical challenges to overcome to measure trace metal isotope ratios in seawater, which are common to each of the five trace metals described here (Fe, Ni, Cu, Zn, Cd). First, akin to concentration measurements (see Sect. 3.5.1.1), the sea salt matrix means that samples cannot be measured directly by mass spectrometry. Instead, the matrix must be separated and the trace metal of interest pre-concentrated. Second, mass spectrometric techniques for measuring trace metal concentrations (see Sect. 3.5.1.1) rely on measuring the whole metal pool or the most abundant isotope of each metal. For isotopic ratio measurement, however, the minor abundance isotopes must also be measured. These minor isotopes are typically present at even lower concentrations (often >50x lower), compounding the problems of clean handling and sample volumes needed. For example, Fe has four stable isotopes with abundances of 6% (^{54}Fe), 92% (^{56}Fe), 2% (^{57}Fe) and 0.3% (^{58}Fe). Thus, in a typical deep ocean seawater sample of 0.5 nmol L^{-1} Fe, ^{54}Fe is only present at only 30 pmol L^{-1}, and just 1.5 pmol L^{-1} are ^{58}Fe. In order to accurately measure a 56/54 or 58/54 Fe ratio, Fe must therefore be pre-concentrated cleanly from much larger volumes of sample (e.g. 1–20 L; e.g. Conway et al. 2013; Lacan, et al. 2008) than those typically needed for concentration analysis (e.g. 30 mL; see Sect. 3.5.1.1). Such large volume requirements mean that it has been historically challenging to make clean measurements at high spatial resolution in the ocean. Typically, chemical methods to prepare samples for isotope analysis involve two to three purification stages, with one stage removing the bulk of the sea salt matrix (see below) and later stages purifying the element of interest from any minor elements which are potential isobaric interferences, as well as removing any remaining salts (see below). A range of approaches is described here, but whichever method is used, the key requirements are for clean (ideally procedural blank concentrations of ≤1 ng) and quantitative recovery (ideally ~100%) of the trace metal of interest.

A further challenge with trace metal isotope analyses is that, as with early concentration methods, most early isotopic methods focused on a single metal isotope system of interest (e.g. just Cd; Lacan et al. 2006; Ripperger and Rehkämper 2007). For multiple trace metal isotope measurements, separate aliquots of seawater were needed, meaning even larger total volumes of clean seawater were required. More recently, however, methods have been developed that can process the same sample for multiple metals in seawater for later isotopic analysis (e.g. Fe, Zn and Cd by Conway et al. 2013; or Ni, Cu and Zn by Takano et al. 2017) (see Box 3.11). These multiple element approaches have played an important role in facilitating the rapid application of multiple isotopic systems as tracers of oceanic processes as part of GEOTRACES.

Sea Salt Matrix Removal Stage

The first step of any seawater processing method for isotope analysis is to pre-concentrate the metal of interest cleanly from 1–4+ L down to a small volume for analysis (<1 mL), while removing the seawater matrix (see Box 3.11; Fig. 3.21). Due to the large volumes (and thus the larger total amount of salts) needed for isotopic compared to concentration measurements, automated processing techniques such as the SeaFAST are not yet commercially available, although in development (e.g. Field et al. 2019). Instead, pre-concentration methods have focused on in-house methods developed by different laboratory groups. These methods include using chelating resins, organic solvent separation (e.g. Thompson et al., 2013; Ellwood et al., 2014) or co-precipitation with Mg or Al (e.g. Bermin et al. 2006; Cameron and Vance 2014; De Jong et al. 2007; Staubwasser et al. 2013; Xue et al. 2012) to cleanly separate the trace metal(s) of interest from seawater. For the former, studies have made use of Qiagen NTA, Nobias PA-1, Chelax-100 or Bio-Rad AG1-X8 ion-exchange resins to extract metals from seawater; resin beads are either added directly to seawater, shaken and filtered out, or seawater is pumped through resin-packed columns (e.g. Abouchami et al. 2011; Bermin et al. 2006; Conway et al. 2013; John and Adkins 2010; Lacan, et al. 2008; Ripperger and Rehkämper 2007; Vance et al. 2016; Xue et al. 2012). The transition metal cations (or metal bromide or chloride complexes if anion resin) 'stick' to the functional groups of the resin and are thus removed from the seawater leaving the salts behind. At the time of writing, most isotope groups have moved to make use of Nobias PA-1 techniques (e.g. Archer et al. 2020; Ellwood et al. 2020; Sieber et al. 2021; Yang et al. 2020), principally due to the low blank and the high extraction efficiency of this resin for multiple trace metals. However, other resins continue to be used for single elements (e.g. NTA for Cu by Baconnais et al. 2019; AG1-X8 for Cd by Xie et al. 2015).

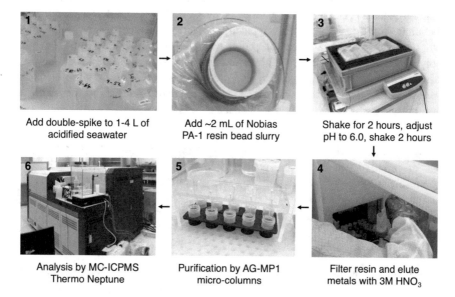

Fig. 3.21 Step-by-step scheme for multiple trace metal isotope ratio analysis (photograph credits: Tim Conway, University of Florida, USA)

Box 3.11: Chemical and Analytical Scheme for Multiple Trace Metal Isotope Ratio Analysis

The figure below shows a simplified schematic of the typical chemical method that is being used at ETH Zürich (Switzerland) and the University of South Florida (USA) to process seawater for Fe, Cd and Zn isotope ratios (follows Conway et al. 2013, and Sieber et al. 2021) (Fig. 3.21). Double spikes of ^{57}Fe-^{58}Fe, ^{64}Zn-^{67}Zn and ^{111}Cd-^{113}Cd are typically added prior to processing to allow for correction of both procedural and instrumental isotopic fractionation (see Box 3.15 for more explanation of double spikes).

Following extraction from a water sample, the trace metals can then be eluted from resins or redissolved from precipitates into a small volume of HNO_3 or HCl for further purification steps (e.g. on micro-columns, see below) and then analysis (see Sect. 3.5.2.3). For marine particles or sediments, following a suitable leaching or digestion process (see Sect. 3.4.3), samples can be redissolved for similar purification. However, care must be taken to establish that any leaching process does not fractionate isotope ratios (e.g. Revels et al. 2015).

Purification Stage

Although the first stage described above removes most of the salt from a seawater sample, the functional groups of ion-exchange resins often bind a range of elements, meaning that the resulting acidic solution contains not just the metal of interest, but often many other transition metals and traces of remnant salts such as Ca and Mg (e.g. for Nobias PA-1; Sohrin et al. 2008; see Box 3.12; Fig. 3.22). If a precipitation technique is used, the precipitate typically contains other elements as well. These other elements cause a range of problematic isobaric (e.g. ^{58}Ni on ^{58}Fe) and polyatomic (e.g. $^{40}Ca^{16}O$ on ^{56}Fe) interferences during mass spectrometric analysis, which adversely affect the accuracy of isotope ratios or the ionization of elements (see Sect. 3.5.1.1. for more background on interferences). As such, further purification of the sample is required prior to analysis of isotope ratios.

Box 3.12: Elution of Different Transition Metals from AGMP-1 Resin

The figure below shows an example of a second-stage AGMP-1 column purification scheme, showing the separate elution of Cu, Fe, Zn and Cd from major salts and interfering elements when acid strength and type through the resin are varied (Conway et al. 2013), with volume being cumulative throughout the column scheme (Fig. 3.22).

The goal of the second purification step is thus to achieve only the single metal of interest being dissolved in a small volume of weak acid (e.g. 0.5–1 mL of 0.2 M HNO_3), ready for isotope analysis. This purification is typically achieved by taking the sample through a single (or several) anion-exchange resin-packed columns,

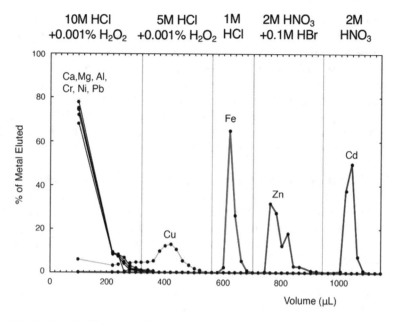

Fig. 3.22 Elution of different transition metals from an AGMP-1 resin (modified from Conway et al. 2013)

usually filled AGMP-1 or AG1-X8 anion-exchange resins (from Bio-Rad). AGMP-1 is perhaps the most widely used resin for this purpose (e.g. Archer and Vance 2004), although AG1-X8 has been widely used for Cd purification, usually followed by an additional Eichrom TruSpec column to further separate Cd from Sn and Mo (e.g. Abouchami et al. 2011; Ripperger and Rehkämper 2007; Xue et al. 2012). In each case, the functionality of these anion-exchange resins relies on metals having different distribution coefficients—that metals 'stick' variably to the resin—as the strength or type of acid is varied (e.g. Strelow 1980). Such AGMP-1 techniques were first applied for separation of Cu, Zn or Fe from rocks for isotope analysis (e.g. Beard and Johnson 1999; Maréchal et al. 1999). For seawater, where the transition metal concentrations are much lower, and thus blanks must be kept suitably low, studies have modified these techniques to work with smaller volumes of resin. Archer and Vance (2004) developed an AGMP-1 column separation scheme for Cu and Zn that lowered the blank contamination of the method by an order of magnitude (down to 0.5–1 ng) compared to earlier techniques. AGMP-1 micro-column techniques have since been developed to purify Fe or Cd and then Cu, Zn, Fe and Cd from seawater for isotope analysis (e.g. see Box 3.12; Conway et al. 2013; John and Adkins 2010; Lacan et al. 2006; Takano et al. 2013). These micro-columns have typical blanks of less than 0.2 ng per element (e.g. Conway et al. 2013).

When using AGMP-1 resin (see Box 3.12), the sample from the first purification stage is typically dried down (evaporated to dryness), so that it can be dissolved in a specific reagent (>5 M HCl) and then 'loaded' on the AGMP-1 resin column. Different metals are then 'eluted' from the resin column by changing the acid strength or acid type (see Box 3.12 for an example column elution scheme). Such procedures have been shown to successfully separate Cu, Fe, Zn and Cd from elements such as Ca, Mg, Ni, Cr, Sn, etc. (Conway et al. 2013). Nickel, however, elutes from the columns with elements such as Na, Al, Ca and Mg (see Box 3.12), and so further purification steps (e.g. Nobias PA-1 or multiple AGMP-1 columns) are typically needed to remove interfering elements before isotope analysis of Ni is possible (Archer et al. 2020; Wang et al. 2019; Yang et al. 2020).

3.5.2.3 Analytical Procedures for Trace Metal Isotope Analysis

Here, we provide a brief overview of the details of trace metal isotope analysis, with a focus on the bioactive metals measured in seawater, often using Fe as an example. For a fuller discussion on the detail of isotope systematics and details of isotope ratio analysis for the bioactive metals, we point the reader elsewhere (e.g. Dauphas et al. 2017; Johnson et al. 2020; Moynier et al. 2017; Rehkämper et al. 2012; Teng et al. 2017). The concepts reviewed in the following sections apply to all transition metals discussed here, although we note each element has its own quirks and interferences to account for during both chemistry and analysis.

Isotope Ratio Basics, Nomenclature and 'Zero' Isotope Standards

Following chemical purification, transition metal isotope ratios are measured by mass spectrometer. However, the natural range of mass-dependent variability for transition metal isotope ratios (e.g. $^{56}Fe/^{54}Fe$), while measurable, is typically very small, usually at the 0.001 or permil (‰) level (e.g. Horner et al. 2021; Johnson and Beard 1999). Further, instrumental 'mass' bias, or the systematic loss of lighter ions during analysis, means that it is difficult to measure absolute isotope ratios at this level of precision. Thus, isotope ratios are typically measured relative to a measured standard RM, usually in 'delta notation' (see Box 3.13). For example, Fe isotope ratios are typically expressed as:

$$\delta^{56}Fe = \left[\frac{\left(^{56}Fe/^{54}Fe\right)_{sample}}{\left(^{56}Fe/^{54}Fe\right)_{IRMM-014}} - 1 \right] \times 1000$$

In this case the $^{56}Fe/^{54}Fe$ of the sample is not expressed as an absolute isotope ratio, but rather relative to the $^{56}Fe/^{54}Fe$ of the international Fe isotope standard IRMM-014 (an isotope zero standard) (Taylor et al. 1993) that is measured by the same mass spectrometer. For each element, data is expressed in the form of a common isotope pair (or range of pairs) and typically relative to one international 'zero' standard (see Table 3.2). It is worth noting that some groups report in other notations (e.g. epsilon units; Table 3.2), which is calculated by replacing the ×1000 in the delta equation with ×10,000, meaning that, for example, $\varepsilon^{114/110}Cd$ is equivalent to $10 \times \delta^{114}Cd$. For converting between isotope ratio pairs, typically a

Table 3.2 Typical notation and isotope zero standards for trace metal isotope ratios in seawater. Where multiple notations exist, the common notation and isotope pairs (as included in the GEOTRACES Intermediate Data Product) are shown in bold. Ranges for deep seawater (expressed in the common notation) are from Horner et al. (2021); for isotope standards see Abouchami et al. (2013); Dauphas et al. (2017); Moynier et al. (2017); and Elliott and Steele (2017). A more detailed discussion of conversion between isotope ratios and notation for Cd can be found in Rehkämper et al. (2012)

Element	Notation	Isotope pair(s)	Isotope zero standard	Range in deep seawater (‰)
Fe	δ^{56}Fe, δ^{57}Fe	^{56}Fe/^{54}Fe, ^{57}Fe/^{54}Fe	IRMM-014	−2.4 to +1.5
Ni	δ^{60}Ni	^{60}Ni/^{58}Ni	NIST SRM 986	+1.2 to +1.5
Cu	δ^{65}Cu	^{65}Cu/^{63}Cu	NIST SRM 976	+0.6 to +0.8
Zn	δ^{66}Zn	^{66}Zn/^{64}Zn	JMC-Lyon	−0.2 to +0.6
Cd	δ^{114}Cd, δ^{112}Cd, $\varepsilon^{114/110}$Cd, $\varepsilon^{112/110}$Cd	^{114}Cd/^{110}Cd, ^{112}Cd/^{110}Cd	NIST SRM 3108	+0.2 to +0.4

simple mass difference approach can be applied; for example, $\delta^{56/54}$Fe $= \sim 2/3 \times \delta^{57/54}$Fe, or $\delta^{114/110}$Cd $= \sim 2 \times \delta^{112/110}$Cd.

For some 'metallic' elements such as B or Li, there is large enough (~10‰) mass-dependent isotope variability in nature to allow isotope ratios to be measured by relatively imprecise instruments such as single-detector SF-ICP-MS instruments (e.g. precision of +0.5‰; Misra et al. 2014). However, for the transition metals, which display much lower levels of natural isotopic variability (e.g. 1–4‰; Horner et al. 2021), higher precision (<0.2‰) is needed. Such levels of precision require 'multi-collector' (MC) mass spectrometers, which operate by collecting a series of ions simultaneously into multiple 'faraday cup' ion-counting detectors, allowing isotope abundances to be measured simultaneously rather than sequentially, as is the case for a single collector SF-ICP-MS (e.g. Walder and Freedman 1992). MC instruments take two forms, classified depending on how the sample is introduced and ionized, either ICP- (inductively coupled plasma) or TI- (thermal ionization) MS. MC-ICP-MS instruments introduce a sample via an ionized plasma (as described in Sect. 3.5.1.1), while for MC-TI-MS a sample is loaded on a metal 'filament' which is then heated and the sample vaporized/ionized (see Johnson and Beard 1999; Schmitt et al. 2009). Of the two, the most widely applied technique in seawater literature for trace metal isotope ratios is MC-ICP-MS, making use of one of three instruments: the Thermo Scientific Neptune (9–10 laboratory groups; Fe, Ni, Cu, Zn and Cd), Nu Instruments Plasma I or II (6 groups; mostly Cd) or the GV Instruments IsoProbe (1 group; Zn). One group utilizes TIMS (Thermo Fisher Triton instrument) to measure Cd isotope ratios (e.g. Abouchami et al. 2011). It is worth noting that for Cd, which is the only isotope system to be measured in aliquots of the same deep seawater sample by TI-MS and both Nu and Thermo MC-ICP-MS, excellent agreement has been shown, meaning that data from all three methods are inter-comparable (e.g. Boyle et al. 2012). For a more detailed discussion of the comparison of MC-ICP-MS and TI-MS for Cd isotope analysis, see Rehkämper et al. (2012).

Box 3.13: Transition Metal Isotope Standards

The choice of isotope standard reference material (RM) and 'zero' standards are established by consensus between laboratory groups and can take the form of in-house shared materials (e.g. JMC-Lyon; Moynier et al. 2017) or commercially available standard RM (e.g. NIST-3108 Cd; Abouchami et al. 2013), with the isotope ratio difference between different isotope standards established by consensus (e.g. Abouchami et al. 2013; Archer et al., 2017). For in-house shared materials, investigators typically acquire an aliquot via personal communication with the original laboratory. For commercial RM, which are often designed as concentration standards, it is important to acquire the correct batch. In both cases, the limited availability of such RM often means that isotope zero standards can become depleted and must be replaced by others. This depletion has already occurred with JMC-Lyon for Zn, with AA-ETH being suggested as a replacement solution (Archer et al., 2017). In this case, it is encouraged to measure ratios relative to AA-ETH but report ratios relative to JMC-Lyon using the multi-laboratory established offset of 0.28‰ between the standards so that new data can most easily be compared with literature data (Archer et al., 2017).

MC-ICP-MS Analytical Techniques and Mass Bias Correction Techniques

For simplicity, here we focus on the measurement of trace metal isotope ratios by MC-ICP-MS, as used by most trace metal isotope groups, and specifically give a brief overview of the systematics of the Neptune Plus as used at the University of South Florida (for a general introduction of the mechanism of sample introduction by ICP-MS, see Sect. 3.5.1.1). The concepts discussed here for Neptune MC-ICP-MS analysis are generally applicable to the Nu Plasma instruments, and mass bias correction techniques are typically similar for MC-ICP-MS and TI-MS. We do not provide detailed discussion of TIMS, but instead point the reader to Schmitt et al. (2009), Abouchami et al. (2011) and Rehkämper et al. (2012) for further reading.

For seawater samples, where metal concentrations are low (especially surface waters), the main concern for isotope analysis is to maximise signal size in order to reduce uncertainty on isotope ratios (e.g. John and Adkins 2010). Compounding this issue for some elements (e.g. Fe), which have isobaric interferences that are large and similar in mass ($^{56}Fe^+$ and $^{40}Ar^{16}O^+$), 'high'-resolution (HR) mode must be used on the Neptune MC-ICP-MS (Weyer and Schwieters 2003). While HR mode allows successful measurement of the Fe isotope peaks separate from their Argide interferences, it also comes at a cost – approximately six to seven times reduction in signal size compared to low resolution (LR; Weyer and Schwieters 2003). As such, every attempt is made to boost signal size, which can be done in two ways –

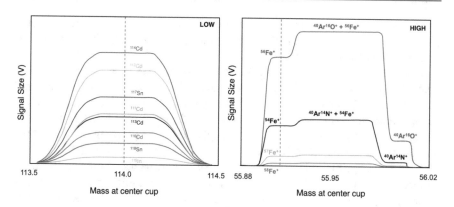

Fig. 3.23 Peak alignment for measurement of trace metal isotope ratios on the Thermo Neptune MC-ICP-MS in low-resolution mode (left) and high-resolution mode (right) (Cartoon credit: Tim Conway, University of Florida, USA). Dashed vertical lines indicate measurement mass on each peak

either by dissolving the sample in the smallest volume of acid possible (often <0.5 mL) or by optimising the MC-ICP-MS introduction system. For the latter, there are two aspects which are helpful: (i) choosing the optimal interface cone combination and (ii) choosing the right introduction system. Neptune MC-ICP-MS methods often boost signal size (i.e. V per ng mL^{-1}) by making use of the high-sensitivity combination of Jet sampler and X-skimmer interface cones (e.g. Conway et al. 2013). Most MC-ICP-MS methods also use a desolvation system, such as the commercially available Cetac Aridus I/II, ESI Apex Q or ESI Apex Omega (e.g. Bermin et al. 2006; Conway et al. 2013; Sieber et al. 2021), as a suitable introduction system. Desolvation systems use a series of heated and cooled spray chambers, together with added Ar or N$_2$ gases (or both), to boost the signal size of the element of interest while minimising oxide interferences. It should be noted, however, that specific care must be taken when customising the introduction system set-up, since both desolvators and high-sensitivity cones can induce larger interference formation (e.g. CdH on Cd or ZnH on Zn; Archer et al., 2017; Sieber et al., 2019) or mass bias effects (Archer and Vance 2004; Bermin et al. 2006), compared to regular H-cones or a spray chamber. For example, larger mass bias effects attributed to the Aridus desolvator have led to δ^{65}Cu being more commonly measured using a spray chamber and H cones (e.g. Bermin et al. 2006; Takano et al. 2013; Little et al. 2014).

Once the introduction system is set up and the plasma lit, the instrument must be tuned for optimal performance, including signal size, signal stability and peak shape, by tuning the gas flows, torch position and lenses. Peak shape is tuned to generate wide, flat-topped peaks in LR and sharply resolved peaks in HR (see Box 3.14, Fig. 3.23). This tuning is akin to what is needed for concentration measurements (Sect. 3.5.1.1). In addition, the Faraday detectors or 'cups' must also be positioned to detect the isotopes of interest, with cup positions collectively termed a 'cup configuration'. Lastly, a specific measurement mass on the peaks must be chosen (see Box 3.14). Similar to the metal concentration measurements, this set-up requires extensive training of the analysts.

Once peaks are aligned and measurement position has been chosen, each cup reports the raw voltage of a single isotope during analysis. The raw voltages must then be corrected for both instrumental background and isobaric interferences before isotope ratios are calculated simply as the ratio of corrected voltages. Instrumental background on the Neptune MC-ICP-MS is typically corrected for in two ways: first, the gain and baseline function are used to correct instrumental noise on the detectors (prior to an analytical session), and second the instrumental background or 'blank' is corrected by subtracting the voltage from each detector in a solution of the acid used for sample dissolution, prior to sample measurement (or sometimes the average of two blank analyses before and after sample). A careful sequence of rinsing between samples is also necessary to ensure that the introduction system returns to background values. As with concentration measurements by ICP-MS (see Sect. 3.5.1.1), isobaric interferences are corrected for mathematically.

Background and isobaric interference corrected isotope ratios must then be further corrected for the inaccuracy caused by instrumental mass bias, which is the effect of preferentially 'losing' lighter isotopes during ICP-MS analysis leading to ratios biased towards heavy isotopes (see Johnson et al. 2020 for a recent discussion). Although the causes of mass bias are not completely understood, instrumental mass bias is systematic and can be corrected for using one of two empirically derived mass bias equations (e.g. Rehkämperab and Halliday 1998). The size of instrumental mass bias in MC-ICP-MS is also typically specific to each instrument and introduction system and predictably increases with mass difference between isotope ratios, but decreases with atomic mass (Johnson et al. 2020). For measuring transition metal isotope ratios via MC-ICP-MS or TIMS, instrumental mass bias is typically corrected in one of three ways, either by sample-standard bracketing, by doping with a second element, or by 'double-spiking' (see Bermin et al. 2006 for extended discussion of the choice for Cu and Zn in seawater). In the first case, an isotope standard reference solution is analysed immediately before and after each sample in an analytical session, and the sample isotope ratio is expressed relative to these two 'bracketing' standards (e.g. for Fe; John and Adkins 2010). Sample-standard bracketing is used routinely for elements where a double spike is not possible (e.g. Cu), but may struggle to account for very rapid changes in mass bias from sample to sample. While care is taken to matrix-match samples to standards for MC-ICP-MS, any change in the sample matrix (e.g. presence of organics) may also induce mass bias effects which are not accounted for in the standards. Doping samples with a second element to correct for instrumental mass bias has also been used for transition metal isotope analysis (e.g. Zn for Cu; Maréchal et al. 1999), often combined with sample-standard bracketing for seawater samples (e.g. Takano et al. 2013; Yang et al. 2020).

Box 3.14: Peak Alignment for Measurement of Trace Metal Isotope Ratios by MC-ICP-MS

The following text gives a brief peak alignment example for isotope ratio measurements on the Thermo Neptune MC-ICP-MS (Fig. 3.23). For the low-resolution mode (e.g. for Zn and Cd), the cups (denoted by different colours in Fig. 3.23) are aligned so that the broad flat-topped peaks line up and the isotope voltages are simultaneously measured at the centre of each peak (red dashed line; left panel; Fig. 3.23). Interference peaks (in this case Sn and In on Cd) are also measured to facilitate correction of isobaric interferences. For the high-resolution mode (e.g. Fe), the cups are aligned so that the sharp, flat-topped compound peaks are lined up along the left side of the peak (right panel; Fig. 3.23). A measurement mass on the left hand 'shoulder 'of the peak, which is free from polyatomic interferences, is chosen (red dashed line; Fig. 3.23). Note the much smaller mass range in high-resolution mode in the figure below. For more details on high-resolution measurements, see Weyer and Schwieters (2003).

The double spike technique (Dodson 1963; Russell et al. 1978) involves the addition of an unnatural spike of two isotopes of the element being measured (e.g. ^{57}Fe and ^{58}Fe; Johnson and Beard 1999; Lacan, et al. 2008) to the natural sample, prior to chemical processing and analysis. Any fractionation of isotope ratios in the natural sample during processing or analysis will similarly affect the spike isotopes, and so mass bias can be corrected for mathematically using a series of equations in a three-dimensional data reduction scheme (see Siebert et al. 2001 for more details). Addition of a double spike also allows for the calculation of precise concentration data by isotope dilution (see Box 3.15), provided that the spike amount and the sample weight are known. A double spike technique requires three isotope ratios (e.g. 56/54, 57/54 and 58/54 for Fe; Dodson 1963), and so while it can be used for Ni, Fe, Zn and Cd, it is not suitable for Cu that only has two naturally occurring isotopes. Use of a double spike is advantageous because it accounts for matrix effects or rapid mass bias changes that can cause problems for sample-standard bracketing, as well as accounts for any fractionation of isotope ratios during chemical processing. Use of a double spike is not without challenges, however, as the spike must be made (by purchasing high-purity single spikes as metal ingots or compounds, dissolving and mixing them) and calibrated (see Box 3.15). If a multi-element technique is being used, multiple double spikes must be 'cleaned' to prevent contamination from the spike solution biasing other metal isotope ratios (e.g. Sun et al. 2021). Typically, such cleaning methods will involve scaled-up versions of the metal purification approach used for seawater samples, but may require adjustments (e.g. Sieber et al., 2019).

Different laboratory groups use slightly different approaches for analytical protocols, but a typical analytical session at USF would begin by analysing a series of mixtures of the 'natural' zero standard and the calibrated double spike at a range of concentrations and ratios to check that the same delta value (ideally $0 \pm 0.1‰$) is obtained for the zero standard across the range of intended sample compositions. For example, for Fe, these solutions would be made from IRMM-014 and the double spike and typically take the form of 5:1, 2:1, 1:1, 1:2 and 1:5 (natural/spike ratio) and 5:10, 10:20, 25:50, 50:100 and 100:200 (natural ng/g: spike ng/g) concentrations. After this, groups of 5–6 seawater samples are analysed, with an IRMM-014:double spike mixture measured before and after each group. Sample $\delta^{56}Fe$ values are then calculated using the double spike iterative technique (Siebert et al. 2001) and subsequently expressed relative to the average (or 'zero') of the $\delta^{56}Fe$ of the two mixtures.

Box 3.15: Double Spike Calibration

Once the double spike is purified, the three isotope ratios required for double spike calculations must be established in both the natural zero standard and the spike. The isotope ratios of the spike and natural metal must then be entered into the double spike calculation scheme, which can then be used to correct for instrumental mass bias in samples. Establishment or 'calibration' of the isotope ratios in the spike and natural metal (usually in the zero standard) is not trivial, however. The calibration process typically requires measurement of both natural and spike solutions and then mathematical optimisation of the spike composition by ensuring a set of natural-double spike mixture solutions generate a delta of 0‰ in the double spike calculation scheme. The spike calibration process can either be carried out once, after preparation of the spike, or more frequently if required. For use in concentration measurements in samples via isotope dilution, the concentration of the spike must also be first established by inverse isotope dilution with a concentration RM.

Uncertainty (Precision and Accuracy) on Trace Metal Isotope Ratios

Uncertainty on isotope measurement by MC-ICP-MS is typically expressed as a combination of 'accuracy' and 'reproducibility' (precision). Accuracy is influenced by mass bias and interference corrections, with insufficiently corrected isobaric or polyatomic interferences leading to significant inaccuracy (e.g. several ‰) on isotope ratios. Precision can be expressed as the internal statistics of a single MC-ICP-MS measurement, where uncertainty arises principally from a combination of detector noise, ion counting statistics and plasma flicker within the MC-ICP-MS (e.g. John and Adkins 2010), or external precision that also includes any effects during sampling, chemical extraction and interference correction, as well as day-to-day run variability. A large and/or variable procedural blank to signal ratio can drive inaccuracy and reduce precision. Different groups express accuracy and precision on isotope ratios using different approaches, which are summarised here. In all cases, however, it is crucial to robustly establish both types of uncertainty in order to assess data quality, before any data variability can be interpreted.

Accuracy of $\delta^{56}Fe$, $\delta^{60}Ni$, $\delta^{63}Cu$, $\delta^{66}Zn$ and $\delta^{114}Cd$ in seawater can be assessed in several ways: (1) measurement of the isotope ratio of a standard reference material (e.g. Abouchami et al. 2013); (2) 'doping' a standard reference material into low-metal seawater and then comparing obtained values to literature values (e.g. Bermin et al. 2006; Conway et al. 2013); or (3) comparing isotope measurements obtained by different groups in reference seawater (e.g. SAFe) or at GEOTRACES crossover stations (e.g. Boyle et al. 2012; Conway et al., 2016). The challenges of collection of sufficient volume for multiple laboratories have restricted intercomparison, but efforts have taken place in the Southern Ocean, North and South Atlantic and Southwest Pacific (Boyle et al. 2012; Xue, et al. 2013; Xie, et al. 2019; Ellwood et al. 2020), and multiple groups have now published Cd and Zn isotope ratios for the SAFe standards (e.g. Cd is summarised in Sieber, et al. 2019). These exercises generally show good agreement between groups, but highlight a need for more intercomparison, especially for low-concentration surface samples for elements like Cd (Janssen et al. 2019).

To establish reproducibility on isotope measurements (also referred to as analytical precision), there are several approaches which can be taken. The simplest is to look at the internal statistics of a single ICP-MS analysis (the 2x standard error; 2SE), which depends largely on the concentration for sample limited analyses—for example, for a set of North Atlantic seawater samples, 2SE internal error varied from ~0.01‰ for samples with high concentrations of Fe (>1 nM in 1 L) to 0.1‰ for low-concentration surface samples (~0.1 nM in 1 L; Conway et al. 2013). However, this estimation of precision does not consider within run or run-to-run uncertainty, or uncertainties associated with sample processing. A fuller assessment of precision (external precision) is to calculate the 2× standard deviation (2SD) of the isotope ratio of a sample by measuring multiple aliquots of samples that have been through complete chemical processing and ideally measured over multiple analytical sessions (e.g. Bermin et al. 2006; Xue et al. 2012; Ellwood et al. 2020). For Fe, external precision can be on the order of 0.04 to 0.07‰ (Conway et al., 2016 Ellwood et al. 2020). For seawater, however, this approach is usually prohibited by the volume requirements of large numbers of replicate samples (ideally 30+ samples). Instead, assessment of external precision on isotope ratios for seawater samples commonly uses the 2SD of the isotope ratio of replicate measurements of an isotope standard RM over multiple analytical periods. Such external precision is typically greater than internal precision and on the order of 0.03–0.09‰ depending on element and laboratory. For example, while the internal 2SE of a $\delta^{56}Fe$ measurement at USF by Neptune MC-ICP-MS can be as low as 0.01 or 0.02‰, the 2SD of measurement of the NIST 3126 Fe reference material is 0.04‰, equivalent to the 0.04‰ established from replicate analyses of seawater at USF (Sieber et al. 2021). As such, such estimates of external precision should typically be quoted as the best estimate of uncertainty on a seawater measurement, unless the internal error is larger (and then the internal error should be used; e.g. Sieber et al. 2021).

3.5.3 Trace Metal Speciation Measurement Techniques

Biological effects and the geochemical behaviour (solubility, reactivity, residence times) of trace metals in the ocean are highly dependent on the physical and chemical speciation of the metal of interest. Thus, measurements of dissolved metal concentrations alone may not yield sufficient information for understanding the fate of trace metals in the ocean (Gledhill and Buck 2012; Tessier and Turner 1995; Vraspir and Butler 2009). For marine chemists, the redox speciation and organic complexation (including the free ion concentration) of bioactive metals in the dissolved fraction are of particular interest to better understand the toxicity, bioavailability and geochemical behaviour of trace metals in natural waters (Achterberg et al. 2018; Hirose 2006). Chemical speciation refers to the specific chemical form of an element, while speciation analysis is an analytical process for identifying and/or measuring the quantities of a chemical species of interest. Though redox speciation plays an important role in the cycling and fate of dissolved marine trace metals, redox speciation is beyond the scope of this chapter, but some redox species can be quantified in seawater samples (e.g. Boye et al. 2006; Bruland et al. 2014; Gledhill and van den Berg 1995; Han and Pan 2021; Oldham et al. 2021; Padan et al. 2019; Sander et al. 2009; Stumm and Morgan 1996). Here, we focus solely on the chemical speciation, i.e. complexation, including free metal ion concentration (M^+, with M being the metal of interest) and the organic complexation of trace metals (ML, with L being the organic ligand) in the marine environment (see Fig. 3.1). Inorganic metal complexation (MY, with Y referring to, e.g. OH^-, CO_2^{-3}, Cl^-, etc., Fig. 3.1) also exists but is not discussed in detail here.

The speciation of many dissolved metals, namely, Fe but also Cu, Co, Mn, Zn and Ni, in marine systems is controlled by complexation processes with organic binding ligands, consisting of organic low molecular weight compounds up to large macromolecules (Donat et al. 1994; Ellwood and Van den Berg 2000; Gledhill and Buck 2012; Saito and Moffett 2001; Vraspir and Butler 2009; Whitby et al. 2018). The organic ligands can form stable complexes with various metals, keeping them in solution – thus both possibly inhibiting their biological uptake and/or reducing their adsorption to particles and thereby removal from the water column (Gledhill and Buck 2012; Vraspir and Butler 2009). For Fe, however, ligands can be a blessing, facilitating biological Fe uptake in Fe-limited environments such as the HNLC regions (Hassler et al. 2011b). Examples of organic metal-binding ligands are humic substances derived from terrestrial humus, as well as thiols, siderophores and exo-polymeric substances (e.g. Hassler et al. 2011a; b; Hirose 2006; Velasquez et al. 2016; Velasquez et al. 2011; Vraspir and Butler 2009). Most organic ligands are, however, yet to be identified and their contributions quantified. Similarly, the processes, sources, dynamics and driving factors of organic metal-binding ligand complexation in the ocean are still poorly understood and are thus an area of ongoing research (e.g. Buck et al. 2016; Campos and van den Berg 1994; Hartland et al. 2019; Hirose 2006; Kleint et al. 2016; Laglera et al. 2011; Laglera et al. 2020; Sander et al. 2007; Whitby et al. 2020; Whitby et al. 2018).

The quantitative characterization of metal complexation with organic ligands is currently largely carried out via the estimation of ambient organic ligand concentrations ([L]) and conditional stability constants ($\log K_{ML}^{cond}$, with M being the metal of interest) of the metal-ligand complex via electrochemical techniques, i.e. voltammetry (Henze 1990; Hirose 2006; Vraspir and Butler 2009). Conditional stability constants reflect the strength of a complex between a metal ion and an organic ligand at a specific temperature, pressure, pH and ionic strength. These conditional stability constants are complex specific and are thus an important parameter to indirectly identify the character of organic ligands and their ecological roles in the marine environment (Hirose 2006; Vraspir and Butler 2009) (see Box 3.16 for more information on the characterization of ligands). For instance, $\log K_{ML}^{cond}$ are commonly divided into various classes; in the case of Cu-binding ligands, two ligand classes are usually distinguished, i.e. L_1 referring to strong ligands ($\log K1_{CuL,Cu2+}^{cond} > 12$ and up to 16) and L_2 referring to weak ligands ($\log K2_{CuL,Cu2+}^{cond} < 12$) (Croot and Johansson 2000; Vraspir and Butler 2009, Bundy et al., 2013; Muller and Batchelli, 2013, Whitby et al. 2018). However, ligand classes with considerably lower $\log K2_{CuL,Cu2+}^{cond}$ have been quantified in some studies of coastal and estuarine settings (e.g. Heller and Croot 2015; Sander et al. 2015). Information about the nature of naturally observed ligand classes can be obtained by comparison to measured $\log K_{ML}^{cond}$ of known organic molecules studied in the laboratory. Such a comparative approach can give a preliminary indication of the prevalence of different marine organic ligands, for example, actively and/or passively produced by phytoplankton and/or originating from humic substances. The following sections will focus on the description of the quantitative and semi-qualitative voltammetric techniques which are widely used for trace metal speciation analysis in the scientific community.

Box 3.16: Molecular Characterization of Metal-Binding Organic Ligands

Molecular characterization and quantification of naturally occurring marine organic ligands in the ocean, while largely an emerging discipline, can be done via excitation-emission matrix (EEM) fluorescence combined with parallel factor analysis (PARAFAC) (Yamashita et al. 2011), nuclear magnetic resonance (NMR) (Hertkorn et al. 2006; Rehman et al. 2017), high-performance liquid chromatography mass spectrometry (HPLC-MS) (Boiteau et al. 2016; Mawji et al. 2008; Velasquez et al. 2016) and Fourier transform ion cyclotron resonance mass spectrometry (FT-ICR-MS) (Pohlabeln and Dittmar 2015). Studies making use of such techniques have provided new insights, for example, into the distribution and form of certain organic binding molecules for Fe, Ni and Cu in the Pacific Ocean (e.g. Boiteau and Repeta 2015; Boiteau et al. 2016; Bundy et al., 2018; Boiteau et al., 2019), or to investigate siderophores associated with nitrogen fixers (Gledhill et al., 2019). However,

(continued)

Box 3.16 (continued)
most of the methods for characterising and quantifying marine organic-binding compounds are still in development or refinement or limited to a small number of laboratories and yet to be widely applied. As such, we do not provide a detailed discussion here but point readers to the references above.

3.5.3.1 Voltammetric Techniques

Voltammetry can be used for measuring total dissolved concentrations of various trace metals, but the efficiency of quantifying metal concentrations using voltammetric methods lags behind ICP-MS techniques and is thus no longer extensively used. However, voltammetric methods are nowadays still used to quantify the speciation of various metals in marine waters both onboard ship and in the laboratory (Campos and van den Berg 1994; Croot et al. 1999; Gerringa et al. 2015; Gledhill and van den Berg 1995; Plavšić et al. 2009; Rue and Bruland 1995; Ružić 1982; Saito and Moffett 2001; Van den Berg 1982; Van Den Berg 1986). While other methods for metal speciation determination in aquatic environments such as ion-selective electrodes (ISE), diffusive gradients in thin films (DGT; see Sect. 3.3.1.2) and ion-exchange techniques (IET) exist (Florence, 1986; Gerringa et al. 1991; Achterberg and Braungardt, 1999), voltammetry is the most used method for metal speciation analysis due to its selectivity, its suitability for ultra-low-concentration levels, particularly in saline waters (detection limit of 10^{-9} to 10^{-12} mol L^{-1}) (Campos and van den Berg 1994; Croot et al. 1999; Gledhill and van den Berg 1995; Rue and Bruland 1995; Ružić 1982; Van den Berg 1982), and the fact that this method needs no sample treatment such as salt removal. Voltammetric methods also have the advantage that they permit the discrimination between the free ionic form of a metal and a metal ion complexed with an organic ligand, including calculation of [L] and $\log K_{ML}^{cond}$ of the prevalent metal ligand complexes (Gerringa et al. 2014; Hudson et al. 2003; Omanović et al. 2010; Omanović et al. 2015; Pižeta et al. 2015; Rue and Bruland 1995; Ružić 1982). These parameters can then be used to evaluate the reactivity (e.g. bioavailability, toxicity, geochemical behaviour) of a trace metal of interest and thus lead to interpretations of its fate in the water column.

The free and labile metal ion is usually defined as the ionic form of a metal, e.g. Cu^{2+} for Cu and Fe^{3+} and Fe^{2+} for Fe, but can also include all forms of the metal which are reactive or labile and can be detected by a particular analytical method (Henze 1990). Thus, in practice, these forms often also include metals complexed by inorganic ligands (Bruland et al. 2014). For voltammetry, the reactive species of a metal is specifically the fraction that is readily reactive at a working electrode (Henze 1990) usually a mercury-based electrode. The mercury (Hg) electrode was developed and defined by Jaroslav Heyrovsky in the early 1920s (Zuman 2001) and further refined by Matson et al. (1965). However, due to Hg being a chemical of major health concern, Hg electrodes are currently being progressively replaced with other materials such as carbon or noble metals (gold, silver, platinum) which are

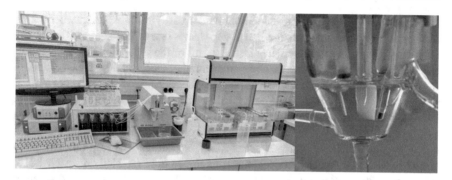

Fig. 3.24 Photograph of an automated voltammetry set-up (left) and a three-electrode voltammetric cell (right) (photograph credits: Dario Omanović, Rudjer Boskovic Institute, Croatia)

often modified with films, nanostructured materials or reagents, to improve the selectivity and/or sensitivity of the analysis (Borrill et al. 2019; Worsfold et al. 2019c). Voltammetry is a method based on current(redox)-potential-response measurements in a three-electrode system—the working electrode, the reference electrode and the counter electrode, which is also known as the auxiliary electrode (Han and Pan 2021; Henze 1990). The registered analytical signal is the change in the current-potential behaviour during a redox reaction of a metal of interest at the stationary working electrode, which itself is immersed into an electro-active solution, e.g. a saltwater sample (Henze 1990). The reference electrode is used as a reference for measuring and controlling the potential of the working electrode (Henze 1990). The counter electrode serves as a source or sink of electrons to balance the current observed at the working electrode (Henze 1990).

The electron transfer between redox species at the working electrode and the counter electrode generates a current which provides (1) quantitative information on the analyte of interest, that is, the signal is proportional to the metal concentration, and (2) qualitative information of the analyte since the potential where the signal is detected is analyte specific (Achterberg et al. 2018; Borrill et al. 2019; Han and Pan 2021; Henze 1990). The latter allows voltammetry to be used for mono-elemental or multi-elemental determination. Many laboratories use a three-electrode configuration that is composed of a hanging mercury drop electrode (HMDE) as the working electrode, an Ag|AgCl|3 M KCl reference electrode and a glassy carbon or platinum counter electrode. In general, the basic instrumentation for voltammetric measurements consists of a potentiostat (device that controls the potential between the electrodes while measuring the resulting current flow, i.e. the signal); a three-electrode cell, as described above; and a computer for automated measurements and data acquisition (Fig. 3.24) (e.g. Achterberg et al. 2018; Borrill et al. 2019; Han and Pan 2021; Henze 1990). The obtained signal is used to estimate [L] and $\log K_{ML}^{cond}$, which together with the metal concentration can be used to calculate the free metal ion concentration of the analyte of interest.

Stripping voltammetry (SV) is a subdivision of voltammetry and constitutes the most widely used voltammetric technique in electroanalytical chemistry since it is the most sensitive electrochemical technique currently available (Achterberg et al. 2018; Borrill et al. 2019; Han and Pan 2021; Henze 1990). The high sensitivity and selectivity of stripping voltammetry are the result of the separation of the analytical technique into two steps, that is, a pre-concentration step and a so-called stripping step. The pre-concentration step consists of an electrochemical deposition of a metal species onto or into the working electrode at a constant potential, necessary to isolate the metal of interest from the matrix. This first step can involve either an anodic or a cathodic potential. This step is followed by the second 'stripping' step during which the analyte of interest is 'stripped' back into the solution. The resulting current is proportional to the analyte concentration in the sample (Achterberg et al. 2018; Borrill et al. 2019; Han and Pan 2021; Henze 1990). Depending on the reduction or oxidation of analytes during the potential sweep, SV can be classed as either anodic SV (ASV; oxidation with reductive pre-concentration step) or cathodic SV (CSV; reduction with oxidative pre-concentration step) (see Box 3.17). Currently, competitive ligand exchange-adsorptive cathodic striping voltammetry (CLE-AdCSV) is the most widely used voltammetric method for metal speciation analysis, especially in seawater matrices (Han and Pan 2021). The principle of this method is the addition of a well-characterized artificial ligand (AL), establishing a competitive equilibrium between the AL and the natural ligands (L) in a seawater sample (Achterberg et al. 2018; Borrill et al. 2019; Han and Pan 2021; Henze 1990; Rue and Bruland 1995). The AL forms an electrochemically active complex with the metal of interest, and after titration with increasing concentrations of M, the natural [L] and $\log K_{ML}^{cond}$ can be determined. The main advantage of the technique is its greater sensitivity compared to conventional ASV and CSV and that the $[M^+]$ can be easily determined based on the M-AL concentration (Borrill et al. 2019; Han and Pan 2021; Henze 1990; Ružić 1982; Van den Berg 1982). Various synthetic ligands are available. For Fe, salicylaldoxime (SA; Abualhaija and van den Berg 2014; Campos and van den Berg 1994; Rue and Bruland 1995), 1-nitroso-2-naphthol (NN; Gledhill and van den Berg 1994), thiazolylazo-p-cresol (TAC; Croot and Johansson 2000) and dihydroxynaphthalene (DHN; Laglera et al. 2011; van den Berg 2006) are commonly used ligands (Han and Pan 2021). Depending on the ligand groups of interest, with respect to their $\log K_{ML}^{cond}$, the analyst can choose the appropriate detection window (D), with D being defined as the product of the concentration of AL and the conditional stability constant of M-AL (e.g. Apte et al. 1988; Laglera and Filella 2015).

The following section describes in brief the analytical aspects and procedures of CLE-AdCSV used by various marine laboratories for the analysis of Cu and Fe speciation. For a more detailed description of the analytical steps and procedures, the reader is referred to available literature belonging to the specific application including Rue and Bruland (1995) and Campos and van den Berg (1994).

Box 3.17: Metal Determination: ASV versus CSV

Both CSV and ASV are similar but differ in the nature and direction of the pre-concentration and stripping steps, as well as in the metals that can be determined (Han and Pan 2021; Worsfold et al. 2019c). ASV is employed to detect metal species that can be reduced, accumulated on and then reoxidized from the surface of the working electrode under appropriate potentials (Han and Pan 2021). Metals that can be determined by ASV are Pb, Cd and Zn in natural waters. ASV-labile metal fractions include free metal ions, inorganically bound metals and weakly organically bound metals (Achterberg et al. 2018; Borrill et al. 2019; Han and Pan 2021; Henze 1990; Van Den Berg 1986), and thus the labile fraction obtained by ASV generally corresponds well with the concentration of bioavailable metals (Han and Pan 2021). CSV is widely used for the determination and speciation analysis of more than 30 elements that cannot be reduced on electrode surface and determined by ASV, such as Fe, Co and Ni (Han and Pan 2021).

3.5.3.2 Voltammetric Analysis of Metal Complexation by Ligand Titration Using CLE-AdCSV

Currently, the quantification of metal speciation ([L] and $\log K_{ML}^{cond}$) in discrete samples is widely undertaken using a titration approach with an artificial ligand (AL), i.e. CLE-AdCSV. The CLE-AdCSV method can be implemented for fresh or thawed unacidified seawater samples. Briefly, a seawater sample is divided into typically 10–15 sub-samples, which should be pipetted into trace metal clean and previously conditioned voltammetric cups. Preconditioning of voltammetric cups with a matrix matched seawater sample, including metal additions and buffer, is necessary to avoid adsorption of the metal of interest on the cup wall prior to the measurements. First, the samples are generally buffered to a method specific pH using a particular buffer. For Fe and Cu, the pH has to be buffered to ambient seawater pH (~8.04) using, e.g. a borate buffer or a HEPES buffer ((4-(2-hydroxyethyl)-1-piperazineethanesulfonic acid; a zwitterionic sulfonic acid buffering agent), to ensure efficient electrodeposition at the Hg drop (Borrill et al. 2019). Lately, however, some methods have been introduced that quantify metal speciation at natural pH without the need for buffer addition (Sanvito and Monticelli 2021). This step is followed by the addition of the AL of choice. Note that the added AL must be in great excess (factor 1000 approximately) of the metal concentration of interest to obtain reliable results. Typically, metal concentrations are quantified in sub-samples using ICP-MS prior to the voltametric measurements. Before the AL addition, increasing amounts of the metal of interest are added to the vials. After the addition of AL, buffer and metal, an equilibration period (often overnight owing to slow kinetics) is applied during which the naturally present ligands compete (i.e. equilibration) with the synthetic ligand for the metal of interest in a controlled manner. Once equilibrated, the metal speciation in the samples can be determined via CSV (see Box 3.18 and 3.19).

Box 3.18: Forward and Reverse Titration

It should be noted that the metal titration approach described here, also termed forward titration, is suitable only for conditions where the natural organic ligand concentrations exceed the dissolved metal concentrations in the sample, which is the case for most oceanic conditions (Achterberg et al. 2018). However, in estuarine, coastal and hydrothermal waters that receive significant metal inputs, metal concentrations can exceed the natural ligand concentrations and then a reverse titration rather than a forward titration has to be applied, by increasing the concentration of the added ligand rather than the metal concentration (Hawkes et al. 2013).

Prior to analysis, the sample may be de-aerated with an inert gas (N_2 or argon) for at least 2 min to remove reactive dissolved oxygen from solution in order to reduce interferences of the reactive oxygen in the voltammogram and to improve the sensitivity of the voltammetric determination (e.g. Campos and van den Berg 1994; Henze 1990; Ružić 1982; Van den Berg 1982). No purge (e.g. Buck et al. 2007) or compressed air is also used by some laboratories for some CLE-AdCSV methods including for Fe, e.g. with SA (Abualhaija and van den Berg 2014) or DHN (Sanvito and Monticelli 2021). No purge or compressed air is used in order to equilibrate the solution with the above standing atmospheric pressure and/or avoid carbon dioxide removal and thus changes in sample pH (Sanvito and Monticelli 2021). A deposition potential is then applied to the working electrode to cause the analyte of interest to be adsorbed on the Hg drop. The solution is commonly stirred during deposition to maximize electrode-analyte contact and to decrease the diffusion layer of the Hg drop. The selection of the deposition potential depends upon the analyte of interest and the redox process (reduction or oxidation). Deposition times can be increased to acquire more sensitivity (distinctive peaks) or decreased to avoid electrode saturation issues of high metal samples. Subsequently, the sample is scanned in a certain potential range (cathodic or anodic) to strip the analyte of interest from the electrode while measuring the resulting current. For quality assurance, each measurement should be repeated at least two times, each with a freshly generated Hg drop. The pH of each sample should be monitored in sample aliquots (the pH electrode must not be immersed into the sample as it may introduce contamination), preferably before the titration, to ensure reliable results—the efficiency of the electrodeposition at the Hg drop is pH dependent. Once the titrations are completed, data transformation allows the calculation of [L] and $\log K_{ML}^{cond}$, in addition to the free aqueous metal concentration (Omanović et al. 2015; Pižeta et al. 2015; Sander et al. 2011; Wells et al. 2013). Data transformation can be facilitated using software packages such as proMCC developed by Omanović et al. (2015) or the R-based software package developed by Gerringa et al. (2014). Both system operation and data fitting require extensive training of the analyst.

Box 3.19: Limitations of Voltammetric Methods

While voltammetry remains an indispensable tool for evaluating metal speciation, this method suffers from several limitations that can lead to significant misrepresentation of metal complexation parameters in seawater samples. Some of the major issues and uncertainties of the electrochemical method are listed below.

1. The voltammetric technique works under the assumption that natural organic ligands exist in well-defined classes, which can be operationally distinguished and measured independently from each other (Town and Filella 2000). However, in complex and heterogeneous natural saline waters, it is much more likely that a 'ligand soup' exists without specific partitioning into different ligand classes (Town and Filella 2000).
2. It is largely assumed that metal-ligand complexation processes occupy a 1: 1 metal-ligand stoichiometry (Omanović et al. 2010), even though ligands with multiple binding sites could be expected and might be more realistic (Tipping, 1998).
3. Organic compounds, sulphides in hydrothermal samples and multiple prevalent metals may shift or distort the stripping peaks for the analyte of interest. These problems can often be minimized by adjusting the deposition time, changing the deposition potential or adding a pre-treatment step. For instance, sulphide-rich hydrothermal samples can undergo a pretreatment to remove acid volatile sulphides (AVS) from the sample solutions, which improves the electrochemical measurement of the analyte of interest (Sander et al. 2007; Kleint et al. 2016; Cotte et al., 2018).

3.5.3.3 Data Quality Control for Trace Metal Speciation Measurements

At present, no reference material exists for the *speciation* of trace metals. Thus, method validation is currently limited to measuring the repeatability and reproducibility of the procedure, i.e. by measuring some samples in duplicate. In addition, it is recommended that the accuracy of the voltammetric method be evaluated by measuring [L] of UV-oxidized seawater samples spiked with a known amount of artificial ligand or humic acid using the procedure described above (e.g. Gerringa et al. 2021a, b; Whitby et al. 2020, 2018).

3.6   Considerations of Data Quality, Inter-Comparability and Accessibility

Data produced by the marine trace metal community is critical to improve our current understanding of the biogeochemistry of the ocean and its variability, to force models of ocean biogeochemistry and thus enhance predictions for future ocean changes and to help design effective data-driven management and mitigation strategies (Worsfold et al. 2019). Further, biogeochemical data, including time series data, are becoming more and more critical in informing climate modelers and climate-relevant decisions of governments and the civic society, and thus, there is an urgent need for good practice and good quality data using the FAIR principles (i.e. data has to be findable, accessible, interoperable and reusable) (Tanhua et al. 2019; Worsfold et al. 2019). Now, national funding agencies and journals often require that data is publicly deposited in online repositories, with attached metadata and data quality assessments. Data quality assurance (accuracy, reproducibility; see Sects. 3.5.1.4 and 3.5.2.3) is particularly important for trace metal research as many commonly used methods and analytical techniques operate at or close to their limit of detection, especially in open ocean waters (Worsfold et al. 2019). Since many techniques are designed or customised by individual laboratories, metadata (information on, e.g. used methods and procedures and availability of ancillary data) is also essential.

In order to achieve the best possible trace metal data, GEOTRACES recommends intercalibration exercises including designated 'crossover' stations between expeditions, the analysis of RM alongside samples and the practice of reporting overall uncertainty as well as the internal instrumental precision (Cutter 2013; Worsfold et al. 2019). Crossover stations are when multiple different expeditions occupy the same geographic location and sample the water column, albeit at different times. Such crossovers are essential for combining multiple expedition datasets into usable global GEOTRACES products. Crossover stations allow inter-comparison of the fidelity of sampling systems and/or the analytical techniques used to measure metals of interest by different laboratories/groups on different expeditions (e.g. Middag et al. 2015a, b). The fidelity of analytical techniques can also be assessed by different laboratories on aliquots of the same sample using different techniques (e.g. Middag et al. 2015a, b; Jensen et al. 2020). Such intercomparison exercises, especially when using different sampling systems, test not only inter-laboratory precision but also any uncertainty resulting from sample collection and handling. For example, Jensen et al. (2019) compared Zn from full water column profiles in the Arctic Ocean analysed by three different techniques by three laboratories (two SeaFAST ICP-MS and one FIA) and found a RSD of 6% for the average of the three methods, similar to the stated internal precision of those individual techniques. Similarly, Middag et al. (2015a, b) found that for the full water column of the North Atlantic, two laboratories using ICP-MS largely statistically agreed within stated analytical precision, except for the surface samples which was attributed to temporal variation between station occupations. Strong agreement has also been found between laboratories in trace metal isotope intercomparison exercises (e.g. Boyle et al. 2012; Conway et al. 2013), but the field is less mature than that of trace metal concentrations due to the smaller number of laboratories that routinely make these measurements, and the degree of agreement still depends on the element being studied. Voltammetric techniques are also seldom intercalibrated owing to the time-consuming analysis technique and the small number of laboratories routinely measuring metal speciation. However, published intercomparison studies showed that consensus values could be largely obtained between different laboratories and using different voltametric methods, but that the number of identified ligand classes present and the groups of ligands resolved in a target sample need more attention in future work (e.g. Bruland et al. 2000; Buck et al. 2016, 2012; Pižeta et al. 2015).

Going forward, continuing to include information on laboratory precision and accuracy, combined with metadata and intercomparison between different laboratories and sampling systems, will enhance the ability of the trace metal community to maintain consistent data quality and to compare datasets obtained by different investigators using different instruments, analytical protocols and/or techniques (Worsfold et al. 2019). In fact, GEOTRACES requires evidence of data quality and intercomparison to be assessed by a committee before data can be included in a publicly available GEOTRACES data product, to ensure synthesized global datasets (Schlitzer et al. 2018). Such close attention to quality control, combined with free public accessibility, means that GEOTRACES datasets can serve as a reliable baseline to assess future change. GEOTRACES data and derived insights will likely be an invaluable tool to develop diagnostic and predictive models on the biogeochemical cycles that ultimately drive life on our planet that is increasingly perturbed by anthropogenic influences.

Acknowledgements We thank Amber Annett, Hajime Obata, Patrick Laan, Gert van Dijken, Dario Omanović, Jessica Fitzsimmons, Amir Mohammadi, Adam Hartland, Antonio Tovar Sánchez, Phoebe Lam, Mathijs van Manen and Nils Strackbein (MARUM QUEST Team) for providing us with pictures and figures for the chapter. We are also very grateful to Hein de Baar, Sylvia Sander, Loes Gerringa, Ken Bruland, Dieter Garbe-Schönberg, Matthias Sieber and Pier van der Merwe for proofreading sections of the manuscript and/or providing useful insights.

References

Abadie C et al (2017) Iron isotopes reveal distinct dissolved iron sources and pathways in the intermediate versus deep Southern Ocean. Proc Natl Acad Sci 114(5):858–863

Abouchami W et al (2011) Modulation of the Southern Ocean cadmium isotope signature by ocean circulation and primary productivity. Earth Planet Sci Lett 305(1–2):83–91

Abouchami W et al (2013) A common reference material for cadmium isotope studies–NIST SRM 3108. Geostand Geoanal Res 37(1):5–17

Abualhaija MM, van den Berg CMG (2014) Chemical speciation of iron in seawater using catalytic cathodic stripping voltammetry with ligand competition against salicylaldoxime. Mar Chem 164:60–74

Achterberg EP, Braungardt C (1999) Stripping voltammetry for the determination of trace metal speciation and in-situ measurements of trace metal distributions in marine waters. Anal Chim Acta 400(1–3):381–397

Achterberg EP et al (2013) Natural iron fertilization by the Eyjafjallajökull volcanic eruption. Geophys Res Lett 40(5):921–926

Achterberg EP, Gledhill M, Zhu K (2018) Voltammetry—cathodic stripping. Elsevier

Adams WJ, Chapman PM (2007) Assessing the hazard of metals and inorganic metal substances in aquatic and terrestrial systems. CRC Press

Allan IJ et al (2008) Chemcatcher® and DGT passive sampling devices for regulatory monitoring of trace metals in surface water. J Environ Monit 10(7):821–829

Anderson RF (2020) GEOTRACES: accelerating research on the marine biogeochemical cycles of trace elements and their isotopes. Annu Rev Mar Sci 12(1):49–85

Anderson RF et al (2014) GEOTRACES: changing the way we explore ocean chemistry. Oceanography 27(1):50–61

Aparicio-González A, Duarte CM, Tovar-Sánchez A (2012) Trace metals in deep ocean waters: A review. J Mar Syst 100:26–33

Apte SC, Gardner MJ, Ravenscroft JE (1988) An evaluation of voltammetric titration procedures for the determination of trace metal complexation in natural waters by use of computers simulation. Anal Chim Acta 212:1–21

Apte SC, Batley G, and Maher WA (2002) Monitoring of trace metals and metalloids in natural waters. Environmental Monitoring Handbook

Archer C, Vance D (2004) Mass discrimination correction in multiple-collector plasma source mass spectrometry: an example using Cu and Zn isotopes. J Anal At Spectrom 19(5):656–665

Archer C, Andersen MB, Cloquet C, Conway TM, Dong S, Ellwood M et al (2017) Intercalibration of a proposed new primary reference standard AA-ETH Zn for zinc isotopic analysis. J Anal At Spectrom 32(2):415–419

Archer C et al (2020) The oceanic biogeochemistry of nickel and its isotopes: new data from the South Atlantic and the Southern Ocean biogeochemical divide. Earth Planet Sci Lett 535: 116118

Arrigo KR (2005) Marine microorganisms and global nutrient cycles. Nature 437(7057):349–355

Asadi S et al (2019) Aerosol emission and superemission during human speech increase with voice loudness. Sci Rep 9(1):1–10

Baconnais I et al (2019) Determination of the copper isotope composition of seawater revisited: a case study from the Mediterranean Sea. Chem Geol 511:465–480

Beard BL, Johnson CM (1999) High precision iron isotope measurements of terrestrial and lunar materials. Geochim Cosmochim Acta 63(11–12):1653–1660

Bell J, Betts J, Boyle E (2002) MITESS: a moored in situ trace element serial sampler for deep-sea moorings. Deep-Sea Res I Oceanogr Res Pap 49(11):2103–2118

Berger CJM et al (2008) Application of a chemical leach technique for estimating labile particulate aluminum, iron, and manganese in the Columbia River plume and coastal waters off Oregon and Washington. J Geophys Res Oceans 113:C2

Bergquist BA, Wu J, Boyle EA (2007) Variability in oceanic dissolved iron is dominated by the colloidal fraction. Geochim Cosmochim Acta 71(12):2960–2974

Bermin J et al (2006) The determination of the isotopic composition of Cu and Zn in seawater. Chem Geol 226(3–4):280–297

Biller DV, Bruland KW (2012) Analysis of Mn, Fe, Co, Ni, Cu, Zn, Cd, and Pb in seawater using the Nobias-chelate PA1 resin and magnetic sector inductively coupled plasma mass spectrometry (ICP-MS). Mar Chem 130–131:12–20

Bishop JKB, et al. (1985) A multiple-unit large-volume in situ filtration system for sampling oceanic particulate matter in mesoscale environments

Bishop JKB, Lam PJ, Wood TJ (2012) Getting good particles: accurate sampling of particles by large volume in-situ filtration. Limnol Oceanogr Methods 10(9):681–710

Boiteau RM, Repeta DJ (2015) An extended siderophore suite from Synechococcus sp. PCC 7002 revealed by LC-ICPMS-ESIMS. Metallomics 7(5):877–884

Boiteau RM et al (2016) Structural characterization of natural nickel and copper binding ligands along the US GEOTRACES Eastern Pacific Zonal Transect. Front Mar Sci 3:243

Boiteau RM, Till CP, Coale TH, Fitzsimmons JN, Bruland KW, Repeta DJ (2019) Patterns of iron and siderophore distributions across the California current system. Limnol Oceanogr 64(1):376–389

Borman CJ et al (2009) The use of flow-injection analysis with chemiluminescence detection of aqueous ferrous iron in waters containing high concentrations of organic compounds. Sensors 9(6):4390–4406

Borrill AJ, Reily NE, Macpherson JV (2019) Addressing the practicalities of anodic stripping voltammetry for heavy metal detection: a tutorial review. Analyst 144(23):6834–6849

Bowie AR, Lohan MC (2009) Determination of iron in seawater. In: Practical guidelines for the analysis of seawater. CRC Press, pp 247–270

Bowie AR et al (2002) Real-time monitoring of picomolar concentrations of iron(II) in marine waters using automated flow injection-chemiluminescence instrumentation. Environ Sci Technol 36(21):4600–4607

Bowie AR et al (2003) Shipboard analytical intercomparison of dissolved iron in surface waters along a north–south transect of the Atlantic Ocean. Mar Chem 84(1–2):19–34

Bowie AR, Sedwick PN, Worsfold PJ (2004) Analytical intercomparison between flow injection-chemiluminescence and flow injection-spectrophotometry for the determination of picomolar concentrations of iron in seawater. Limnol Oceanogr Methods 2(2):42–54

Boyd PW et al (2017) Biotic and abiotic retention, recycling and remineralization of metals in the ocean. Nat Geosci 10(3):167–173

Boye M et al (2006) The chemical speciation of iron in the north-East Atlantic Ocean. Deep-Sea Res I Oceanogr Res Pap 53(4):667–683

Boyle EA, Sclater F, Edmond JM (1976) On the marine geochemistry of cadmium. Nature 263(5572):42–44

Boyle EA et al (2012) GEOTRACES IC1 (BATS) contamination-prone trace element isotopes Cd, Fe, Pb, Zn, Cu, and Mo intercalibration. Limnol Oceanogr Methods 10(9):653–665

Brewer PG, Spencer DW, Robertson DE (1972) Trace element profiles from the GEOSECS-II test station in the Sargasso Sea. Earth Planet Sci Lett 16(1):111–116

Brewer PG (1975) Minor elements in sea water. In: Riley JP, Skirrow G (eds) Chemical oceanography. Academic Press, Waltham, pp 415–496

Bruland KW (1983) CHAPTER 45—trace elements in sea-water. In: Riley JP, Chester R (eds) Chemical oceanography. Academic Press, pp 157–220

Bruland KW, Franks RP (1983) Trace metals in seawater. Chem Oceanogr 8:157–220

Bruland KW, Lohan MC (2003) Controls of trace metals in seawater. Treatise Geochem 6:625

Bruland KW, Knauer GA, Martin JH (1978) Zinc in north-East Pacific water. Nature 271(5647):741–743

Bruland KW et al (1979) Sampling and analytical methods for the determination of copper, cadmium, zinc, and nickel at the nanogram per liter level in sea water. Anal Chim Acta 105:233–245

Bruland KW, Donat JR, Hutchins DA (1991) Interactive influences of bioactive trace metals on biological production in oceanic waters. Limnol Oceanogr 36(8):1555–1577

Bruland KW et al (2000) Intercomparison of voltammetric techniques to determine the chemical speciation of dissolved copper in a coastal seawater sample. Anal Chim Acta 405(1–2):99–113

Bruland KW et al (2003) Treatise on geochemistry. Oceans Mar Geochem 6:23–48

Bruland KW, Middag R, Lohan MC (2014) 8.2—controls of trace metals in seawater A2—Holland, Heinrich D. In: Turekian KK (ed) Treatise on geochemistry, 2nd edn. Elsevier, Oxford, pp 19–51

Buck KN et al (2007) Dissolved iron speciation in two distinct river plumes and an estuary: implications for riverine iron supply. Limnol Oceanogr 52(2):843–855

Buck KN et al (2012) The organic complexation of iron and copper: an intercomparison of competitive ligand exchange-adsorptive cathodic stripping voltammetry (CLE-ACSV) techniques. Limnol Oceanogr Methods 10(7):496–515

Buck KN, Gerringa LJA, Rijkenberg MJA (2016) An intercomparison of dissolved iron speciation at the Bermuda Atlantic Time-series Study (BATS) site: results from GEOTRACES crossover station a. Front Mar Sci 3:262

Buma AGJ et al (1991) Metal enrichment experiments in the Weddell-scotia seas: effects of iron and manganese on various plankton communities. Limnol Oceanogr 36(8):1865–1878

Bundy RM, Barbeau KA, Buck KN (2013) Sources of strong copper-binding ligands in Antarctic Peninsula surface waters. Deep-Sea Res II Top Stud Oceanogr 90:134–146

Bundy, R.M., Boiteau, R.M., McLean, C., Turk-Kubo, K.A., McIlvin, M.R., Saito, M.A., Van Mooy, B.A. and Repeta, D.J., 2018. Distinct siderophores contribute to iron cycling in the mesopelagic at station ALOHA. Frontiers in Marine Science, 5, p.61.

Cameron V, Vance D (2014) Heavy nickel isotope compositions in rivers and the oceans. Geochim Cosmochim Acta 128:195–211

Campos AM, van den Berg CMG (1994) Determination of copper complexation in sea water by cathodic stripping voltammetry and ligand competition with salicylaldoxime. Anal Chim Acta 284(3):481–496

Capodaglio G, Barbante C, Cescon P (2001) Trace metals in Antarctic Sea water. In: Environmental contamination in Antarctica. Elsevier, pp 107–154

Caroli S, Cescon P, and Walton DWH (2001) Trace metals in Antarctic sea water. Environmental Contamination in Antarctica: A Challenge to Analytical Chemistry:107

Chester R (1990) Trace elements in the oceans. In: Marine geochemistry. Springer, pp 346–421

Cleanroom-Technology (2011) Datex study into human particle shedding. GIT Verlag

Clough R et al (2015) Uncertainty contributions to the measurement of dissolved co, Fe, pb and V in seawater using flow injection with solid phase preconcentration and detection by collision/reaction cell—quadrupole ICP–MS. Talanta 133:162–169

Conway TM, John SG (2014a) The biogeochemical cycling of zinc and zinc isotopes in the North Atlantic Ocean. Glob Biogeochem Cycles 28(10):1111–1128

Conway TM, John SG (2014b) Quantification of dissolved iron sources to the North Atlantic Ocean. Nature 511(7508):212–215

Conway TM et al (2013) A new method for precise determination of iron, zinc and cadmium stable isotope ratios in seawater by double-spike mass spectrometry. Anal Chim Acta 793:44–52

Conway TM et al (2021) A decade of progress in understanding cycles of trace elements and their isotopes in the oceans. Chem Geol:120381

Conway TM, John SG, Lacan F (2016) Intercomparison of dissolved iron isotope profiles from reoccupation of three GEOTRACES stations in the Atlantic Ocean. Mar Chem 183:50–61

Cotte L, Omanović D, Waeles M, Laës A, Cathalot C, Sarradin PM, Riso RD (2018) On the nature of dissolved copper ligands in the early buoyant plume of hydrothermal vents. Environ Chem 15 (2):58–73

Croot PL, Johansson M (2000) Determination of iron speciation by cathodic stripping voltammetry in seawater using the competing ligand 2-(2-Thiazolylazo)-p-cresol (TAC). Electroanalysis: an international journal devoted to fundamental and practical aspects of. Electroanalysis 12(8):565–576

Croot PL, Moffett JW, Luther III GW (1999) Polarographic determination of half-wave potentials for copper-organic complexes in seawater. Mar Chem 67(3–4):219–232

Cullen JT, Sherrell RM (1999) Techniques for determination of trace metals in small samples of size-fractionated particulate matter: phytoplankton metals off Central California. Mar Chem 67(3):233–247

Cunliffe M, Wurl O (2014) Guide to best practices to study the ocean's surface. Plymouth, UK, Marine Biological Association of the United Kingdom for SCOR, 118 pp. (Occasional Publications of the Marine Biological Association of the United Kingdom). https://doi.org/10.25607/OBP-1512

Cutter GA (2013) Intercalibration in chemical oceanography—getting the right number. Limnol Oceanogr Methods 11(7):418–424

Cutter GA, Bruland KW (2012) Rapid and noncontaminating sampling system for trace elements in global ocean surveys. Limnol Oceanogr Methods 10:425–436

Cutter GA, et al. (2017) Sampling and sample-handling protocols for GEOTRACES cruises, Version 3

Dauphas N, John SG, and Rouxel O (2017) 11 iron isotope systematics. Non-Traditional Stable Isotopes:415–510

Davison W, Zhang H (1994) In situ speciation measurements of trace components in natural waters using thin-film gels. Nature 367(6463):546–548

de Baar HJW et al (1990) On iron limitation of the Southern Ocean: experimental observations in the Weddell and Scotia Seas. Mar Ecol Prog Ser:105–122

de Baar HJW (1994) von Liebig's law of the minimum and plankton ecology (1899–1991). Prog Oceanogr 33(4):347–386. https://doi.org/10.1016/0079-6611(94)90022-1

De Baar HJW et al (2008) Titan: a new facility for ultraclean sampling of trace elements and isotopes in the deep oceans in the international Geotraces program. Mar Chem 111(1–2):4–21

de Baar HJW, van Heuven SMAC, Middag R (2018) Ocean biochemical cycling and trace elements. In: White WM (ed) Encyclopedia of geochemistry: a comprehensive reference source on the chemistry of the earth. Springer International Publishing, Cham, pp 1–21

de Jong JTM et al (1998) Dissolved iron at subnanomolar levels in the Southern Ocean as determined by ship-board analysis. Anal Chim Acta 377(2–3):113–124

De Jong J et al (2007) Precise measurement of Fe isotopes in marine samples by multi-collector inductively coupled plasma mass spectrometry (MC-ICP-MS). Anal Chim Acta 589(1):105–119

Dodson MH (1963) A theoretical study of the use of internal standards for precise isotopic analysis by the surface ionization technique: part I-general first-order algebraic solutions. J Sci Instrum 40(6):289

Donat JR, Lao KA, Bruland KW (1994) Speciation of dissolved copper and nickel in South San Francisco Bay: a multi-method approach. Anal Chim Acta 284(3):547–571

Douthitt CB (2008) The evolution and applications of multicollector ICPMS (MC-ICPMS). Anal Bioanal Chem 390(2):437–440

Ellwood MJ, Nodder SD, King AL, Hutchins DA, Wilhelm SW, Boyd PW (2014) Pelagic iron cycling during the subtropical spring bloom, east of New Zealand. Mar Chem 160:18–33

Ebling AM, Landing WM (2015) Sampling and analysis of the sea surface microlayer for dissolved and particulate trace elements. Mar Chem 177:134–142

Elliott T, Steele RCJ (2017) The isotope geochemistry of Ni. Rev Mineral Geochem 82(1):511–542

Ellwood MJ, Van den Berg CMG (2000) Zinc speciation in the northeastern Atlantic Ocean. Mar Chem 68(4):295–306

Ellwood MJ et al (2018) Insights into the biogeochemical cycling of iron, nitrate, and phosphate across a 5,300 km South Pacific zonal section (153 E–150 W). Glob Biogeochem Cycles 32(2):187–207

Ellwood MJ et al (2020) Distinct iron cycling in a Southern Ocean eddy. Nat Commun 11(1):1–8

EPA, US (1996) Sampling ambient water for trace metals at EPA water quality criteria levels. Office of Water Engineering and Analysis Division Washington, EPA Method 1669

EPA (1997) Determination of trace elements in water by preconcentration and inductively coupled plasma-mass spectrometry. U.S. Environmental Protection Agency Office of Water Office of Science and Technology Engineering and Analysis Division (4303)

Falkner KK, Edmond JM (1990) Gold in seawater. Earth Planet Sci Lett 98(2):208–221

Farley KJ, Morel FM (1986) Role of coagulation in the kinetics of sedimentation. Environ Sci Technol 20(2):187–195

Field MP, Conway TM, Summers BA, Saetveit N, and Sakowski JC (2019) Automated processing of seawater samples for iron isotope ratio determination. Analytical Poster Note, Goldschmidt 2019, Elemental Scientific, Omaha, Nebraska, USA

Fitzsimmons JN, Boyle EA (2012) An intercalibration between the GEOTRACES GO-FLO and the MITESS/vanes sampling systems for dissolved iron concentration analyses (and a closer look at adsorption effects). Limnol Oceanogr Methods 10(6):437–450

Fitzsimmons JN, Boyle EA (2014a) Assessment and comparison of Anopore and cross flow filtration methods for the determination of dissolved iron size fractionation into soluble and colloidal phases in seawater. Limnol Oceanogr Methods 12(4):246–263

Fitzsimmons JN, Boyle EA (2014b) Both soluble and colloidal iron phases control dissolved iron variability in the tropical North Atlantic Ocean. Geochim Cosmochim Acta 125:539–550

Fitzsimmons JN et al (2015a) The composition of dissolved iron in the dusty surface ocean: an exploration using size-fractionated iron-binding ligands. Mar Chem 173:125–135

Fitzsimmons JN et al (2015b) Partitioning of dissolved iron and iron isotopes into soluble and colloidal phases along the GA03 GEOTRACES North Atlantic Transect. Deep-Sea Res II Top Stud Oceanogr 116:130–151

Floor GH et al (2015) Combined uncertainty estimation for the determination of the dissolved iron amount content in seawater using flow injection with chemiluminescence detection. Limnol Oceanogr Methods 13(12):673–686

Florence TM (1986) Electrochemical approaches to trace element speciation in waters. A review. Analyst 111(5):489–505

Garbe-Schönberg (2006) KIPS -a new multiport valve-based all-Teflon fluid sampling system for ROVs. Geology

Gasparon M (1998) Trace metals in water samples: minimising contamination during sampling and storage. Environ Geol 36(3):207–214

GEOTRACES-Group (2006) GEOTRACES Science Plan. Scientific Committee on Oceanic Research, Baltimore, Maryland

Gerringa LJA, Van der Meer J, Cauwet G (1991) Complexation of copper and nickel in the dissolved phase of marine sediment slurries. Mar Chem 36(1–4):51–70

Gerringa LJA et al (2014) A critical look at the calculation of the binding characteristics and concentration of iron complexing ligands in seawater with suggested improvements. Environ Chem 11(2):114–136

Gerringa LJA et al (2015) Organic complexation of iron in the West Atlantic Ocean. Mar Chem 177:434–446

Gerringa LJA et al (2020) Dissolved trace metals in the Ross Sea. Front Mar Sci 7:874

Gerringa LJA, Rijkenberg MJA, Slagter HA, Laan P, Paffrath R, Bauch D et al (2021a) Dissolved Cd, Co, Cu, Fe, Mn, Ni, and Zn in the Arctic Ocean. J Geophys Res Oceans 126(9): e2021JC017323

Gerringa LJA et al (2021b) Comparing CLE-AdCSV applications using SA and TAC to determine the Fe binding characteristics of model ligands in seawater. Biogeosci Discuss:1–40

Gillain G, et al. (1982) Sampling techniques and analytical methods. Distribution, Transport and Fate of Heavy Metals in the Belgian Coastal Marine Environment, Actions de Recherche Concertées-Geconcerteerde Onderzoeksacties, Oceanologie, final report 2:13–39

Gledhill M, Buck KN (2012) The organic complexation of iron in the marine environment: a review. Front Microbiol 3:69

Gledhill M, van den Berg CMG (1994) Determination of complexation of iron (III) with natural organic complexing ligands in seawater using cathodic stripping voltammetry. Mar Chem 47(1): 41–54

Gledhill M, van den Berg CMG (1995) Measurement of the redox speciation of iron in seawater by catalytic cathodic stripping voltammetry. Mar Chem 50(1–4):51–61

Gledhill M, Basu S, Shaked Y (2019) Metallophores associated with Trichodesmium erythraeum colonies from the Gulf of Aqaba. Metallomics 11(9):1547–1557

Goldberg, M.M. 1996 Trace metal clean room. National Environmental Research Laboratory, United States Environmental Protection Agency EPA/600/R-96/018

Goldhaber SB (2003) Trace element risk assessment: essentiality vs. toxicity. Regul Toxicol Pharmacol 38(2):232–242

Grand MM et al (2016) Determination of trace zinc in seawater by coupling solid phase extraction and fluorescence detection in the lab-on-valve format. Anal Chim Acta 923:45–54

Grand MM et al (2019) Developing autonomous observing systems for micronutrient trace metals. Front Mar Sci 6:35

Hales B, Takahashi T (2002) The pumping SeaSoar: a high-resolution seawater sampling platform. J Atmos Ocean Technol 19(7):1096–1104

Han H, Pan D (2021) Voltammetric methods for speciation analysis of trace metals in natural waters. Trends Environ Anal Chem:e00119

Hartland A et al (2019) Aqueous copper bioavailability linked to shipwreck-contaminated reef sediments. Sci Rep 9(1):1–13

Hassler CS et al (2011a) Exopolysaccharides produced by bacteria isolated from the pelagic Southern Ocean—role in Fe binding, chemical reactivity, and bioavailability. Mar Chem 123(1–4):88–98

Hassler CS et al (2011b) Saccharides enhance iron bioavailability to Southern Ocean phytoplankton. Proc Natl Acad Sci 108(3):1076–1081

Hawkes JA et al (2013) Characterisation of iron binding ligands in seawater by reverse titration. Anal Chim Acta 766:53–60

Heller MI, Croot PL (2015) Copper speciation and distribution in the Atlantic sector of the Southern Ocean. Mar Chem 173:253–268

Henderson GM, Achterberg EP, Bopp L (2018) Changing trace element cycles in the 21st Century Ocean. Elements 14(6):409–413

Henze G (1990) Application of polarographic and voltammetric techniques in environmental analysis. In: Metal speciation in the environment. Springer, pp 391–408

Hertkorn N et al (2006) Characterization of a major refractory component of marine dissolved organic matter. Geochim Cosmochim Acta 70(12):2990–3010

Hirose K (2006) Chemical speciation of trace metals in seawater: a review. Anal Sci 22(8): 1055–1063

Homoky WB et al (2016) Quantifying trace element and isotope fluxes at the ocean–sediment boundary: a review. Philos Trans R Soc A Math Phys Eng Sci 374(2081):20160246

Homoky WB et al (2021) Iron colloids dominate sedimentary supply to the ocean interior. Proc Natl Acad Sci 118(13):e2016078118

Honeyman BD, Santschi PH (1989) A Brownian-pumping model for oceanic trace metal scavenging: evidence from Th isotopes. J Mar Res 47(4):951–992

Horner TJ, Little SH, Conway TM, Farmer JR, Hertzberg JE, Janssen DJ, Lough AJM, McKay JL, Tessin A, Galer SJG, Jaccard SL (2021) Bioactive trace metals and their isotopes as paleoproductivity proxies: an assessment using GEOTRACES-era data. Glob Biogeochem Cycles 35(11):e2020GB006814

Hudson RJM, Rue EL, Bruland KW (2003) Modeling complexometric titrations of natural water samples. Environ Sci Technol 37(8):1553–1562

Hunter P (2008) A toxic brew we cannot live without: micronutrients give insights into the interplay between geochemistry and evolutionary biology. EMBO Rep 9(1):15–18

Hutchins DA, Boyd PW (2016) Marine phytoplankton and the changing ocean iron cycle. Nat Clim Change 6(12):1072–1079

Jackson SL et al (2018) Determination of Mn, Fe, Ni, Cu, Zn, Cd and Pb in seawater using offline extraction and triple quadrupole ICP-MS/MS. J Anal At Spectrom 33(2):304–313

Janssen DJ et al (2019) Particulate cadmium stable isotopes in the subarctic Northeast Pacific reveal dynamic Cd cycling and a new isotopically light Cd sink. Earth Planet Sci Lett 515:67–78

Jensen LT et al (2019) Biogeochemical cycling of dissolved zinc in the Western Arctic (Arctic GEOTRACES GN01). Glob Biogeochem Cycles 33(3):343–369

Jensen LT et al (2020) Assessment of the stability, sorption, and exchangeability of marine dissolved and colloidal metals. Mar Chem 220:103754

John SG, Adkins JF (2010) Analysis of dissolved iron isotopes in seawater. Mar Chem 119(1–4): 65–76

John SG, Conway TM (2014) A role for scavenging in the marine biogeochemical cycling of zinc and zinc isotopes. Earth Planet Sci Lett 394:159–167

Johnson CM, Beard BL (1999) Correction of instrumentally produced mass fractionation during isotopic analysis of Fe by thermal ionization mass spectrometry. Int J Mass Spectrom 193(1): 87–99

Johnson KS et al (2007) Developing standards for dissolved iron in seawater. EOS Trans Am Geophys Union 88(11):131–132

Johnson C, Beard B, Weyer S (2020) Iron geochemistry: an isotopic perspective. Springer

Karl DM, Lukas R (1996) The Hawaii Ocean time-series (HOT) program: background, rationale and field implementation. Deep-Sea Res II Top Stud Oceanogr 43(2–3):129–156

Kleint C et al (2016) Voltammetric investigation of hydrothermal iron speciation. Front Mar Sci 3: 75

Klunder MB et al (2011) Dissolved iron in the Southern Ocean (Atlantic sector). Deep-Sea Res Part Ii-Topical Studies in Oceanography 58(25–26):2678–2694

Klunder MB et al (2012) Dissolved iron in the Arctic Ocean: important role of hydrothermal sources, shelf input and scavenging removal. J Geophys Res Oceans 117:C4

Knutsson J (2013) Passive sampling for monitoring of inorganic pollutants in water. Chalmers University of Technology

Koprivnjak JF et al (2009) Chemical and spectroscopic characterization of marine dissolved organic matter isolated using coupled reverse osmosis–electrodialysis. Geochim Cosmochim Acta 73(14):4215–4231

Lacan F et al (2006) Cadmium isotopic composition in the ocean. Geochim Cosmochim Acta 70(20):5104–5118

Lacan F, Radic A, Jeandel C, Poitrasson F, Sarthou G, Pradoux C, Freydier R (2008) Measurement of the isotopic composition of dissolved iron in the open ocean. Geophys Res Lett 35(24)

Lagerström ME et al (2013) Automated on-line flow-injection ICP-MS determination of trace metals (Mn, Fe, Co, Ni, Cu and Zn) in open ocean seawater: application to the GEOTRACES program. Mar Chem 155:71–80

Laglera LM, Filella M (2015) The relevance of ligand exchange kinetics in the measurement of iron speciation by CLE–AdCSV in seawater. Mar Chem 173:100–113

Laglera LM, Battaglia G, van den Berg CMG (2011) Effect of humic substances on the iron speciation in natural waters by CLE/CSV. Mar Chem 127(1–4, 134):–143

Laglera LM et al (2020) Iron organic speciation during the LOHAFEX experiment: iron ligands release under biomass control by copepod grazing. J Mar Syst 207:103151

Lam PJ, Ohnemus DC, Auro ME (2015) Size-fractionated major particle composition and concentrations from the US GEOTRACES North Atlantic zonal transect. Deep-Sea Res II Top Stud Oceanogr 116:303–320

Landing WM, Bruland KW (1981) The vertical distribution of iron in the Northeast Pacific. EOS: Transactions of the American Geophysical Union 62:906

Lane TW, Morel FMM (2000) A biological function for cadmium in marine diatoms. Proc Natl Acad Sci 97(9):4627–4631

Lane TW et al (2005) A cadmium enzyme from a marine diatom. Nature 435(7038):42–42

Lee J-M et al (2011a) Analysis of trace metals (Cu, Cd, Pb, and Fe) in seawater using single batch nitrilotriacetate resin extraction and isotope dilution inductively coupled plasma mass spectrometry. Anal Chim Acta 686(1–2):93–101

Lee W-J, et al. (2011b) Spherical PF resin beads prepared from phenol-liquefied Bambusa dolichoclada with suspension polymerization

Little SH et al (2014) The oceanic mass balance of copper and zinc isotopes, investigated by analysis of their inputs, and outputs to ferromanganese oxide sediments. Geochim Cosmochim Acta 125:673–693

Little SH et al (2016) Key role of continental margin sediments in the oceanic mass balance of Zn and Zn isotopes. Geology 44(3):207–210

Lohan MC, Aguilar-Islas AM, Bruland KW (2006) Direct determination of iron in acidified (pH 1.7) seawater samples by flow injection analysis with catalytic spectrophotometric detection: application and intercomparison. Limnol Oceanogr Methods 4(6):164–171

Maréchal CN, Télouk P, Albarède F (1999) Precise analysis of copper and zinc isotopic compositions by plasma-source mass spectrometry. Chem Geol 156(1–4):251–273

Martin JH, Gordon RM (1988) Northeast Pacific iron distributions in relation to phytoplankton productivity. Deep Sea Res Part A Oceanographic Research Papers 35(2):177–196

Martin JH, Bruland KW, and Broenkow WW (1976) Cadmium transport in the California current [Water pollutants]

Martin JH, Gordon RM, Fitzwater SE (1990) Iron in Antarctic waters. Nature 345(6271):156–158

Martin JH et al (1994) Testing the iron hypothesis in ecosystems of the equatorial Pacific Ocean. Nature 371(6493):123–129

Matamoros V (2012) 1.13 – Equipment for water sampling including sensors. In: Pawliszyn J (ed) Comprehensive sampling and sample preparation. Academic Press, Oxford, pp 247–263

Matson WR, Roe DK, Carritt DE (1965) Composite graphite-mercury electrode for anodic stripping voltammetry. Anal Chem 37(12):1594–1595

Mawji E et al (2008) Collision-induced dissociation of three groups of hydroxamate siderophores: ferrioxamines, ferrichromes and coprogens/fusigens. Rapid Commun Mass Spectrom 22(14): 2195–2202

Mawji E, Schlitzer R, Dodas EM, Abadie C, Abouchami W, Anderson RF, Baars O, Bakker K, Baskaran M, Bates NR, Bluhm K (2015) The GEOTRACES intermediate data product 2014. Mar Chem 177:1–8

McDonnell AMP et al (2015) The oceanographic toolbox for the collection of sinking and suspended marine particles. Prog Oceanogr 133:17–31

Measures CI et al (2008) A commercially available rosette system for trace metal—clean sampling. Limnol Oceanogr Methods 6(9):384–394

Michaels AF, Knap AH (1996) Overview of the US JGOFS Bermuda Atlantic time-series study and the Hydrostation S program. Deep-Sea Res II Top Stud Oceanogr 43(2–3):157–198

Middag R et al (2011) Dissolved manganese in the Atlantic sector of the Southern Ocean. Deep Sea Res Part Ii-Topical Studies in Oceanography 58(25–26):2661–2677

Middag R et al (2015a) Intercomparison of dissolved trace elements at the Bermuda Atlantic time series station. Mar Chem 177(Part 3):476–489

Middag R et al (2015b) Dissolved aluminium in the ocean conveyor of the West Atlantic Ocean: effects of the biological cycle, scavenging, sediment resuspension and hydrography. Mar Chem 177(Part 1):69–86

Middag R et al (2020) The distribution of nickel in the West-Atlantic Ocean, its relationship with phosphate and a comparison to cadmium and zinc. Front Mar Sci 7:105

Mills MM et al (2004) Iron and phosphorus co-limit nitrogen fixation in the eastern tropical North Atlantic. Nature 429(6989):292–294

Minami T et al (2015) An off-line automated preconcentration system with ethylenediaminetriacetate chelating resin for the determination of trace metals in seawater by high-resolution inductively coupled plasma mass spectrometry. Anal Chim Acta 854:183–190

Misra S et al (2014) Determination of δ11B by HR-ICP-MS from mass limited samples: application to natural carbonates and water samples. Geochim Cosmochim Acta 140:531–552

Moffett JW, German CR (2020) Distribution of iron in the Western Indian Ocean and the eastern tropical south pacific: an inter-basin comparison. Chem Geol 532:119334

Moigne L, Frédéric AC et al (2014) Sequestration efficiency in the iron-limited North Atlantic: implications for iron supply mode to fertilized blooms. Geophys Res Lett 41(13):4619–4627

Moore CM et al (2006) Iron limits primary productivity during spring bloom development in the Central North Atlantic. Glob Chang Biol 12(4):626–634

Moore CM et al (2009) Large-scale distribution of Atlantic nitrogen fixation controlled by iron availability. Nat Geosci 2(12):867–871

Moore CM et al (2013) Processes and patterns of oceanic nutrient limitation. Nat Geosci 6(9): 701–710

Morel FMM, Milligan AJ, Saito MA (2014) 8.5—marine bioinorganic chemistry: the role of trace metals in the oceanic cycles of major nutrients. In: Holland HD, Turekian KK (eds) Treatise on geochemistry, 2nd edn. Elsevier, Oxford, pp 123–150

Morrison, Archie TIII, John D Billings, and Kenneth W Doherty (2000) The McLane WTS-LV: a large volume, high accuracy, oceanographic sampling pump. OCEANS 2000 MTS/IEEE Conference and Exhibition. Conference Proceedings (Cat. No. 00CH37158), 2000. Vol. 2, pp. 847–852. IEEE

Moynier F et al (2017) The isotope geochemistry of zinc and copper. Rev Mineral Geochem 82(1): 543–600

Muller FL, Batchelli S (2013) Copper binding by terrestrial versus marine organic ligands in the coastal plume of River Thurso, North Scotland. Estuar Coast Shelf Sci 133:137–146

Mueller AV, et al. (2018) A novel high-resolution trace metal clean sampler for stationary, profiling, and mobile deployment. 2018 OCEANS-MTS/IEEE Kobe Techno-Oceans (OTO), 2018, pp. 1–8. IEEE

Murray JW et al (1992) EqPac: a process study in the central equatorial Pacific. Oceanography 5(3): 134–142

Nehme C (2020) HVAC DESIGN FOR CLEAN ROOMS: Charles Nehme

Nielsdóttir MC et al (2009) Iron limitation of the postbloom phytoplankton communities in the Iceland Basin. Glob Biogeochem Cycles 23:3

Nishioka J, et al. (2020) Iron supply from the marginal seas and its influence on biological production in the North Pacific Ocean. Ocean Sciences Meeting 2020, 2020. AGU

Noble AE et al (2020) A review of marine water sampling methods for trace metals. Environ Forensic 21(3–4):267–290

Obata H et al (2017) Dissolved iron and zinc in Sagami Bay and the Izu-Ogasawara Trench. J Oceanogr 73(3):333–344

Ohnemus DC et al (2014) Laboratory intercomparison of marine particulate digestions including piranha: a novel chemical method for dissolution of polyethersulfone filters. Limnol Oceanogr Methods 12(8):530–547

Okamura K et al (2013) Development of a 128-channel multi-water-sampling system for underwater platforms and its application to chemical and biological monitoring. Methods Oceanogr 8: 75–90

Oldham VE et al (2021) Inhibited manganese oxide formation hinders cobalt scavenging in the Ross Sea. Glob Biogeochem Cycles 35(5):e2020GB006706

Olesik JW (2014) 15.17—inductively coupled plasma mass spectrometers. In: Holland HD, Turekian KK (eds) Treatise on geochemistry (second edition). Elsevier, Oxford, pp 309–336

Omanović D et al (2010) Significance of data treatment and experimental setup on the determination of copper complexing parameters by anodic stripping voltammetry. Anal Chim Acta 664(2):136–143

Omanović D, Garnier C, Pižeta I (2015) ProMCC: an all-in-one tool for trace metal complexation studies. Mar Chem 173:25–39

Pađan J et al (2019) Improved voltammetric methodology for chromium redox speciation in estuarine waters. Anal Chim Acta 1089:40–47

Pađan J et al (2020) Organic copper speciation by anodic stripping voltammetry in estuarine waters with high dissolved organic matter. Front Chem:8

Patterson CC (1965) Contaminated and natural lead environments of man. Arch Environ Health 11(3):344–360

Patterson C, Settle D, Glover B (1976) Analysis of lead in polluted coastal seawater. Mar Chem 4(4):305–319

Pilson MEQ (2012) An introduction to the chemistry of the sea. Cambridge University Press

Pižeta I et al (2015) Interpretation of complexometric titration data: an intercomparison of methods for estimating models of trace metal complexation by natural organic ligands. Mar Chem 173:3–24

Planquette H, Sherrell RM (2012) Sampling for particulate trace element determination using water sampling bottles: methodology and comparison to in situ pumps. Limnol Oceanogr Methods 10(5):367–388

Plavšić M et al (2009) Determination of the copper complexing ligands in the Krka river estuary. Fresenius Environ Bull 18(3):327–334

Pohlabeln AM, Dittmar T (2015) Novel insights into the molecular structure of non-volatile marine dissolved organic sulfur. Mar Chem 168:86–94

Price NM, Morel FMM (1990) Cadmium and cobalt substitution for zinc in a marine diatom. Nature 344(6267):658–660

Protti P (2001) Introduction to modern voltammetric and polarographic analisys techniques. AMEL srl:p10

Rapp I et al (2017) Automated preconcentration of Fe, Zn, Cu, Ni, Cd, Pb, Co, and Mn in seawater with analysis using high-resolution sector field inductively-coupled plasma mass spectrometry. Anal Chim Acta 976:1–13

Rauschenberg S, Twining BS (2015) Evaluation of approaches to estimate biogenic particulate trace metals in the ocean. Mar Chem 171:67–77

Rehkämper M et al (2012) Natural and anthropogenic Cd isotope variations. In: Handbook of environmental isotope geochemistry. Springer, pp 125–154

Rehkämperab M, Halliday AN (1998) Accuracy and long-term reproducibility of lead isotopic measurements by multiple-collector inductively coupled plasma mass spectrometry using an external method for correction of mass discrimination. Int J Mass Spectrom 181(1–3):123–133

Rehman ZU et al (2017) Advanced characterization of dissolved organic matter released by bloom-forming marine algae. Desalin Water Treat 69:1–11

Resing JA et al (2015) Basin-scale transport of hydrothermal dissolved metals across the South Pacific Ocean. Nature 523(7559):200–203

Revels BN et al (2015) Fractionation of iron isotopes during leaching of natural particles by acidic and circumneutral leaches and development of an optimal leach for marine particulate iron isotopes. Geochim Cosmochim Acta 166:92–104

Richter R (2003) Clean chemistry: techniques for the modern laboratory. Milestone Press Shelton, CT

Rijkenberg MJA et al (2014) The distribution of dissolved iron in the West Atlantic Ocean. PLoS One 9(6):e101323

Rijkenberg MJA et al (2015) "PRISTINE", a new high volume sampler for ultraclean sampling of trace metals and isotopes. Mar Chem 177(Part 3):501–509

Rijkenberg MJA et al (2018) Dissolved Fe in the deep and upper Arctic Ocean with a focus on Fe limitation in the Nansen Basin. Front Mar Sci 5:88

Riley GA, Van Hemert D, Wanngersky PJ (1965) Organic aggregates in surface and DEEP waters of the Sargasso Sea 1. Limnol Oceanogr 10(3):354–363

Riley JS et al (2012) The relative contribution of fast and slow sinking particles to ocean carbon export. Glob Biogeochem Cycles 26:1

Ripperger S, Rehkämper M (2007) Precise determination of cadmium isotope fractionation in seawater by double spike MC-ICPMS. Geochim Cosmochim Acta 71(3):631–642

Ripperger S et al (2007) Cadmium isotope fractionation in seawater—a signature of biological activity. Earth Planet Sci Lett 261(3–4):670–684

Rolison JM et al (2018) Iron isotope fractionation during pyrite formation in a sulfidic Precambrian Ocean analogue. Earth Planet Sci Lett 488:1–13

Rue EL, Bruland KW (1995) Complexation of iron (III) by natural organic ligands in the central North Pacific as determined by a new competitive ligand equilibration/adsorptive cathodic stripping voltammetric method. Mar Chem 50(1–4):117–138

Russell WA, Papanastassiou DA, Tombrello TA (1978) Ca isotope fractionation on the earth and other solar system materials. Geochim Cosmochim Acta 42(8):1075–1090

Ružić I (1982) Theoretical aspects of the direct titration of natural waters and its information yield for trace metal speciation. Anal Chim Acta 140(1):99–113

Ryan-Keogh TJ et al (2013) Spatial and temporal development of phytoplankton iron stress in relation to bloom dynamics in the high-latitude North Atlantic Ocean. Limnol Oceanogr 58(2): 533–545

Saito MA, Moffett JW (2001) Complexation of cobalt by natural organic ligands in the Sargasso Sea as determined by a new high-sensitivity electrochemical cobalt speciation method suitable for open ocean work. Mar Chem 75(1–2):49–68

Saito MA, Goepfert TJ, Ritt JT (2008) Some thoughts on the concept of colimitation: three definitions and the importance of bioavailability. Limnol Oceanogr 53(1):276–290

Saito MA et al (2013) Slow-spreading submarine ridges in the South Atlantic as a significant oceanic iron source. Nat Geosci 6(9):775–779

Sander S et al (2005) Effect of UVB irradiation on Cu2+–binding organic ligands and Cu2+ speciation in alpine lake waters of New Zealand. Environ Chem 2(1):56–62

Sander SG et al (2007) Organic complexation of copper in deep-sea hydrothermal vent systems. Environ Chem 4(2):81–89

Sander SG, Hunter K, and Frew R (2009) 14-sampling and measurements of trace metals in seawater. Practical Guidelines for the Analysis of Seawater 305

Sander SG et al (2011) Numerical approach to speciation and estimation of parameters used in modeling trace metal bioavailability. Environ Sci Technol 45(15):6388–6395

Sander SG, Buck KN, Wells M (2015) The effect of natural organic ligands on trace metal speciation in San Francisco Bay: implications for water quality criteria. Mar Chem 173:269–281

Sanderson MP et al (1995) Primary productivity and trace-metal contamination measurements from a clean rosette system versus ultra-clean go-Flo bottles. Deep-Sea Res II Top Stud Oceanogr 42(2–3):431–440

Santschi PH (2018) Marine colloids, agents of the self-cleansing capacity of aquatic systems: historical perspective and new discoveries. Mar Chem 207:124–135

Sanvito F, Monticelli D (2021) Exploring bufferless iron speciation in seawater by competitive ligand equilibration-cathodic stripping voltammetry: does pH control really matter? Talanta 229:122300

Schintu M, Marrucci A, Marras B (2014) Passive sampling technologies for the monitoring of organic and inorganic contaminants in seawater. In: Current environmental issues and challenges. Springer, pp 217–237

Schlitzer R, Anderson RF, Dodas EM, Lohan M, Geibert W, Tagliabue A, Bowie A, Jeandel C, Maldonado MT, Landing WM, Cockwell D (2018) The GEOTRACES intermediate data product 2017. Chem Geol 493:210–223

Schmitt A-D, Galer SJG, Abouchami W (2009) High-precision cadmium stable isotope measurements by double spike thermal ionisation mass spectrometry. J Anal At Spectrom 24(8):1079–1088

Sclater FR, Boyle E, Edmond JM (1976) On the marine geochemistry of nickel. Earth Planet Sci Lett 31(1):119–128

Sedwick PN et al (2015) A zonal picture of the water column distribution of dissolved iron (II) during the US GEOTRACES North Atlantic transect cruise (GEOTRACES GA03). Deep-Sea Res II Top Stud Oceanogr 116:166–175

Seewald JS et al (2002) A new gas-tight isobaric sampler for hydrothermal fluids. Deep-Sea Res I Oceanogr Res Pap 49(1):189–196

Seyitmuhammedov K (2021) Biogeochemical cycling of trace metals in shelf regions: the importance of cross-shelf exchange. University of Otago

Shah SB (2021) Heavy metals in scleractinian corals. Springer Nature

Sherrell RM, Boyle EA (1992) The trace metal composition of suspended particles in the oceanic water column near Bermuda. Earth Planet Sci Lett 111(1):155–174

Sieber M, Conway TM, de Souza GF, Hassler CS, Ellwood MJ, Vance D (2019) High-resolution Cd isotope systematics in multiple zones of the Southern Ocean from the Antarctic Circumnavigation Expedition. Earth Planet Sci Lett 527:115799

Sieber M et al (2021) Isotopic fingerprinting of biogeochemical processes and iron sources in the iron-limited surface Southern Ocean. Earth Planet Sci Lett 567:116967

Siebert C, Nägler TF, Kramers JD (2001) Determination of molybdenum isotope fractionation by double-spike multicollector inductively coupled plasma mass spectrometry. Geochem Geophys Geosyst 2:7

Sohrin Y, Bruland KW (2011) Global status of trace elements in the ocean. TrAC Trends Anal Chem 30(8):1291–1307

Sohrin Y et al (2008) Multielemental determination of GEOTRACES key trace metals in seawater by ICPMS after preconcentration using an ethylenediaminetriacetic acid chelating resin. Anal Chem 80(16):6267–6273

Sonstadt E (1872) On the presence of gold in sea water. Chem News 26(671):159–161

Staubwasser M et al (2013) Isotope fractionation between dissolved and suspended particulate Fe in the oxic and anoxic water column of the Baltic Sea. Biogeosciences 10(1):233–245

Strady E et al (2008) PUMP–CTD-system for trace metal sampling with a high vertical resolution. A test in the Gotland Basin. Baltic Sea Chemosphere 70(7):1309–1319

Strelow FEW (1980) Improved separation of iron from copper and other elements by anion-exchange chromatography on a 4% cross-linked resin with high concentrations of hydrochloric acid. Talanta 27(9):727–732

Stumm W, Morgan JJ (1996) Aquatic chemistry: chemical equilibria and rates in natural waters, paperback. Wiley, New York

Sun M, Archer C, and Vance D (2021) New methods for the chemical isolation and stable isotope measurement of multiple transition metals, with application to the earth sciences. Geostandards and Geoanalytical Research

Sunda W (2012) Feedback interactions between trace metal nutrients and phytoplankton in the ocean. Front Microbiol 3:204

Tagliabue A et al (2016) How well do global ocean biogeochemistry models simulate dissolved iron distributions? Glob Biogeochem Cycles 30(2):149–174

Tagliabue A et al (2017) The integral role of iron in ocean biogeochemistry. Nature 543(7643): 51–59

Takano S et al (2013) Determination of isotopic composition of dissolved copper in seawater by multi-collector inductively coupled plasma mass spectrometry after pre-concentration using an ethylenediaminetriacetic acid chelating resin. Anal Chim Acta 784:33–41

Takano S et al (2017) A simple and rapid method for isotopic analysis of nickel, copper, and zinc in seawater using chelating extraction and anion exchange. Anal Chim Acta 967:1–11

Tanhua T et al (2019) Ocean FAIR data services. Front Mar Sci 6:440

Taylor PDP et al (1993) Stable isotope analysis of iron by gas-phase electron impact mass spectrometry. Anal Chem 65(21):3166–3167

Teng F-Z, Dauphas N, Watkins JM (2017) Non-traditional stable isotopes: retrospective and prospective. Rev Mineral Geochem 82(1):1–26

Tessier A, Turner DR (1995) Metal speciation and bioavailability in aquatic systems. Wiley, Chichester

Tessier A, Campbell PGC, Bisson MJAC (1979) Sequential extraction procedure for the speciation of particulate trace metals. Anal Chem 51(7):844–851

Thompson CM, Ellwood MJ, Wille M (2013) A solvent extraction technique for the isotopic measurement of dissolved copper in seawater. Anal Chim Acta 775:106–113

Tipping E (1998) Humic ion-binding model VI: an improved description of the interactions of protons and metal ions with humic substances. Aquat Geochem 4(1):3–47

Tonnard M et al (2020) Dissolved iron in the North Atlantic Ocean and Labrador Sea along the GEOVIDE section (GEOTRACES section GA01). Biogeosciences 17(4):917–943

Tovar-Sánchez A (2012) 1.17 – Sampling approaches for trace element determination in seawater. In: Pawliszyn J (ed) Comprehensive sampling and sample preparation. Academic Press, pp 317–334, ISBN 9780123813749. https://doi.org/10.1016/B978-0-12-381373-2.00017-X

Tovar-Sánchez A et al (2014) Spatial gradients in trace metal concentrations in the surface microlayer of the Mediterranean Sea. Front Mar Sci 1:79

Tovar-Sánchez A et al (2020) Characterizing the surface microlayer in the Mediterranean Sea: trace metal concentrations and microbial plankton abundance. Biogeosciences 17(8):2349–2364

Town RM, Filella M (2000) A comprehensive systematic compilation of complexation parameters reported for trace metals in natural waters. Aquat Sci 62(3):252–295

Trull TW, et al. (2010) The Australian integrated marine observing system Southern Ocean time series facility. OCEANS'10 IEEE SYDNEY, 2010, pp. 1–7. IEEE

Turk J (2001) Field guide for surface water sample and data collection. USDA Forest Service, Washington (DC), p 67

Twining BS, Baines SB (2013) The trace metal composition of marine phytoplankton. Annu Rev Mar Sci 5(1):191–215

Twining BS et al (2015a) Metal contents of phytoplankton and labile particulate material in the North Atlantic Ocean. Prog Oceanogr 137:261–283

Twining BS et al (2015b) Comparison of particulate trace element concentrations in the North Atlantic Ocean as determined with discrete bottle sampling and in situ pumping. Deep-Sea Res II Top Stud Oceanogr 116:273–282

US-Geological-Survey 2006 Chapter A4. Collection of water samples. Report, 09-A4

Ussher SJ et al (2010) Distribution of size fractionated dissolved iron in the Canary Basin. Mar Environ Res 70(1):46–55

Van den Berg CMG (1982) Determination of copper complexation with natural organic ligands in seawater by equilibration with MnO2 I. Theory Mar Chem 11(4):307–322

Van Den Berg CM (1986) Determination of copper, cadmium and lead in seawater by cathodic stripping voltammetry of complexes with 8-hydroxyquinoline. J Electroanal Chem Interfacial Electrochem 215(1–2):111–121

van der Merwe P et al (2019) The autonomous clean environmental (ACE) sampler: a trace-metal clean seawater sampler suitable for open-ocean time-series applications. Limnol Oceanogr Methods 17(9):490–504

Van Dorn WG (1956) Large-volume water samplers. EOS Trans Am Geophys Union 37(6): 682–684

van den Berg CM (2006) Chemical speciation of iron in seawater by cathodic stripping voltammetry with dihydroxynaphthalene. Anal Chem 78(1):156–163

van Hulten M et al (2017) Manganese in the West Atlantic Ocean in the context of the first global ocean circulation model of manganese. Biogeosciences 14(5):1123–1152

Vance D et al (2016) The oceanic budgets of nickel and zinc isotopes: the importance of sulfidic environments as illustrated by the Black Sea. Philos Trans R Soc A Math Phys Eng Sci 374(2081):20150294

Vance D et al (2017) Silicon and zinc biogeochemical cycles coupled through the Southern Ocean. Nat Geosci 10(3):202–206

Velasquez I et al (2011) Detection of hydroxamate siderophores in coastal and sub-Antarctic waters off the South Eastern Coast of New Zealand. Mar Chem 126(1–4):97–107

Velasquez IB et al (2016) Ferrioxamine siderophores detected amongst iron binding ligands produced during the remineralization of marine particles. Front Mar Sci 3:172

Vink S et al (2000) Automated high resolution determination of the trace elements iron and aluminium in the surface ocean using a towed fish coupled to flow injection analysis. Deep-Sea Res I Oceanogr Res Pap 47(6):1141–1156

Von Damm KL et al (1985) Chemistry of submarine hydrothermal solutions at 21° N, East Pacific rise. Geochim Cosmochim Acta 49(11):2197–2220

Vraspir JM, Butler A (2009) Chemistry of marine ligands and siderophores. Annu Rev Mar Sci 1:43

Waeber T, Marie-Louise SS, Slaveykova V (2012) Trace metal behavior in surface waters: emphasis on dynamic speciation, sorption processes and bioavailability. Arch Sci 65:119–142

Walder AJ, Freedman PA (1992) Communication. Isotopic ratio measurement using a double focusing magnetic sector mass analyser with an inductively coupled plasma as an ion source. J Anal At Spectrom 7(3):571–575

Wang R-M et al (2019) Zinc and nickel isotopes in seawater from the Indian sector of the Southern Ocean: the impact of natural iron fertilization versus Southern Ocean hydrography and biogeochemistry. Chem Geol 511:452–464

Weber T et al (2018) Biological uptake and reversible scavenging of zinc in the global ocean. Science 361(6397):72–76

Wells ML, Goldberg ED (1992) Marine submicron particles. Mar Chem 40(1):5–18

Wells M, Buck KN, Sander SG (2013) New approach to analysis of voltammetric ligand titration data improves understanding of metal speciation in natural waters. Limnol Oceanogr Methods 11(9):450–465

Weyer S, Schwieters JB (2003) High precision Fe isotope measurements with high mass resolution MC-ICPMS. Int J Mass Spectrom 226(3):355–368

Whitby H et al (2018) Copper-binding ligands in the NE Pacific. Mar Chem 204:36–48

Whitby H et al (2020) A call for refining the role of humic-like substances in the oceanic iron cycle. Sci Rep 10(1):1–12

Wilde, Franceska D, and Dean B Radtke (1998) Handbooks for Water-resources Investigations: National field manual for the collection of water-quality data. Field measurements: US Department of the Interior, US Geological Survey

Worsfold PJ et al (2013) Flow injection analysis as a tool for enhancing oceanographic nutrient measurements—a review. Anal Chim Acta 803:15–40

Worsfold PJ et al (2014) Determination of dissolved iron in seawater: a historical review. Mar Chem 166:25–35

Worsfold PJ et al (2019) Estimating uncertainties in oceanographic trace element measurements. Front Mar Sci 5:515

Worsfold, Paul, et al. 2019c Encyclopedia of analytical science: Elsevier

Wu JF (2007) Determination of picomolar iron in seawater by double Mg(OH)(2) precipitation isotope dilution high-resolution ICPMS. Mar Chem 103(3–4):370–381

Wurl O (2009) Sampling and sample treatments. In: Practical guidelines for the analysis of seawater. CRC Press, pp 13–44

Wuttig K et al (2019) Critical evaluation of a seaFAST system for the analysis of trace metals in marine samples. Talanta 197:653–668

Wyatt NJ et al (2014) Biogeochemical cycling of dissolved zinc along the GEOTRACES South Atlantic transect GA10 at 40 S. Glob Biogeochem Cycles 28(1):44–56

Xie RC et al (2015) The cadmium–phosphate relationship in the western South Atlantic—the importance of mode and intermediate waters on the global systematics. Mar Chem 177:110–123

Xue Z et al (2012) A new methodology for precise cadmium isotope analyses of seawater. Anal Bioanal Chem 402(2):883–893

Xue Z, Rehkämper M, Horner TJ, Abouchami W, Middag R, van de Flierd T, de Baar HJ (2013) Cadmium isotope variations in the Southern Ocean. Earth Planet Sci Lett 382:161–172

Xie RC, Rehkämper M, Grasse P, van de Flierdt T, Frank M, Xue Z (2019) Isotopic evidence for complex biogeochemical cycling of Cd in the eastern tropical South Pacific. Earth Planet Sci Lett 512:134–146

Yamamoto A et al (2019) Glacial CO2 decrease and deep-water deoxygenation by iron fertilization from glaciogenic dust. Clim Past 15(3):981–996

Yamashita Y et al (2011) Assessing the spatial and temporal variability of dissolved organic matter in Liverpool Bay using excitation–emission matrix fluorescence and parallel factor analysis. Ocean Dyn 61(5):569–579

Yang S-C et al (2020) A new purification method for Ni and Cu stable isotopes in seawater provides evidence for widespread Ni isotope fractionation by phytoplankton in the North Pacific. Chem Geol 547:119662

Zhang H, Davison W (1995) Performance characteristics of diffusion gradients in thin films for the in situ measurement of trace metals in aqueous solution. Anal Chem 67(19):3391–3400

Zuman P (2001) Electrolysis with a dropping mercury electrode: J. Heyrovsky's contribution to electrochemistry. Crit Rev Anal Chem 31(4):281–289

Radionuclides as Ocean Tracers

4

Valentí Rodellas, Montserrat Roca-Martí, Viena Puigcorbé, Maxi Castrillejo, and Núria Casacuberta

Contents

V. Rodellas (✉)
Institut de Ciència i Tecnologia Ambientals (ICTA-UAB), Universitat Autònoma de Barcelona, Bellaterra, Spain
e-mail: valenti.rodellas@uab.cat

M. Roca-Martí
Department of Oceanography, Dalhousie University, Halifax, NS, Canada

Department of Marine Chemistry and Geochemistry, Woods Hole Oceanographic Institution, Woods Hole, MA, USA

V. Puigcorbé
Department of Marine Biology and Oceanography, Institut de Ciències del Mar (ICM-CSIC), Barcelona, Spain

Centre for Marine Ecosystems Research, School of Science, Edith Cowan University, Joondalup, WA, Australia

M. Castrillejo
Department of Physics, Imperial College London, London, UK

Laboratory of Ion Beam Physics, ETH Zürich, Zurich, Switzerland

N. Casacuberta
Physical Oceanography Institute of Biogeochemistry and Pollutant Dynamics, Department of Environmental System Science, ETH Zürich, Zurich, Switzerland

© The Author(s), under exclusive license to Springer Nature Switzerland AG 2023
J. Blasco, A. Tovar-Sánchez (eds.), *Marine Analytical Chemistry*,
https://doi.org/10.1007/978-3-031-14486-8_4

Abstract

Radionuclides, both from natural and anthropogenic origin, are powerful ocean tracers that provide key information on fluxes, pathways and time scales of marine processes. Their added value compared to hydrographic parameters (e. g., temperature and salinity) relies either on their known rates of radioactive decay and production or on their time-variable releases from sources. Both aspects introduce a temporal dimension that allows quantifying rates or time scales of marine processes. Their wide range of half-lives and historical inputs, together with their physicochemical characteristics allow tracing a broad spectrum of marine processes. This chapter aims at providing an overview of the application of radionuclides as tracers of ocean processes. The chapter is structured in four sections: The first part reviews the main principles of radioactivity and the origin of radionuclides. The second section introduces the key aspects that allow using radionuclides as ocean tracers, followed by (third section) three real and contemporary instructive examples that cover different marine processes and require radionuclides with specific properties: i) thorium-234/uranium-238 (^{234}Th/^{238}U) pair to quantify the biological pump, ii) radium (Ra) isotopes to estimate the magnitude of submarine groundwater discharge, and iii) iodine-129 (^{129}I) to investigate the large-scale ocean circulation. Finally, this chapter provides a summary of the different methods and techniques used to measure radionuclides in seawater.

Keywords

Radionuclides · Isotopes · Ocean tracers · Biological pump · Submarine groundwater discharge · Ocean circulation · Radionuclide measurement techniques

4.1   Introduction

We live in a "radioactive world" where radioactive isotopes (radionuclides) are present in all the environmental compartments (atmosphere, lithosphere, cryosphere, biosphere, and hydrosphere). It is thus not surprising that the ocean contains large amounts and a great variety of radionuclides, most of them derived from natural sources but also from recent anthropogenic activities. However, the concentrations of radionuclides in the ocean are typically too small to pose a radiological risk for human or marine life. Hence, why do we care about radionuclides in the ocean? Radionuclides have been widely investigated in oceanography because they can reveal crucial details about marine processes, i.e., they can be used as *ocean tracers*. Ocean tracers are natural or anthropogenic compounds, elements, or isotopes that are present in small but measurable amounts in the marine environment, such that variations in their abundances or distribution can be used to obtain information on ocean processes. The added value of radionuclides as ocean tracers relies either on their known rates of radioactive decay and production or on their time-variable releases from sources, which both introduce a temporal dimension that allows quantifying temporal scales (e.g., residence times) and rates (e.g., fluxes) of processes.

This chapter aims at providing an overview of the use of radionuclides as tracers of processes occurring in the marine environment. The first two sections review the basic principles of radioactive decay and the use of radionuclides as ocean tracers. The third section provides three instructive examples of how radionuclides trace relevant ocean processes: the thorium-234/uranium-238 (^{234}Th/^{238}U) pair as a tracer of the biological pump, radium (Ra) isotopes as tracers of submarine groundwater discharge, and iodine-129 (^{129}I) as a tracer of large-scale ocean circulation. Finally, a fourth section is dedicated to summarize the different methods and techniques used to measure radionuclides in seawater.

4.1.1 Basic Concepts of Radioactivity

The use of radionuclides as tracers of ocean processes requires a basic understanding of the main principles related to radioactivity and radioactive decay. This section is thus an introduction to basic concepts such as isotopes; radioactive decay; radionuclides; alpha, beta and gamma decay; activity; half-life; decay chain; or secular equilibrium.

4.1.1.1 Nuclear Instability and Types of Radioactive Decay

Isotopes are atoms that belong to the same element in the periodic table (i.e., same number of protons or same atomic number, Z), but that differ in the number of neutrons (N) and the mass number (i.e., sum of the number of protons and neutrons, A). For example, ^{12}C, ^{13}C, and ^{14}C are three isotopes of the element carbon, which have in all cases 6 protons but a different number of neutrons: 6, 7, and 8, respectively. The relationship between the number of protons and neutrons in the atom nucleus defines the stability of each isotope. When the protons and neutrons in the nucleus are configured in an unstable way (i.e., too many/few neutrons to bind the protons together), the isotope will spontaneously decay into a more stable configuration during the process known as *radioactive decay*. Unstable isotopes are known as *radionuclides*. Over 340 different isotopes have been found in nature and about 80 of them are radioactive (Masarik 2009). Indeed, all elements with atomic numbers higher than 80 ($Z > 80$) have radioactive isotopes, and all isotopes of elements with $Z > 83$ are radioactive.

While elements are illustrated in the periodic table, a common way of representing all the isotopes and their nuclear stability is by means of the table of nuclides (also known as Segrè chart or Karlsruhe nuclide chart; Fig. 4.1). This table is a two-dimensional graph of the isotopes of the elements in which the x-axis represents the number of neutrons, and the y-axis represents the number of protons in the nucleus. All isotopes in the same row (i.e., same number of protons) are isotopes of the same element, but only a small number of those are stable. The line

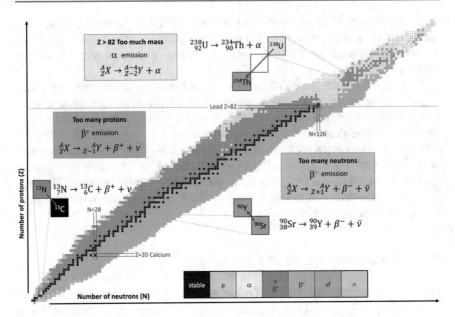

Fig. 4.1 Table of nuclides showing all known isotopes arranged according to the number of protons as a function of the number of neutrons in the nucleus. Black squares represent stable isotopes, and the other colors indicate the primary type of radioactive decay of each radionuclide: emission of protons (p), alpha particles (α), electron capture (ε) or beta-plus (β^+), beta-minus (β^-), spontaneous fission (sf) and neutron emission (n). The types of radioactive decay α, β^+, and β^- are illustrated with its general equation and one example (^{238}U, ^{13}N, and ^{90}Sr, respectively). ^{48}Ca ($Z = 20$, $N = 28$) and ^{208}Pb ($Z = 82$, $N = 126$) are highlighted as examples of nuclides that have unusual stability because they have a "magic" number of protons and neutrons (i.e., widely recognized "magic" numbers are 2, 8, 20, 28, 50, 82, and 126)

that connects all the stable isotopes is known as the valley of stability (Fig. 4.1, black color). The stability declines to both sides of this valley when decreasing or increasing the number of neutrons.

Unstable atomic nuclei find stability through radioactive decay, which consists of the loss and emission of energy and/or mass. The types of decay depend on the mass of the unstable nuclei and the relationship between neutrons and protons. Radioactive decay involves the emission of subatomic particles (e.g., electrons, protons, neutrons, and lighter particles), photons (electromagnetic radiation), or both. The most common types of radioactive decay are alpha, beta, and gamma (Fig. 4.2) and are summarized below:

- When isotopes are heavy ($Z > 82$), the atomic nucleus will likely emit an *alpha* (α) particle (Fig. 4.1, yellow color), which consists of two protons and two neutrons. When a radionuclide decays via alpha decay it produces a lighter nuclide, with the mass number reduced by 4 and the atomic number by 2, and thus located further down and left in the table of nuclides with respect to the original nuclide (example of ^{238}U in Fig. 4.1).

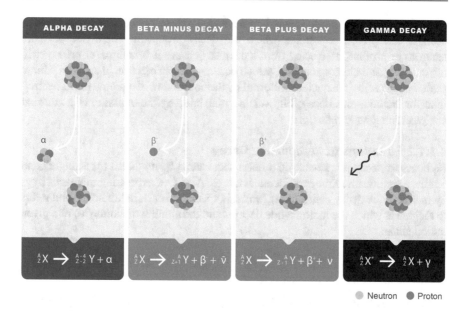

Fig. 4.2 Schematic representation of the most common decay modes: alpha, beta-minus, beta-plus, and gamma decay (designed by gemmasola.com)

- If the isotopes have an excess of neutrons compared to protons (Fig. 4.1, right side of the valley of stability), the decay mode will likely be *beta-minus*. In the beta-minus decay, the nucleus emits an electron (β^-) and an antineutrino ($\bar{\upsilon}$) in a process that converts a neutron into a proton. This decay mode produces a nuclide with one extra proton and one neutron less than the original nuclide (moving up and left in the table of nuclides; example of ^{90}Sr in Fig. 4.1).
- If the isotopes have an excess of protons compared to neutrons (Fig. 4.1, left side of the valley of stability), the decay mode will likely be *beta-plus*. In the beta-plus decay, the nucleus emits a positron (β^+) and a neutrino (υ) in a process that converts a proton into a neutron. The transition to stability in this decay mode is by producing a nuclide with one neutron more and one proton less than the original nuclide (moving down and right in the table of nuclides; example of ^{13}N in Fig. 4.1).
- When a radioactive nucleus decays by the emission of an alpha or beta particle, the resulting nucleus is usually left in an excited state, and it can decay to a lower energy state by emitting a *gamma ray* (γ), a highly energetic photon. This radioactive mode is not represented in Fig. 4.1 because it only implies the emission of electromagnetic radiation and does not change the number of protons or neutrons of the original nuclide.

Aside from alpha, beta, and gamma, there are other less-common decay modes: spontaneous fission, which can occur only in very heavy elements ($Z > 92$), with the heavy nucleus splitting into two lighter nuclei and releasing neutrons in the process; neutron and proton emissions, which are decay processes where one or more neutrons or protons are ejected from a nucleus because it is neutron or proton rich, respectively; and electron capture, which occurs when an electron in an atom's inner shell is captured by the nucleus followed by the conversion of a proton into a neutron (like for the beta-plus decay; Fig. 4.1) accompanied by the emission of a neutrino and electromagnetic radiation.

4.1.1.2 Equations of Radioactive Decay

Radioactive decay is a stochastic process that cannot be predicted for a single atom but for a significant number of identical atoms. A way of expressing radioactivity is by means of the decay constant (λ), which is specific to each radionuclide and refers to the rate at which the radionuclide decays, and the number of atoms (N) of a given radionuclide:

$$-\frac{dN}{dt} = \lambda \cdot N \tag{4.1}$$

In this equation, $-dN/dt$ denotes *activity* ($A = \lambda \cdot N$). Activity is expressed in becquerels (Bq), which is the unit of radioactivity in the International System of Units and is defined as the number of radioactive transformations per second (i.e., nucleus decays per second or dps). Activity is also commonly expressed in units of disintegrations per minute (dpm; 60 dpm = 1 Bq). The older unit of radioactivity was the curie (Ci), which is roughly the number of decays that take place in 1 g of ^{226}Ra per second and corresponds to $3.7 \cdot 10^{10}$ Bq.

As obtained by developing Eq. (4.1), the temporal evolution of the number of atoms of a radionuclide can be described by an exponential decay:

$$N = N_0 \cdot e^{-\lambda \cdot t} \tag{4.2}$$

where N_0 is the initial number of atoms at time 0 (t_0), t is the time spanned from t_0, and N is the number of atoms left at time t.

The behavior described by Eq. (4.2) implies that if a radionuclide is incorporated and subsequently isolated in an environment with no exchange with the surroundings and no additional production, the abundance of this radionuclide will decrease exponentially only due to its radioactive decay. The exponential decay of three different radioactive isotopes (^{228}Ra, ^{137}Cs, and ^{209}Po) is represented in Fig. 4.3. In all cases, N_0 is assumed to be 100 atoms, and the shape of the exponential curve is defined by the decay constant of each isotope. As illustrated in Fig. 4.3, the higher the decay constant (λ) the faster the radionuclide decays (e.g., ^{228}Ra), whereas radionuclides with lower decay constants (e.g., ^{209}Po) will remain longer in the system due to its slower decay rate.

The decay constant of a radionuclide can also be related to its *half-life* ($T_{1/2}$), which is a much more intuitive concept that refers to the length of time after which

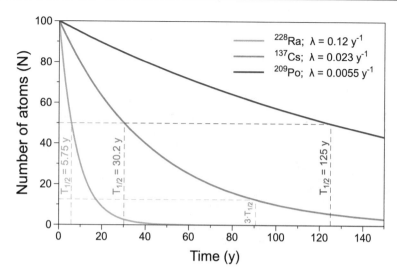

Fig. 4.3 Temporal evolution of the abundance of a radionuclide (expressed in number of atoms, N) as a function of time. Three radionuclides with significantly different decay constants (and half-lives) are represented (^{228}Ra, ^{137}Cs, ^{209}Po). The initial number of atoms (N_0) is assumed to be 100 for all the radionuclides. As exemplified for ^{137}Cs, after three half-lives ($3 \cdot T_{1/2}$), there is only 12.5% of the initial number of atoms remaining in the system

half of the initial number of atoms have decayed. Numerically, the half-life can be estimated from the decay constant as follows:

$$\frac{N_0}{2} = N_0 \cdot e^{-\lambda \cdot T_{\frac{1}{2}}} \tag{4.3}$$

which results in:

$$T_{\frac{1}{2}} = \frac{\ln(2)}{\lambda} \tag{4.4}$$

Half-lives of radionuclides can span from microseconds to millions of years. Following the examples in Fig. 4.3, the half-lives of ^{228}Ra, ^{137}Cs, and ^{209}Po correspond to the time needed to reduce the initial number of atoms from 100 (N_0) to 50 ($N_0/2$) and can be calculated from Eq. (4.4) (5.75 years, 30.2 years, and 125 years, respectively). The number of atoms remaining after one half-life is 50%, after 2 half-lives 25%, after 3 half-lives 12.5%, etc. After 7 half-lives, the percentage of nuclides remaining is <1%, and thus it can be considered that almost all atoms have decayed. For example, ^{228}Ra needs about 40 years (seven times its $T_{1/2}$) to disintegrate "almost" completely (Fig. 4.3).

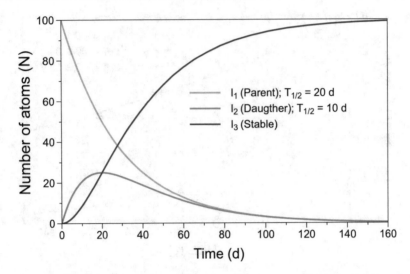

Fig. 4.4 Temporal evolution of the abundance of three radionuclides belonging to the same decay chain as a function of time. In this example, it is assumed that I_1 ($T_{1/2} = 20$ days) decays to I_2, I_2 ($T_{1/2} = 10$ days) decays to I_3, and I_3 is stable. We consider that at $t = 0$ there are 100 atoms of the parent radionuclide (I_1) and none of the daughter radionuclide (I_2) and stable isotope (I_3)

4.1.1.3 Decay Chains

Most radionuclides do not decay directly to a stable form, but rather to another unstable nuclide (i.e., another radionuclide). When a radionuclide (I_1, or parent isotope) decays into another radionuclide (I_2, or daughter isotope), the latter also decays into another isotope (stable or not) forming a *decay chain*. In this case, the number of atoms of a daughter isotope remaining in the system after a certain time is the result of not only its own decay but also its production by the decay of the parent isotope. In other words, the number of atoms of the daughter radionuclide will depend on both the decay constant of the parent radionuclide (λ_{I1}) and its own decay constant (λ_{I2}). The temporal evolution of the number of atoms of both radionuclides (parent I_1 and daughter I_2) can be estimated using the Bateman equations, which can be propagated to account for all the daughter radionuclides:

$$N_{I1} = N_{I1,0} \cdot e^{-\lambda_{I1} \cdot t} \tag{4.5}$$

$$\frac{dN_{I2}}{dt} = -\lambda_{I2} \cdot N_{I2} + \lambda_{I1} \cdot N_{I1} \tag{4.6}$$

$$N_{I2} = \frac{\lambda_{I1}}{\lambda_{I2} - \lambda_{I1}} \cdot N_{I1,0} \cdot \left(e^{-\lambda_{I1} \cdot t} - e^{-\lambda_{I2} \cdot t}\right) + N_{I2,0} \cdot e^{-\lambda_{I2} \cdot t} \tag{4.7}$$

where $N_{I1,0}$ and $N_{I2,0}$ are the atoms of the parent and daughter at time 0. In Fig. 4.4 we represent the decay of a parent (I_1) to a radioactive daughter (I_2), which subsequently decays into another daughter (I_3), which in this case is stable. In this example, we set the initial number of atoms of both daughters (I_2 and I_3) to 0, and thus the number of atoms of the stable daughter (I_3) at a given time can be calculated as:

$$N_{I3} = N_{I1,0} - N_{I1} - N_{I2} \tag{4.8}$$

The Bateman equations can also be rewritten in terms of activity, by considering that $A = \lambda \cdot N$:

$$A_{I1} = A_{I1,0} \cdot e^{-\lambda_{I1} \cdot t} \tag{4.9}$$

$$A_{I2} = \frac{\lambda_{I2}}{\lambda_{I2} - \lambda_{I1}} \cdot A_{I1,0} \cdot \left(e^{-\lambda_{I1} \cdot t} - e^{-\lambda_{I2} \cdot t}\right) + A_{I2,0} \cdot e^{-\lambda_{I2} \cdot t} \tag{4.10}$$

In a decay chain, it often happens that the parent radionuclide decays at a much slower rate than its daughter ($\lambda_{I1} \ll \lambda_{I2}$). If this is the case and the radionuclides are produced or lost only by radioactive decay, the activity of both the parent and the daughter radionuclide will be the same after more than seven half-lives of the daughter radionuclide. This phenomenon is called *secular equilibrium* and can be numerically explained from Eq. (4.10), if λ_{I1} is considered negligible compared to λ_{I2}, resulting in the following equation:

$$A_{I2} = A_{I1,0} \cdot \left(1 - e^{-\lambda_{I2} \cdot t}\right) + A_{I2,0} \cdot e^{-\lambda_{I2} \cdot t} \tag{4.11}$$

After seven times the half-life of the daughter radionuclide (I_2), the exponential terms are almost 0, so the equation simplifies to $A_{I2} = A_{I1,0}$, which is similar to A_{I1} because of the slow decay rate of the parent radionuclide. The concept of secular equilibrium, which occurs frequently in nature, is crucial to use certain radionuclides as tracers.

4.1.2 Why Do We Find Radionuclides in the Environment?

Some of the radionuclides present on Earth were already produced during the formation of our planet (nucleogenesis) about 4500 million years ago, the so-called *primordial radionuclides*. Most of these primordial radionuclides have already decayed, and only the ones having half-lives comparable or longer than the Earth's age are nowadays naturally present in the environment. Other radionuclides are continuously produced through different processes involving nuclear reactions: *natural decay series radionuclides*, produced from the radioactive decay of primordial radionuclides; *cosmogenic radionuclides*, produced by the interaction of cosmic radiation with stable isotopes; and *anthropogenic radionuclides*, originating from man-made reactions occurring in nuclear reactors or detonations (Fig. 4.5). Radionuclides are thus originated either from the Earth's crust (i.e., primordial

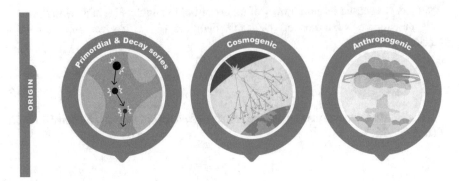

Fig. 4.5 Origin (sources) of radionuclides in the environment (designed by gemmasola.com)

radionuclides and natural decay series radionuclides) or from extraterrestrial (i.e., cosmogenic radionuclides) or artificial sources (i.e., anthropogenic radionuclides) and then distributed between environmental compartments depending on their properties. Nowadays, both natural and anthropogenic radionuclides are found in the atmosphere, the hydrosphere (e.g., in the ocean, rivers, or groundwater), the cryosphere, the lithosphere (e.g., in rocks, sediments, or soils), and the biosphere.

4.1.2.1 Primordial and Natural Decay Series Radionuclides

Primordial radionuclides were produced during the Earth's nucleogenesis, but only few dozens of them are still present in our planet due to their long half-life ($\gg 10^8$ years). Primordial radionuclides with relatively short half-lives compared to the Earth's age (e.g., ^{237}Np, $T_{1/2} = 2.1 \cdot 10^6$ years) are already extinct. Some of the surviving primordial radionuclides decay directly into stable isotopes, such as ^{40}K ($T_{1/2} = 1.3 \cdot 10^9$ years) and ^{87}Rb ($T_{1/2} = 4.8 \cdot 10^{10}$ years), which mainly produce stable ^{40}Ca and ^{87}Sr, respectively. Other primordial radionuclides decay through a chain of radionuclides with shorter half-lives into a stable isotope (the natural decay series radionuclides). This is the case of the three naturally occurring radioactive decay chains, the often-called uranium and thorium decay series. Each one of these series originates with a primordial radionuclide (^{238}U, ^{232}Th, or ^{235}U) with a long half-life ($4.5 \cdot 10^9$ years, $1.4 \cdot 10^{10}$ years, and $7.0 \cdot 10^8$ years, respectively) and ends with a stable isotope of Pb (Fig. 4.6). In between, there is a series of radionuclides that include isotopes of Th, Ra, Rn, Po, and Pb, among others, with half-lives ranging from microseconds to hundreds of thousands of years. If not subjected to chemical or physical separation, each of these series reaches a state of secular equilibrium (see Sect. 4.1.1.3). In the environment, however, it is common to observe disequilibrium among radionuclides of the same decay chain (e.g., between ^{238}U and ^{234}Th; see Sect. 4.3.1), and this disequilibrium is used to obtain information on the environmental processes that produced such partitioning.

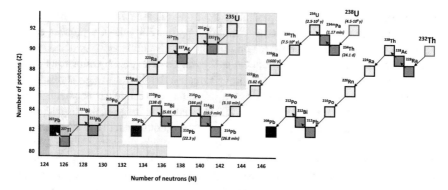

Fig. 4.6 Schematic diagram of the natural decay series (^{238}U, ^{232}Th, and ^{235}U) superimposed to a zoomed view of the table of nuclides (Fig. 4.1). Notice that ^{238}U and ^{232}Th decay chains are moved to the right for a better representation, but the position of the parent radionuclide is indicated in red (^{238}U) and blue (^{232}Th) squares. Only half-lives of the ^{238}U decay chain radionuclides are represented

4.1.2.2 Cosmogenic Radionuclides

Cosmogenic radionuclides are produced by cosmic radiation that penetrates the Earth from space and interacts with atmospheric gases or solids onto the Earth's surface (e.g., rocks). This cosmic radiation includes both extraterrestrial particles that strike the Earth (largely protons) and secondary particles generated by the interaction of cosmic radiation with atmospheric atoms. Most cosmogenic radionuclides are thus produced in the atmosphere and can enter the soils and the hydrosphere through fallout. Examples of cosmogenic radionuclides are isotopes of Be, C, Al, Cl, Ca, and I, with half-lives ranging from seconds to millions of years. The best-known cosmogenic radionuclide is ^{14}C ($T_{1/2} = 5730$ years), produced from the interaction of a cosmogenic neutron with atmospheric ^{14}N. ^{14}C rapidly reacts with atmospheric oxygen to form carbon dioxide (CO_2). ^{14}C is commonly used to date biological materials (e.g., plants, corals) that incorporate CO_2 (with ^{14}C) or to trace ocean circulation, since ocean waters are a major sink of atmospheric CO_2.

4.1.2.3 Anthropogenic Radionuclides

Anthropogenic radionuclides are those that are not naturally present in the environment (e.g., ^{137}Cs) or whose natural levels have been largely exceeded by anthropogenic sources (e.g., ^{129}I). They are mainly produced by human activities during nuclear energy production or detonations of nuclear weapons. The most common anthropogenic radionuclides are the fission products of ^{235}U produced either via bomb testing or released as waste from the nuclear energy industry. Examples of anthropogenic radionuclides include ^{90}Sr ($T_{1/2} = 28.8$ years), ^{137}Cs ($T_{1/2} = 30.2$ years), or ^{129}I ($T_{1/2} = 1.6 \cdot 10^7$ years). In addition, anthropogenic radionuclides can also be produced from the interaction of neutrons released from nuclear weapon explosions with atmospheric atoms (e.g., ^3H, $T_{1/2} = 12.3$ years), a process called neutron capture. Civil and military activities have thus deliberately or accidentally introduced large quantities of radionuclides into the environment by numerous ways, including releases to the atmosphere through nuclear detonations (e.g., atmospheric nuclear tests from 1954 to 1963) or as a consequence of nuclear accidents (e.g., Chernobyl in 1986 and Fukushima in 2011), leading to large-scale dispersion of radionuclides in soils and the hydrosphere through fallout. Other sources of anthropogenic radionuclides to the marine environment include discharges from nuclear fuel processing and power plants into rivers or the coastal ocean (e.g., Sellafield plant, UK); leaks from storage and processing facilities into the ground subsequently transported by groundwaters (e.g., Hanford facility, USA); or the dumping of nuclear waste directly into the ocean (practiced until the 1990s). More information on anthropogenic radionuclides is provided in Sect. 4.3.3.

4.2   Radionuclides as Ocean Tracers

Studying the movement of ocean waters, the fluxes of particulate or dissolved material (e.g., nutrients, metals, contaminants) supplied by external sources, and their internal cycling in the sea is crucial to better understand marine biogeochemistry, the global carbon cycle, and the role of the ocean in the Earth's climate. Most of these processes are usually difficult to measure, quantify, or constrain, and they have historically been studied by combining a set of different tools (e.g., models, observational techniques, ocean tracers, moorings). In this regard, radionuclides have emerged as powerful tools to obtain key information on pathways, fluxes, and temporal scales of ocean processes. The value of radionuclides as ocean tracers relies in their widespread distribution in the marine environment, their well-known rates of radioactive decay and production, their often well-constrained input sources, and their wide range of half-lives and physicochemical characteristics that allow tracing a broad spectrum of marine processes. In this section, we will introduce the key aspects that need to be considered when using radionuclides as tracers of a given process (Sect. 4.2.1), and we will highlight the five most common ocean processes that are investigated using radionuclides (Sect. 4.2.2).

4.2.1   Which Radionuclides Can Trace a Given Process?

Ocean processes operate over a wide range of temporal and spatial scales and often involve complex interactions with other environmental compartments such as the atmosphere or the lithosphere. In order to choose the right tracer, it is key to find the radionuclide with the appropriate input source, physicochemical behavior, and half-life for the targeted process (Fig. 4.7).

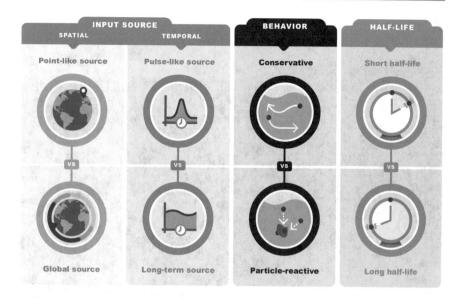

Fig. 4.7 Schematic representation of the main tracer characteristics that should be considered when choosing a suitable radionuclide to study a given ocean process (designed by gemmasola. com)

4.2.1.1 Input Source

Radionuclides found in the ocean can be produced by any of the origins described in Sect. 4.1.2 (i.e., primordial and natural decay series, cosmogenic, and anthropogenic radionuclides). These radionuclides present in the Earth's compartments are then introduced to the ocean through different pathways: *atmospheric inputs*, including gas exchange and wet/dry deposition; *riverine discharge*, including radionuclides dissolved in river waters and transported with riverine particles; *submarine groundwater discharge*, including radionuclides dissolved in continental subsurface waters that ultimately enter the ocean; *sediment inputs*, supplying radionuclides to the water column via diverse processes such as diffusion, resuspension, bioturbation, and other forms of remobilization of radionuclides from sediments; *anthropogenic inputs*, referring to those radionuclides that reached the ocean either by releases from the nuclear industry, from testing of nuclear weapons or nuclear accidents, or from industries that enhance the concentration of natural radionuclides (e.g., oil extraction industry, phosphate industry); and in situ *production*, referring to the direct production of radionuclides from the decay of parent radionuclides already present in the ocean (Fig. 4.8).

The pathways or sources supplying radionuclides to the ocean can vary in space and time. The *spatial component* of a source refers to the area in which radionuclides are introduced to the marine environment. Sources can range from *point-like* to *global* (Fig. 4.7). Point-like sources are constrained to a specific area, while global sources are distributed everywhere in the marine environment and thus are not area-specific. Examples of point-like sources include riverine discharge or liquid waste released from nuclear reprocessing plants directly into the ocean. Examples of global

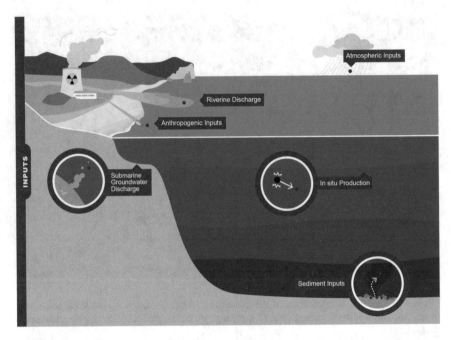

Fig. 4.8 Representation of the different pathways or sources supplying radionuclides to the ocean (designed by gemmasola.com)

sources include in situ production, cosmogenic production, or stratospheric fallout from large nuclear detonations. Yet, global sources do not necessarily have a uniform distribution. For example, the distribution of the global fallout from nuclear weapon tests varies across different latitudes due to the locations where the tests took place and the dynamics of atmospheric circulation.

The *temporal component* of a source refers to the temporal evolution of the input of radionuclides to the marine environment. Sources can range from *pulse-like* to *long-term* (Fig. 4.7). Pulse-like sources last only for a certain period of time (e.g., a nuclear accident), while long-term sources are permanent in time (e.g., in situ production and riverine discharge). Yet, long-term sources are not necessarily constant. For example, the input of radionuclides through riverine discharge can show strong seasonal fluctuations.

The source of cosmogenic radionuclides is mainly considered global and long-term because these radionuclides are continuously supplied to the ocean from the atmosphere, although there are input fluctuations in both space and time. Generally speaking, the sources of primordial and natural decay series radionuclides are also considered global and long-term. However, some radionuclides from the natural decay chains are supplied or removed at ocean interfaces (e.g., sediment-water, land-ocean, or air-water interfaces) and, therefore, their source can be considered point-like. That would be the case of Ra isotopes, which get enriched in groundwater and inflow into the ocean in a relatively constrained area. Finally, anthropogenic radionuclides can have point-like or global input sources, and although they are usually pulse-like, some inputs can be sustained over years or decades (e.g., inputs from nuclear reprocessing plants). The temporal evolution of the releases of radionuclides in a specific location defines the input function, which is an important term particularly for the use of anthropogenic radionuclides as tracers of the temporal scales or rates of marine processes (see Sects. 4.2.2.1 and 4.3.3 for further explanations).

4.2.1.2 Physicochemical Behavior

From the physicochemical perspective, radionuclides can be either conservative or particle-reactive following their behavior in the ocean (Fig. 4.7).

- *Conservative* radionuclides show none or little interaction with marine particles or with biology, and they are removed from the water column only through radioactive decay and physical processes (e.g., outgassing). The distribution of conservative radionuclides will thus be governed by physical transport processes including ocean circulation, mixing, and diffusion, and hence, they will be ideal to study such processes. Examples of conservative radionuclides include ^{236}U, ^{129}I (e.g., Wefing et al. 2021), and ^{3}H used as ocean circulation tracers (e.g., Jenkins and Smethie 1996); Ra isotopes used to trace land-ocean interaction processes (e.g., Garcia-Orellana et al. 2021); or ^{222}Rn used to quantify ocean-atmosphere gas exchange (e.g., Rutgers van der Loeff et al. 2014).
- *Particle-reactive* radionuclides have a high affinity for particles and, therefore, can be removed from the water column by scavenging onto marine particles. As a result, particle-reactive radionuclides can be used as tracers to study the cycling and fate of particles in the ocean. Examples of particle-reactive radionuclides include ^{234}Th, ^{210}Po, and ^{90}Y used to quantify particle export fluxes in the open ocean (e.g., Cochran and Masqué 2003), or ^{210}Pb used to date sediments (e.g., Arias-Ortiz et al. 2018).

In some cases, particularly for those radionuclides of anthropogenic origin, the physicochemical form in which they enter the marine environment can determine their behavior in seawater. For example, oxidized Pu can behave conservatively, while reduced Pu is highly particle-reactive. The environmental conditions can also influence the reactivity of a certain radionuclide. For example, while Cs is considered conservative in the open ocean, ^{137}Cs serves as a tracer to study particle mixing and sediment accumulation in coastal environments characterized by high particle loads (e.g., Smith and Ellis 1982).

Finally, particle-reactive radionuclides but also some radionuclides that are soluble in seawater (e.g., ^{14}C) can interact to significant extent with biology and thus become tracers of important marine biogeochemical cycles. Marine organisms can uptake radionuclides directly from the water (via adsorption, absorption across body surfaces, or a combination of both) or via ingestion of food, and release them as exudates or upon remineralization of the organic matter. The intake of these radionuclides may lead to their bioaccumulation in marine organisms, particularly for radionuclides of elements with known biological functions (e.g., ^{14}C). The fact that some radionuclides can be incorporated into organisms allows to further expand the use of radionuclides to study a wide variety of biologically mediated marine processes.

4.2.1.3 Half-Life

The final key aspect to consider is the half-life of the radionuclide (Fig. 4.7). In contrast to stable isotopes, radionuclides decay at well-constrained rates and this intrinsic characteristic of each radionuclide provides insights into the temporal context of the investigated process. Radionuclides can thus be used as "clocks," being instrumental to determine the temporal scale (e.g., ages, residence times, transit times) and rates (e.g., fluxes) of processes. It is therefore essential to select a radionuclide presenting a half-life of adequate length for the time scale of the process being evaluated.

- Radionuclides with a *short half-life* (e.g., few days) can be ideal to study processes occurring over short time scales (e.g., water movement in a coastal bay), but decay too quickly to address processes with time scales of months, years, or decades (e.g., large-scale circulation). Radionuclides with very short half-lives in comparison to their parents can be very useful to trace environmental processes that can break the expected secular equilibrium between the parent and daughter radionuclides (see Sects. 4.2.2.2 and 4.3.1). In these cases, the half-life of the daughter radionuclide should be comparable to the temporal scale of the environmental process being studied. Examples of short-lived radionuclides used as ocean tracers include ^{90}Y ($T_{1/2} = 64.6$ h; used to quantify pulses of particle export (e.g., Orlandini et al. 2003)) or ^{224}Ra ($T_{1/2} = 3.66$ days; used to quantify water age and mixing rates in coastal waters (e.g., Moore 2000)).
- Radionuclides with a *long half-life* (>years) are suitable to investigate long time scale processes such as ocean circulation, but may overlook seasonal to shorter processes such as phytoplankton blooms. Examples of long-lived radionuclides include ^{230}Th ($T_{1/2} = 7.5 \cdot 10^4$ years), which is used in paleoceanography to understand changes in past ocean circulation (e.g., Bradtmiller et al. 2014), or ^{14}C ($T_{1/2} = 5730$ years) used to date sediments that have been deposited thousands of years ago (e.g., Hajdas et al. 2021). It is worth noting that some radionuclides decay at such slow rates that they can be considered "stable" tracers despite being radioactive (e.g., ^{236}U, $T_{1/2} = 2.3 \cdot 10^7$ years). While their millions of years' long half-life precludes their use as radioactive "clocks," they can still provide key information about time scales if they have time varying input functions (see Sect. 4.3.3.3).

4.2.2 What Are the Most Common Ocean Processes Studied Using Radionuclides?

There is a long list of radionuclides that have been used as ocean tracers. Here we will briefly describe the five most common processes studied using radionuclides in

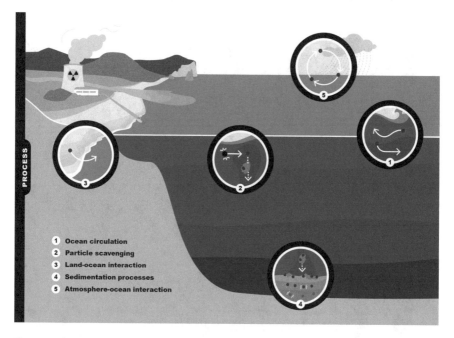

Fig. 4.9 Representation of the main ocean processes commonly studied using radionuclides as tracers (designed by gemmasola.com)

the marine environment (Fig. 4.9, Table 4.1), while emphasizing the key attributes that radionuclides should have to trace such processes.

4.2.2.1 Ocean Circulation

Understanding global ocean circulation patterns and the movement and mixing of water masses is crucial to comprehend critical aspects of climate regulation and marine life (see Sect. 4.3.3). The ideal radionuclide to trace ocean circulation should be conservative in seawater and should have a constrained (well-known) input function. For radionuclides with long-term inputs, their radioactive decay can be used as a built-in-clocks; to estimate transport times, water ages, or rates of water mass mixing. For instance, the short-lived radium isotopes ^{224}Ra ($T_{1/2} = 3.66$ days) and ^{223}Ra ($T_{1/2} = 11.3$ days), which are produced from the naturally occurring decay series, have been applied to constrain water transport and mixing rates in coastal areas, since these isotopes are supplied from the continent (long-term and point-like sources) and have half-lives comparable to the time scales of coastal mixing processes (Moore 2000) (see Sect. 4.3.2). For radionuclides with a half-life much longer than the time scale of a given circulation process, the fluctuations of their inputs to the ocean with time (i.e., time-variable input function) can be used to shed light on the temporal context. This is the case of the so-called transient tracers, which include several anthropogenic radionuclides that in spite of having very long half-lives (e.g., millions of years), their time-varying release from nuclear activities after the 1950s

Table 4.1 Examples of radionuclides used to trace different ocean processes. Key properties of the radionuclides (half-life, behavior, input source) and their origin are also indicated

Main application	Radionuclide	Half-life	Behavior	Input	Origin
Ocean circulation					
Ocean circulation (basin scale)	^{236}U	$2.3 \cdot 10^7$ years	Conservative	Point-like and global. Pulse-like and long-term	Anthropogenic radionuclide: supplied from both global fallout and nuclear reprocessing plants
Ocean circulation (basin scale)	^{129}I	$1.6 \cdot 10^7$ years	Conservative	Point-like. Long-term	Anthropogenic radionuclide: supplied from nuclear reprocessing plants
Mixing in the coastal ocean (~days)	^{224}Ra	3.66 days	Conservative	Point-like. Long-term	Decay-series: supplied from ^{228}Th in sediments, rocks
Particle scavenging					
Particle scavenging (~weeks)	^{234}Th	24.1 days	Particle-reactive	Global. Long-term	Decay-series: supplied from ^{238}U in water
Particle scavenging (~days)	^{90}Y	64.6 h	Particle-reactive	Global. Long-term	Anthropogenic radionuclide: supplied from ^{90}Sr
Particle scavenging (~months)	^{210}Po	138 days	Particle-reactive	Global. Long-term	Decay-series: supplied from ^{210}Pb in water, atmosphere
Land-ocean interaction					
Submarine groundwater discharge (regional scale)	^{228}Ra	5.75 years	Conservative	Point-like. Long-term	Decay-series: supplied from ^{232}Th in sediments, rocks
Submarine groundwater discharge (local scale)	^{222}Rn	3.82 days	Gas, conservative	Point-like. Long-term	Decay-series: supplied from ^{226}Ra in sediments, rocks, water
Sedimentation processes					
Geochronology (~30 k years)	^{14}C	5730 years	Incorporated into C cycle and organisms	Global. Long-term	Cosmogenic + anthropogenic radionuclide (bomb testing)
Sedimentation rates (~100 years)	^{210}Pb	22.3 years	Particle-reactive	Global. Long-term	Decay-series: supplied from ^{222}Rn in atmosphere, water

Atmosphere-ocean interaction				
Atmospheric deposition	^{7}Be	53.3 days	Particle-reactive	Global. Long-term	Cosmogenic: interaction cosmic rays with atmospheric O and N
Atmospheric deposition	^{232}Th	$1.4 \cdot 10^{10}$ years	Particle-reactive	Global. Long-term	Primordial radionuclide
Air-sea gas exchange	^{222}Rn	3.82 days	Gas, conservative	Point-like. Long-term	Decay-series: supplied from ^{226}Ra in sediments, rocks, water

allows the study of recent ocean circulation processes. For example, ^{236}U ($T_{1/2} = 2.3 \cdot 10^7$ years) can be used alone or in combination with other conservative long-lived radionuclides (e.g., ^{129}I, $T_{1/2} = 1.6 \cdot 10^7$ years) to trace North Atlantic waters circulating in the Arctic Ocean (e.g., Casacuberta et al. 2018; Castrillejo et al. 2018).

4.2.2.2 Particle Scavenging

Particle scavenging in the ocean is an essential part of many biogeochemical cycles relevant to marine life, and it is also a key mechanism that allows the ocean to remove CO_2 from the atmosphere, thus impacting climate (see Sect. 4.3.1). The study of particle cycling in the ocean requires the use of radionuclides that are particle-reactive and can therefore be used as tracers of particles and their associated chemical species. Particle-reactive radionuclides can be used to study various processes related to particle dynamics in the ocean, including aggregation and disaggregation, the remineralization of particulate organic matter, or the sinking of particles from the ocean surface to depth (e.g., Cochran and Masqué 2003; Lam and Marchal 2015). Particle-reactive radionuclides used to trace scavenging processes are generally sourced by in situ production from the decay of their parents (global and long-term source), which are typically conservative in ocean waters. Another key aspect is that the parent decays at a much lower rate than its daughter. As a result, secular equilibrium (see Sect. 4.1.1.3) between the two would be expected if there were no particles in the ocean. However, the movement of particles in the ocean by gravitational sinking or by lateral transport, creates a radioactive disequilibrium between the parent and daughter radionuclides that is used to study such processes. For instance, the highly particle-reactive ^{234}Th ($T_{1/2} = 24.1$ days) is widely used as a particle tracer, and it is continuously produced by the decay of ^{238}U ($T_{1/2} = 4.5 \cdot 10^9$ years), which is dissolved in ocean waters and has a much longer half-life (see Sect. 4.3.1). Other radionuclides used as particle tracers include ^{90}Y ($T_{1/2} = 64.6$ h) produced from the anthropogenic radionuclide ^{90}Sr ($T_{1/2} = 28.8$ years), or ^{210}Pb ($T_{1/2} = 22.3$ years) and ^{210}Po ($T_{1/2} = 138$ days), which are produced by ^{226}Ra ($T_{1/2} = 1600$ years) and ^{210}Pb, respectively, from the ^{238}U decay chain. In the latter example, both ^{210}Pb and ^{210}Po get adsorbed onto particle surfaces, but Po is also incorporated into particles possibly as an analog of a biologically active element (i.e., sulfur; Stewart et al. 2008). As a consequence, Po is preferentially removed by sinking particles relative to Pb, allowing the use of the ^{210}Po/^{210}Pb pair as a tracer of particle export even though in this case both radionuclides are particle-reactive. The half-life of the daughter radionuclide determines the time scale of the processes that can be traced by a given radionuclide pair, ranging from days (^{90}Y) and weeks (^{234}Th), to months (^{210}Po), and decades (^{210}Pb).

4.2.2.3 Land-Ocean Interaction

The supply of solutes and solid material across the land-ocean interface strongly controls the chemical composition of coastal seawater, the global cycles of many elements (e.g., macro- and micronutrients), as well as the productivity of coastal ecosystems (see Sect. 4.3.2). These compounds are delivered through a variety of processes including the riverine discharge of solutes and solids, the weathering or diffusion from deposited sediments, or the discharge of groundwater to the coastal ocean. In this regard, the flux of solutes between land and ocean can be quantified by using radionuclides that are both sourced by long-term processes occurring across the land-ocean interface and conservative in seawater. Ra isotopes (^{224}Ra, ^{223}Ra, ^{226}Ra, and ^{228}Ra) are the most commonly applied radionuclides to evaluate land-ocean interaction processes, such as fluxes derived from submarine groundwater discharge or sediment inputs (Garcia-Orellana et al. 2021). Ra isotopes are appropriate tracers of these coastal processes because they are primarily and continuously produced in rocks or sediments, and thus, they are supplied to the ocean by land-ocean (i.e., solid-water) interactions. Moreover, Ra isotopes behave conservatively in seawater and have a wide range of half-lives, allowing to trace coastal processes on various spatial and temporal scales (see Sect. 4.3.2). Another example of tracers of land-ocean interaction processes is ^{222}Rn ($T_{1/2} = 3.82$ days), which is a radioactive gas continuously produced by the decay of ^{226}Ra mainly present in aquifer solids and sediments. As a consequence, water that has been in contact with sediments becomes enriched in ^{222}Rn allowing the use of this radionuclide to trace inputs of waters that have interacted with solids (e.g., groundwater discharge, exchange of seawater across the sediment-water interface) (e.g., Burnett and Dulaiova 2006; Rodellas et al. 2018).

4.2.2.4 Sedimentation Processes

Marine sediments are a major reservoir for particle-reactive elements. Hence, they are natural archives that allow the reconstruction and evaluation of changes that occurred in the marine environment in the past. To that end, a detailed knowledge of sedimentation processes and the ability to identify the age of the deposited sediments are required. This can be achieved by studying the depth distribution of radionuclides contained in the sediments. Radioactive tracers of sedimentation processes have generally a global and long-term source. They are also radionuclides that either attach onto biogenic or lithogenic particles or are incorporated into marine organisms, sinking with them and accumulating on the seabed. For example, the ^{210}Pb present in the water column (produced by the decay of ^{222}Rn both in the atmosphere and in the water column) is scavenged by sinking particles on their transit to the ocean interior. Another example is ^{14}C, a radionuclide from atmospheric origin that enters the oceanic carbon cycle by air-sea exchange, is incorporated into organisms (e.g., carbonate shells), and is ultimately deposited on the seabed. Once particles or organisms reach the sediments and "isolate" from the water column, the scavenged or incorporated radionuclides begin to decrease exponentially as a consequence of radioactive decay. This exponential decrease is determined by the decay constant of each radionuclide (i.e., "built-in clock") and can be used to determine the age of a given sediment layer or to estimate sedimentation rates. As mentioned before, radionuclides are suitable tracers for examining processes taking place over time scales comparable to their half-lives (from one to seven $T_{1/2}$; see Sect. 4.1.1.2). For instance, ^{14}C ($T_{1/2} = 5730$ years) is effective for time scales in the order of thousands of years (e.g., Lougheed et al. 2020), while

^{210}Pb ($T_{1/2} = 22.3$ years) is widely used as a tracer of sedimentation processes for the past ~20–150 years (e.g., Arias-Ortiz et al. 2018). Other short-lived particle-reactive radionuclides are suitable to quantify sedimentation rates from several months (e.g., ^{7}Be; ^{234}Th; e.g., Lecroart et al. 2005) to a decade (e.g., ^{228}Th; e.g., Smoak et al. 1999). Anthropogenic radionuclides produced by bomb testing (e.g., ^{137}Cs) and deposited onto the ocean via global fallout (i.e., a pulse-like source) also represent a useful tool to identify specific sediment layers: a peak of ^{137}Cs in a sediment profile can be assumed to represent the period of 1954–1963, when most of the atmospheric nuclear tests were performed, or it can help to pinpoint when a nuclear accident happened, such as the Chernobyl nuclear power plant accident in 1986 (e.g., Paradis et al. 2017).

4.2.2.5 Atmosphere-Ocean Interaction

Radionuclides have also been successfully applied to trace the bidirectional exchange between the atmosphere and the ocean. This is of relevance because atmospheric deposition and air-sea gas exchange processes are key pathways that connect the ocean and the atmosphere, impacting global biogeochemical cycles and climate.

Atmospheric deposition can be an important source of nutrients (e.g., iron) to the marine environment and impacts marine life, particularly in areas where primary production is limited by the lack of nutrients (e.g., Jickells et al. 2005). Estimates of atmospheric deposition can be determined by using particle-reactive radionuclides that are produced in the atmosphere, such as the cosmogenic radionuclide ^{7}Be ($T_{1/2} = 53.3$ days). ^{7}Be is supplied to the ocean surface by wet (rain and snow) and dry deposition attached to aerosols (Young and Silker 1980). Once in the ocean, ^{7}Be does not have a strong affinity for particulate matter and can therefore be used as a conservative tracer in particle-poor waters (Andrews et al. 2008). As a result, measurements of ^{7}Be in the upper water column allow the calculation of the atmospheric ^{7}Be flux. This ^{7}Be flux together with aerosol chemical analysis can then be used to determine the atmospheric supply of biologically important chemical species, such as trace metals, to the water column (e.g., Kadko et al. 2015). Another example of a tracer of atmospheric deposition is the long-lived primordial radionuclide ^{232}Th ($T_{1/2} = 1.4 \cdot 10^{10}$ years). The input of ^{232}Th to the open ocean is dominantly derived from dust deposition, in contrast to the shorter-lived Th radionuclides (^{230}Th, ^{228}Th, and ^{234}Th), which are primarily produced in seawater. Despite its long half-life, the ^{232}Th present in surface waters is used as a proxy for recent dust deposition due to the high particle reactivity of Th and its subsequent removal by particle scavenging (e.g., Hsieh et al. 2011). On the other hand, ^{232}Th in marine sediments is used to determine dust deposition over millennial and glacial-interglacial time scales (e.g., Kienast et al. 2016).

Air-sea gas exchange is a key component of the global cycle of greenhouse gases such as CO_2, N_2O, or CH_4. Under the current context of rapid climate change, it is imperative to quantify air-sea fluxes of greenhouse gases to allow projections of possible future climate scenarios. In this regard, radioactive gases have been used as tracers of fluxes between the atmosphere and the ocean. For instance, ^{14}C produced during bomb testing and incorporated as CO_2 in upper ocean waters can be used to evaluate the air-sea gas exchange of soluble gases (e.g., Sweeney et al. 2007). Another example is ^{222}Rn, produced in the water column by the decay of ^{226}Ra. Without any gas exchange, ^{222}Rn in the water column would reach secular equilibrium with ^{226}Ra. However, the upper water column usually shows a deficiency of ^{222}Rn relative to the expected secular equilibrium that accounts for the ^{222}Rn lost by sea-air gas exchange and thus allows determining such fluxes (e.g. Bender et al., 2011; Rutgers van der Loeff et al. 2014).

4.3 Case Studies for the Application of Radionuclides as Ocean Tracers

Natural and anthropogenic radionuclides have been used to provide information on key ocean processes, including water movement, fluxes across interfaces, or the movement of particles and solutes in the water column (Fig. 4.9). To illustrate the potential of radionuclides in marine studies and to highlight the characteristics that allow using them, here we will provide three comprehensive examples: the pair $^{234}Th/^{238}U$ to trace the biological pump, Ra isotopes to quantify the magnitude of submarine groundwater discharge, and ^{129}I to study ocean circulation. We have selected these three examples to cover different marine processes and because they rely on different properties of the radionuclides: (a) the disequilibrium between parent (^{238}U) and daughter (^{234}Th) radionuclides that provides information on the process causing this partitioning, (b) known half-lives of Ra isotopes that allow constraining removal rates in a mass balance, and (c) the characteristic input function of ^{129}I that serves as a fingerprint of water masses.

4.3.1 The $^{234}Th/^{238}U$ Pair as a Tracer of the Biological Pump

The thorium-234/uranium-238 ($^{234}Th/^{238}U$) pair is extensively used as an ocean tracer of particle export, particularly to obtain quantitative information on the biological pump and the marine carbon cycle. This section aims to provide a brief overview of the biological pump and to describe, in a practical way, how the $^{234}Th/^{238}U$ pair can be used to quantify particle export fluxes.

4.3.1.1 What Is the Biological Pump and Why Is It Important?

The biological pump is a major mechanism by which the ocean removes CO_2 from the atmosphere and sequesters it in the deep ocean. In the surface of the ocean, photosynthetic organisms convert dissolved CO_2 into particulate organic carbon (POC), and a series of biological and physical processes transfer a portion of this carbon to the ocean interior (Volk and Hoffert 1985). As POC is exported through the water column, it is consumed and eventually converted back to CO_2, a process known as remineralization. The depth at which POC is remineralized depends on a variety of factors (e.g., particle sinking, bacterial activity, etc.) and determines for

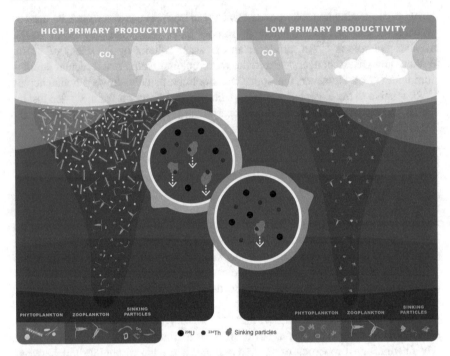

Fig. 4.10 The ^{234}Th/^{238}U pair as a tracer of the biological pump. Two scenarios are illustrated: one representative of a productive area with higher export flux associated with sinking particles (left) and a low productive area with lower export flux (right). The productive scenario shows enhanced export of ^{234}Th (red dots) associated with sinking particles relative to the less productive scenario. Purple dots represent atoms of ^{238}U (designed by gemmasola.com)

how long carbon is stored or sequestered in the ocean (Kwon et al. 2009). In general terms, the deeper the organic matter is exported, the longer this carbon will remain isolated from the atmosphere. There are several export pathways for organic matter to reach the deep ocean, but here we will focus on what is generally considered the largest one: the gravitational sinking of particles (Boyd et al. 2019) (Fig. 4.10).

The magnitude of POC export driven by the biological pump varies widely across spatial and temporal scales, and measurements of particle export are sparse and heterogeneously distributed in the ocean. As a result, estimates of POC export in the global ocean range by more than a factor of 2 (e.g., Boyd and Trull 2007; Henson et al. 2011), limiting our ability to quantify the ocean carbon sink in the present and to predict future changes. Apart from carbon, the biological pump modulates the distribution of many other elements in the ocean, including major nutrients and trace metals that are essential for phytoplankton growth, particle-reactive elements, or toxic compounds. Therefore, the study of particle export in the ocean is of broad interest not only for constraining the cycling and fate of carbon but also other chemical species that are biogeochemically important or relevant to humans.

Measuring the export of POC associated with the sinking of particles is challenging. The only direct way to measure sinking particle fluxes in the ocean is by deploying sediment traps during relatively short periods of time during oceanographic expeditions or by using moored sediment traps at ocean observatories. Therefore, indirect methods are essential to increase the number of observations across the ocean and decrease the uncertainty associated with the magnitude of the ocean carbon sink. One of the most common indirect methods for quantifying the magnitude of sinking particle fluxes in the ocean is the use of parent-daughter radionuclide pairs. As summarized in Sect. 4.2.2.2, these pairs of radionuclides have two key characteristics: (a) the daughter radionuclide has a *shorter half-life* than its parent, which means that if both radionuclides are added to and removed from the system only by radioactive decay, parent and daughter would be in secular equilibrium (i.e., their activities would be the same), and (b) the daughter radionuclide has *high affinity* for particles (i.e., it is the particle tracer), whereas its parent is less particle-reactive or characterized by a conservative behavior in seawater. This difference in particle reactivity between the parent and daughter radionuclides leads to the preferential scavenging and removal of the daughter radionuclide from seawater via particle scavenging. The result is a deviation from secular equilibrium that allows the quantification of particle export in the ocean.

4.3.1.2 Why Is the ^{234}Th/^{238}U Pair an Ideal Tracer of Particle Export?

The radioactive disequilibrium between ^{234}Th and ^{238}U in oceanic surface waters was first reported by Bhat et al. (1968), but it was not until the 1980s that the scavenging of ^{234}Th was shown to be a function of primary production and proposed to trace particle export. Indeed, the application of this pair as a proxy for POC export as we know it today commenced in the 1990s (Buesseler et al. 1992) and has been widely used since then (Le Moigne et al. 2013; Puigcorbé et al. 2020), motivated by important methodological advances in analyzing ^{234}Th in seawater as well as the need to quantify POC export at high spatial and temporal resolution.

The ^{234}Th/^{238}U pair ticks all the boxes of the ideal tracer to quantify particle export and the strength of the biological pump:

- First, ^{234}Th has a much shorter half-life than its parent ^{238}U: ^{234}Th is a short-lived radionuclide ($T_{1/2} = 24.1$ days) that is constantly produced in seawater by alpha decay of the longer-lived ^{238}U ($T_{1/2} = 4.5 \cdot 10^9$ years). In addition, ^{234}Th has a half-life that is suitable to study the biological pump over time scales of several days to weeks, which are similar to the biological and physical processes that influence sinking particle fluxes in the upper ocean.
- Second, Th and U have contrasting chemical properties in seawater and, therefore, behave very differently: Th has one of the strongest affinities for particle surfaces among all elements (International Atomic Energy Agency 1985), whereas U is conservative in well-oxygenated seawater and is tightly coupled to salinity. As a consequence, ^{234}Th attaches to particles, and, if these particles sink, ^{234}Th sinks with them (Fig. 4.10), breaking the secular equilibrium between ^{234}Th and ^{238}U.

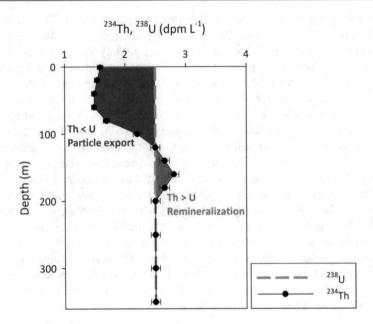

Fig. 4.11 Typical ^{234}Th (black circles) and ^{238}U (vertical dashed gray line) activity profile vs. depth. Dark blue area shows ^{234}Th deficit (^{234}Th/^{238}U activity ratios <1) indicative of particle export. The orange area shows ^{234}Th excess (^{234}Th/^{238}U activity ratios >1) indicative of remineralization and/or particle disaggregation

- And lastly, the methods to quantify the activity of ^{234}Th and ^{238}U in seawater are well-established and relatively simple (see Box 4.3. of Sect. 4.4.1.3).

Figure 4.11 shows the typical depth profile of ^{234}Th and ^{238}U in seawater. In the upper tens or hundreds of meters of the water column, ^{234}Th activities are lower than those expected from radioactive decay of ^{238}U due to adsorption and removal of ^{234}Th by sinking particles, creating a deficit of ^{234}Th in the upper water column (i.e., ^{234}Th/^{238}U activity ratios <1). With increasing water depth and decreasing particle concentration, activities of ^{234}Th increase towards secular equilibrium with ^{238}U (i.e., ^{234}Th/^{238}U activity ratios = 1). In subsurface waters, excess of ^{234}Th (i.e., ^{234}Th/^{238}U activity ratios >1) can often be identified due to remineralization or disaggregation of ^{234}Th-carrying particles exported from overlying waters.

4.3.1.3 How to Quantify Particle Export Using the ^{234}Th/^{238}U Pair?

In this section we will describe, step by step, how the ^{234}Th/^{238}U pair can be used as a tracer of POC export by taking a study conducted in the Northwest Atlantic Ocean as an example (Puigcorbé et al. 2017). We will provide a practical example of flux calculations using data from the Demerara Plain (station PE33; ~13°N, ~53°W) on June 26, 2010, and then will show the distribution of POC export observed by Puigcorbé et al. (2017) across the Northwest Atlantic based on the ^{234}Th/^{238}U method.

Step 1: Quantification of ^{234}Th Fluxes Due to Particle Scavenging

To use the ^{234}Th/^{238}U pair as a tracer of POC export, first of all, the disequilibrium between ^{234}Th and ^{238}U in the upper water column needs to be well defined. To do that, one has to collect seawater samples (2–4 L each) at high vertical resolution in the upper hundreds of meters of the water column to determine ^{234}Th activities following the radiochemical method briefly described in Box 4.3. of Sect. 4.4.1.3. In contrast, ^{238}U activities can be directly inferred from salinity data (e.g., Owens et al. 2011). ^{234}Th and ^{238}U activities are typically given in units of disintegrations per minute per liter of water (dpm L^{-1}; 1 Bq = 60 dpm).

Once we obtain the depth profile of ^{234}Th and ^{238}U, the next step is to calculate ^{234}Th export fluxes associated with sinking particles at a reference depth or depth horizon (see example in Box 4.1). The magnitude of the flux of ^{234}Th at a given depth will depend on the magnitude of the disequilibrium between ^{234}Th and ^{238}U observed in overlying waters. ^{234}Th fluxes can be described by a suite of scavenging models of varying complexity in the representation of particle and sorption dynamics (Savoye et al. 2006). Here we will focus on the simplest model, the one-box scavenging model, which does not differentiate between particulate and dissolved ^{234}Th and, therefore, can be used when total (dissolved + particulate) activities of ^{234}Th in the water column are measured. The one-box model represents the change of total ^{234}Th activity with time as the balance between continuous production of ^{234}Th from ^{238}U in seawater, radioactive decay and removal of ^{234}Th by particle export, and the inputs or outputs of ^{234}Th by physical transport processes:

$$\frac{\partial A_{Th}}{\partial t} = A_U \, \lambda_{Th} - A_{Th} \, \lambda_{Th} - P_{Th} + V \qquad (4.12)$$

where A_U and A_{Th} are the ^{238}U and ^{234}Th total activities integrated from surface waters down to the depth horizon of interest (dpm m^{-2}), λ_{Th} is the radioactive decay constant of ^{234}Th (0.02876 day^{-1}), P_{Th} is the net particle export flux of ^{234}Th (dpm m^{-2} day^{-1}), and V is the sum of the advective and diffusive fluxes (dpm m^{-2} day^{-1}). Traditionally, the depth horizon has been fixed at 100 or 150 m, regardless of the conditions encountered at the study site at the time of sampling. However, more recently, fluxes have been increasingly assessed relative to a biogeochemically relevant reference depth (e.g., base of the euphotic zone) to improve comparisons across different oceanic regions and help understand the mechanistic controls on the biological pump (Buesseler et al. 2020b).

In the open ocean, the net particle export flux of ^{234}Th is usually determined by assuming steady-state conditions ($\frac{\partial A_{Th}}{\partial t} = 0$) and neglecting physical transport processes ($V = 0$). In this scenario, P_{Th} can be calculated by integrating the deficit of ^{234}Th with respect to ^{238}U from surface waters to the depth horizon of interest and multiplying it by the decay constant of ^{234}Th:

$$P_{Th} = \lambda_{Th}(A_U - A_{Th}) \qquad (4.13)$$

Box 4.1 ^{234}Th Flux Calculations

The depth profile of ^{234}Th and ^{238}U activities in the upper 250 m of the water column at station PE33 (Puigcorbé et al. 2017) is represented in the following figure. At this station, a deficit of ^{234}Th relative to ^{238}U was evident in the upper 100 m of the water column at the time of sampling indicating the removal of ^{234}Th from seawater due to particle export. Secular equilibrium between ^{234}Th and ^{238}U was reached at 100 m.

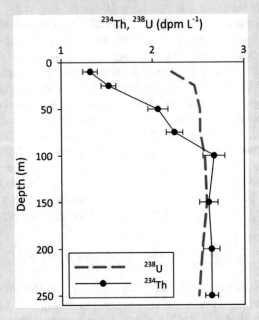

In this specific case, 5 samples were collected between the surface and 100 m. The data from those five depths, which are necessary to calculate the ^{234}Th flux, is represented in the table below. First, the ^{234}Th deficit at each of these depths has to be calculated by subtracting the ^{234}Th activity from the ^{238}U activity. Then, the ^{234}Th deficit is integrated down to 100 m by dividing the upper 100 m of the water column into 5 layers, given the sampling depth resolution at this station. The thickness of each layer is determined using a midpoint integration method. The final ^{234}Th flux at 100 m is then obtained by multiplying the sum of the integrated ^{234}Th deficit in all layers by the ^{234}Th decay constant: 51,625 dpm m^{-2} · 0.02876 day^{-1} = 1485 dpm m^{-2} day^{-1}.

(continued)

Box 4.1 (continued)

The associated uncertainty of this ^{234}Th flux is calculated by propagating the uncertainties of the activities of ^{234}Th and ^{238}U. Hence, the final ^{234}Th flux at 100 m would be reported as 1490 ± 140 dpm m^{-2} day^{-1}.

Depth (m)	^{234}Th activity (dpm L^{-1})	^{238}U activity (dpm L^{-1})	^{234}Th deficit (dpm L^{-1})	Layer thickness (m)	Integrated ^{234}Th deficit (dpm m^{-2})
10	1.32 ± 0.08	2.21 ± 0.04	0.89 ± 0.09	17.5	$15{,}575 \pm 1565$
25	1.52 ± 0.08	2.46 ± 0.04	0.94 ± 0.09	20.0	$18{,}800 \pm 1789$
50	2.06 ± 0.11	2.52 ± 0.04	0.46 ± 0.12	25.0	$11{,}500 \pm 2926$
75	2.24 ± 0.09	2.52 ± 0.04	0.28 ± 0.10	25.0	7000 ± 2462
100	2.67 ± 0.12	2.57 ± 0.04	-0.10 ± 0.13	12.5	-1250 ± 1581
				100	**$51{,}625 \pm 4772$**
				Flux at 100 m (dpm m^{-2} day^{-1})	**1485 ± 137**

The use of Eq. (4.13) has been shown to provide accurate flux estimates during specific temporal windows in study areas where physical transport or spatial variability in ^{234}Th activities is minimal. However, when the study site can be reoccupied over time (i.e., sampling of the same water masses), potential temporal changes in ^{234}Th activities ($\frac{\partial A_{Th}}{\partial t} \neq 0$) due to changing particle export conditions, such as during the development of a phytoplankton bloom, can be captured using a non-steady-state model. Moreover, when ^{234}Th is measured at an appropriate vertical and spatial resolution, the physical transport term can also be included in the model (i.e., V in Eq. (4.12); (e.g., Savoye et al. 2006; Resplandy et al. 2012)) by considering the impact of advection and diffusion on the measured ^{234}Th activities in seawater, which can be important in dynamic regimes (e.g., upwelling areas or eddy fields).

Step 2: Conversion from ^{234}Th Fluxes to POC Fluxes

In order to quantify export fluxes of POC (P_{POC}) or any other element of interest, the ^{234}Th flux has to be multiplied by the ratio of POC (or the element of interest) to ^{234}Th associated with sinking particles at the chosen depth horizon:

$$P_{POC} = P_{Th} \left(\frac{POC}{Th} \right)_{Sinking\ particles} \tag{4.14}$$

POC/^{234}Th ratios vary regionally, temporally, and with depth depending on multiple factors such as particle source, sinking velocity, and food-web dynamics (Buesseler et al. 2006; Puigcorbé et al. 2020) (Fig. 4.12). To accurately estimate export fluxes, it is essential to use the POC/^{234}Th ratio that corresponds to the composition of sinking material at a given location, time, and depth. These ratios are usually estimated by measuring POC and ^{234}Th in particles collected using

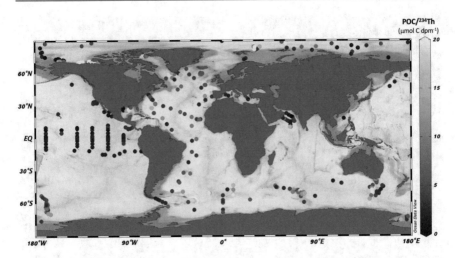

Fig. 4.12 Compilation of POC/^{234}Th ratios measured in >50 μm particles collected using large volume filtration via in situ pumps deployed at 100 ± 10 m (adapted from Puigcorbé et al. 2020). The color scale is cut off at 20 μmol C dpm^{-1}, and therefore maximum values correspond to >20 μmol C dpm^{-1}. (Figure prepared using Ocean Data View (Schlitzer 2021))

sediment traps and/or particles of a specific size range obtained by filtration. When sediment traps cannot be deployed, particles larger than >50 μm obtained by large volume filtration are usually considered to be the particle size class that best represents sinking particles. The determination of this parameter is often the greatest source of uncertainty in the application of ^{234}Th as a proxy for POC export. Hence, collecting as many particulate POC/^{234}Th ratio profiles as possible together with total ^{234}Th profiles during a given observation period is encouraged, as well as using different particle sampling methods to set upper and lower limits for POC export.

Example: Changes of POC Export Fluxes Across the Northwest Atlantic

By following the process described above and using Eqs. (4.13) and (4.14), Puigcorbé et al. (2017) estimated the magnitude of POC export fluxes along a transect in the Northwest Atlantic from 64°N to the equator from early May to early July in 2010. The transect covered five oceanic domains (subpolar, temperate, oligotrophic, riverine, equatorial) with contrasting physical and biogeochemical characteristics, including a wide range of chlorophyll-a concentrations (often used as a proxy for phytoplankton biomass; Fig. 4.13).

As mentioned before, the magnitude of POC export varies strongly over spatial and temporal scales, and those changes are strongly related to biological factors, such as phytoplankton biomass (Fig. 4.13), food-web structure, or bacterial remineralization. Therefore, the contrasting biogeochemical regions sampled lead to differences in the magnitude of ^{234}Th export, POC/^{234}Th ratios, and, therefore, POC export fluxes (Fig. 4.14).

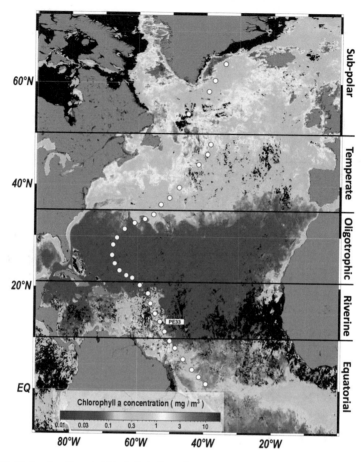

Fig. 4.13 Station locations (white circles) along the northern GEOTRACES GA02 transect in the Northwest Atlantic overlaying the mean chlorophyll-*a* concentrations (mg m^{-3}) derived from MODIS AQUA remote sensing data (http://oceancolor.gsfc.nasa.gov) at the sampling time (i.e., May for subpolar and temperate domains, May–June for oligotrophic domain, June for riverine domain, and June–July for equatorial domain). Black horizontal lines divide the five oceanic domains. Station PE33, used as example in Box 4.1, is shown within the riverine domain. (adapted from Puigcorbé et al. 2017)

Figure 4.14 shows ^{234}Th fluxes at 100 m ranging from negligible in the oligotrophic domain to >2700 dpm m^{-2} day^{-1} in the subpolar domain. Large variability in POC/^{234}Th ratios at 100–150 m was also observed, especially in dynamic domains where storms, seasonal phytoplankton blooms, or external inputs of nutrients have a different impact across stations. These ^{234}Th fluxes and POC/^{234}Th ratios resulted in POC fluxes ranging from negligible in the oligotrophic domain to 28 mmol C m^{-2} day^{-1} in the subpolar domain at 100 m (Fig. 4.14). The highest average POC fluxes were found in the riverine domain (12 ± 5 mmol C m^{-2} day^{-1}), suggesting that the Amazon River outflow resulted in enhanced particle export in

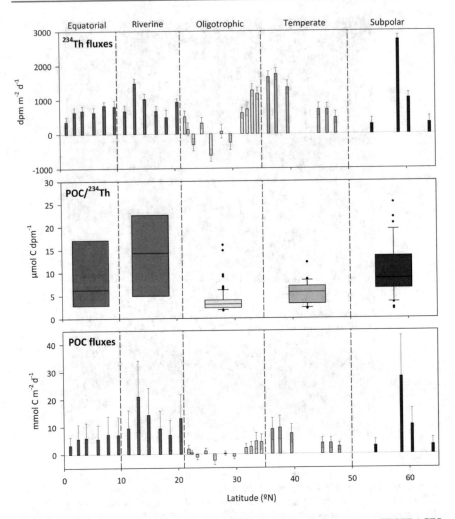

Fig. 4.14 ^{234}Th and POC export fluxes estimated at 100 m along the northern GEOTRACES GA02 transect. POC/^{234}Th ratios are from a compilation of 12 studies conducted in the North Atlantic Ocean obtained from large (>51 to 53 μm) or sediment trap particles collected at 100–150 m. The average POC/^{234}Th ratio in each domain was used to estimate the POC fluxes at the stations sampled along the northern GA02 transect (data from Puigcorbé et al. 2017)

this region. The subpolar domain also presented high average POC fluxes (11 ± 12 mmol C m^{-2} day^{-1}), although highly variable, most likely due to strong weather conditions and patchiness of phytoplankton blooms. The temperate and equatorial domains showed similar average POC fluxes (6.0 ± 3.0 and 5.7 ± 1.5 mmol C m^{-2} day^{-1}, respectively), whereas the oligotrophic domain presented by far the lowest POC fluxes (1.2 ± 2.2 mmol C m^{-2} day^{-1}) along the transect linked to low primary productivity in the area (see low chlorophyll-*a* concentrations in Fig. 4.13). This general distribution of POC export across domains is consistent with previous studies conducted in the North Atlantic (see details in Puigcorbé et al. 2017).

In this study, the use of the radionuclide pair ^{234}Th/^{238}U allowed the quantification of the magnitude of POC export fluxes across a variety of bioregions in the Northwest Atlantic Ocean. This type of large-scale assessment is difficult to be achieved with direct methods of particle flux measurements (i.e., sediment traps) and provides important quantitative information that is critical to understand the links between bioregion-specific characteristics and carbon export.

4.3.1.4 Closing Remarks

The ^{234}Th/^{238}U pair has been used for several decades to quantify the biological pump and has shed valuable insights into particle export, particularly in the open ocean, over a range of spatial scales. International programs, such as GEOTRACES (https://www.geotraces.org), have allowed the measurement of isotopes in the ocean over large spatial scales, including the ^{234}Th/^{238}U pair (see example in Sect. 4.3.1.3). On the other hand, programs such as EXPORTS (https://oceanexports.org) have recently used ^{234}Th/^{238}U to quantify sinking particle fluxes at localized areas with measurements at an unprecedented high resolution (Buesseler et al. 2020a). With the need to reduce the uncertainties of POC export in the global ocean and the increasing number of interdisciplinary programs devoted to studying the biological pump, it is an exciting time to use ^{234}Th/^{238}U as an ocean tracer. The ^{234}Th/^{238}U method can provide one of the highest spatial resolutions of export estimates, but like any other method, it has its limitations. Therefore, combining ^{234}Th data with other particle export methods, such as sediment traps or parent-daughter radionuclide pairs with different time scales (e.g., ^{210}Po/^{210}Pb), is particularly useful to improve our understanding of particle export and their associated elements in the ocean. Interdisciplinary field campaigns, along with the development of new technologies, provide a unique opportunity to expand observations in the ocean and put export flux data into the context of the multiple components of the biological pump. This will be key to characterize the spatial and temporal variability of the biological pump and decrease the uncertainty associated with ocean carbon sink estimates at regional and global scales in the present and the future.

4.3.2 Ra Isotopes as Tracers of Submarine Groundwater Discharge

Radium (Ra) isotopes have been widely used as tracers of land-ocean interactions, particularly to quantify the magnitude of fluxes of water and solutes supplied by groundwater to the coastal ocean (the so-called submarine groundwater discharge, SGD). In this section we will briefly describe the basic concepts of SGD, explain the principles that make Ra isotopes ideal tracers of SGD, and provide a practical example to highlight the relevance of using these radionuclides in coastal oceanography.

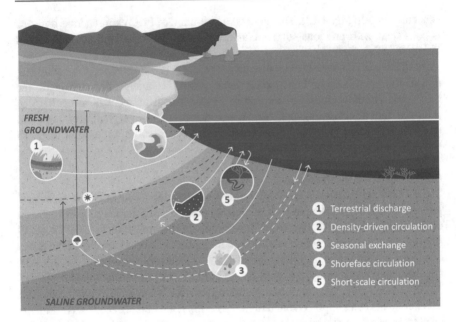

FRESH
GROUNDWATER

SALINE GROUNDWATER

1 Terrestrial discharge
2 Density-driven circulation
3 Seasonal exchange
4 Shoreface circulation
5 Short-scale circulation

Fig. 4.15 Conceptual diagram representing the discharge of groundwater to the coastal ocean, i.e., SGD. The different icons represent the different mechanisms that are responsible for SGD fluxes to the ocean: (1) terrestrial SGD (usually the discharge of fresh groundwater); (2–5) marine SGD, including density-driven seawater circulation (2), seasonal exchange of seawater (3), shoreface seawater circulation driven by tidal inundation and wave set-up (4), and short-scale recirculation driven by multiple processes such as bioirrigation and wave and tidal pumping (5). (adapted from Garcia-Orellana et al. 2021)

4.3.2.1 What Is Submarine Groundwater Discharge and Why Is It Important?

The coastal ocean plays a key role in modulating the transfer of solutes (e.g., nutrients, metals, contaminants) from land to the ocean. Quantifying their sources across the land-ocean interface is thus essential to understand the biogeochemical cycles in the coastal zone and in the open ocean. While riverine inputs have long been documented as a major source of freshwater, sediments, and solutes to the coastal ocean, the importance of groundwater has been often overlooked. This is mainly because the discharge of groundwater occurs below the water surface and cannot be easily observed or measured. Over the past two decades, this invisible subsurface water flow has received increased attention, and SGD is now recognized as a major supplier of solutes to coastal waters (Santos et al. 2021).

The discharge of groundwater to the coastal ocean commonly referred to as SGD (Fig. 4.15) includes both fresh groundwater originated from the recharge inland (terrestrial SGD) and seawater that infiltrates into the coastal aquifer and returns to the ocean with a different chemical composition (marine SGD) (Moore 2010). Inputs of terrestrial SGD to the ocean are expected to occur where an aquifer with a positive hydraulic gradient is connected with the sea through permeable rocks or sediments, a condition that is met in most coastal areas. This terrestrial component of SGD represents in most cases a minor fraction of total SGD, although it can be locally important and constitute the main source of freshwater to specific coastal sites (Luijendijk et al. 2020). The major component of SGD on a global scale is the circulation of seawater through the coastal aquifer (marine SGD), which is driven by several mechanisms occurring in most world's shorelines (e.g., wave action, tidal pumping, bioirrigation) (Santos et al. 2012; Kwon et al. 2014). Both terrestrial and marine SGD usually have significantly greater concentrations of nutrients and trace elements than seawater because of anthropogenic and natural sources to groundwater and the biogeochemical transformations occurring in the coastal aquifer (Slomp and Van Cappellen 2004). Therefore, even small SGD inputs can represent a large contribution of solutes (e.g., nutrients, metals, contaminants) to the sea, sometimes rivaling that of riverine sources. Recent studies provide increasing evidence that groundwater is an important source of solutes to bays and coves, salt marshes, coral reefs, or entire ocean basins (e.g., Rodellas et al. 2014; Cho et al. 2018). These SGD-driven inputs of solutes can have profound impacts for coastal ecosystems, such as supporting bacterial communities and primary producers, triggering algal blooms, or promoting eutrophication of coastal waters or hypoxia events (Lecher et al. 2018; Ruiz-González et al. 2021). Hence, understanding the magnitude of SGD as well as the fluxes of solutes supplied by this process is of great importance for coastal and open ocean biogeochemical cycles.

While SGD is a key process occurring in most of the world's shorelines, it is usually difficult to detect and quantify because the discharge occurs below the water surface and it is often patchy and variable with time. In this context, geochemical tracers of SGD have emerged as powerful tools to quantify the magnitude of SGD, as well as the associated fluxes of dissolved compounds. An ideal tracer of SGD should (a) be highly enriched in coastal groundwater relative to seawater, so that SGD constitutes a relevant source of the tracer to the coastal ocean; (b) behave conservatively in seawater; and (c) integrate the SGD signal over a broad area and time period, smoothing the spatiotemporal heterogeneities of the process. The four isotopes of Ra (^{226}Ra, ^{228}Ra, ^{223}Ra, ^{224}Ra), as well as radon (^{222}Rn), meet these constraints reasonably well, and these naturally occurring radionuclides are today the most widely applied tracers of SGD (Garcia-Orellana et al. 2021). These radionuclides have been applied to quantify SGD in a wide range of marine environments, including sandy beaches, bays, coastal lagoons, estuaries, salt marshes, ocean basins, and the global ocean. In the following subsections, we will provide some insights on the use of Ra isotopes to quantify SGD.

4.3.2.2 Why Are Ra Isotopes Ideal Tracers of SGD?

There are four naturally occurring isotopes of Ra present in the environment: ^{226}Ra ($T_{1/2} = 1600$ years), ^{228}Ra ($T_{1/2} = 5.75$ years), ^{223}Ra ($T_{1/2} = 11.3$ days) and ^{224}Ra ($T_{1/2} = 3.66$ days). Ra isotopes belong to the natural decay series, and thus they are continuously produced by the decay of their immediate Th parents (Fig. 4.6), which are insoluble and highly particle-reactive, so they are bound to the geological matrix (i.e., particles, sediments, rocks). Ra isotopes are more mobile than their Th parents, particularly in brackish to saline environments where Ra bound in the surface of solids is removed via cationic exchange with other cations (e.g., Na^+, K^+, Ca^{2+}) and remains in solution as a dissolved divalent cation (Ra^{2+}) or forming stable complexes with anionic ligands (e.g., $ClRa^+$, $RaSO_4$, $RaCO_3$). The contrasting chemical behavior between Ra and Th is responsible for the partitioning of these isotopes in aquatic environments, allowing a continuous Ra enrichment of waters in contact with sediments or rocks. This enrichment of Ra in groundwater supplied by the decay of Th in the aquifer solids constitutes the basis for using Ra isotopes as tracers of SGD.

Owing to their properties, Ra isotopes are considered ideal tracers to quantify SGD fluxes of water and associated solutes. The key attributes that make Ra isotopes powerful tracers of SGD are:

- Ra isotopes are significantly *enriched in coastal groundwater* relative to seawater (up to 1–2 orders of magnitude), as a consequence of the continuous production of Ra from the Th bound to aquifer solids. Therefore, even small inputs of SGD to the coastal ocean can be recognized by a strong Ra signal.
- Ra isotopes in coastal areas are usually primarily *sourced from SGD*. Ra anomalies in the coastal ocean can thus be mainly attributed to SGD inputs, allowing to estimate SGD-driven fluxes even though the other potential Ra sources (e.g., riverine inputs, sediment diffusion) are not precisely constrained.
- Ra *behaves conservatively* in the ocean on the time scales of coastal mixing processes. As a consequence, Ra is mainly removed from the coastal system via transport offshore (mixing with open ocean waters with low Ra activities) and radioactive decay, which can be easily constrained through the well-known decay constants of the different Ra isotopes.
- Ra isotopes have a *wide range of half-lives*, ranging from 3.66 days to 1600 years. Since isotopes are best suited to evaluate processes occurring at temporal scales comparable to their half-lives, the availability of four Ra isotopes allows examining SGD over a variety of time scales, from few days (e.g., ^{224}Ra to estimate SGD in a coastal bay) to tens of years (e.g., ^{228}Ra to estimate SGD in an ocean basin). Ra isotopes can also be combined to estimate water ages or mixing rates because all Ra isotopes are affected identically by mixing processes, but their removal rate is significantly different depending on their half-life (Moore 2000). The different Ra isotopes can additionally be used to infer the relative contribution of SGD inflowing from different systems (e.g., aquifers) because different systems are likely to have different isotopic signatures (i.e., ratios between Ra isotopes) (e.g., Diego-Feliu et al. 2021).
- SGD estimates derived from Ra isotopes *integrate the SGD signal* over a broad area and time period. This allows integrating the inherent spatiotemporal heterogeneities of the process and overcoming the limitations of point-like measurements.

- Ra isotopes can be *accurately measured* in seawater samples using well-established methods: chiefly, the RaDeCC system and gamma detectors (see Sect. 4.4.1).

4.3.2.3 How to Quantify SGD Using Ra Isotopes?

There are different ways of quantifying SGD by using Ra isotopes, but the simplest and most widely applied strategy is based on conducting a mass balance of Ra in the study area, in which all the potential sources and sinks of Ra are considered. The mass balance allows determining the Ra flux supplied by SGD as the difference between inputs and outputs. Finally, if the Ra concentration in SGD is well constrained, the Ra flux can be converted into a volumetric SGD flow. In this section we will briefly describe the two steps to determine SGD via Ra isotopes and present a practical example of SGD determination in the Mediterranean Sea.

Step 1: Determination of Ra Fluxes Supplied by SGD: The Ra Mass Balance

If a marine system (e.g., ocean basin, bay, gulf, coastal lagoon, etc.) is assumed to be in steady state, total Ra losses should be balanced by continuous additions of Ra into the system. By characterizing all the Ra sinks and sources (with the only exception of SGD, which is the unknown targeted process), the flux of Ra supplied by SGD (F_{SGD}; in dpm day^{-1} (or Bq day^{-1})) can be quantified as the difference between total outputs (F_{out}) and inputs (F_{in}):

$$F_{SGD} = F_{out} - F_{in} \qquad (4.15)$$

Determining the SGD-driven flux of any Ra isotope requires thus first identifying all the relevant sources and sinks of the isotope and then appropriately constraining their fluxes. Potential sources (F_{in}) and sinks (F_{out}) of Ra isotopes in the coastal ocean are shown in Fig. 4.16. Aside from SGD, Ra sources include riverine discharge or inputs from other surface water bodies (including Ra dissolved in river water and remobilized from riverine particles), atmospheric input from wet or dry deposition, inputs from seafloor sediments via diffusion, bioturbation or resuspension, inputs from offshore or adjacent water systems, and the production from the decay of their Th parents in the water column. On the contrary, Ra is mainly lost from the system by radioactive decay and export of Ra offshore or to adjacent water bodies. Additionally, Ra can also be supplied to or removed from the water column through a variety of biological or chemical processes (e.g., particle scavenging, coprecipitation with other compounds or minerals), which are often referred to as internal cycling (Fig. 4.16).

Step 2: Determination of the Fluxes of Water and Nutrients Supplied by SGD

The flux of Ra supplied by SGD to the system under study (determined from the mass balance in Step 1) needs to be converted into a volumetric water flow to provide meaningful information on the magnitude of SGD. To do so, one needs to characterize the concentration of Ra isotopes in groundwater discharging to the ocean, a term that is often referred to as the *SGD endmember*. Constraining the Ra

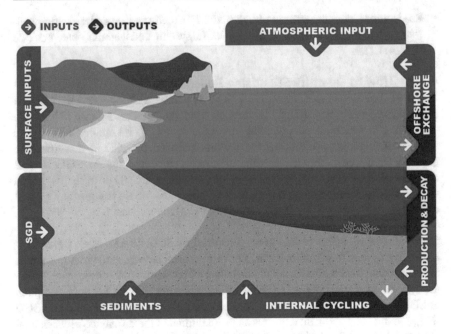

Fig. 4.16 Generalized box model summarizing all the processes supplying or removing Ra isotopes from the coastal ocean (adapted from Garcia-Orellana et al. 2021)

concentration in the SGD endmember is not a simple task because Ra concentrations can vary significantly at each study site. Consequently, a representative value for this endmember is usually constrained by measuring Ra concentrations in a large number of groundwater samples, including near-shore piezometers and wells, porewaters collected at the seafloor, or direct measurements of SGD actually discharging to the system via springs or using benthic chambers (Cook et al. 2018). Once both the SGD-driven Ra flux (F_{SGD}; in dpm day^{-1}) and the Ra concentration in the SGD endmember (C_{Ra-SGD}; in dpm m^{-3}) are characterized, the flow of SGD (Q_{SGD}; in m^3 day^{-1}) can be simply estimated as follows:

$$Q_{SGD} = \frac{F_{SGD}}{C_{Ra-SGD}} \tag{4.16}$$

In a similar manner, fluxes of nutrients (or other solutes) supplied by SGD to the coastal ocean can be estimated by multiplying the Ra-derived water flux (Q_{SGD}) by the nutrient concentration in groundwater discharging to the ocean (i.e., the nutrient SGD endmember). However, as in the case of the Ra endmember, it is often difficult to determine a nutrient concentration representative of SGD because nutrients are modified in the coastal aquifer as a consequence of several biogeochemical transformations. A commonly used approach to overcome this limitation consists in collecting samples that cover the whole spectrum of possibilities (e.g., different salinities, different geologies, including wells and porewaters). This allows constraining the possible variability of this term, providing a lower and an upper limit estimate of the fluxes of nutrients supplied by SGD.

Example: SGD to the Mediterranean Sea

We will illustrate the application of Ra isotopes following an example in which ^{228}Ra was used to quantify SGD in the entire Mediterranean Sea (based on Rodellas et al. 2015). The Mediterranean Sea is a semi-enclosed basin and is considered one of the most oligotrophic seas in the world (Bethoux et al. 1998). It thus relies on external inputs of nutrients to sustain its marine productivity. Atmospheric deposition and riverine runoff have been traditionally considered the main sources of nutrients to the Mediterranean Sea (Guerzoni et al. 1999; Ludwig et al. 2009), but the role of SGD as a conveyor of nutrients to the Mediterranean Sea had been largely overlooked.

In this study, ^{228}Ra was used instead of other Ra isotopes because its half-life ($T_{1/2} = 5.75$ years) is considerably lower than the residence time of the upper Mediterranean Sea waters (~100 years). The decay of ^{228}Ra is thus a dominant sink term in this basin. Since the ^{228}Ra lost due to radioactive decay can be easily estimated from the ^{228}Ra inventory in the system and its decay constant, using ^{228}Ra for the Mediterranean Sea allows accurately constraining the total ^{228}Ra removal terms. Beyond radioactive decay, additional sinks of ^{228}Ra are particle scavenging (almost negligible) and the export of Mediterranean waters to the Atlantic Ocean and Black Sea at the boundaries of this basin (through the straits of Gibraltar and Dardanelles, respectively).

Assuming that ^{228}Ra concentrations in the Mediterranean Sea are in steady state, total ^{228}Ra outputs should be balanced by continuous inputs of this isotope. Major sources of Ra isotopes to the Mediterranean Sea (aside from SGD) include riverine discharge, atmospheric deposition, supply from seafloor sediments, as well as inputs from the Atlantic Ocean and the Black Sea. In the case of this study, the mass balance was restricted to the upper water column (<600 m depth) because the main ^{228}Ra inputs occur at the uppermost layers. In doing so, ^{228}Ra released from deep sediments can be neglected, but vertical exchange fluxes need to be considered.

Quantifying the source and sink terms in the mass balance (Eq. 4.15) requires determining the ^{228}Ra concentration in the different water masses (e.g., Atlantic, Black Sea, and deep waters) and sources (e.g., Ra dissolved in rivers, desorbed from riverine and atmospheric particles, in sediments), as well as the ^{228}Ra inventory in the study area (i.e., the total amount of ^{228}Ra in the upper Mediterranean Sea). This inventory can be obtained by a detailed characterization of the ^{228}Ra distribution in the water column of the basin, which was achieved by collecting more than 100 samples throughout the Mediterranean Sea during several oceanographic cruises and complemented with additional data from the literature. Details on sampling for ^{228}Ra and its measurement by gamma spectrometry are presented in Box 4.4. in Sect. 4.4.1.3.

The distribution of ^{228}Ra in the Mediterranean Sea is shown in Fig. 4.17. The highest ^{228}Ra concentrations are observed in coastal areas, which is consistent with Ra inputs from continental margins. ^{228}Ra concentrations measured along the

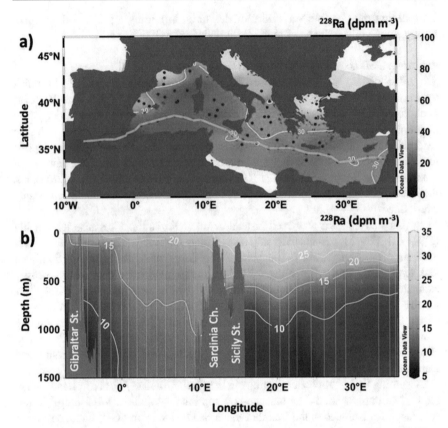

Fig. 4.17 Distribution of ^{228}Ra concentrations (dpm m^{-3}) in the Mediterranean Sea: (**a**) ^{228}Ra concentrations in surface waters, where dots indicate the sampled stations; (**b**) ^{228}Ra concentrations in the upper 1500 m of the water column in a W-E transect (light gray line in panel **a**) across the Mediterranean Sea (see details in Rodellas et al., 2015). (Figures prepared using Ocean Data View (Schlitzer 2021))

Mediterranean Sea mainly reflect the thermohaline circulation of the basin: relatively fresh Atlantic waters with low ^{228}Ra concentrations enter through the Gibraltar Strait within the upper 100–200 m of the water column. Surface waters move eastwards and continuously receive ^{228}Ra from the continental margins at a greater rate than radioactive decay or other forms of removal, resulting in ^{228}Ra concentrations generally increasing from west to east. Intermediate waters (found at 200–600 m) are normally formed in the Eastern Mediterranean basin and circulate westwards towards the Strait of Gibraltar. Since at intermediate depths there are no other major sources of ^{228}Ra, ^{228}Ra concentrations in these intermediate waters generally decrease from east to west. ^{228}Ra concentrations in deep waters (>600 m) are considerably lower, confirming that the main ^{228}Ra inputs occur in the upper layers.

Once the ^{228}Ra distribution in the Mediterranean Sea is well understood, the total amount of ^{228}Ra in the upper (<600 m) Mediterranean Sea can be estimated (total inventory of $35 \cdot 10^{15}$ dpm). The different input and output fluxes of ^{228}Ra to the Mediterranean Sea were quantified combining direct measurements and data available in the literature, as shown in Table 4.2. Total outputs and inputs are largely dominated by the radioactive decay (>60% of total outputs) and the sediment flux from the continental shelf and slope (~40% of total inputs). The results show a difference of $(1.8 \pm 1.2) \cdot 10^{15}$ dpm year^{-1} between the total loss of ^{228}Ra (F_{out} in Eq. 4.15) and the inputs from the evaluated sources (F_{in} in Eq. 4.15), which can only be balanced by an additional source of ^{228}Ra from the continent. This flux is thus equivalent to the flux of ^{228}Ra supplied by SGD (F_{SGD} in Eq. 4.15).

To convert the SGD-derived ^{228}Ra flux into an SGD flow, the ^{228}Ra concentration in the SGD inflowing to the Mediterranean Sea needs to be characterized ($C_{Ra\text{-}SGD}$; Eq. 4.16). In the Mediterranean basin, several studies have

Table 4.2 Terms in the ^{228}Ra mass balance for the upper Mediterranean Sea, including the method of estimation, the estimated fluxes, and the proportion that these fluxes represent of the total outputs or inputs. Details on flux estimates for each term are reported in Rodellas et al. (2015)

Term	Calculation	^{228}Ra flux (10^{15}) dpm year^{-1}	^{228}Ra flux % outputs
Decay	Total ^{228}Ra inventory multiplied by decay constant ($\lambda = 0.120$ year^{-1})	4.2 ± 0.7	64
Advection to deep waters	^{228}Ra concentration in the different areas multiplied by water transfer rates	1.95 ± 0.13	30
Particle scavenging	Total ^{228}Ra inventory multiplied by Ra removal rate (~0.002 year^{-1})	0.07 ± 0.01	1.1
Outflow (Gibraltar)	^{228}Ra concentration in Gibraltar outflowing waters multiplied by outflow rates	0.30 ± 0.03	4.5
Outflow (Dardanelles)	^{228}Ra concentration in Dardanelles outflowing waters multiplied by outflow rates	0.06 ± 0.01	0.9
Total outputs		**6.6 ± 0.7**	**100**
Rivers	^{228}Ra concentration in rivers (dissolved + in particles) multiplied by water and sediment fluxes	0.40 ± 0.29	6.0
Atmospheric dust	Dust input multiplied by maximum ^{228}Ra desorption from particles	0.08 ± 0.03	1.2
Sediments	Continental shelf and slope areas (<600 m) multiplied by ^{228}Ra diffusive fluxes (differentiating fine-grained, coarse-grained, and slope sediments)	2.7 ± 0.9	41
Advection from deep waters	^{228}Ra concentration in deep waters multiplied by water transfer rates	1.1 ± 0.2	16
Inflow (Gibraltar)	^{228}Ra concentration in Atlantic Ocean multiplied by inflow rates	0.37 ± 0.03	5.7
Inflow (Dardanelles)	^{228}Ra concentration in the Black Sea multiplied by inflow rates	0.14 ± 0.01	2.1
Total inputs		**4.7 ± 1.0**	**72**
SGD	Difference between total outputs and inputs	**1.8 ± 1.2**	**28**

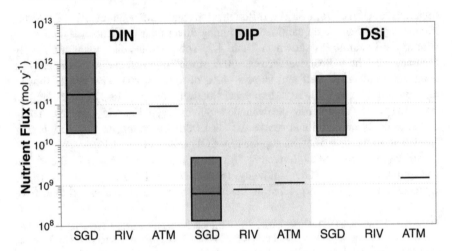

Fig. 4.18 Comparison of nutrient inputs (*DIN* dissolved inorganic nitrogen, *DIP* dissolved inorganic phosphorus, *DSi* dissolved silica) to the Mediterranean Sea supplied by SGD, rivers (RIV), and atmospheric deposition (ATM) (adapted from Rodellas et al. 2015)

reported ^{228}Ra data in SGD (45 different locations), with concentrations spanning a wide range (130–72,000 dpm m^{-3}; see references in Rodellas et al. 2015). A ^{228}Ra concentration of 640–2200 dpm m^{-3} covers the range between the first and third quartiles of all the data in SGD from the Mediterranean Sea, accounting for the expected variability in SGD and excluding extreme values. Using this range as C_{Ra-SGD} (Eq. 4.16), the resulting SGD flow to the Mediterranean Sea ranges from $0.2 \cdot 10^{12}$ to $4.3 \cdot 10^{12}$ m^3 year^{-1}. The magnitude of SGD appears to be comparable or most likely larger than the riverine discharge to the Mediterranean Sea ($\sim 0.3 \cdot 10^{12}$ m^3 year^{-1}; Ludwig et al. 2009).

Considering the estimated magnitude of SGD to the entire basin and the high concentrations of nutrients usually found in coastal groundwater, SGD could represent a relevant source of nutrients to the Mediterranean Sea. The flux of bioavailable nutrients (dissolved inorganic nitrogen (DIN), phosphorus (DIP) and silica (DSi)) supplied by SGD can be estimated by multiplying the SGD flow by the nutrient concentrations reported in SGD studies along the Mediterranean Sea coastline (see references in Rodellas et al. 2015). This results in SGD-driven fluxes of $(20–1300) \cdot 10^9$ mol year^{-1} of DIN, $(0.09–4.0) \cdot 10^9$ mol year^{-1} of DIP, and $(13–570) \cdot 10^9$ mol year^{-1} of DSi (Fig. 4.18). Fluxes of nutrients from SGD are comparable or even higher than those supplied by atmospheric and riverine sources, revealing that SGD is likely a major external source of DIN, DIP, and DSi to the Mediterranean Sea (Fig. 4.18). Importantly, unlike atmospheric inputs (distributed throughout the basin) and riverine discharge (restricted to river mouth surroundings), SGD is essentially ubiquitous in coastal areas (Trezzi et al. 2016). Thus, the relative significance of SGD as a source of nutrients may be even higher for most of the

Mediterranean coastal areas, where SGD can represent by far the dominant source of dissolved nutrients and the major sustainment for coastal ecosystems.

The use of Ra isotopes ([228]Ra, in this case) allowed constraining the magnitude of SGD and its associated nutrient inputs to the entire Mediterranean Sea, a process that had been largely overlooked. These radioactive tracers revealed that SGD might play a relevant role on coastal biogeochemical cycles in the Mediterranean Sea.

4.3.2.4 How to Quantify Transport Time Scales Using Ra Isotopes?

In addition to SGD studies, Ra isotopes have also been of particular value in quantifying residence times of solutes in coastal environments and mixing processes between coastal waters and the open ocean (Moore 2000; Charette et al. 2016). The retention of solutes in coastal ecosystems and their transport between the coastal and open ocean continuum are key parameters directly related to the quality of the coastal ecosystem, the vulnerability to pollution threats (e.g., accumulation of contaminants, risk of eutrophication), and the structure and productivity of ocean ecosystems.

Activities (or concentrations) of Ra isotopes are expected to decrease offshore mainly as a consequence of mixing with low-Ra offshore waters and radioactive decay (different removal rates depending on the half-life of each Ra isotope). On the contrary, the ratios between the activities of any pair of Ra isotopes (i.e., activity ratios) are expected to change only as a consequence of radioactive decay, since other processes, such as mixing, will not affect the activity ratios. Thus, Ra activity ratios can provide key information on water residence times and mixing rates in the coastal ocean.

There is a suite of models of varying complexity that allow determining transport time scales in the coastal ocean using a combination of Ra isotopes (e.g., Moore 2015; Lamontagne and Webster 2019). Here we will focus on the simplest approach, which allows estimating the apparent ages of coastal waters, i.e., the time elapsed since a water parcel isolates from the source of Ra. In an ideal system where all the Ra inputs occur at the shoreline and the ocean has negligible Ra concentrations, activity ratios of different pairs of Ra isotopes are expected to decrease with time as shown in Fig. 4.19. Thus, by comparing the activity ratios of a pair of Ra isotopes in the source (AR_{In}) and in any water parcel (AR_{SW}), its apparent age (τ) can be estimated as follows (Moore 2000):

$$\tau = \frac{\ln{(AR_{In})} - \ln{(AR_{SW})}}{\lambda_s - \lambda_l} \tag{4.17}$$

where λ_s and λ_l are the decay constants of the shorter-lived and longer-lived Ra isotope, respectively. The activity ratio of any pair of Ra isotopes can be used to estimate water ages, provided that the half-life of the shorter-lived Ra isotope is appropriate for the time scales expected in the study area. Imagine that the [224]Ra/[223]Ra activity ratio of a water parcel is 1/3 (33.3%) of the activity ratio of the Ra source (see Fig. 4.19). In this case, the water age of the parcel can be estimated to be 8.5 days by using Eq. (4.17). For instance, Sanial et al. (2015)

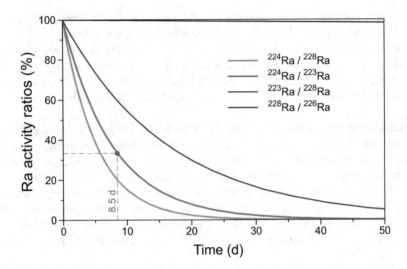

Fig. 4.19 Change of activity ratio of a pair of Ra isotopes as a function of the time a water parcel has been disconnected from the Ra source (water age or apparent age, see Eq. 4.17). Ra activity ratios are expressed in % (i.e., activity ratio in the source, $AR_{In} = 100\%$). Notice that the $^{228}Ra/^{226}Ra$ activity ratio only decreases by 2% in 50 days, and thus these isotopes cannot be used to estimate water ages on a short temporal scale

used $^{224}Ra/^{223}Ra$ activity ratios to estimate young water ages (4–8 days) in the northern Kerguelen Plateau, whereas Kipp et al. (2019) combined ^{228}Ra and ^{226}Ra to estimate residence times on the order of ~20 years in intermediate waters of the western Arctic Ocean.

4.3.2.5 Closing Remarks

Submarine groundwater discharge (SGD) has historically been largely overlooked in coastal oceanography mainly because it is difficult to identify and quantify. The development of approaches based on Ra isotopes has decisively contributed to characterize the magnitude of the fluxes supplied by SGD to the coastal ocean and has made possible the identification of SGD as a major source of nutrients to coastal ecosystems. Despite the growing understanding of this process, there are key science questions related to SGD that remain open. There is, for instance, the need to improve the understanding of nutrient transformations in coastal aquifers or to predict how ongoing climate change will modify the quantity and quality of SGD (see Santos et al. 2021). Approaches based on Ra isotopes are crucial to contribute to answering these research questions. However, to further build the understanding of SGD at local and global scales, these approaches should be combined with other methods and techniques (e.g., other tracers, hydrogeological modeling, benthic chambers). Indeed, since SGD is a process occurring at the land-ocean interface and involves traditionally independent disciplines (hydrogeology, biogeochemistry, oceanography, microbiology, etc.), the fragmentation of points of view has hindered the understanding of this process. Interdisciplinary efforts that combine multiple approaches and provide a holistic view of SGD are thus crucial to foster the understanding of SGD.

4.3.3 Anthropogenic Radionuclides as Tracers of Ocean Circulation

Anthropogenic radionuclides (also known as artificial radionuclides) have been widely used as tracers of ocean circulation since they were first incorporated into the environment in the 1950s. In this section we will introduce some of the key circulation processes that can be traced by means of anthropogenic radionuclides in the Arctic Ocean and the Subpolar North Atlantic (SPNA). The focus will be on the artificial iodine-129 (^{129}I) as a tool to understand the pathways, temporal changes, and transport time scales of circulation in these important regions.

4.3.3.1 What Is Ocean Circulation and Why Is It Important?

Ocean circulation is the large-scale movement of waters in the ocean basins, and it plays a fundamental role in modulating Earth's climate. Heat and CO_2 are critical climate properties that are constantly exchanged between the sea surface and the atmosphere. In coastal areas, essential nutrients and pollutants are also introduced to the surface of the ocean constituting a vital resource, or a threat, for marine life. All properties taken up by oceans are transported by marine currents and water masses within a few hours to months in a regional domain or over longer time scales across different ocean basins. By means of ocean circulation, heat, carbon, and nutrients are sequestered down to abyssal depths where they are stored for long periods of time. Therefore, the ocean plays a crucial role in controlling global budgets of nutrients, heat, and carbon. Among all oceans, there are two regions that are particularly important in the current context of climate change: the Arctic Ocean and the SPNA (IPCC 2021). The Arctic region is warming at least twice as fast as the rest of the planet, and its sea ice extent has significantly declined over the past 40 years (https://arctic.noaa.gov/). Recent studies have demonstrated that the inflow of Atlantic waters bringing heat to the polar region is one of the major causes for the melting and thinning of the Arctic sea ice (Polyakov et al. 2017). Consequently, Arctic changes in heat and freshwater fluxes can modify the water density at afar locations and contribute to a possible slowdown of the Atlantic Meridional Overturning Circulation (AMOC) in the SPNA (Sévellec et al. 2017). As it happens, anthropogenic radionuclides emerge as ideal tools to trace key components of the AMOC in the SPNA and the circulation of Atlantic Waters in the Arctic Ocean (Smith et al. 2011; Casacuberta et al. 2018; Castrillejo et al. 2018; Wefing et al. 2021).

4.3.3.2 Why Are Anthropogenic Radionuclides Ideal Tracers of Ocean Circulation?

Anthropogenic radioactivity refers to a wide range of radionuclides associated with nuclear activities conducted worldwide since the 1950s that have entered the ocean either by point-like or global sources and in a pulse-like or long-term fashion

(Fig. 4.7). Their pathways to the marine environment are either through direct inputs (e.g., liquid waste discharge, riverine inputs) or atmospheric deposition (Fig. 4.8), depending on their sources. Those having a conservative behavior in seawater and a long half-life might be suitable to study ocean circulation. What makes anthropogenic radionuclides especially different compared to the naturally occurring ones explained previously (i.e., ^{238}U, ^{234}Th, and Ra isotopes) is that (a) they did not exist in the environment before their introduction by human activities, or their natural levels were generally negligible compared to man-made inputs (i.e., of artificial origin); and (b) they have not reached steady state in the oceans because of their recent introduction (i.e., these radionuclides are only present in relatively young waters). Additionally, the use of anthropogenic radionuclides as ocean circulation tracers relies on knowing the sources and the amounts released by each source.

Sources of Anthropogenic Radionuclides to the Ocean

The largest source of many anthropogenic radionuclides to the ocean is the testing of *nuclear weapons* in the atmosphere that caused the global dispersion and deposition of radioactivity, known as *global fallout* (Fig. 4.20). The peak of radioactive deposition in the ocean surface took place during and immediately after the 1960s when thermonuclear weapons injected radioactive debris above the atmosphere's tropopause dispersing radionuclides worldwide. This can be observed taking ^{90}Sr as an example (Fig. 4.20, inset plot 1 for global fallout), where the peak represents the greatest deposition due to the testing of weapons with the highest explosion yield (see also Box 4.2). The vast amounts of ^{90}Sr and ^{137}Cs, among other radionuclides, fingerprinted surface seawaters globally, thus serving as powerful tracers of ocean circulation (Povinec et al. 2005).

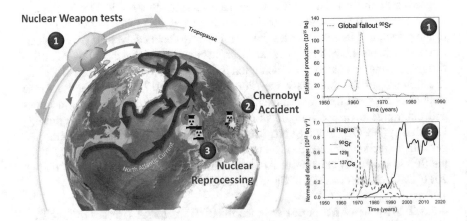

Fig. 4.20 Principal sources of anthropogenic radionuclides in the Arctic-North Atlantic region. The blue arrows represent the transport pathways of liquid waste released from European nuclear fuel reprocessing plants as they join the schematic, large-scale circulation represented by the red arrows

Another source of anthropogenic radioactivity is related to *nuclear accidents* (Fig. 4.20; see e.g., Chernobyl). The accidents at the Chernobyl (1986) and Fukushima Dai-ichi (2011) nuclear power plants resulted in relatively large inputs of anthropogenic radionuclides into the marine environment. To put a quantitative example for ^{137}Cs, whereas weapon tests released about $9.5 \cdot 10^{17}$ Bq to the atmosphere, the Chernobyl and Fukushima accidents added $1.0 \cdot 10^{17}$ and $1.5 \cdot 10^{16}$ Bq, respectively (Steinhauser 2014; Buesseler et al. 2017). Although the two accidents had very different characteristics (land vs. ocean deposition), both produced a large pulse- and point-like release that imprinted the atmosphere and nearby seas (i.e., the Black Sea and Mediterranean Sea by Chernobyl and the Pacific Ocean by Fukushima).

Box 4.2 Example of Weapon Test ^{90}Sr as Tracer of Water Circulation in the North Atlantic

The figure below represents the distribution of ^{90}Sr across the western Atlantic Ocean in the 1970s as part of the GEOSECS program (https://odv.awi.de/data/ocean/geosecs/). The presence of ^{90}Sr is derived from the testing of nuclear weapons. Units are reported in decays per minute (dpm) in 100 L of seawater (1 dpm 100L^{-1} translates to 0.17 Bq m^{-3}). The highest concentrations of ^{90}Sr were observed in surface waters of the northern Atlantic Ocean, coinciding with greater deposition rates of global fallout at the northern hemisphere compared to the southern hemisphere (UNSCEAR 2000). One of the highlights of this figure is the penetration of ^{90}Sr to greater depths via dense water formation in the high latitudinal North Atlantic. The creation of North Atlantic Deep Waters is a crucial component of the AMOC (Frajka-Williams et al. 2019).

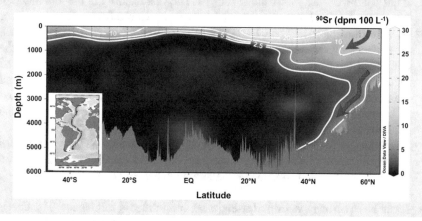

The third major input of anthropogenic radioactivity into the oceans originates from the *nuclear fuel reprocessing plants* (Fig. 4.20). These are facilities dedicated to recover U and Pu from spent nuclear fuel for reuse in reactors, which are authorized to release limited amounts of the low-level radioactive waste that is produced during reprocessing activities. The release of nuclear waste occurs both as gaseous discharges and as liquid effluents through pipes into the ocean (Povinec et al. 2005). The latter are of utmost interest to oceanographers as they label marine waters with a characteristic radionuclide composition. There are currently two main nuclear reprocessing plants: Sellafield, formerly called Windscale, which discharges to the Irish Sea, and Cap La Hague, which discharges to the English Channel (Fig. 4.20). Discharges from these facilities occurred since the 1950–1960s in a time-dependent fashion for ^{137}Cs, ^{90}Sr, and ^{129}I, among many others (Fig. 4.20, inset plot 3 only shown for La Hague).

Once the abovementioned sources are well-constrained in time and space, any radionuclide behaving conservatively in seawater and having a half-life long enough to be tracked in the following decades after its supply to the ocean can be explored as a potential tracer of ocean circulation. The three isotopes mostly used for that purpose are ^{137}Cs, ^{90}Sr, and ^{129}I (Livingston and Jenkins 1983; Povinec et al. 2005).

4.3.3.3 How Can ^{129}I Help Study the Circulation in the Arctic Ocean and the SPNA?

The use of ^{129}I as a tracer of ocean circulation processes in the Arctic and North Atlantic Oceans emerged after the 1990s, following the large increase of releases from the nuclear reprocessing industry to the English Channel and Irish Sea. The main attributes that make ^{129}I a powerful tracer to study ocean circulation are:

- It is *conservative* in seawater. The long residence time of iodine in the open ocean (ca. 10^5 years) suggests a nearly conservative behavior. And, because ^{129}I has a very long half-life of about $1.6 \cdot 10^7$ years, its oceanic distribution is mainly shaped by ocean circulation.
- It has *point-like, well-constrained* sources. The vast majority of the ^{129}I found in the Arctic and North Atlantic Oceans has been introduced from a combination of releases from Sellafield and La Hague. Because of the close location to each other, these two sources are usually considered a single source located at the outflow of the North Sea (Fig. 4.20). Sellafield and La Hague have introduced much more ^{129}I (>6000 kg) than nuclear weapon tests (100–150 kg globally) and the Chernobyl accident (<7 kg) (He et al. 2013); thus, the plume of reprocessed ^{129}I can easily be detected by its high concentrations that are well above the expected background levels (sum of global fallout, Chernobyl, and natural).
- Its *discharges* from the nuclear reprocessing plants (mostly La Hague) *increased sharply* after the 1990s (Fig. 4.20; see inset plot 3 for La Hague) making ^{129}I a very valuable tool to estimate circulation time scales of waters in the Arctic and SPNA (e.g., Smith et al. 2005, 2011).
- It can be *accurately measured* using mass spectrometric techniques. The development of accelerator mass spectrometry (AMS) and its further optimization in the 1980s allow today the measurement of this isotope at low concentrations (about 10^5 atoms L^{-1}) in less than 500 mL of seawater (see Box 4.6 in Sect. 4.4.2).

In the following subsections, we will present three examples in which ^{129}I can be used as a powerful tool to unravel circulation pathways and temporal changes in the Arctic Ocean. We will also demonstrate how transport time scales of water masses in the SPNA can be reconstructed by means of this tracer. In all cases, the basis is that the signal of ^{129}I exiting the North Sea is then transported northwards to the Nordic Seas, therefore imprinting the Atlantic waters that either circulate through the Arctic, the SPNA, or both regions.

Example 1: Pathways of Atlantic Waters in the Arctic Ocean

Between 2012 and 2016, several expeditions in the Arctic Ocean included the sampling of water for the analysis of ^{129}I. Figure 4.21 shows the distribution of ^{129}I concentrations for the surface layer (Fig. 4.21a) and the Atlantic layer (Fig. 4.21b). Although the Atlantic layer normally occupies depths between 200 and 1000 m, here we only represent the upper branch of it (240–300 m), also known as Fram Strait Branch Water (FSBW).

In these plots we observe that the surface water generally shows higher concentrations (up to $1.2 \cdot 10^{10}$ atoms L^{-1}) compared to the Atlantic layer (up to $6.0 \cdot 10^9$ atoms L^{-1}). The reason for this is that waters entering with the Norwegian Coastal Current (NCC) carry a greater signal of the nuclear reprocessing plants, compared to the deeper waters carried by the North Atlantic Current (NAC) (Fig. 4.20) which only contained the global fallout signal, i.e., very low ^{129}I (Smith et al. 2011). The high surface concentrations entering from the Barents Sea opening recirculate within the Eurasian Basin, where concentrations up to $6.0 \cdot 10^9$ atoms L^{-1} are still observed over the North Pole, thus indicating a rather advective

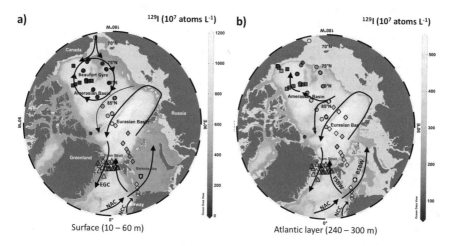

Fig. 4.21 Distribution of ^{129}I in the Arctic Ocean between 2012 and 2016 for the (**a**) surface layer and (**b**) the Atlantic mid-depth layer (240–300 m), also known as Fram Strait Branch Water (FSBW). *EGC* East Greenland Current, *NAC* North Atlantic Current, *NCC* Norwegian Coastal Current, *BSBW* Barents Sea Branch Water, *FSBW* Fram Strait Branch Water. (Figures prepared using Ocean Data View (Schlitzer 2021))

flow of this Atlantic surface branch (Fig. 4.21a) (Casacuberta et al. 2018; Wefing et al. 2021). The lowest ^{129}I concentrations ($<1.0 \cdot 10^9$ atoms L^{-1}) are observed in the Amerasian Basin and are representative of Pacific waters entering the Arctic Ocean through the Bering Strait. Similar to the NAC, Pacific waters only contain global fallout ^{129}I. The contrast of high/low ^{129}I in Arctic polar surface waters is therefore a powerful tool to mark the Atlantic-Pacific water front (Karcher et al. 2012).

Finally, the distribution of ^{129}I concentrations at mid-depths (Fig. 4.21b) suggests that Atlantic waters (and more precisely the FSBW) penetrate quite far from their entrance gate into the Arctic Ocean. The signal of ^{129}I reaches the Amerasian Basin, and its distribution suggests new pathways of circulation that were not described previously (Smith et al. 2021).

Example 2: Changes of Surface Circulation in the Arctic Ocean

The ^{129}I concentrations in the Arctic Ocean have not always been the same. The primary reason is that the ^{129}I discharge rate from the nuclear reprocessing plants also varied over time. However, the ^{129}I distribution can additionally change due to variations in the circulation patterns, something that has been revealed by both observations and models (Karcher et al. 2012; Smith et al. 2021). In Fig. 4.22, ^{129}I concentrations at polar surface waters in 1987, 1997, and 2015 from the NAOSIM model output illustrate these shifts in circulation patterns, which are associated with two main atmospheric patterns (anticyclonic and cyclonic), which alternate at 5–7-year intervals (Morison et al. 2012; Proshutinsky et al. 2015).

During anticyclonic regimes (Fig. 4.22a, c), high sea-level atmospheric pressure dominated over the Arctic Ocean driving upper ocean water (and sea ice) clockwise. The atmosphere was cold and dry, and the freshwater flux (composed of river, sea ice melt, and Pacific water) from the Arctic to the subarctic seas was reduced (Proshutinsky et al. 2015). During these regimes, the Beaufort Gyre is expanded, hindering the circulation of surface Atlantic waters through the Canada Basin (Karcher et al. 2012). By contrast, during cyclonic circulation regimes (Fig. 4.22b), low sea-level atmospheric pressure dominated driving sea ice and upper ocean waters counterclockwise. Under that situation, the Arctic atmosphere was relatively warm and humid, and the freshwater flux from the Arctic Ocean towards the subarctic seas intensified. In this regime, the Beaufort Gyre is reduced in size allowing for surface Atlantic waters carrying ^{129}I to penetrate into the Canada Basin.

The intensity and duration of the two atmospheric circulation regimes (anticyclonic and cyclonic) are not always predictable, and repeated observations and simulations of oceanic ^{129}I (Figs. 4.21 and 4.22) have proved the relationship of changes in Arctic ocean circulation to shifts in atmospheric patterns. During the expeditions in 2012–2016, the Arctic was under anticyclonic circulation conditions, and ^{129}I revealed the location of the Pacific-Atlantic front over the Lomonosov Ridge (Fig. 4.22c). On the contrary, in the mid-1990s (under strong cyclonic conditions), the Pacific-Atlantic front was further into the Amerasian Basin (Fig.4.22b), and the mid-depth Atlantic layer was confined to the Russian and Canadian shelves as a boundary current (Smith et al. 2021).

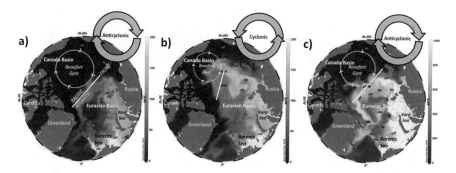

Fig. 4.22 Distribution of ^{129}I (units are in 10^7 atoms L^{-1}) at the surface from an Arctic regional ocean model (NAOSIM) in (**a**) 1987, when an anticyclonic regime dominated; (**b**) 1997, when a strong cyclonic regime dominated; and (**c**) 2015, again with anticyclonic patterns. (Figure kindly provided by M. Karcher)

Example 3 Circulation Time Scales to the Deep Labrador Sea in the Subpolar Region

The SPNA hosts the transformation of northward-flowing warm subtropical waters carried by the NAC (Fig. 4.23a) into southward-flowing, colder, and denser waters that convect down to abyssal depths (also known as ventilation process). Southward-flowing waters are enriched with reprocessing-derived ^{129}I, which is supplied from the Arctic and Nordic Seas via the Greenland-Scotland passages (Fig. 4.23a). On the contrary, waters from lower latitudes carried by the NAC only contain global fallout ^{129}I (i.e., low concentrations), thus making this tracer ideal to identify the transport of deep waters formed in the SPNA (Edmonds et al. 2001; Smith et al. 2005; Alfimov et al. 2013; Castrillejo et al. 2018, Castrillejo et al., 2022).

Figure 4.23a shows the distribution of ^{129}I observed at 2900 m depth in the SPNA between 2014 and 2018. The ^{129}I concentrations were well above $8.0 \cdot 10^8$ atoms L^{-1} in the Labrador Sea in comparison to the eastern part of the section that showed tracer values in the global fallout range. The deep layers of the western SPNA are occupied by the Denmark Strait Overflow Water (DSOW), a water mass formed in the Nordic Seas that ventilates the abyssal depths of the Atlantic Ocean. The ^{129}I concentrations in DSOW (flowing near the seabed) have rapidly increased in time as observed from the time series in the central Labrador Sea (Fig. 4.23b), and this temporal evolution can be used to infer transit times of water from the nuclear reprocessing plants to the deep Labrador Sea (Edmonds et al. 2001).

The first step to calculate transit times is to construct a tracer *input function* from the documented discharges from Sellafield and La Hague (Fig. 4.24a). The ^{129}I input function (black solid line) is obtained by mixing and diluting the ^{129}I discharged from La Hague and Sellafield with a known volume of water (carrying global fallout ^{129}I only) passing by the English Channel and the Irish Sea, respectively. The historical records for ^{129}I concentrations at a reference location of 60°N can be reconstructed either using a transfer factor of 25 Bq m^{-3}/PBq year^{-1} (Smith et al. 1999) or a mixing model (Christl et al. 2015; Wefing et al. 2021). Both cases apply a

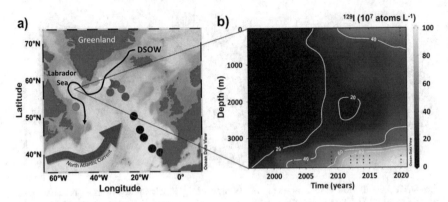

Fig. 4.23 (**a**) Distribution of ^{129}I concentrations at 2900 m depth in 2014–2018. (**b**) Temporal evolution of ^{129}I concentrations in the central Labrador Sea observed between the 1990s and the 2020s (data compiled in Castrillejo et al. 2018). *DSOW* Denmark Strait Overflow Water. (Base figures prepared using Ocean Data View (Schlitzer 2021))

delay of 4 years (for Sellafield) and 2 years (for La Hague) following the time of discharge, and the two approaches lead to very similar results. The input function therefore represents the dynamics of ^{129}I concentrations exiting the North Sea and considering homogeneous mixing of the two nuclear reprocessing plants discharges.

The second step is to compare the ^{129}I input function to the mean ^{129}I concentrations observed in the deep (2900 m) Labrador Sea, representing the DSOW (red dots in Fig. 4.24b). A dilution and a time delay to the ^{129}I input function (black solid line) is applied so that it matches the mean ^{129}I concentrations in DSOW (red dots). The time shift from the original input function to the observed data corresponds to the transit time of waters from the exit at the North Sea (60°N) to the deep Labrador Sea. In this particular case, the rapid increase in DSOW ^{129}I concentrations observed in the late 1990s probably corresponds to the arrival of the first tracer front that left the North Sea approximately 5 years earlier (dotted line in Fig. 4.24b), whereas the second increase observed in the 2010s would correspond to the arrival of a delayed and more advective second front that needed about 15 years to travel from the exit of the North Sea to the deep Labrador Sea (dashed line in Fig. 4.24b). Although this example here represents a very simple approach of the use of tracer time series to understand circulation time scales, it gives a general impression of the potential of this tracer to provide new insights into circulation pathways and time scales. Indeed, based on these findings and additional ^{129}I observations in the Arctic and Nordic Seas, different authors suggested that waters reaching the deep Labrador Sea may follow at least two different pathways (Fig. 4.20): one short loop that flows into the northwestern Nordic Seas and then returns south through the Denmark Strait into the SPNA and a second longer loop that circulates into the Arctic basins before reaching the deep Labrador Sea (Smith et al. 2005; Castrillejo et al. 2018). Other more sophisticated ways of calculating circulation times are the use of transit time distributions, which consider that a water parcel is composed by

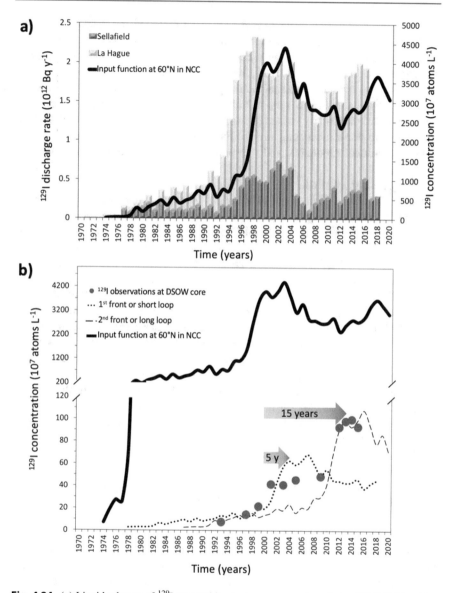

Fig. 4.24 (**a**) Liquid releases of ^{129}I reported by nuclear reprocessing plants of Sellafield and La Hague (Wefing et al. 2021) and the input function of ^{129}I at the exit of the North Sea at 60°N in the Norwegian Coastal Current (NCC). (**b**) Comparison between ^{129}I observations in the Labrador Sea and three different input functions. The solid black input function is the same as in (**a**). The dotted and dashed black input functions are estimated for the Denmark Strait Overflow Water (DSOW) core in the central Labrador Sea. The dotted line represents the arrival of the first front after following a short circulation loop of about 5 years, while the dashed line corresponds to the second tracer front that would have circulated following a longer circulation loop of about 15 years. (Figures adapted from Castrillejo et al. 2018 and Smith et al. 2005)

the mixing of waters with different tracer ages. For those willing to deepen in tracer oceanography, they are encouraged to learn about different ways of calculating time scales and mixing regimes, using a combination of tracers and applying different numerical models (Smith et al. 2011; Wefing et al. 2021).

Tracking the spread of the radioactive plume of ^{129}I from nuclear reprocessing plant discharges provides a very valuable tool to investigate ocean processes in regions that are particularly vulnerable to climate change. The three examples of this subsection provide instrumental information about the potential of ^{129}I to understand pathways, circulation patterns, and transport time scales of ocean water masses.

4.3.3.4 Closing Remarks

The use of artificial radionuclides as tracers of ocean circulation started shortly after the peak of nuclear weapon tests in the 1960s with the first comprehensive and global ocean expeditions (i.e., GEOSECS) taking place along the western Atlantic and Pacific Oceans. However, after the 1970s, the emissions of anthropogenic gases such as CFCs and SF_6 provided a cheaper tool to understand circulation and ventilation processes (Jenkins and Smethie 1996). The use of anthropogenic radionuclides (mostly ^{137}Cs) remained alive only in the domain affected by the nuclear reprocessing plant releases (i.e., Nordic Seas and Arctic Ocean). The big revival of anthropogenic radionuclides as ocean tracers has come with advancements in isotopic counting techniques. New developments and optimization of AMS and atom trap trace analysis are opening a new front in tracer oceanography. Long-lived radionuclides such as ^{236}U, ^{14}C, and ^{39}Ar can be now counted at a level of a few atoms per liter or ratios to their abundant isotope that are lower than 10^{-13} (Casacuberta et al. 2014a; Ebser et al. 2018). These achievements are bringing today new tools to physical oceanographers, adding a crucial piece of the puzzle to understanding ocean circulation, its changes due to global warming, and its future role in modulating Earth's climate.

4.4   Measurement of Radionuclides

The application of radionuclides as tracers of ocean processes relies on an accurate quantification of their concentration in marine samples (e.g., seawater, particles, sediments) using appropriate detection techniques. The oldest but (still) most common techniques to quantify radionuclides in environmental samples are based on the detection of radiation produced during the decay of radionuclides (alpha, beta, or gamma). They are commonly referred to as counting or radiometric techniques and allow to precisely quantify very low concentrations of radionuclides at relatively low costs. In some instances, mass spectrometric techniques are also used to measure radionuclides, especially for those whose half-life is too long to measure their decay accurately or when the amount of sample available is very small. Mass spectrometric techniques are based on the separation of atoms (and isotopes) according to their mass, which usually involves a greater infrastructural investment. In this section, we will provide an overview of the most commonly applied techniques to quantify radionuclides in marine samples. For a more comprehensive description, we encourage readers to consult the more extensive reviews by Hou and Roos (2008), Suckow (2009), Knoll (2010), and L'Annunziata (2013).

4.4.1 Radiometric Techniques

4.4.1.1 Basic Concepts of Radiometric Techniques

Radiometric techniques are based on detecting the particles and energy released during radioactive decay. As shown in Sect. 4.1.1.1, the decay of radionuclides usually produces the emission of charged particles (i.e., alpha or beta), followed by emission of highly energetic electromagnetic radiation (gamma rays). These particles and rays have enough energy to create pairs of ions when interacting with the atoms of any material (i.e., producing negatively charged electrons and positively charged atoms with a missing electron) and are therefore collectively referred to as ionizing radiation. Most of the radiometric techniques are indeed based on detecting the creation of pairs of charged particles produced by the interaction of ionizing radiation with the material of the detector (Fig. 4.25). If a suitable electric field is applied, the ions created can be converted into an electronic signal, which is then amplified and subsequently counted as a pulse. Some of the detectors produce a simple pulse each time an ionizing radiation interacts with the detector, regardless of the number of ion pairs directly created by the radiation. Other detectors produce

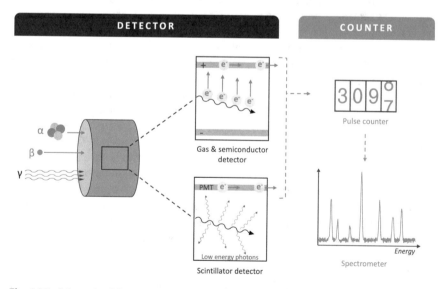

Fig. 4.25 Schematic of the counting process. Alpha, beta, and/or gamma radiation interact with the detector material, ultimately producing electrons that will be converted to a pulse that is counted. Some systems "only" count the interactions, whilst others additionally allow conducting spectrometry. The most commonly used detectors (gas, semiconductor, and scintillation) are represented in this figure

different pulses depending on the number of ion pairs initially created by the radiation arriving at the detector, which is proportional to the radiation type and energy. As a result, the latter detectors provide information on the nature of the ionizing radiation, allowing to conduct spectrometry and thus distinguish the radioactive composition of a sample (Fig. 4.25).

It is important to notice that different types of radiation interact differently with matter and thus require different radiometric techniques. Since alpha and beta are charged particles, they interact with the detector primarily through electromagnetic forces, repelling or attracting electrons from atoms by virtue of their charge, resulting in their ionization. Gamma radiation can also cause the ionization of atoms in the detector, but in this case produced by the transfer of energy from electromagnetic waves to matter. Below we will summarize the most relevant properties of these three types of radioactive decay from the perspective of their measurement:

Alpha particles of a certain radionuclide are emitted with a very well-defined energy. Thus, counting techniques able to identify and distinguish energy from different alpha particles (i.e., alpha spectrometry) have the great advantage of being able to identify the radionuclide from which the alpha particle is emitted and measure different radionuclides simultaneously. Alpha particles usually have high energy (in the order of several megaelectronvolts, MeV), but their energy is rapidly lost due to their charge and high mass, resulting in particles that can only travel a few centimeters in air. As a consequence, the detection of alpha radiation requires specific methods to prevent self-absorption within the sample and energy loss before interaction with the detector. Examples of these methods include extracting the targeted element by a physicochemical process prior to counting, using detectors that surround the sample, or measuring the sample in a vacuum chamber.

Beta particles are lighter, their energy is lower (tens to hundreds of kiloelectron volts, keV) relative to alpha particles, but they can have a range of up to few meters in air. Unlike alpha particles, the electron (or positron) released during the beta decay does not have a characteristic emission energy, but a broad distribution of energies extending to a defined maximum energy. Therefore, unlike for alpha and gamma radiation, the lack of a well-defined energy emission for beta particles makes it difficult to distinguish between different beta-emitting radionuclides. Consequently, the detection of beta radiation also requires specific methods to ensure that the analyst knows which radionuclide is emitting the pulses measured by the detector. Examples of these methods include treating the sample prior to counting to eliminate the influence of non-targeted radionuclides, or shielding radiation that does not come from the sample itself (i.e., background radiation).

Gamma radiation consists of electromagnetic waves or photons with very high energy (i.e., short wavelengths) ranging from keV to MeV. Gamma radiation is produced due to the reorganization of the radionuclide's nucleus after alpha, beta, or spontaneous fission decay. As a consequence, gamma rays with different energies can be emitted from the same radionuclide, since the nucleus reorganization is not necessarily the same after each decay event. However, these energies are discrete and very well defined, and the probabilities for the different energy emissions are well known. Thus, detectors that can measure the energy of gamma radiation and quantify the energy spectra of gamma emissions (i.e., gamma spectrometry) can be used to identify the radionuclide that emitted the radiation and quantify its abundance. This allows measuring a number of gamma-emitting radionuclides simultaneously. Gamma rays can usually penetrate a considerable distance in matter and, as a result, the radionuclides of interest do not need to be separated from the sample matrix before counting.

4.4.1.2 General Properties of Radiation Detectors

Before reviewing the different types of detectors, one needs to be familiar with four basic concepts of radiation detectors used to quantify radionuclides in environmental samples: background, temporal resolution, energy resolution, and detection efficiency (Fig. 4.26).

Background Radioactivity is a common phenomenon in the environment. Therefore, when dealing with environmental samples with relatively low radionuclide contents, it is important to consider that both the detector itself and its surroundings can represent additional sources of radioactivity. To ensure an appropriate quantification of the radionuclides of interest, it is crucial to distinguish the radioactivity produced by the sample from background sources. A common method consists in placing the detector inside a shielding (Fig. 4.26), the material and size of which would depend on the type of radiation to be analyzed. For instance, considering the short range of alpha particles, background radiation has a minimal influence on alpha measurements, and thus shielding is often not necessary. On the contrary, a thick shielding of lead is usually necessary to reduce the gamma background and conduct measurements of environmental samples with low-level activities.

Temporal Resolution When radiation interacts with the detector, the energy of the radiation is transferred to the detector through ionization, and this energy is converted into electric pulses that are subsequently quantified. After a given decay event, the detector system often needs a few micro- to milliseconds to recover and be able to record another decay event (e.g., remove ions from previous interaction, recover the electric field, etc.). The temporal resolution of a detector refers to the time needed for a detector to differentiate and record two successive decay events. Therefore, detectors with relatively long (bad) temporal resolutions (Fig. 4.26) might not be appropriate for high activity samples, since a significant number of decay events occurring in the sample might not be recorded.

Energy Resolution Some radiation detectors allow conducting spectrometry and can therefore differentiate the energy of the radiation emitted by each radionuclide. The energy resolution of a detector refers to the capacity of the detector to distinguish pulses at different energies, and it is a key property to conduct spectrometry. In general terms, the narrower the energy range of a peak is, the better the energy resolution. Detectors with bad energy resolution (i.e., wide peaks) could lead to the overlapping of peaks from different radionuclides and thus make spectrometry difficult (Fig. 4.26).

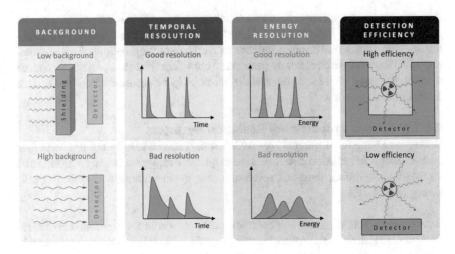

Fig. 4.26 Schematic representation of the general properties of radiation detectors

Detection Efficiency Only a fraction of the ionizing radiation emitted by a sample reaches the detector, and not all the radiation reaching the detector produces a measurable pulse. The relation between the number of particles or photons that are detected compared to those that are emitted by the sample is known as the efficiency of the detection system. Detection efficiency depends on the detector characteristics and the type and energy of the radiation emitted (e.g., low-energy radiation can be absorbed before interacting with the detector). The configuration of the detector, such as the distance between the sample and the detector or the active area of the detector, will also impact the detection efficiency (Fig. 4.26). Having an accurate and precise calibration of the detection efficiency of the detector is crucial to conduct quantitative measurements.

4.4.1.3 Types of Radiation Detectors

While most of the radiometric techniques are generally based on the principles described above, there are several types of detectors optimized to either analyze different types of radiation or to fit specific purposes (e.g., dosimeters, spectrometers). In this section, we will briefly describe the three most commonly applied detectors to quantify the radioactive content of environmental samples: gas detectors, semiconductor detectors, and scintillation detectors, with a special focus on those that are used to measure the ocean tracers described in Sect. 4.3.1 and 4.3.2 (see reviews from Hou and Roos (2008), Suckow (2009) and L'Annunziata (2013) for more details).

Gas Detectors

The general principle of gas detectors is based on the interaction of ionizing radiation with gas. The detector consists of a cylindrical metallic tube filled with gas (usually a noble gas) with a thin wire in the center that acts as an anode. When the incident radiation enters the cylinder, it interacts with the gas and ionizes it, resulting in the production of ion pairs. By applying a specific voltage between the wire (anode) and the tube wall (cathode), the electrons and positive ions drift towards the anode and cathode, respectively, which produces a pulse that can be amplified and recorded using an electrical current measuring device (Fig. 4.25). The number of ion pairs created as a result of the interaction of the incident radiation with the gas depends on both the radiation (type and energy) and the voltage applied between anode and cathode. There are three main types of gas detectors differentiated according to the voltage of operation (summarized below from lowest to highest voltage):

- **Ionization chambers** operate under low voltages (~100 to 150 V) (i.e., low electric fields between anode and cathode), but these voltages are high enough to allow all the ion pairs directly created by the incident radiation (primary ionizations) to be collected at the electrodes. Since the number of primary ionizations produced depends on the type of radiation (i.e., alpha, beta, or gamma) and it is proportional to its energy, the amount of charge reaching the electrodes is directly related to the energy and type of the incident radiation. Therefore, ionization chambers can be used for spectrometry.
- **Proportional counters** operate at higher voltages than ionization chambers (~200 to 800 V). By applying higher electric fields, the ion pairs directly created by the incident radiation (primary ions) are accelerated towards the electrodes and produce additional ionizations by collision with the atoms of the gas (secondary ionizations) on their path towards the electrodes. These secondary ionizations allow amplifying the electric signal, but the pulse generated is still proportional to the number of primary ionizations produced by the incident radiation, allowing to conduct spectrometry.
- **Geiger-Müller (GM) detectors** are one of the oldest and most commonly applied types of radiation detectors. Due to the high voltage applied in GM (~800 to 2000 V), the primary ions produced by the incident radiation ionize additional atoms of gas during their drift towards the electrodes. This produces secondary ions that in turn produce additional ion pairs by collision, eventually leading to avalanches of ion pairs known as Townsend avalanches. The amplification factor can reach up to 10^{10}, which is the number of ionizations that a single primary ionization can create. Due to the large avalanche induced by any primary ionization, the presence of ions from previous ionizations typically lasts tens to hundreds of microseconds, and during this time the system is insensitive to new ionization events and is thus not able to detect new incident radiation. Hence, a limitation of this type of detectors is that they have a relatively poor temporal resolution and therefore are not useful to measure high activity rates. On the other hand, GM detectors are simple and insensitive to small voltage fluctuations, which makes them very useful for general measurement of low-level radiation. However, it is important to consider that the pulses measured in GM detectors are independent of the nature of the incident radiation (i.e., alpha, beta, or gamma) and its energy, which means that GM cannot be used for spectrometry. Therefore,

to ensure that the detector is quantifying the radionuclide of interest, samples often need to be treated (e.g., chemical separation) prior to counting to isolate the radionuclide of interest. Additionally, repeated counting can be done to monitor the decay curve and compare it against the decay constant of the radionuclide of interest.

Box 4.3 Quantifying ^{234}Th in Seawater Using Geiger-Müller Detectors

^{234}Th is a particle-reactive radionuclide widely used as tracer of particle export in the open ocean (Sect. 4.3.1) which is generally measured with low-level beta GM multicounters. Each multicounter system is a gas flow unit which can measure five samples simultaneously. These detectors are operated at a voltage between 1100 and 1400 V and are filled with 99% argon and 1% isobutane or propane. The design of low-level beta GM multicounters includes two types of shielding that allow the measurement of low activity samples: a thick lead shield ($\sim$10 cm; see figure below) that blocks external radiation and reduces background to very low levels, and a guard counter and an electronic system (called discriminator/anticoincidence module) that allow detecting cosmic ray background that penetrates the lead shield and discriminate it from the radioactive emissions from the samples.

To quantify ^{234}Th in seawater, first of all, samples need to be pre-concentrated. Typically, total (i.e., dissolved + particulate) ^{234}Th activities are measured from 2 to 4 L of seawater (Clevenger et al. 2021) usually collected using sampling bottles attached to a rosette. Briefly, samples are immediately acidified and spiked with ^{230}Th, which is used as a chemical yield tracer (Pike et al. 2005), and processed following the manganese dioxide (MnO_2) precipitation technique (Buesseler et al. 2001). The precipitates are subsequently filtered through 25 mm quartz microfiber filters. These filters are then dried and mounted for counting under one layer of plastic film and two layers of standard aluminum foil. The aluminum foil is used to block low-energy beta emitters that have been precipitated along with ^{234}Th, including the low-energy beta emissions from ^{234}Th itself ($E_{max} = 0.27$ MeV). Instead, ^{234}Th is measured via detection of its high-energy beta-emitting daughter, ^{234m}Pa ($T_{1/2} = 1.17$ min, $E_{max} = 2.19$ MeV), which is in secular equilibrium with ^{234}Th (Rutgers van der Loeff et al. 2006). Samples have to be recounted after, at least, 6 half-lives of ^{234}Th ($\sim$5 months) to determine the non-^{234}Th activity of high-energy beta emitters collected by the MnO_2 precipitate and measured during the first counting. The last step is to determine the chemical recovery of Th by measuring the yield tracer (^{230}Th) by inductively coupled plasma mass spectrometry (see Sect. 4.4.2).

(continued)

Box 4.3 (continued)

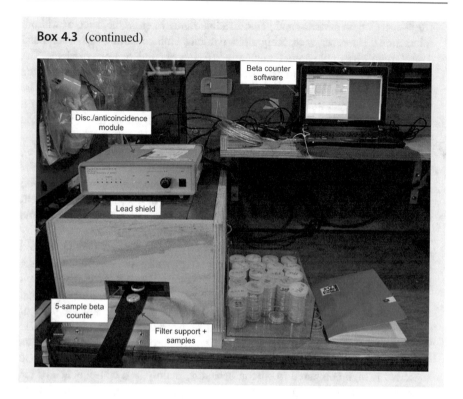

Semiconductor Detectors

Semiconductor detectors are systems based on semiconductor crystalline solids, as its name implies. Semiconductors are materials with electrical conductivities in between insulators and conductors, and they allow the flow of electrons (preferentially in one direction) if enough energy is supplied. The functioning of semiconductor detectors is indeed comparable to a classic gas detector (Fig. 4.25) but using a semiconductor solid instead of a noble gas: radiation interacts with solids and generates electron-hole pairs along the path of radiation through the detector. The amount of energy required to create an electron-hole pair is specific, and the number of electron-hole pairs generated is proportional to the energy of the radiation interacting with the semiconductor. Under the influence of an electric field, electrons and holes are swiped away to the electrodes, producing a pulse that can be measured. These counters allow accurately characterizing the energy of the incident radiation, and therefore they are perfectly suited to conduct spectrometry. There are two main types of semiconductor detectors:

- **Silicon detectors** are the state-of-the-art method for the detection of alpha particles, mainly because of their high energy resolution (ideal for spectrometry) and relatively high detection efficiency. They are very small crystals, allowing easy detection of alpha particles without interference from background radiation (high-energy beta particles and gamma radiation might cross the detector without interaction). However, measuring alpha particles emitted from an environmental sample requires sample pretreatment due to the short range of alpha particles. The sample sometimes needs to be dissolved if it is a solid matrix (e.g., marine sediments), and the target radionuclide needs to be chemically separated. The radionuclide of interest is finally deposited as a thin layer onto the surface of a small metal disk. To maximize detector efficiency for alpha particles, the disk is placed only a few millimeters from the silicon detector in a vacuum chamber.

- **Germanium detectors** contain a single high purity germanium crystal (HPGe), which can have much larger sizes than silicon crystals. The larger size of the crystal, together with its higher density and atomic number, allows the absorption of gamma radiation. The main advantage of HPGe detectors is their excellent energy resolution and relatively good detection efficiency, providing gamma spectra that permit the quantification of a number of gamma-emitting radionuclides simultaneously. The geometry of the detector can also be adapted to optimize the efficiency of the measurement (e.g., well-type detectors, where the sample is positioned in a small hole in the detector crystal), particularly relevant for low activity samples such as those commonly measured in marine water and sediments. Major drawbacks of this method are the high price and the need of cooling the detector (with liquid nitrogen or a cryocooler) to avoid leakage current (i.e., current generated by thermal energy, producing electric noise that masks the current generated by radiation). In addition, the background in HPGe detectors is high compared to silicon detectors, mainly as a consequence of the cosmic radiation and Earth's natural radioactivity, usually requiring shielding to analyze environmental samples.

Box 4.4 Quantifying ^{228}Ra and ^{226}Ra in Seawater Using HPGe Semiconductor Detectors

^{228}Ra and ^{226}Ra are widely applied tracers of submarine groundwater discharge (Sect. 4.3.2), and they are commonly measured using germanium semiconductor detectors.

^{228}Ra and ^{226}Ra are typically at such low levels in seawater that their measurement requires pre-concentration from large water samples. This pre-concentration is commonly achieved by filtering 50–350 L of seawater through fibers (or other filters) impregnated with MnO_2, which quantitatively adsorb Ra isotopes from water. Ra isotopes adsorbed in these fibers are then transferred to small vials adapted for gamma counting through either chemical leaching followed by coprecipitation (Moore and Reid 1973) or ashing the fiber and packing it into a vial (Charette et al. 2001). The vial is sealed to prevent ^{222}Rn outgassing from the sample and stored for a minimum of 3 weeks before counting to ensure the secular equilibrium between ^{226}Ra and its daughter radionuclides. After this aging time, vials are ready to be counted in a low-background HPGe detector to quantify ^{228}Ra and ^{226}Ra.

(continued)

Box 4.4 (continued)

In this detector, the pulses generated by the interaction of gamma radiation with the semiconductor crystal are electronically stored in a multichannel analyzer that generates a spectrum of the different energies detected (see figure below). Since ^{228}Ra and ^{226}Ra in a closed system (i.e., the sealed vials) will reach secular equilibrium with their radioactive daughters, the activity of these isotopes can be measured through the activity of ^{228}Ac and ^{214}Pb, respectively (Fig. 4.6), which emit gamma radiation with an energy of 911.6 and 351.9 keV, respectively. The area of the peaks generated by these radionuclides at these specific emission lines is proportional to their activity, and thus it can be used to determine the activity of ^{228}Ra and ^{226}Ra in seawater samples (see figure below).

Considering the low activities of ^{228}Ra and ^{226}Ra in marine samples and the relatively high levels of background radiation, the design of germanium detectors includes a thick lead shield surrounding the detector (see figure below). In addition, the vials containing the samples are often placed in well-type detectors, a configuration that significantly increases the detection efficiency.

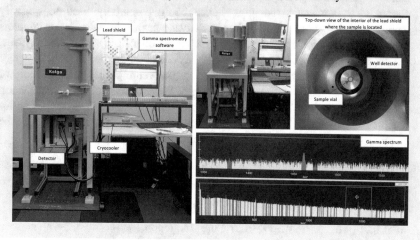

Scintillation Detectors

Scintillation counting is a well-established and popular technique for the detection and quantification of radioactivity. The basic principle behind scintillation techniques is slightly different than that of gas and semiconductor detectors: instead of detecting the charge created by ionizing radiation, scintillation counters detect light photons emitted by certain materials when they absorb radiation (Fig. 4.25). These optically visible photon emissions need to be converted into an electric signal, which is achieved by coupling the scintillation material to a photomultiplier tube (PMT). The amplified electron signal is measured and the magnitude and number of electric pulses can be related to the radiation type/energy and the abundance of each radionuclide. This counting technique therefore allows conducting spectrometry.

There are two major types of scintillation detectors based on the material that is used to absorb the radiation and produce the scintillation:

- **Liquid scintillation detectors** are based on scintillators made of organic materials dissolved in suitable solvents. The method of liquid scintillation counting (LSC) usually involves placing a liquid sample containing the radionuclide of interest into a small vial and adding a special cocktail that contains the scintillator. This method is particularly useful to directly determine radionuclide concentrations in water samples (e.g., ^{222}Rn in submarine groundwater; Lefebvre et al. 2013) or in digested/dissolved solids (e.g., ^{226}Ra in digested marine sediments; Villa et al. 2005). Since the radionuclide of interest is introduced into the scintillation cocktail, the detection efficiency for most alpha and beta decays is close to 100%. However, the liquid is not dense enough to stop high-energy electromagnetic radiation. An advantage of LSC is its better temporal resolution compared to gas detectors, which allows quantifying samples with much higher activities.
- **Solid scintillation detectors** are made of high density and high atomic number inorganic crystalline materials (such as thallium-activated sodium iodide (NaI) or bismuth germanate (BGO)). These detectors have the capability to stop high-energy electromagnetic radiation, hence being commonly used to quantify gamma-emitting radionuclides. The main advantage of solid scintillation counters is their high detection efficiency due to the large dimensions of the solid crystals. On the contrary, the energy resolution of this detection method is relatively poor in comparison to semiconductor detectors, preventing the quantification of radionuclides emitting with similar energies. For that reason, solid scintillation detectors are generally used to measure the total radiation of a sample or for mapping experiments. While less common, solid scintillation detectors can also be applied to the detection of particulate radiation such as alpha and beta.

Box 4.5 Quantifying ^{224}Ra and ^{223}Ra in Seawater Using a RaDeCC System

^{224}Ra and ^{223}Ra (often referred to as short-lived Ra isotopes because of their short half-lives, 3.66 and 11.3 days) are frequently used in oceanography as tracers of submarine groundwater discharge and transport time scales (Sect. 4.3.2). The reference technique to quantify these short-lived Ra isotopes is the RaDeCC system, which is a solid scintillation detector (Moore and Arnold 1996).

To measure the activities of ^{224}Ra and ^{223}Ra with the RaDeCC system, seawater samples are filtered through fibers impregnated with MnO_2 (see Box 4.4). These fibers are then placed into cartridges that are connected to the instrument. The RaDeCC system consists essentially of (a) a large volume scintillation flask with the inside part coated with silver-activated zinc sulfide (ZnS) that allows detecting alpha-decay events produced in the flask; (b) a photomultiplier tube to convert the flashes into an electrical signal; (c) an

(continued)

Box 4.5 (continued)

electronic gateway system, which registers counts and splits the registered events into different channels; and (d) a pump, which continuously circulates helium through the sample and brings it to the scintillation chamber. Since ^{224}Ra and ^{223}Ra continuously produce Rn isotopes (which are gases), the circulation of helium carries Rn isotopes to the scintillation chamber, and thus the system detects the alpha-decay events produced by the daughters of these Ra isotopes (Fig. 4.6). Given that the half-lives of ^{220}Rn and ^{216}Po produced by ^{224}Ra are significantly different than the half-lives of ^{219}Rn and ^{215}Po produced by ^{223}Ra, the RaDeCC system can use the time elapsed between two subsequent decay events to classify them in three different channels: one channel counts decay events from the ^{223}Ra decay chain; a second channel counts events from the ^{224}Ra decay chain; and the third channel records all events (see more details in Moore and Arnold (1996) and Diego-Feliu et al. (2020)). Therefore, while this system does not allow distinguishing the energy of the alpha-decay events (it does not conduct spectrometry), it allows distinguishing different Ra isotopes according to the half-lives of their daughters.

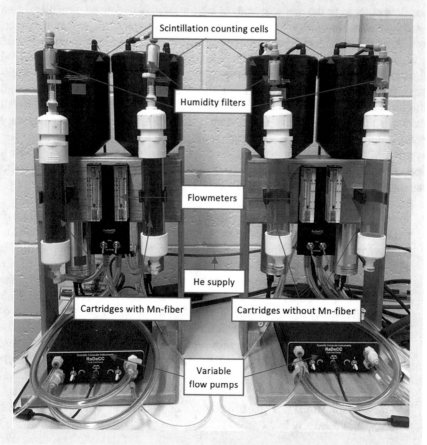

4.4.2 Mass Spectrometric Techniques

A relatively recent development in radioanalytical technologies has been the application of mass spectrometric systems. *Mass spectrometry* (MS) is particularly useful in the analysis of long-lived radionuclides (e.g., ^{129}I in Sect. 4.3.3) or small volume samples, because the low activity of the sample prevents its measurement through radiometric techniques. Since MS is also applied to measure the concentration of stable isotopes and elements in environmental samples, here we will only provide a general overview of these methods. Details on mass spectrometric techniques can be found elsewhere (Hou and Roos 2008; Suckow 2009).

Contrary to the radiometric techniques described above, mass spectrometers do not detect and analyze radioactive decay, but instead they directly quantify the number of atoms. The basic principle of MS consists on generating charged particles (ions) from the atoms or molecules in the samples, which are then accelerated through a semi-circular path and sorted by using a magnetic or electric field that classifies the ions according to their mass (mass-to-charge ratio), and the relative abundance of the ions of different masses is then counted. These principles dictate the main components of a MS: (a) an *ion source* that generates charged particles and accelerates them to the analyzer with as little spread in energy as possible; (b) an *analyzer*, where normally a magnetic field forces the particles around a corner, deflecting the beam and separating the particles according to their momentum and charge state (with the heavy particles being found on the outer side of the curve and the lighter ones on the inner side of the curve); and (c) a *detector*, where the mass-separated charged particles (ion beams) are collected and counted. Importantly, all this process needs to take place in vacuum to reduce energy-scattering collisions. All the mass spectrometers have these same characteristics, although they can vary considerably between different types. Here we will summarize the three main types of MS used for the determination of radionuclides in marine samples, with a special focus on accelerator mass spectrometry, the most sensitive technique used to measure long-lived radionuclides such as ^{129}I (see Sect. 4.3.3):

- **Inductively coupled plasma mass spectrometry (ICP-MS)** is nowadays one of the most common analytical techniques, and it is routinely used, for example, to measure isotopes of Pu, Np, Th, U and Pa in marine samples (Kenna et al. 2012; Mas et al. 2012). Samples need to be in aqueous solutions, requiring thus a previous digestion/dissolution in case of solid samples. In the ICP-MS, the chemical compounds in the sample solution are then decomposed into their atomic constituents in an inductively coupled argon plasma at temperatures of approximately 6000–8000 K to ensure a high degree of ionization (>90% for most elements). The positively charged ions are transferred into the high vacuum of the mass spectrometer, separated by mass filters (quadrupole or magnetic sector mass analyzers) and finally measured by an ion detector. This instrument has the capability to determine radionuclides down to fg (10^{-15} g) levels in small samples and allows the analysis of aqueous samples directly and rapidly (in a few minutes) at a relatively low cost.

- **Thermal ionization mass spectrometry (TIMS)** is an extensively used technique to measure radionuclide concentrations and isotopic compositions (e.g., Cs and U isotopes; Henderson 2002; Zhu et al. 2020). In thermal ionization the sample is placed directly on a small filament by evaporation of the solution, and ions are produced by electrically heating this filament. The most frequently applied technique in TIMS works with two heated filaments (one for evaporation of the sample and the other for ionization of evaporated atoms) which are arranged opposite to each other. The ions created on the filament are accelerated across an electrical potential gradient and focused into a beam via a series of slits and electrostatically charged plates. This ion beam then passes through a magnetic field which disperses the original beam into separate beams on the basis of their mass-to-charge ratio. These mass-resolved beams are then directed into collectors where the ion beams are converted into voltage. Comparison of voltages corresponding to individual ion beams yields precise isotope ratios. TIMS allows measuring radionuclides down to ng (10^{-9} g) with high precision. However, the throughput for this method is smaller than other mass spectrometers, and the cost of the measurement is also higher than for ICP-MS.
- **Accelerator mass spectrometry (AMS)** is the most sensitive counting method for long-lived radionuclides, and it is the state-of-the-art technique for the measurement of a long list of ocean tracers, including ^{14}C, ^{129}I, ^{236}U, and Pu isotopes, among others (Christl et al. 2013). AMS combines two mass spectrometers coupled to a tandem accelerator (see Box 4.6). The main principle is that sample material is introduced in the ion source and negative ions can be extracted and accelerated at low energies (in the order of 10 keV). At the tandem accelerator, ions are then accelerated to high energies (up to several MeV), reaching speeds at several percent of the speed of light (10^7 m s^{-1}). This happens in two main steps within the tandem accelerator: first, negative ions are accelerated by the positive high voltage, and, then, after conversion to positive ions at the stripper, positive ions are accelerated again by repulsion from the same high voltage. The *stripper* is therefore a key element of the tandem accelerator, which usually consists of a canal filled with gas (e.g., He or Ar) at low pressure, causing the rupture of any molecular bond. These positively charged ions (now only atomic ions exist) are then further discriminated by mass-to-charge ratio using a combination of high-energy filters (i.e., magnets and an electrostatic analyzer) in several steps. The abundant stable isotope is measured in the Faraday cups (located before and after the tandem accelerator; see Box 4.6) as a current generated by the charged particles (in the order of microamperes, µA). Individual ions of the non-abundant radioactive isotope are detected by single-ion counting in the gas ionization chamber. The special strength of AMS among the mass spectrometric methods is its power to separate a rare isotope from a much more abundant neighboring mass ("abundance sensitivity," e.g., ^{14}C from ^{12}C, or ^{129}I from ^{127}I) and the isobar (isotopes of different elements but same atomic weight) suppression (e.g., ^{14}C from ^{14}N, ^{129}I from ^{129}Xe). The output of the AMS is generally the atomic ratio of the rare isotope over the abundant one (e.g., $^{14}C/^{12}C$, $^{129}I/^{127}I$, $^{236}U/^{238}U$) at great sensitivities down to 10^{-15} for milligram-size samples. The

small sample size increases the risk of contamination and, thus, a careful sample treatment is necessary prior to measurement. Although AMS provides the greatest sensitivity, there are only about 50 AMS facilities worldwide, and their operational costs are higher than for any other mass spectrometry technique.

Box 4.6 Quantifying ^{129}I in Seawater Using Accelerator Mass Spectrometry

Anthropogenic iodine (^{129}I) is a powerful tracer of ocean circulation, especially in the Arctic and North Atlantic regions (Sect. 4.3.3), which is measured using AMS due to its low concentrations in seawater. Before AMS analysis, seawater samples have to be processed in the laboratory so that the final precipitates mostly contain iodine isotopes.

Nowadays, the determination of ^{129}I in seawater can be done in less than 500 mL of sample, and since iodine is mostly conservative in seawater, there is no need for filtration before sample storage in plastic bottles (preferably opaque bottles to minimize the volatilization of iodine as gaseous I_2). Prior to radiochemistry, samples are weighted and spiked with a known amount of ^{127}I, exceeding about three orders of magnitude the ^{127}I naturally present in seawater. The extraction of both iodine isotopes is done using different methods that include ion exchange resins or solvent extraction techniques (e.g., Casacuberta et al. 2014b; see left picture below). Finally, the extracted iodine solution is precipitated as silver iodide (AgI), centrifuged and dried at 80 °C prior to its pressing into sample holders. The most abundant ^{127}I isotope and the rare ^{129}I are detected quasi simultaneously in the Faraday cups and the gas-filled ionization chamber, respectively. Each sample is measured in cycles of a few minutes, together with blanks and ^{129}I/^{127}I standards. The final ^{129}I/^{127}I atom ratio for each sample is corrected by background counts and standard nominal values, and the concentrations of ^{129}I are calculated from the known amount of ^{127}I added to the sample.

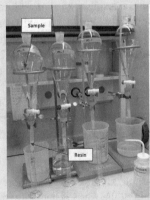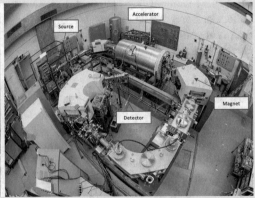

Dedication

This chapter is dedicated to Jordi Garcia-Orellana (1971-2022), our teacher, mentor and friend. Jordi stimulated in all of us the interest to study and understand the environment with the use of radionuclides. His love for science and nature made him a unique scientist and person. He will be remembered for his passion, enthusiasm, generosity and humor. Gràcies per tant, Jordi. Et trobem a faltar.

Acknowledgments We thank J. Garcia-Orellana and P. Masqué for introducing us to the amazing world of environmental radioactivity. VR acknowledges the Beatriu de Pinós postdoctoral program of the Catalan Government (2019-BP-00241) and the financial support of the Spanish Government through the "Maria de Maeztu" programme for Units of Excellence (CEX2019-000940-M) and the Catalan Government (MERS; 2017 SGR – 1588). MRM acknowledges the Ocean Frontier Institute International Postdoctoral Fellowship Program. VP acknowledges the support from "la Caixa" Foundation (ID 100010434; code 105183) and from the European Union's Horizon 2020 research and innovation programme (Marie Skłodowska-Curie grant agreement No 847648). VP also acknowledges the 'Severo Ochoa Centre of Excellence' accreditation (CEX2019-000928-S). MC was partly funded by the ETH Career Seed Grant (SEED-06 19-2). NC funding came from the European Research Council (ERC Consolidator GAP-101001451).

References

Alfimov V, Aldahan A, Possnert G (2013) Water masses and [129]I distribution in the Nordic seas. Nucl Instrum Methods Phys Res Sect B Beam Interact Mater Atoms 294:542–546. https://doi.org/10.1016/J.NIMB.2012.07.042

Andrews JE, Hartin C, Buesseler KO (2008) [7]Be analyses in seawater by low background gamma-spectroscopy. J Radioanal Nucl Chem 2771(277):253–259. https://doi.org/10.1007/S10967-008-0739-Y

Arias-Ortiz A, Masqué P, Garcia-Orellana J et al (2018) Reviews and syntheses: ^{210}Pb-derived sediment and carbon accumulation rates in vegetated coastal ecosystems—setting the record straight. Biogeosciences 15:6791–6818. https://doi.org/10.5194/BG-15-6791-2018

Bender ML, Kinter S, Cassar N, Wanninkhof R (2011) Evaluating gas transfer velocity parameterizations using upper ocean radon distributions. J Geophys Res 116:C02010. https://doi.org/10.1029/2009JC005805

Bethoux J, Morin P, Chaumery C, Connan O (1998) Nutrients in the Mediterranean Sea, mass balance and statistical analysis of concentrations with respect to environmental change. Mar Chem 63:155–169

Bhat SG, Krishnaswamy S, Lal D et al (1968) [234]Th/[238]U ratios in the ocean. Earth Planet Sci Lett 5: 483–491. https://doi.org/10.1016/S0012-821X(68)80083-4

Boyd PW, Trull TW (2007) Understanding the export of biogenic particles in oceanic waters: is there consensus? Prog Oceanogr 72:276–312. https://doi.org/10.1016/j.pocean.2006.10.007

Boyd PW, Claustre H, Levy M et al (2019) Multi-faceted particle pumps drive carbon sequestration in the ocean. Nature 568:327–335. https://doi.org/10.1038/s41586-019-1098-2

Bradtmiller LI, McManus JF, Robinson LF (2014) [231]Pa/[230]Th evidence for a weakened but persistent Atlantic meridional overturning circulation during Heinrich Stadial 1. Nat Commun 51(5):1–8. https://doi.org/10.1038/ncomms6817

Buesseler KO, Bacon MP, Cochran JK, Livingston HD (1992) Carbon and nitrogen export during the JGOFS North Atlantic Bloom experiment estimated from [234]Th:[238]U disequilibria. Deep Sea Res Part A Oceanogr Res Pap 39:1115–1137. https://doi.org/10.1016/0198-0149(92)90060-7

Buesseler KO, Benitez-Nelson C, Rutgers van der Loeff M et al (2001) An intercomparison of small- and large-volume techniques for thorium-234 in seawater. Mar Chem 74:15–28. https://doi.org/10.1016/S0304-4203(00)00092-X

Buesseler KO, Benitez-Nelson CR, Moran SB et al (2006) An assessment of particulate organic carbon to thorium-234 ratios in the ocean and their impact on the application of ^{234}Th as a POC flux proxy. Mar Chem 100:213–233. https://doi.org/10.1016/j.marchem.2005.10.013

Buesseler K, Dai M, Aoyama M et al (2017) Fukushima Daiichi–derived radionuclides in the ocean: transport, fate, and impacts. Ann Rev Mar Sci 9:173–203. https://doi.org/10.1146/ANNUREV-MARINE-010816-060733

Buesseler KO, Benitez-Nelson CR, Roca-Martí M et al (2020a) High-resolution spatial and temporal measurements of particulate organic carbon flux using thorium-234 in the northeast Pacific Ocean during the EXport Processes in the Ocean from RemoTe Sensing field campaign. Elem Sci Anth 8:030. https://doi.org/10.1525/elementa.030

Buesseler KO, Boyd PW, Black EE, Siegel DA (2020b) Metrics that matter for assessing the ocean biological carbon pump. Proc Natl Acad Sci 117:9679–9687. https://doi.org/10.1073/PNAS.1918114117

Burnett WC, Dulaiova H (2006) Radon as a tracer of submarine groundwater discharge into a boat basin in Donnalucata, Sicily. Cont Shelf Res 26:862–873. https://doi.org/10.1016/j.csr.2005.12.003

Casacuberta N, Christl M, Lachner J et al (2014a) A first transect of ^{236}U in the North Atlantic Ocean. Geochim Cosmochim Acta 133:34–46. https://doi.org/10.1016/J.GCA.2014.02.012

Casacuberta N, Christl M, Vockenhuber C (2014b) Chemistry of ^{129}I in seawater samples: setting up the ^{129}I radiochemistry at LIP/EAWAG

Casacuberta N, Christl M, Vockenhuber C et al (2018) Tracing the three Atlantic branches entering the Arctic Ocean with ^{129}I and ^{236}U. J Geophys Res Ocean 123:6909–6921. https://doi.org/10.1029/2018JC014168

Castrillejo M, Casacuberta N, Christl M et al (2018) Tracing water masses with ^{129}I and ^{236}U in the subpolar North Atlantic along the GEOTRACES GA01 section. Biogeosciences 15:5545–5564. https://doi.org/10.5194/BG-15-5545-2018

Castrillejo M, Casacuberta N, Vockenhuber C, Lherminier P (2022) Rapidly increasing artificial iodine highlights pathways of Iceland-Scotland Overflow Water and Labrador Sea Water. Front Mar Sci 9. https://www.frontiersin.org/articles/10.3389/fmars.2022.897729. https://doi.org/10.3389/fmars.2022.897729

Charette MA, Buesseler KO, Andrews JE (2001) Utility of radium isotopes for evaluating the input and transport of groundwater-derived nitrogen to a Cape Cod estuary. Limnol Oceanogr 46:465–470

Charette MA, Lam PJ, Lohan MC et al (2016) Coastal ocean and shelf-sea biogeochemical cycling of trace elements and isotopes: lessons learned from GEOTRACES. Philos Trans A Math Phys Eng Sci 374(2081):20160076

Cho H-M, Kim G, Kwon EY et al (2018) Radium tracing nutrient inputs through submarine groundwater discharge in the global ocean. Sci Rep 8:2439. https://doi.org/10.1038/s41598-018-20806-2

Christl M, Vockenhuber C, Kubik PW et al (2013) The ETH Zurich AMS facilities: performance parameters and reference materials. Nucl Instrum Methods Phys Res Sect B Beam Interact Mater Atoms 294:29–38. https://doi.org/10.1016/J.NIMB.2012.03.004

Christl M, Casacuberta N, Vockenhuber C et al (2015) Reconstruction of the ^{236}U input function for the Northeast Atlantic Ocean: implications for ^{129}I/^{236}U and ^{236}U/^{238}U-based tracer ages. J Geophys Res Ocean 120:7282–7299. https://doi.org/10.1002/2015JC011116

Clevenger SJ, Benitez-Nelson CR, Drysdale J et al (2021) Review of the analysis of ^{234}Th in small volume (2–4 L) seawater samples: improvements and recommendations. J Radioanal Nucl Chem 329:1–13. https://doi.org/10.1007/S10967-021-07772-2/FIGURES/10

Cochran JK, Masqué P (2003) Short-lived U/Th series radionuclides in the ocean: tracers for scavenging rates, export fluxes and particle dynamics. Rev Mineral Geochem 52(1):461–492

Cook PG, Rodellas V, Stieglitz TC (2018) Quantifying surface water, porewater and groundwater interactions using tracers: tracer fluxes, water fluxes and endmember concentrations. Water Resour Res 54(3):2452–2465. https://doi.org/10.1002/2017WR021780

Diego-Feliu M, Rodellas V, Alorda-Kleinglass A et al (2020) Guidelines and limits for the quantification of Ra isotopes and related radionuclides with the radium delayed coincidence counter (RaDeCC). J Geophys Res Ocean 125:e2019JC015544. https://doi.org/10.1029/2019JC015544

Diego-Feliu M, Rodellas V, Saaltink MW et al (2021) New perspectives on the use of ^{224}Ra/^{228}Ra and ^{222}Rn/^{226}Ra activity ratios in groundwater studies. J Hydrol 596:126043. https://doi.org/10.1016/j.jhydrol.2021.126043

Ebser S, Kersting A, Stöven T et al (2018) ^{39}Ar dating with small samples provides new key constraints on ocean ventilation. Nat Commun 91(9):1–7. https://doi.org/10.1038/s41467-018-07465-7

Edmonds HN, Zhou ZQ, Raisbeck GM et al (2001) Distribution and behavior of anthropogenic ^{129}I in water masses ventilating the North Atlantic Ocean. J Geophys Res Ocean 106:6881–6894. https://doi.org/10.1029/1999JC000282

Frajka-Williams E, Ansorge IJ, Baehr J et al (2019) Atlantic meridional overturning circulation: observed transport and variability. Front Mar Sci 6:260. https://doi.org/10.3389/FMARS.2019.00260/BIBTEX

Garcia-Orellana J, Rodellas V, Tamborski J et al (2021) Radium isotopes as submarine groundwater discharge (SGD) tracers: review and recommendations. Earth Sci Rev 220:103681. https://doi.org/10.1016/j.earscirev.2021.103681

Guerzoni S, Chester R, Dulac F et al (1999) The role of atmospheric deposition in the biogeochemistry of the Mediterranean Sea. Prog Oceanogr 44:147–190. https://doi.org/10.1016/S0079-6611(99)00024-5

Hajdas I, Ascough P, Garnett MH et al (2021) Radiocarbon dating. Nat Rev Methods Prim 11(1):1–26. https://doi.org/10.1038/s43586-021-00058-7

He P, Aldahan A, Possnert G, Hou XL (2013) A summary of global ^{129}I in marine waters. Nucl Instrum Methods Phys Res Sect B Beam Interact Mater Atoms 294:537–541, ISSN 0168-583X. https://doi.org/10.1016/j.nimb.2012.08.036

Henderson GM (2002) Seawater (^{234}U/^{238}U) during the last 800 thousand years. Earth Planet Sci Lett 199:97–110. https://doi.org/10.1016/S0012-821X(02)00556-3

Henson SA, Sanders R, Madsen E et al (2011) A reduced estimate of the strength of the ocean's biological carbon pump. Geophys Res Lett 38:L04606. https://doi.org/10.1029/2011GL046735

Hou X, Roos P (2008) Critical comparison of radiometric and mass spectrometric methods for the determination of radionuclides in environmental, biological and nuclear waste samples. Anal Chim Acta 608:105–139. https://doi.org/10.1016/J.ACA.2007.12.012

Hsieh Y-T, Henderson GM, Thomas AL (2011) Combining seawater ^{232}Th and ^{230}Th concentrations to determine dust fluxes to the surface ocean. Earth Planet Sci Lett 312:280–290. https://doi.org/10.1016/J.EPSL.2011.10.022

International Atomic Energy Agency (1985) Sediment Kd's and concentration factors for radionuclides in the marine environment, Vienna, Austria

IPCC (2021) AR6 climate change 2021: the physical science basis

Jenkins WJ, Smethie WM (1996) Transient tracers track ocean climate signals. Oceanus 39:29–32

Jickells TD, An ZS, Andersen KK et al (2005) Global iron connections between desert dust, ocean biogeochemistry, and climate. Science (80-) 308:67–71. https://doi.org/10.1126/SCIENCE.1105959

Kadko D, Landing WM, Shelley RU (2015) A novel tracer technique to quantify the atmospheric flux of trace elements to remote ocean regions. J Geophys Res Ocean 120:848–858. https://doi.org/10.1002/2014JC010314

Karcher M, Smith JN, Kauker F et al (2012) Recent changes in Arctic Ocean circulation revealed by iodine-129 observations and modeling. J Geophys Res Ocean 117:8007. https://doi.org/10.1029/2011JC007513

Kenna TC, Masqué P, Mas JL et al (2012) Intercalibration of selected anthropogenic radionuclides for the GEOTRACES Program. Limnol Oceanogr Methods 10:590–607. https://doi.org/10.4319/lom.2012.10.590

Kienast SS, Winckler G, Lippold J et al (2016) Tracing dust input to the global ocean using thorium isotopes in marine sediments: ThoroMap. Global Biogeochem Cycles 30:1526–1541. https://doi.org/10.1002/2016GB005408

Kipp LE, Kadko DC, Pickart RS et al (2019) Shelf-basin interactions and water mass residence times in the Western Arctic Ocean: insights provided by radium isotopes. J Geophys Res Ocean 124:3279–3297. https://doi.org/10.1029/2019JC014988

Knoll G (2010) Radiation detection and measurement. John Wiley & Sons, New York

Kwon EY, Primeau F, Sarmiento JL (2009) The impact of remineralization depth on the air–sea carbon balance. Nat Geosci 2:630–635. https://doi.org/10.1038/ngeo612

Kwon EY, Kim G, Primeau F et al (2014) Global estimate of submarine groundwater discharge based on an observationally constrained radium isotope model. Geophys Res Lett 41:8438–8444. https://doi.org/10.1002/2014GL061574

L'Annunziata MF (2013) Handbook of radioactivity analysis, 3rd edn. Elsevier, Amsterdam

Lam PJ, Marchal O (2015) Insights into particle cycling from thorium and particle data. Annu Rev Mar Sci 7:159–184. https://doi.org/10.1146/annurev-marine-010814-015623

Lamontagne S, Webster IT (2019) Cross-shelf transport of submarine groundwater discharge tracers: a sensitivity analysis. J Geophys Res Ocean 124:453–469. https://doi.org/10.1029/2018JC014473

Le Moigne FAC, Henson SA, Sanders RJ, Madsen E (2013) Global database of surface ocean particulate organic carbon export fluxes diagnosed from the [234]Th technique. Earth Syst Sci Data 5:295–304. https://doi.org/10.5194/essd-5-295-2013

Lecher A, Mackey K, Lecher AL, Mackey KRM (2018) Synthesizing the effects of submarine groundwater discharge on marine biota. Hydrology 5:60. https://doi.org/10.3390/hydrology5040060

Lecroart P, Schmidt S, Jouanneau J-M, Weber O (2005) Be-7 and Th-234 as tracers of sediment mixing on seasonal time scale at the water-sediment interface of the Thau Lagoon. Radioprotection 40:S661–S667. https://doi.org/10.1051/RADIOPRO:2005S1-097

Lefebvre K, Barbecot F, Ghaleb B et al (2013) Full range determination of [222]Rn at the watershed scale by liquid scintillation counting. Appl Radiat Isot 75:71–76. https://doi.org/10.1016/J.APRADISO.2013.01.027

Livingston HD, Jenkins WJ (1983) Radioactive tracers in the sea. In: Oceanography. Springer, New York, pp 163–191. https://doi.org/10.1007/978-1-4612-5440-9_11

Lougheed BC, Ascough P, Dolman AM et al (2020) Re-evaluating ^{14}C dating accuracy in deep-sea sediment archives. Geochronology 2:17–31. https://doi.org/10.5194/GCHRON-2-17-2020

Ludwig W, Dumont E, Meybeck M, Heussner S (2009) River discharges of water and nutrients to the Mediterranean and Black Sea: major drivers for ecosystem changes during past and future decades? Prog Oceanogr 80:199–217. https://doi.org/10.1016/j.pocean.2009.02.001

Luijendijk E, Gleeson T, Moosdorf N (2020) Fresh groundwater discharge insignificant for the world's oceans but important for coastal ecosystems. Nat Commun 11:1–12. https://doi.org/10.1038/s41467-020-15064-8

Mas JL, Villa M, Hurtado S, García-Tenorio R (2012) Determination of trace element concentrations and stable lead, uranium and thorium isotope ratios by quadrupole-ICP-MS in NORM and NORM-polluted sample leachates. J Hazard Mater 205–206:198–207. https://doi.org/10.1016/J.JHAZMAT.2011.12.058

Masarik J (2009) Origin and distribution of radionuclides in the continental environment. In: Radioactivity in the environment, vol 16. Elsevier, Amsterdam, pp 1–25. https://doi.org/10.1016/S1569-4860(09)01601-5

Moore WS (2000) Ages of continental shelf waters determined from [223]Ra and [224]Ra. J Geophys Res C Ocean 105:22117–22122

Moore WS (2010) The effect of submarine groundwater discharge on the ocean. Annu Rev Mar Sci 2:59–88. https://doi.org/10.1146/annurev-marine-120308-081019

Moore WS (2015) Inappropriate attempts to use distributions of ^{228}Ra and ^{226}Ra in coastal waters to model mixing and advection rates. Cont Shelf Res 105:95–100. https://doi.org/10.1016/j.csr.2015.05.014

Moore WS, Arnold R (1996) Measurement of ^{223}Ra and ^{224}Ra in coastal waters using a delayed coincidence counter. J Geophys Res C Ocean 101:1321–1329. https://doi.org/10.1029/95JC03139

Moore WS, Reid DF (1973) Extraction of radium from natural waters using manganese-impregnated acrylic fibers. J Geophys Res 78:8880–8886. https://doi.org/10.1029/JC078i036p08880

Morison J, Kwok R, Peralta-Ferriz C et al (2012) Changing Arctic Ocean freshwater pathways. Nature 481:66–70. https://doi.org/10.1038/nature10705

Orlandini KA, Bowling JW, Pinder IE, Penrose WR (2003) ^{90}Y-^{90}Sr disequilibrium in surface waters: investigating short-term particle dynamics by using a novel isotope pair. Earth Planet Sci Lett 207:141–150. https://doi.org/10.1016/S0012-821X(02)01096-8

Owens SA, Buesseler KO, Sims KWW (2011) Re-evaluating the ^{238}U-salinity relationship in seawater: implications for the ^{238}U–^{234}Th disequilibrium method. Mar Chem 127:31–39. https://doi.org/10.1016/j.marchem.2011.07.005

Paradis S, Puig P, Masqué P et al (2017) Bottom-trawling along submarine canyons impacts deep sedimentary regimes. Sci Rep 71(7):1–12. https://doi.org/10.1038/srep43332

Pike SM, Buesseler KO, Andrews J, Savoye N (2005) Quantification of ^{234}Th recovery in small volume sea water samples by inductively coupled plasma-mass spectrometry. J Radioanal Nucl Chem 2632(263):355–360. https://doi.org/10.1007/S10967-005-0594-Z

Polyakov IV, Pnyushkov AV, Alkire MB et al (2017) Greater role for Atlantic inflows on sea-ice loss in the Eurasian Basin of the Arctic Ocean. Science (80-) 356:285–291. https://doi.org/10.1126/SCIENCE.AAI8204/SUPPL_FILE/POLYAKOV-SM.PDF

Povinec PP, Aarkrog A, Buesseler KO et al (2005) ^{90}Sr, ^{137}Cs and ^{239,240}Pu concentration surface water time series in the Pacific and Indian Oceans—WOMARS results. J Environ Radioact 81:63–87. https://doi.org/10.1016/J.JENVRAD.2004.12.003

Proshutinsky A, Dukhovskoy D, Timmermans ML et al (2015) Arctic circulation regimes. Philos Trans R Soc A Math Phys Eng Sci 373:20140160. https://doi.org/10.1098/RSTA.2014.0160

Puigcorbé V, Roca-Martí M, Masqué P et al (2017) Latitudinal distributions of particulate carbon export across the North Western Atlantic Ocean. Deep Sea Res Part I Oceanogr Res Pap 129:116–130. https://doi.org/10.1016/J.DSR.2017.08.016

Puigcorbé V, Masqué P, Le Moigne FAC (2020) Global database of ratios of particulate organic carbon to thorium-234 in the ocean: improving estimates of the biological carbon pump. Earth Syst Sci Data 12:1267–1285. https://doi.org/10.5194/essd-12-1267-2020

Resplandy L, Martin AP, Le Moigne F et al (2012) How does dynamical spatial variability impact ^{234}Th-derived estimates of organic export? Deep Sea Res Part I Oceanogr Res Pap 68:24–45. https://doi.org/10.1016/j.dsr.2012.05.015

Rodellas V, Garcia-Orellana J, Tovar-Sánchez A et al (2014) Submarine groundwater discharge as a source of nutrients and trace metals in a Mediterranean Bay (Palma Beach, Balearic Islands). Mar Chem 160:56–66. https://doi.org/10.1016/j.marchem.2014.01.007

Rodellas V, Garcia-Orellana J, Masqué P et al (2015) Submarine groundwater discharge as a major source of nutrients to the Mediterranean Sea. Proc Natl Acad Sci U S A 112:3926–3930. https://doi.org/10.1073/pnas.1419049112

Rodellas V, Stieglitz TC, Andrisoa A et al (2018) Groundwater-driven nutrient inputs to coastal lagoons: the relevance of lagoon water recirculation as a conveyor of dissolved nutrients. Sci Total Environ 642:764–780. https://doi.org/10.1016/j.scitotenv.2018.06.095

Ruiz-González C, Rodellas V, Garcia-Orellana J (2021) The microbial dimension of submarine groundwater discharge: current challenges and future directions. FEMS Microbiol Rev 10:1–25. https://doi.org/10.1093/FEMSRE/FUAB010

Rutgers van der Loeff M, Sarin MM, Baskaran M et al (2006) A review of present techniques and methodological advances in analyzing ^{234}Th in aquatic systems. Mar Chem 100:190–212. https://doi.org/10.1016/J.MARCHEM.2005.10.012

Rutgers van der Loeff MM, Cassar N, Nicolaus M et al (2014) The influence of sea ice cover on air-sea gas exchange estimated with radon-222 profiles. J Geophys Res Ocean 119:2735–2751. https://doi.org/10.1002/2013JC009321

Sanial V, van Beek P, Lansard B et al (2015) Use of Ra isotopes to deduce rapid transfer of sediment-derived inputs off Kerguelen. Biogeosci Discuss 12:1415–1430. https://doi.org/10.5194/bg-12-1415-2015

Santos IR, Eyre BD, Huettel M (2012) The driving forces of porewater and groundwater flow in permeable coastal sediments: a review. Estuar Coast Shelf Sci 98:1–15. https://doi.org/10.1016/j.ecss.2011.10.024

Santos IR, Chen X, Lecher AL et al (2021) Submarine groundwater discharge impacts on coastal nutrient biogeochemistry. Nat Rev Earth Environ 2:307–323. https://doi.org/10.1038/s43017-021-00152-0

Savoye N, Benitez-Nelson C, Burd AB et al (2006) ^{234}Th sorption and export models in the water column: a review. Mar Chem 100:234–249. https://doi.org/10.1016/j.marchem.2005.10.014

Schlitzer R (2021) Ocean data view. https://odv.awi.de/

Sévellec F, Fedorov AV, Liu W (2017) Arctic sea-ice decline weakens the Atlantic Meridional overturning circulation. Nat Clim Chang 78(7):604–610. https://doi.org/10.1038/nclimate3353

Slomp CP, Van Cappellen P (2004) Nutrient inputs to the coastal ocean through submarine groundwater discharge: controls and potential impact. J Hydrol 295:64–86. https://doi.org/10.1016/j.jhydrol.2004.02.018

Smith JN, Ellis KM (1982) Transport mechanism for Pb-210, Cs-137 and Pu fallout radionuclides through fluvial-marine systems. Geochim Cosmochim Acta 46:941–954. https://doi.org/10.1016/0016-7037(82)90050-3

Smith JN, Ellis KM, Boyd T (1999) Circulation features in the central Arctic Ocean revealed by nuclear fuel reprocessing tracers from Scientific Ice Expeditions 1995 and 1996. J Geophys Res Ocean 104:29663–29677. https://doi.org/10.1029/1999JC900244

Smith JN, Jones EP, Moran SB et al (2005) Iodine 129/CFC 11 transit times for Denmark Strait Overflow Water in the Labrador and Irminger Seas. J Geophys Res Ocean 110:1–16. https://doi.org/10.1029/2004JC002516

Smith JN, McLaughlin FA, Smethie WM et al (2011) Iodine-129, ^{137}Cs, and CFC-11 tracer transit time distributions in the Arctic Ocean. J Geophys Res Ocean 116:4024. https://doi.org/10.1029/2010JC006471

Smith JN, Karcher M, Casacuberta N et al (2021) A changing Arctic Ocean: how measured and modeled ^{129}I distributions indicate fundamental shifts in circulation between 1994 and 2015. J Geophys Res Ocean 126:e2020JC016740. https://doi.org/10.1029/2020JC016740

Smoak JM, Moore WS, Thunell RC, Shaw TJ (1999) Comparison of ^{234}Th, ^{228}Th, and ^{210}Pb fluxes with fluxes of major sediment components in the Guaymas Basin, Gulf of California. Mar Chem 65:177–194. https://doi.org/10.1016/S0304-4203(98)00095-4

Steinhauser G (2014) Fukushimas forgotten radionuclides: a review of the understudied radioactive emissions. Environ Sci Technol 48:4649–4663. https://doi.org/10.1021/ES405654C/SUPPL_FILE/ES405654C_SI_001.PDF

Stewart G, Fowler SW, Fisher NS (2008) The bioaccumulation of U- and Th-series radionuclides in marine organisms. In: Krishnaswami S, Cochran JK (eds) U-Th series nuclides in aquatic systems. Elsevier Science, Amsterdam, pp 269–305

Suckow A (2009) Analysis of radionuclides. In: Radioactivity in the environment, vol 16. Elsevier, Amsterdam, pp 363–406. https://doi.org/10.1016/S1569-4860(09)01609-X

Sweeney C, Gloor E, Jacobson AR et al (2007) Constraining global air-sea gas exchange for CO_2 with recent bomb ^{14}C measurements. Global Biogeochem Cycles 21(2):GB2015. https://doi.org/10.1029/2006GB002784

Trezzi G, Garcia-Orellana J, Rodellas V et al (2016) Submarine groundwater discharge: a significant source of dissolved trace metals to the North Western Mediterranean Sea. Mar Chem 186:90–100. https://doi.org/10.1016/j.marchem.2016.08.004

UNSCEAR (2000) Sources and effects of ionizing radiation. Vol I: sources. United Nations. Scientific Committee on the effects of atomic radiation. United Nations Publications, New York

Villa M, Moreno HP, Manjón G (2005) Determination of ^{226}Ra and ^{224}Ra in sediments samples by liquid scintillation counting. Radiat Meas 39:543–550. https://doi.org/10.1016/J.RADMEAS. 2004.10.004

Volk T, Hoffert MI (1985) Ocean carbon pumps: analysis of relative strengths and efficiencies in ocean-driven atmospheric CO_2 changes. In: Sundquist ET, Broecker WS (eds) The carbon cycle and atmospheric CO_2: natural variations Archean to present. American Geophysical Union, Washington, DC, pp 99–110

Wefing AM, Casacuberta N, Christl M et al (2021) Circulation timescales of Atlantic water in the Arctic Ocean determined from anthropogenic radionuclides. Ocean Sci 17:111–129. https://doi. org/10.5194/OS-17-111-2021

Young JA, Silker WB (1980) Aerosol deposition velocities on the Pacific and Atlantic Oceans calculated from ^{7}Be measurements. Earth Planet Sci Lett 50:92–104. https://doi.org/10.1016/ 0012-821X(80)90121-1

Zhu L, Xu C, Hou X et al (2020) Determination of ultratrace level ^{135}Cs and ^{135}Cs/^{137}Cs ratio in small volume seawater by chemical separation and thermal ionization mass spectrometry. Anal Chem 92:6709–6718. https://doi.org/10.1021/ACS.ANALCHEM.0C00688/SUPPL_FILE/ AC0C00688_SI_001.PDF

Persistent Organic Contaminants

5

Karina S. B. Miglioranza, Paola M. Ondarza, Sebastián I. Grondona, and Lorena B. Scenna

Contents

Abstract

Persistent organic pollutants (POPs) are toxic compounds of global concern that adversely affect the environment and human health around the world. POPs are forbidden at worldwide level and regulated by the Stockholm Convention, managed by the UNEP. It was created in 2001 and entered into force in 2004 and currently has 184 members. POPs consist mainly of organochlorine pesticides, industrial compounds (PCBs, PBDEs, PFAS, PCN, and SCCPs), and chemical-delivered unintentional by-products of urban and industrial processes (dioxins and furans). They persist in the environment, are adsorbed to organic matter of soils and sediments, bioaccumulate in fatty tissues and

K. S. B. Miglioranza (✉) · P. M. Ondarza · L. B. Scenna
Laboratorio de Ecotoxicología y Contaminación Ambiental, Instituto de Investigaciones Marinas y Costeras, Universidad Nacional de Mar del Plata, CONICET, Mar del Plata, Argentina
e-mail: kmiglior@mdp.edu.ar; pmondar@mdp.edu.ar; lscenna@mdp.edu.ar

S. I. Grondona
Laboratorio de Ecotoxicología y Contaminación Ambiental, Instituto de Investigaciones Marinas y Costeras, Universidad Nacional de Mar del Plata, CONICET, Mar del Plata, Argentina

Instituto de Geología de Costas y del Cuaternario, Universidad Nacional de Mar del Plata, Comisión de Investigaciones Científicas de la Prov. de Buenos Aires, Argentina

© The Author(s), under exclusive license to Springer Nature Switzerland AG 2023
J. Blasco, A. Tovar-Sánchez (eds.), *Marine Analytical Chemistry*,
https://doi.org/10.1007/978-3-031-14486-8_5

biomagnify through the food web, and are transported across international boundaries far from their sources, through moving air masses, water currents, and migratory species. The dynamics of POPs in the environment depends on several factors, such as physicochemical characteristics of POPs and abiotic components, environmental conditions, and properties of the species inhabiting different ecosystems. In terrestrial and aquatic environments, plants, invertebrates, fish, birds, marine mammals, and other organisms can incorporate POPs directly from soils, sediments, and water and also through food chain.

Keywords

Persistent organic pollutants · Bioaccumulation · Biomagnification · Stockholm Convention · Chronic toxicity

5.1   What Are Persistent Organic Contaminants

Persistent organic contaminants are toxic compounds of global concern that adversely affect the environment and human health around the world. They persist in the environment, bioaccumulate in fat tissues through the food web, and are transported across international boundaries far from their sources, through moving air masses, water currents, and migratory species. They can reach regions where they have never been used or produced, such as the Arctic or Antarctic. In this sense, both indigenous people and ecosystems of both areas are particularly at risk considering the impact of this long-range atmospheric transport and potential to bioaccumulation and biomagnification through food web.

This group of compounds, known as persistent organic pollutants (POPs), consists mainly in three general categories, although some of them fit into more than one of this classification:

- Organochlorine pesticides (OCPs) involve pesticides used in agricultural applications and health care.
- Industrial compounds used in several applications.
- Chemical-delivered unintentional by-products of urban and industrial processes.

Twelve POPs were initially listed in the Stockholm Convention. In general, these "legacy" POPs were first produced and/or used several decades ago, and their dangerous properties lead them to being globally banned or restricted since 2004. Then, between 2009 and 2019, 18 new chemicals have been added and listed as POPs. Table 5.1 summarizes the 30 POPs listed in the Stockholm Convention up to January 2021.

Table 5.1 Persistent organic pollutants listed in the Stockholm Convention

Chemical substances
Category = pesticides
Aldrin
Alpha-hexachlorocyclohexane (α-HCH)
Beta-hexachlorocyclohexane (β-HCH)
Chlordane
Chlordecone
Dicofol
Dichlorodiphenyltrichloroethane (DDT)
Dieldrin
Endosulfan
Endrin
Gamma-hexachlorocyclohexane (lindane, γ-HCH)
Heptachlor
Pentachlorophenol (PCP), salts, and esters
Toxaphene
Mirex
Category = industrial
Pentachlorobenzene (PeCB)[a,b]
Decabromodiphenyl ether (deca-BDE)
Hexabromocyclododecane (HBCD)
Polybromodiphenyl ether (PBDE)
Hexachlorobenzene (HCB)[a,b]
Hexachlorobutadiene (HCBD)[b]
Hexabromobiphenyl (HBB)
Perfluorooctanesulfonic acid (PFOS)[a]
Perfluorooctanoic acid (PFOA)
Polychlorinated biphenyls (PCBs)[b]
Polychlorinated naphthalenes (PCN)[b]
Short-chain chlorinated paraffins (SCCPs)
Category = unintentional production
Polychlorinated dibenzo-para-dioxins (PCDDs)
Polychlorinated dibenzofurans (PCDFs)

Additional category: [a]pesticide and [b]unintentional production

5.1.1 Organochlorine Pesticides (OCPs)

These pesticides have at least one ring structure and several chlorine atoms. In relation to their molecular structures, OCPs can be classified into five classes: (1) DDT and metabolites (Dichlorodiphenyldichloroethylene (DDE) and Dichlorodiphenyldichloroethane (DDD)); (2) hexachlorocyclohexanes (α-, β-, and γ-HCH); (3) cyclodienes including aldrin, dieldrin, endrin, heptachlor, heptachlor epoxide, chlordanes (α- and γ-isomers), and endosulfan (α- and β-isomers and endosulfan sulfate); (4) toxaphene; and (5) mirex and chlordecone (Fig. 5.1).

Organochlorine pesticides were introduced in the 1940s and vary mechanisms of toxicity according to their chemical structures. They have been used since the

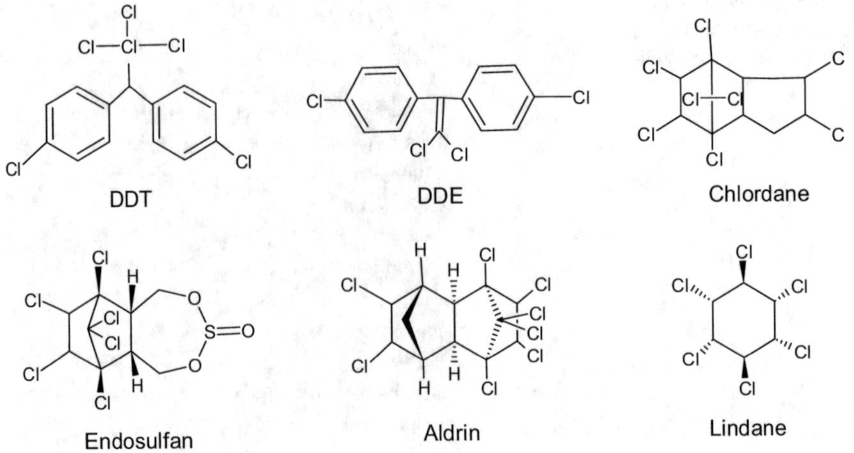

Fig. 5.1 Organochlorine pesticides

mid-twentieth century at worldwide level in agricultural activities and for sanitary vector control of tropical diseases such as malaria. Despite that the production and usage of all these pesticides are forbidden, the most popular insecticide DDT is currently listed in Annex B to the Stockholm Convention with its production and/or use restricted for disease vector control purposes when no equally effective and efficient alternative is available and in accordance with related World Health Organization (WHO) recommendations and guidelines. The main source of exposure to organochlorine pesticides is through eating fatty foods such as milk, dairy products, or fish that are contaminated with these pesticides. It is also possible to pass these pesticides through the placenta to the unborn child or by breastfeeding or also to absorb them through the skin. They are considered as endocrine disruptors, produce alterations in the Central Nervous System (CNS), and trigger carcinogenesis processes, since they are considered as potential generators of cancer by the International Agency for Research on Cancer (IARC) of the World Health Organization. Due to their worldwide distribution and high persistence, these pollutants represent a serious threat to the environment and public health.

5.1.2 Polychlorinated Biphenyls (PCBs)

These compounds were mainly used worldwide between 1929 and the 1980s for industrial purposes (Alcock et al. 1994). They vary in consistency from thin, light-colored liquids to yellow or black waxy solids. PCB characteristics include nonflammability, chemical stability, high boiling point, and electrical insulating properties. All these properties lead to these compounds being used in hundreds of industrial and commercial applications including electrical, heat transfer, and hydraulic equipment, in pigments, dyes, and carbonless copy paper, among other

Fig. 5.2 General structure
of PCBs

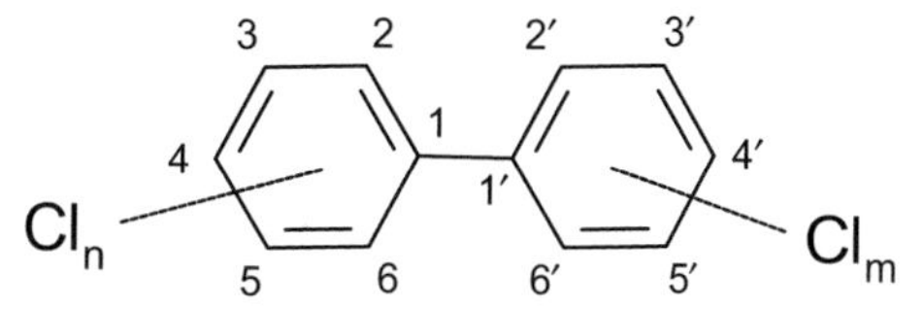

PCBs

applications. PCBs are compounds with high toxicity and have undesirable effects on the environment and on humans. They are most often referred to their commercial trade names, including Aroclor and Kanechlor which represent varied mixtures in their weight percentages of chlorine, indicated by the last two numbers of the four-digit number following the Aroclor name. The most commonly produced Aroclor mixtures were 1016, 1221, 1242, 1248, 1254, and 1260. The PCB molecule consists of 2 connected benzene rings and chlorine atoms that can attach to any or all of 10 different positions (Fig. 5.2) allowing for 209 different congeners and 10 different homologs (McFarland and Clarke 1989).

Individual PCB congeners can be named in two ways, following International Union of Pure and Applied Chemistry (IUPAC) structural name (e.g., 2,2′,5-trichlorobiphenyl) and also with a numbering system originally devised by Ballschmitter and Zell, and adopted by the IUPAC (e.g., PCB 18, or #18), in which PCB 1 is 2-monochlorobiphenyl and PCB 209 is decachlorobiphenyl.

PCBs are chemically very stable, soluble in a wide range of organic solvents, and poorly soluble in water and have low electrical conductivity. Pure PCBs are either liquids or noncrystalline solids at room temperature, and most commercial mixtures are viscous liquids. Vapor pressures of PCBs range from 1.2 Pa for monochlorinated congeners to $7\ 10^6$ Pa for decachlorobiphenyl at 25 °C, and log octanol–water partition coefficients (log K_{OW}) of PCBs are in the range 4.2 (monochlorobiphenyl) to 8.3 (decachlorobiphenyl) at 25 °C.

Once these compounds are released into the environment, PCBs are bioaccumulated into the food chain, due to their high affinity for fatty tissues.

5.1.3 Dioxins and Furans

Polychlorinated dibenzo-p-dioxins (PCDDs) and polychlorinated *dibenzofurans (PCDFs)* are environmental contaminants detectable in almost all compartments of the global ecosystem in trace amounts. Most dioxins and furans are not man-made or produced intentionally, but they are created by some industrial processes or when other chemicals are made. There are 210 different dioxins and furans (Fig. 5.3).

The dioxin considered most toxic is referred to as 2,3,7,8-tetrachlorodibenzo-p-dioxin (TCDD) and is classified to be carcinogenic to humans (Group 1 carcinogen) according to the International Agency for Research on Cancer (IARC). It has also been extensively studied for health effects linked to its presence as a contaminant in

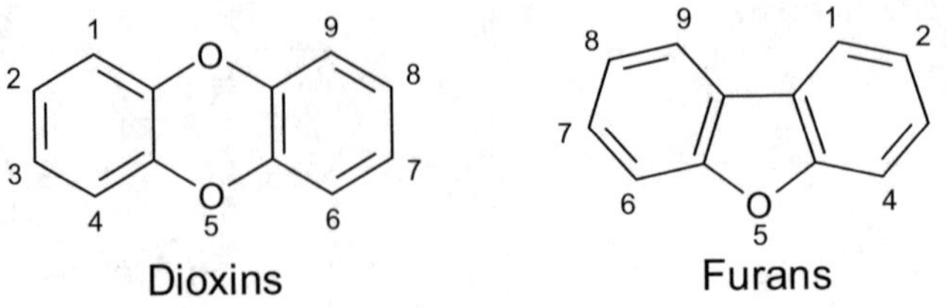

Fig. 5.3 Dioxins and furans

some batches of the herbicide Agent Orange, which was used as a defoliant during the Vietnam War.

Dioxins and furans are unwanted by-products of a wide range of manufacturing processes including incomplete burning of many issues, such as municipal, household, and hospital wastes, plastics and wood, tobacco smoke, fuel including diesel fuel and fuel for agricultural purposes and home heating, electrical power generation, and production of iron and steel. Moreover, smelting, chlorine bleaching of paper pulp, and manufacturing of some herbicides and pesticides represent other possible sources of dioxins.

Although formation of dioxins is local, environmental distribution is global. Dioxins can also be produced from natural processes, such as volcanic eruptions and forest fires. In this sense, most dioxins are introduced to the environment through the air and then attached to small particles that can travel long distances in the atmosphere. Both dioxins and furans are found in very small amounts in the environment, including air, water, soil, and food. Once dioxins enter the body, they are absorbed by fatty tissues for body storage. Moreover, they tend to accumulate in the food chain being people more exposed to dioxins and furans into their bodies through food than through air, water, or soil. It is known that meat, milk products, and fish have higher levels of dioxins and furans than fruit, vegetables, and grains. Long-term storage and improper disposal of PCB-based waste industrial oils may result in dioxin release into the environment and the contamination of human and animal food supplies.

Exposure to dioxins and furans has been associated with a wide range of adverse health effects in laboratory animals and humans. The type and occurrence of these effects typically depend on the level and duration of exposure.

Dioxins are potent endocrine disruptor and are highly lipophilic and extremely stable with a long half-life in humans of 7–9 years (IARC 1997; Needham et al. 1999).

It is known that many countries monitor their food supplies for dioxins in order to get an early detection of contamination and possibility to prevent impacts on a larger scale. Some events related to dioxin contamination have been significant, with dangerous implications. In July 1976, large amounts of dioxans were released during a chemical plant explosion near Seveso, Italy. Therefore, the accident led to the highest known levels of 2,3,7,8-tetrachlorodibenzo-*p*-dioxin (TCDD or dioxin) exposure to a residential population (Mocarelli 2001). A cloud of toxic chemicals was released into the air and eventually contaminated an area of 15 km^2 where 37,000 people lived. Extensive studies in the affected population are continuing to determine long-term human health effects from this incident. Epidemiologic research only became possible because of the rapid establishment of a long-term health surveillance program of the population, to capture information on health effects that could present themselves years after initial exposure.

Another example of dioxin contamination incident occurred in January 1999, where high levels of dioxins were found in poultry and eggs from Belgium. In this context, 500 tons of feed contaminated with PCBs and dioxins was distributed to animal farms in Belgium and to a lesser extent in the Netherlands, France, and Germany. Subsequently, dioxin-contaminated animal-based foods (poultry, eggs, pork, hens, and chicks) were detected in several other countries due to animal feed contaminated with illegally disposed PCB-based waste industrial oil. Estimates of the total number of cancers resulting from this incident range between 40 and 8000.

Only some cases of intentional human poisoning have been reported. The most notable incident was in 2004, in relation to the case of the Ukraine president, Viktor Yushchenko, who was exposed to dioxins and whose face was disfigured by chloracne.

5.1.4 Polybrominated Flame Retardants (BFR)

These compounds are additive that have been used in a variety of consumer products to inhibit or reduce the flammability of combustible products (Covaci et al. 2005). Polybrominated diphenyl ethers (PBDEs) and hexabromocyclododecane (HBCD) represent two important compound classes (Fig. 5.4).

PBDEs have been widely used as a flame retardant in furniture, plastics, cars, textiles, electronic equipment, and building materials (De Wit 2002). They started to be used in 1970, and because of low cost and high compatibility with plastics and fabrics, PBDEs have been applied in a variety of industrial and consumer products. Particularly, the decabrominated diphenyl ether is added to plastics used in electrical

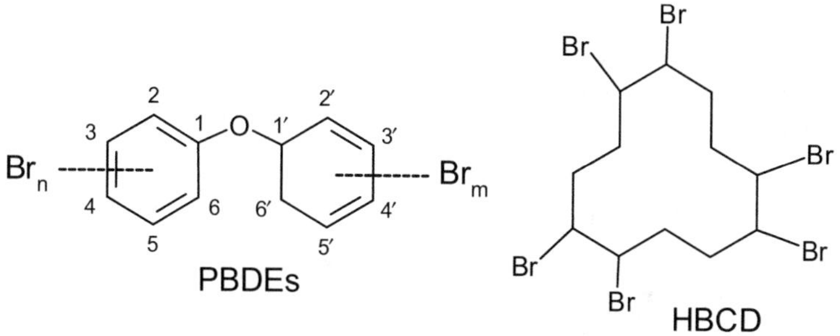

Fig. 5.4 General structure of polybrominated diphenyl ethers (PBDEs) and hexabromocyclododecane (HBCD)

and electronic equipment (computers, TV, and music equipment), to automobile interiors, and for construction and building (i.e., wires, cables, pipes, etc.).

There have been three primarily PBDE commercial formulations, i.e., penta- (over 70% of BDE-47 and BDE-99), octa- (over 40% of BDE-183), and deca- (over 98% of BDE-209) BDEs. Among the three technical mixtures commercially produced, deca-BDE has been the most widely used (80% global demand of 2001). Even though it is not regulated, there is increasing concern for its bioavailability to wildlife and the potential degradation to more toxic and bioaccumulative PBDEs (Stapleton et al. 2006).

5.1.5 Polyfluoroalkyl Substances (PFAS)

Per- and polyfluoroalkyl substances (PFAS) are a group of synthetic chemicals that contain fluorine atoms and have been in use since the 1940s and are found in many consumer products. These compounds include perfluorooctanoic acid (PFOA), perfluorooctanesulfonic acid (PFOS), and many other chemicals (Fig. 5.5).

PFAS owe their properties to the carbon–fluorine bond, which is one of the shortest and strongest known. This property also makes these chemicals highly resistant to breakdown in the environment. PFAS have been manufactured and used in a variety of industries around the globe. These compounds offer characteristics such as heat, stain, and water resistance that are desired by industry and consumers alike. PFOA and PFOS have been the most extensively produced and studied of these chemicals.

PFAS are currently listed in the Annex B of the Stockholm Convention and are registered as restricted compounds; therefore, it is possible that they can be found in many consumer goods such as the following:

- Food packaged in PFAS-containing materials such as pizza boxes, processed with equipment that used PFAS, or grown in PFAS-contaminated soil or water
- Commercial household products, including cookware, nonstick products (e.g., Teflon), stain repellants, polishes, waxes, paints, firefighting foams, and cleaning products
- Drinking water, close to a specific facility (e.g., manufacturer, landfill, and wastewater treatment plant)
- Living organisms, including plants, animals, and humans, where PFAS persist over time

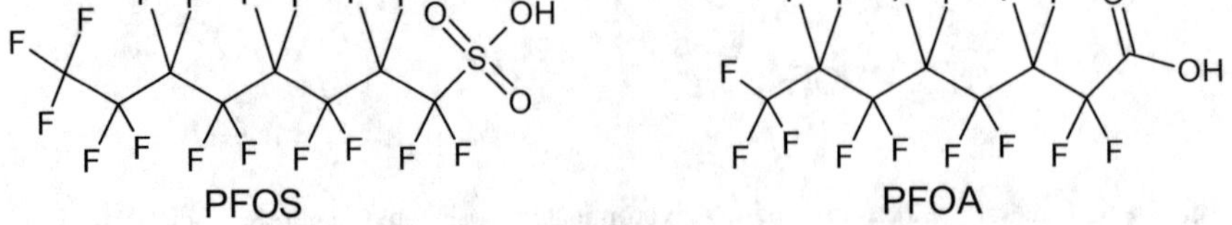

Fig. 5.5 Perfluorooctanesulfonic acid (PFOS) and perfluorooctanoic acid (PFOA)

Although PFOA and PFOS are no longer manufactured in some countries, they are still produced internationally and can be imported in several consumer goods such as leather and apparel, textiles, carpet, paper and packaging, rubber, plastics, and coatings.

Both chemicals are very persistent in the environment and in the human body. Most people have been exposed to PFAS, and some of them can accumulate and remain in the human body for long time periods. There is evidence that exposure to PFAS can lead to adverse human health effects. Studies indicate that PFOA and PFOS can cause reproductive and developmental effects in laboratory animals and also have caused some tumors. Moreover, other evidences indicate that animals exposed to PFAS develop dangerous effects in the liver and the kidney and also in immunological processes.

5.2 What International Actions Have Been Considered to Control POPs

Environmental agreements have the objective to protect the human health and the environment from hazardous chemicals and wastes.

The Stockholm Convention on Persistent Organic Pollutants is a global treaty whose aim is to protect the human health and the environment from highly harmful persistent chemicals that affect the health of humans as well as wildlife. The convention restricts and sometimes eliminates the production, use, importation, and exportation of POPs. These compounds are listed in three annexes with different obligations:

- Annex A – elimination of production and use
- Annex B – restriction of production and use
- Annex C – unintentionally generated and released substances

The convention initially focused on 12 intentionally and unintentionally produced chemicals, called the "dirty dozen" (UNEP 2002). But additionally, the convention has a mechanism to add more compounds; today 30 POPs are covered, 18 more than the initial ones (Fig. 5.6). The 12 initial POPs were chlorinated chemicals and had direct uses such as pesticides, industrial chemicals, and

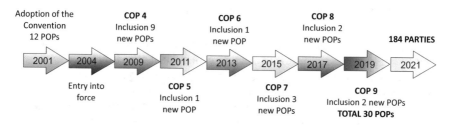

Fig. 5.6 Milestones of the convention 2001–2021

by-products; the more recent POPs are products such as flame retardants and water-repellent chemicals.

The Stockholm Convention is managed by the United Nations Environment Programme (UNEP). The function of parties is to implement obligations of the convention: eliminating or restricting the production and use of the intentionally produced POPs and prohibiting and eliminating production and use or import of POPs and, also, conducting research, identifying areas contaminated with POPs, and providing financial support for the convention. It was created in 2001 and entered into force in 2004 and currently has 184 members. The Stockholm Convention adds an important global dimension to our national and regional efforts to control POPs.

The UNEP has established criteria for identifying new POPs and classified a substance as persistent if it has a half-life exceeding one of the thresholds in the following media: soil or sediment (180 days), surface water (60 days), and air (2 days).

The Basel Convention on the Control of Transboundary Movements of Hazardous Wastes and Their Disposal was created to protect people and the environment from negative effects of improper management of hazardous wastes around the world. This comprehensive global treaty addresses these materials throughout their life cycles, from production and transportation to end use and disposal. It was designed to reduce the movement of these wastes between nations and to prevent the transfer from developed to less developed countries. The convention was opened for signature on March 21, 1989, and entered into force on May 5, 1992. As of the end of March 2021, the convention has 188 members worldwide.

The Rotterdam Convention on the Prior Informed Consent Procedure for Certain Hazardous Chemicals and Pesticides in International Trade is a multilateral treaty that promotes efforts to protect the human health and the environment and allows parties to decide whether to import pesticides and dangerous chemicals listed in the convention. This treaty asks exporters of dangerous chemicals to use proper labeling, include instructions on safe handling, and inform buyers of any known restrictions or prohibitions. Parties can then decide whether to allow or prohibit the importation of chemicals listed in the treaty, and exporting countries are required to ensure that producers within their jurisdiction comply with the convention. The Rotterdam Convention was adopted on September 10, 1998, and entered into force on February 24, 2004, and currently has 164 members.

5.3   Distribution in the Environment

The dynamics and fate of pollutants in the environment depend on several factors, such as physicochemical characteristics of contaminants and abiotic components, environmental parameters, and biochemical, physiological, and behavioral characteristics of the species inhabiting different ecosystems.

POPs are long-lasting chemicals that circulate between different environmental compartments and are capable of traveling long distances, thus becoming ubiquitous global pollutants. Physicochemical characteristics of POPs include high

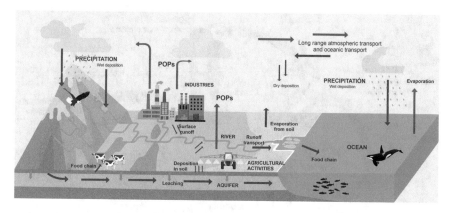

Fig. 5.7 Fate and transport of POPs in the environment and their interactions with the hydrological cycle

hydrophobicity, toxicity, long-range transport, and bioaccumulation. In aquatic and terrestrial systems, they bond strongly to soils and sediments, mainly to organic matter, and are absorbed to fatty tissue and accumulate in the body fat of living organisms, allowing compounds to persist in biota, where the rate of metabolism is low. Because POPs are degraded very slowly, they will be present in the environment for a long time with long half-lives in soils, sediments, air, and biota (Fig. 5.7).

Furthermore, POPs are transported in low concentrations by the movement of fresh and marine waters, and they can be transported long distances in the atmosphere. Therefore, POPs are widely distributed throughout the world, including regions where they have never been used.

Main environmental processes that control the persistence of a compound can be roughly divided into three areas:

- Transportation – volatilization, dilution, and advection
- Partition – sorption and absorption by organisms
- Degradation – biodegradation (aerobic, anaerobic, and metabolism) and abiotic degradation (hydrolysis, photolysis, oxidation, and reduction)

5.3.1 Atmospheric Transport

POPs are cycling between environmental compartments (soil, air, water, snow, and biota) and subject to revolatilization from land and sea surfaces. Cycling of POPs is expected to be influenced by meteorological parameters (wind, temperature, precipitation, and others) that also affect substance phase changes. Atmospheric long-range transport is today considered as major distribution pathways for many POPs. Selected POPs reach even the Arctic or Antarctic, where they are deposited in soil, ice, and sediments and may even accumulate in the food web.

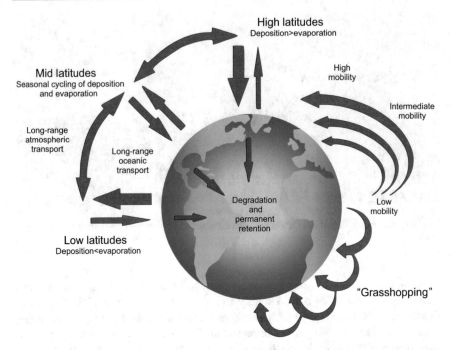

Fig. 5.8 Tracking and distribution of persistent organic pollutants (POPs, adapted from Wania and Mackay 1996)

Effects of global distillation and cold condensation describe the process in which POPs volatilize from warm areas (lower latitudes), migrate, and condense on surfaces such as soil, vegetation, or snow and then migrate again to finally accumulate at higher latitudes with lower temperatures (Fig. 5.8). This process of POP migration, in a series of relatively short jumps, is known as the "grasshopper effect" (Wania and Mackay 1996).

This effect explains why POPs can be found in areas where they have never been used. Once POPs reach polar regions, cold weather and low evaporation rates trap these compounds, leading to enter into the food chain. Furthermore, this ability of POPs to travel long distances and the remobilization of soils may explain why countries that banned the use of POPs are not experiencing a marked decrease in their POP concentrations. The atmosphere has been considered to be the most important and fastest transport route for POPs to surface waters. POPs are divided into gas and aerosol phases and can then be removed from the atmosphere by four main mechanisms: dry deposition of particle-bound pollutants, diffusive gas exchange between the atmosphere and the ocean surface, removal by rain or snow (deposition wet), and direct and indirect photolysis.

Atmospheric monitoring of POPs can be carried out by different pathways. High-volume air samplers are usually used, but they depend on electric energy and, therefore, lead its use to be more difficult due to the accessibility. However, there are other options using passive air samplers (PAS), such as semipermeable membrane devices (SPMD), polyurethane foam disks (PUF), and styrene–divinylbenzene polymeric (XAD) sampler, being all of them artificial passive samplers. Moreover, other options include natural samplers, such as the use of vegetation (lichens, moss, epiphytes, pine needles, and tree bark).

As part of some studies carried out by the monitoring network Latin American Passive Atmospheric Sampling Network (LAPAN), XAD-based passive air samplers (PAS) were used to evaluate organochlorine pesticides (OCPs), polychlorinated biphenyls (PCBs), polybrominated diphenyl ethers (PBDEs), and some currently used pesticides (chlorothalonil, trifluralin, and dichlofluanid) in the atmosphere of Argentinian Patagonia (Miglioranza et al. 2021). Endosulfan, trifluralin, and DDT-related substances were the most prevalent pesticides in the Rio Negro watershed, an intensive agricultural basin settled in Patagonia, Argentina. Contaminant levels decreased toward the south, although a slight increase of total pollutant levels was observed in the most southern sampling site, mainly due to the lower chlorinated PCB congeners.

The use of different plant species has been proved to be very useful for air pollution monitoring, such as trees, shrubs, grass, and also pine needles. Another option is epiphytic plants such as lichens, mosses, and species from the *Tillandsia* genus (Bromeliaceae family) that are more frequently used in air monitoring because they are independent of the substrate. Silva Barni et al. (2019) have used the epiphyte *Tillandsia bergeri* as a feasible biomonitor to OCPs and PCBs along a watershed with activities related to livestock, agricultural, urban, and industrial development. Plants can metabolize different pollutants; therefore, the combined use of natural (plants) and artificial air samplers (XAD-PAS or PUF) results in a very good option, which allows to discriminate efficiently the relative contribution of different contaminants to air pollution.

5.3.2 Soils and Sediments

Soils have been considered as the "most complex matrix" acting as a habitat for living organisms, a carbon sink, a nutrient source, and a hydrological regulator about quality and quantity of water. Owing to the demands for a rapid economic growth, soils are subject to direct and indirect human-induced disturbances. Whether intentionally or unintentionally, organic pollutants (such as pesticides, flame retardants, surfactants, polycyclic aromatic hydrocarbons, PCBs, PCDDs, and PCDFs) end up in the soil as a consequence of human activities. In this sense, POPs can be released directly or indirectly to soil by different routes.

Pesticides may be introduced into soil by common agricultural practices and runoff from gardens, lawns, and roadways. Moreover, they can also reach surface soils by atmospheric deposition of resuspended soil dust, road dust, and air particles (e.g., emissions from traffic, industry, domestic heating, and incineration processes) in wet and/or dry conditions. In the context of an underground scenery, leakages of wastewater pipes may cause dispersion of contaminants. Moreover, abandoned industrial sites, illegal deposits, and accident sites may also contribute to POP release into soil. Therefore, soils can be considered in some cases sinks and also sources of POPs.

Once POPs enter soils and sediments, they may experience a variety of complex physical, chemical, and biological processes that determine their partial or complete degradation or persistence. The extent to which any of these processes occur depends on physicochemical properties of both POPs and soils, as well as on climatic conditions. In this way, they can be metabolized by various chemical and microbial degradation processes. Adsorption and desorption processes, which determine the buffering capacity of soil or sediment, play a major role with respect to POP behavior in the environment.

The organic matter content of soils is a key factor to determine the POP behavior in this matrix, because it can limit the availability and mobility of POPs. Strong adsorption of chemicals can increase their persistence and result in bioaccumulation effects. Sorbed contaminants can also be subject to runoff to surface waters, leaching to aquifers, or being available to organisms. The fate and transport of natural and anthropogenic soil particle-borne organic contaminants are a critical environmental issue, and complex processes are involved.

As a consequence of the high persistence of OCPs, several studies have demonstrated that soils can serve as sink of these pollutants. In this sense, Miglioranza et al. (2013) studied a basin which constitutes the main fruit production area in Argentina, mainly for apples, grapes, peaches, plums, and pears. Very high levels of p,p'-DDE were reported in surface soils, reaching up to 95% of the total pesticides found. Other studies carried out in soils from an organic farm, which have never received direct pesticide application and were settled close to a big horticultural belt, revealed the occurrence of many OCPs, such as DDTs, endosulfans, HCHs, chlordanes, drins, and heptachlors as a result of drift and volatilization from agricultural areas (Gonzalez et al. 2005). In the south of the pampean region, Argentina, studies on an important agricultural belt demonstrated the accumulation of OCP residues in soil as a consequence of the intensive use, low rains, low slope, and drainage design that minimize losses by surface runoff (Andrade et al. 2005).

5.3.3 Surface Water and Groundwater

Freshwater ecosystems play an important role in supplying drinking water, fisheries, and recreation and in maintaining regional ecological balance and sustainable socioeconomic development, but these ecosystems are generally exposed to POP pollution.

Thus, POPs can enter the freshwater system and groundwater through point and nonpoint sources. A point source is a single, identifiable source of pollution, such as a pipe, a direct discharge, or a drain. Industrial wastes are commonly discharged to rivers and the sea. Nonpoint sources of pollution or "diffuse pollution" refer to inputs and impacts which occur over a wide area and are not easily attributed to a single source. The most common nonpoint sources are agricultural runoff, soil leaching, urban runoff, and atmospheric wet and dry depositions that probably lead to concentration relatively uniform but result in sceneries very difficult to manage. In the case of industrial sources of POPs, they are more related to point sources, such as direct discharges to rivers, effluents, or industrial areas with high atmospheric pollution.

According to physicochemical properties, POPs have a high adsorption capacity to organic matter of sediments and particle suspended material. Therefore, in aquatic environments, POPs are preferentially accumulated in bottom sediments becoming to be the source of POPs and gradually are released to aquatic environment.

The main factors that affect bioavailability of POPs in aquatic environments can be divided into two main areas: factors that alter the partitioning of contaminants to sediments (water solubility, K_{ow}) and biological factors that alter the exposure and accumulation of contaminants. In this sense, aquatic organisms incorporate POPs by direct uptake from the water, through the food chain, or by feeding sediments (Ondarza et al. 2014).

The leaching of POPs involves vertical movements of contaminants from surface soils through the soil profile to deeper zones and finally to groundwater, which is vulnerable to pollution. One important factor is the possibility of cosolvation which implies a mobility of POPs in soil enhanced by the presence of another solvent such as chloroform and chlorobenzene, leading to a higher movement of POPs to groundwater.

Groundwater is usually impacted by industry, agriculture, mining, and other human activities, being agricultural practices as one of the main sources for pesticides (Massone et al. 1998; Gonzalez et al. 2012). Many factors influence pesticides and other contaminants leaching to groundwater, but the most important are nature of the soil, physicochemical properties of pollutants, and distance to the aquifer (Grondona et al. 2019). When groundwater becomes contaminated, it is difficult and very expensive to clean up and is a serious threat to water supply.

5.3.4 Organisms

Many factors affect concentrations of POPs in living organisms, such as location and diet. Although POPs are widespread globally, their concentrations vary by orders of magnitude from one site to another. In this sense, sedentary species will be affected by their home range, while migratory species can take POPs from different sites and periods. POPs are bioaccumulative and have the potential to elicit adverse effects on environmental and human health. Due to physicochemical characteristics, they are absorbed to fatty tissues and can be transferred across trophic levels of the food web by processes of bioaccumulation, where POP concentrations are higher in old organisms than younger, and biomagnification where POP concentrations are higher in the predator than the prey (Fig. 5.9). This process leads to a high persistence of POPs in biota, due to a very slow metabolism inside organisms.

In aquatic environments, plants, invertebrates, fish, birds, marine mammals, and other organisms can incorporate POPs directly from sediments and water and also through food chain (Chierichetti et al. 2021; Quadri-Adrogué et al. 2021; Vazquez

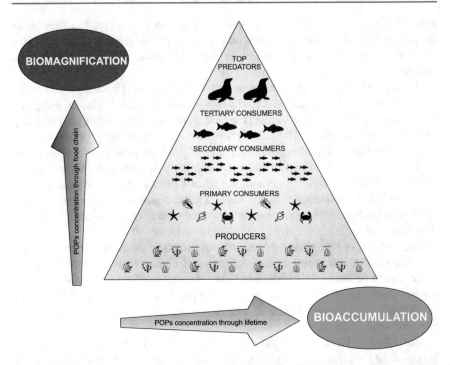

Fig. 5.9 Bioaccumulation and biomagnification processes of persistent organic pollutants (POPs)

et al. 2022). Under laboratory studies, it has been demonstrated that benthic invertebrates can bioaccumulate PBDEs (Leppänen and Kukkonen 2004), and they have a key role in the trophic transfer of these highly hydrophobic chemicals.

Moreover, some floating plants and other aquatic macrophytes are able to absorb POPs directly from the atmosphere. In marine environments, marine mammals accumulate high concentrations of POPs because of their high fat content and because many species of cetaceans and pinnipeds occupy the very top of the marine food chain. Moreover, given that marine mammals suckle their young on high-fat milk, they receive an unhealthy dose of POPs as a result of "lactation transfer" of contaminants from mother to young.

Moreover, PCBs have been found in biota and human tissues and are incorporated mainly through diet, by the consumption of meat, fish, and dairy products (Van den Berg et al. 2006).

The occurrence of microplastics in aquatic environments leads organisms to be exposed to many contaminants. In this sense, pollutants can be leached out from plastics, which can increment their bioavailability to organisms in aquatic habitats. Due to the small size, and mistaken with food, aquatic organisms can ingest microplastics, which represents a long-term threat due to mechanical blockage of the gastrointestinal tract of animals and also due to the high toxicity of adsorbed POPs.

In terrestrial environments, the soil represents the main source of POPs to organisms. Vegetation is the link among atmosphere, soil, and human food supply and plays an important role in the global pollutant cycle (Collins et al. 2006), so it is important to understand those processes by which pollutants enter into these environmental compartments. Plants may accumulate POPs via different pathways, (1) adsorption to the root surface, (2) root uptake and transport to the upper plant tissues, and (3) uptake of airborne vapors by aerial plant parts (stem, leaves, and fruits).

Among plant species, trees account for the majority of global biomass (Trapp et al. 2001) and are dominant constituents of several ecosystems. The wood of stems and the lipophilic cuticles of their leaves provide storage compartments and live cells, especially near the cambium and phloem, being an area for rapid metabolic degradation of anthropogenic chemicals.

It has been studied the uptake and translocation of DDTs in willow plants and how some amendments affected those processes. In this sense, the addition of organic acids increases this pattern and, therefore, affects willow DDT uptake and bioavailability of aged pesticide residues in soils being one important factor to be considered for environmental risk assessment (Mitton et al. 2012).

Regular horticultural practices under organic production include the enrichment of soil with vegetal rests. Among these practices, pine needles are usually added to soils at the beginning of growing periods. It is known that pine needles, rich in waxes, accumulate highly lipophilic pollutants from the atmosphere (Krauthacker et al. 2001). Therefore, those soils enriched with this amendment incorporate those contaminants to soils being available for plant species grown there. It has been demonstrated that species-specific characteristics determine the pollutant mobilization extent and the magnitude of uptake associated with a crop upon particular environmental condition (Gonzalez et al. 2005).

Then, crops can be consumed by cattle, humans, or wildlife, and a transference of POPs occurs by the biomagnification process. Moreover, invertebrates and vertebrates are exposed to POPs, through food chain and also by direct exposure to pollutants. A great variety of invertebrates live within the soil, and particularly, beetles are considered beneficial arthropods in agriculture. Earthworms and arthropods contribute to maintain soil fertility and have been reported that these nontarget organisms can bioaccumulate and metabolize POPs from soils (Miglioranza et al. 1999).

5.3.5 Marine Ecosystems

Oceans play an important role in controlling the environmental transport, fate, and sinks of POPs because these are a global reservoir and ultimate sink of many POPs (Iwata et al. 1993) and constitute a slow but significant medium for their long-range transport. The main sources of POPs to oceans are the atmosphere, rivers, and municipal and industrial wastewater. Due to their global atmospheric transport,

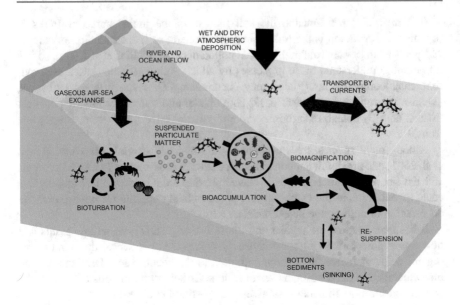

Fig. 5.10 Processes involved in the fate of POPs in the marine environment

POPs can be found in all oceans ranging from the Arctic to the Antarctic and in intertidal or abyssal areas (GESAMP 1990).

When POPs enter the ocean, their physicochemical properties result in the accumulation of these compounds in estuarine and marine sediments that act as a repository of POPs. Furthermore, these contaminants can be accumulated by biota, from waster, sediments, and particle suspended matter. Coastal areas represent the most affected sites where organisms living there receive the highest burden of contaminants.

The fate of POPs in the marine environment depends on different processes (Ilyina et al. 2006, Fig. 5.10):

- Transport by ocean currents via processes of advection and diffusion, being the major pathway for many POPs' dispersion far from emission sources (Li et al. 2004).
- Air–sea exchange is represented by three mechanisms: reversible gaseous exchange, dry particle deposition, and wet deposition.
- Sorption to suspended particulate matter (SPM).
- Deposition in bottom sediments, followed by redistribution due to disturbance of sediment layers by biological activity (bioturbation).
- Resuspension from bottom sediments by erosion processes, including the impact of rivers, effluents, and creeks that outflow on coastal areas.

POPs tend to be bound to particles, which could modify their bioavailability. Benthic invertebrates are particularly exposed to POPs associated to sediments through its ingestion and bioturbation activities, which could result in a trophic transference to higher levels.

Planktonic and benthic invertebrates have been widely used for POP monitoring. Some species indeed showed detrimental effects, modifying their growth, reproduction, mortality, and community composition.

Uptake and depuration processes are dependent on mainly three groups of variables such as biological factors (physiology and ecology characteristics), physicochemical properties of contaminants (persistence, time, route and concentration of exposure, etc.), and environmental conditions (organic matter and bioavailability of xenobiotic, among others). POP passive uptake can usually take place from the water column and, to a lesser extent, through particulate matter, bottom sediments, and food items. Regarding depuration, metabolite products, urine, and feces represent the main elimination ways.

Polychaetes, crabs, equinoderms, and sea cucumbers are considered very sensitive species to the anthropogenic pollution. For instance, multiple taxes were used as biomonitor of POPs in Antarctica, such as bivalves (*Laternula elliptica*), soft coral (*Alcyonium antarcticum*), deposit feeder polychaete (*Flabegraviera mundata*), and herbivore echinoid (*Sterechinus neumayeri*), being PCBs and DDTs as main pollutants found (Palmer et al. 2022). In Argentina, Commendatore et al. (2018) reported OCP, PCB, and PBDE concentrations in sediments and crabs showing a similar pollutant pattern between both matrices. Similar behavior was found by other previous studies in Bahía Blanca estuary, Argentina (Menone et al. 2004, 2006), suggesting that crabs when these are present play a role in the distribution of sediment-bound POPs and they are modifiers of the dynamic of organic pollutants in estuarine areas.

Fish accumulates POPs by ingestion of contaminated food and by contact of their respiratory surfaces and skin with contaminated waters. Those compounds with log K_{ow} values between 3 and 4 are accumulated by fish primarily by the aqueous route, while those compounds with log $K_{ow} > 6$ will likely be incorporated by the dietary route (Hinton et al. 2008). POPs have been found in marine fish from Arctic to Antarctic and from benthophagous to pelagic feeder (Fisk et al. 2001; Weber and Goerke 2003; Ko et al. 2018).

POP accumulation and toxicity depend on several factors such as concentration and bioavailability of the contaminant in the environment and metabolic and physiological traits of each organism. In this way, it was observed that levels of POPs in marine fish, in both teleosts and chondrichthyans, are dependent upon the diet, sex, maturity stage, and habitat type of species (Weber and Goerke 2003; Lanfranchi et al. 2006; Weijs et al. 2015; Chierichetti et al. 2021). Different effects of POPs on marine fish have been recorded. Several of these compounds have the ability to act as endocrine disruptors in coastal fish, for example, by inducing vitellogenesis in males and/or cause premature induction in females. Studies showed that endocrine disturbances tend to impair reproductive functions, such as higher proportion of atretic oocytes in females, reduced gonad size, reduced fecundity, poor hatching success, and decreased fry survival (Matthiessen 2003; Johnson et al. 2013). On the other hand, fish are generally considered to be most sensitive to contaminants during early life stages, which can reduce larval survival and cause a wide range of developmental defects. POPs measured in embryonic and early developmental stages of fish are considered to be primarily of maternal origin. This is a consequence of lipid mobilization occurred during reproduction to support oocyte development, providing a route through POPs that are accumulated in the liver rich in lipids and are transferred to the gonads and eventually to eggs and embryos. Maternal transfer of POPs was also recorded in different species of marine fish (Mull et al. 2013; Horri et al. 2018).

A relevant aspect to take into account is that fish consumption could be an important source of POPs to humans, added to the fact that a high number of marine fish species have become important component of human diet. Moreover, many coastal areas of the world that function as nursery habitat of commercial fish species have been damaged by pollution, which could indeed reduce recruitment and affected coastal marine fisheries.

The main predators in the marine food chain, such as turtles, birds, and mammals, have the potential to receive high levels of POPs due to their biomagnification process (Fig. 5.11).

Sea turtle populations are threatened or are in danger of extinction worldwide, due to certain anthropogenic activities (water quality and habitat deterioration, hunting, and incidental capture, among others). Turtles are very sensitive to chemical pollution through direct contact or diet; however, limited information is available. Blood is widely used as a nondestructive method to assess recently absorbed organic pollutants. POPs reported in sea turtle samples worldwide indicate that OCPs and PCBs are the most frequently detected compounds and are present at higher levels. Among them, p,p'-DDE followed by hexa- and hepta-chlorinated biphenyls (#153, #138, and #180) usually represent main POPs found in marine organisms (Storelli and Zizzo 2014; Camacho et al. 2014; Clukey et al. 2018). POPs can be accumulated in the adipose tissue, which is used as storage or reserve. Then, these lipids are mobilized during migration or reproduction seasons, increasing the bioavailability and distribution of POPs adsorbed, through blood. Along these processes, POP burden can be transferred from the mother to the eggs, especially during the vitellogenesis (Kelly et al. 2008; Guirlet et al. 2010).

Seabirds have been used in numerous studies as bioindicators to study the occurrence and distribution of contaminants in the marine environment (Burger and Gochfeld 2004). They forage in both coastal and pelagic environments, thereby sampling a wide swath of ocean while consuming plankton, fish, squid, and other invertebrates. Particularly, studies on POPs in seabirds have used a wide variety of tissues for their determination. Among them, unhatched eggs are very useful and represent a minimally invasive and nondestructive method, ensuring a negligible impact on individuals and populations under study. Therefore, POP levels in eggs reflect the load of contaminants in female at the time of laying (Mwangi et al. 2016). Other noninvasive methods for seabird POP level studies have been utilizing feces (Rudolph et al. 2016), feathers (Jaspers et al. 2007), blood (Roscales et al. 2010), and preen gland oil (van den Brink et al. 1998; Yamashita et al. 2007).

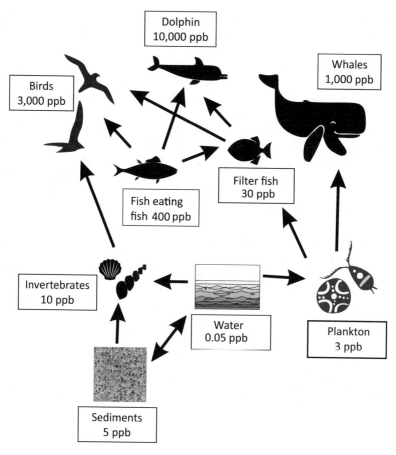

Fig. 5.11 Biomagnification of POPs in a food web *(ppb*, part per billion*)*

Additionally, POP concentrations in organisms are affected by their geographic distribution, foraging area, diet, age, sex, and body condition.

Although physicochemical properties of POPs control much of their environmental behavior, their dynamics and distribution in marine mammals are largely controlled by physiological and biochemical processes.

Marine mammals tend to have large reserves of fat or blubber. Since most POPs have a propensity to accumulate in fat, marine mammals are prone to accumulate high concentrations of POPs since they represent the top of marine food chains, and by the biomagnification process, higher levels are found. The bulk of POP burden is, therefore, associated with blubber in marine mammals, since this energy and insulating tissue represents the key lipid storage in these animals. The main factors controlling lipid distribution in marine mammals are feeding and reproduction. In the case of feeding, POPs accumulated in marine mammals depend on amount of food consumed, efficiency of POP absorption, and POP concentrations in the food. In relation to reproduction, the main factor affecting POP burden is lactation. Females can mobilize up to 90% of the blubber to produce lipid-rich milk to sustain their offspring during nursing. However, males can rely only on slow metabolic degradation to remove POP burden. In this sense, POP levels increase in males with age.

Monitoring of marine mammals provides a holistic view about global trends in POP levels, influenced by some factors such as climate change, industrial and agrochemical pattern uses, and reduced biodiversity in oceans. The habitat use is indicative of POP exposure. In this sense, some individuals move hundreds of kilometers and may be more exposed to global POP background, while others have more limited movements and can be mainly exposed to contaminated sites.

POPs may have a wide range of effects on wild species, such as behavioral changes, hormonal disruption, carcinogenesis, teratogenicity, mutagenic effects, smaller brain size and neurotoxicity, cell and tissue damage, and reproductive problems, among others (Da Cuña et al. 2011; DeLeon et al. 2013; Nossen et al. 2016; Harmon 2015; Silva Barni et al. 2016). Some POPs are known to disrupt endocrine (hormonal) processes, including those involved in breeding and reproduction. Reports are mainly focused on species representatives of the top food chain, such as birds of prey and polar bears. Evidences indicate that POPs may affect levels of estrogen and testosterone, even at relatively low levels of contamination. In this sense, higher levels of POPs in females were associated with lower levels of estrogen, while in males with higher testosterone levels as well as some changes on phenotypic characteristics, potentially reducing the effectiveness of their immune response (Nossen et al. 2016).

Marine pollution caused by plastic debris especially microplastics, which are generally defined as plastic debris less than 5 mm in size, has gained increasing interest in recent years because of their large amount. They are widely detected in marine and coastal environments, including sediments, beaches, seawater, organisms, and even polar regions. Therefore, microplastics are a topic of great concern since their great volume in the ocean and their negative effect on the environment, not only physical but also chemical due to their capacity to adsorb and accumulate contaminants. Thus, they are persistent, universal, widespread, and potential vector of toxic organic compounds to the marine environment. Microplastics and POPs are strongly related, as they interact due to addiction of chemical additives to plastics (PBDEs), through wastewater, urban runoff, or landfill leachate, and after they are released to natural environments. Concentrations of POPs on microplastics depend on many factors, including concentrations and types of pollutants, environmental conditions in surrounding water, and characteristics of microplastics. Pollutants can be desorbed from plastics, which can increment their bioavailability to organisms in aquatic environments. Moreover, microplastics are easily mistaken with food because of their small size, which represents a long-term threat not just by the mechanical blockage of the gastrointestinal tract of animals but also due to the high toxicity of POPs.

5.4   How Are People Exposed to POPs

Humans can be exposed to POPs through diet, occupation, chemical fires, burning of wastes, accidents, and both indoor and outdoor environments (Fig. 5.12). How long people are exposed to POPs is critical to understanding possible effects on humans. It can be a short-term exposure to high concentrations (acute) or a long-term exposure to lower concentrations (chronic).

In the case of exposure to POPs associated to occupational pathway, it could be represented by personal handling pesticides or working in industries where PCBs, PBDEs, and PFAS were used or produced. Farmers are a population especially vulnerable to pesticide exposure through storing, preparing, mixing, and applying these products.

Concerning nonoccupational pathways, they mainly include para-occupational, accidental spills, agricultural drift, residential use, and consumption of contaminated food. Moreover, another route of exposure to POPs is related to contact direct to workers, who introduce these pollutants into their houses with the laundering of work clothes. People living in agricultural areas usually have a higher pesticide exposure compared with urban or suburban residents, due to pesticide drift during application or possible revolatilization from soils. A similar situation was found for people living close to e-waste recycling factories, who are exposed to high PBDE concentrations. Besides, some cleaning practices or pest treatment may affect pollutant levels into residential areas. The ingestion of contaminated food or drinking water represents another exposure way to POPs. Some studies have revealed that food consumption represents the main source (>90 %) of PCDD/PCDF burden in people from Canada (Rawn et al. 2017).

Different human tissues can be considered to evaluate bioaccumulation and effects of POPs in people. As useful indicators of long-time POP exposure, samples of human fat or milk have been measured. For practical reasons, human blood plasma can be used, but it is important to know that noninvasive matrices such as

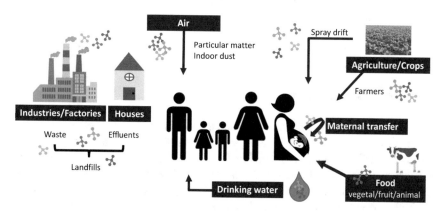

Fig. 5.12 Human exposure to persistent organic pollutants (POPs)

meconium, breast milk, hair, urine, and placenta have been extensively used for these approaches (Tang and Zhai 2017).

Prenatal exposure to POPs is of great concern because fetuses and neonates are vulnerable to these contaminants. The burden of POPs during the pregnancy and early life is mainly transferred from mothers by transplacental delivery and breastfeeding (Vizcaino et al. 2014). Maternal blood, cord blood, placenta, meconium, and breast milk have been used for evaluating this kind of exposure and its adverse effects. Some studies have monitored POPs in different human tissues, mainly associated to the occurrence of OCPs and PCBs (Fernández-Cruz et al. 2020). Health indigenous populations from Arctic or Antarctic have caused special concern about effects by POP levels in their tissues, since they are exposed to food (fish and marine mammals) usually highly contaminated with POPs.

In Argentina, several reports about levels of POPs were associated with OCP accumulation in blood, serum, placenta, mammary fat, and breast milk (Muñoz-de-Toro et al. 2006; Rodriguez et al. 2020).

Endosulfan has been an insecticide widely used in Argentina, mainly on soybean crops and fruit production. Despite its adverse effects, only few studies have been focused on levels of this pesticide in human tissues. In this sense, higher concentrations of α- and β-endosulfan were detected in the placenta from women resident in Argentina (Rodriguez et al. 2020). Another study has related total endosulfan levels in the placenta with higher maternal body mass index before pregnancy (Lopez-Espinosa et al. 2007).

Another study carried out in Arusha, Tanzania, an area where DDT is currently applied to malaria control, showed bioaccumulation of DDTs and some congeners of PCBs (#153, #138, and #180) and PBDEs (i.e., #47, #99, and #100) in the placenta, maternal and cord blood, and breast milk of resident women (Müller et al. 2019). Concentrations of DDT found in placentas were lower than those from North Patagonia in Argentina (an area without malaria control, Rodriguez et al. 2020).

Different samples of human tissues around the world (Europe, Asia, Africa, and Latin America) have evidenced that most frequently POPs detected are the metabolite p,p'-DDE, showing a substantial accumulation and placental transfer to the fetus (Müller et al. 2019; Rodriguez et al. 2020).

5.5 How Do POPs Affect Biota and Human Health

5.5.1 Plants and Animals

POPs can be accumulated in both aquatic and terrestrial plant tissues, where these pollutants can induce toxic effects. The broad strategy of plants when these are confronted with pollutants is including them into standard metabolic cycles. The biotransformation (detoxification) process can be divided into three steps: phase I, which mainly includes transformation by oxidation and hydrolysis; then these metabolites are conjugated in phase II and then compartmentalized in phase III.

POPs promote the enhanced production of reactive oxygen species (ROS) that disrupts the plasmatic membrane structure or metabolic pathways. Moreover, ROS are associated with many effects, such as genotoxicity, lipid and protein peroxidation, and adduct formation. Plants possess both enzymatic and nonenzymatic defense system to reduce these damages. Members of the glutathione-S-transferase (GST) and cytochrome P450 (CYP) enzyme families are involved in diverse biological pathways such as the metabolism of xenobiotics, which enhance the breakdown or conjugation of POPs. Additionally, reduced glutathione (GSH) represents one of the main nonenzymatic defenses, which is involved in the conjugation of contaminants or reducing the ROS species (Mitton et al. 2018).

A wide variety of effects observed in nontarget animal species exposed to POPs include detrimental impacts to biota with photosynthetic symbionts but also reduced growth and reproduction rates of invertebrates and several sublethal effects on physiological or metabolic endpoints.

Among different effects on target pest, OCPs may alter the proper function of the nervous system of insects leading to disorders such as convulsions and paralysis followed by eventual death. Particularly, the insecticide DDT causes uncontrolled, repetitive, and spontaneous discharges along the axon of a neuron, which most often leads to loss of control in the muscles used for breathing and consequently asphyxiation.

Moreover, long-term exposure to OCPs is widely known in mammals, such as rats, to damage the liver, kidneys, the central nervous system, the thyroid, and the bladder. In high doses, dieldrin, heptachlor, and heptachlor epoxide block neurotransmitters in the central nervous system (Narahashi 1992), inducing abnormal brain excitation, headache, confusion, muscle twitching, nausea, and seizures. Laboratory animal studies suggest that heptachlor and heptachlor epoxide cause liver enlargement, liver injuries, and ultimately liver tumors; hexachlorobenzene induces kidney injuries, immunologic abnormalities, reproductive and developmental toxicities, and cancer of the liver and the thyroid.

Literature also provides evidence about effects of PCB exposure to animals (e.g., rats) such as liver damage, immune system suppression, abnormalities in fetal development, enzyme induction, sarcomas, non-Hodgkin lymphomas, and serum lipids.

5.5.2 Human Health Effects and Perception

POPs have been linked to several adverse health effects such as neurotoxicity, endocrine disruption, fetal development, and teratogenicity, among others. The kind, mixture, and levels of POPs are important factors that define human health effects. Following the exposure, POPs can enter the human body through the skin, the respiratory system, and eyes. Diseases related to chronic exposure to POPs include cancer, asthma, dermatitis, endocrine disorders, and reproductive dysfunctions, among others (Fig. 5.13).

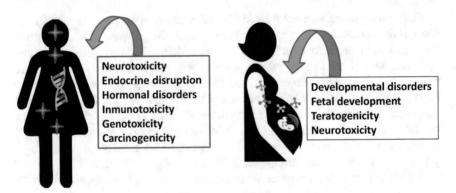

Fig. 5.13 Possible effects of POPs in human health

However, acute poisoning exposure is actually uncommon event and is associated with headaches, stomachaches, vomiting, skin rash, respiratory disorders, and eye irritation, among others (Kalyabina et al. 2021).

Around the world, there are many evidences related to human health effect in workers exposed to POPs for occupational activities, such as a reduction of gestational duration associated with HCB levels in serum of Latin-American pregnant women and their children living in an agricultural area in California (Fenster et al. 2006).

Relatively less is known about nonoccupational exposure effects. It has been observed that α-endosulfan affects both DNA methylation and histone modification enzymes with oncogenic potential in MCF-7 human breast cancer cell line and increases endothelial permeability of human umbilical vein endothelial cells. Levels of DDTs and HCHs in the placenta of nonoccupational exposed women were related with a reduction in birth weight of babies (Anand and Taneja 2020). Even though effects associated with prenatal exposure to pesticides are of increasing concern, despite these possible associations are not well established, some effects were linked with pesticide exposure, like reduction of birth weight, length of gestation, deficiency in cognitive development, growth retardation, and alternations in thyroid and reproductive hormones. Moreover, some studies observed that smoking during pregnancy and maternal ethnicity increase the susceptibility of PCB-153 prenatal exposure and an inverse relationship between PCB concentrations and both free T3 (hormone) and free T4 (hormone) in cord blood suggesting that PCBs may affect the neurologic development of children.

PCBs have been also linked to chronic effects in humans including immune system damage, decreased pulmonary function, bronchitis, and interferences with hormones and cancer. Moreover, other studies indicated that children exposed to PCBs have shown several and serious developmental problems such as low birth weight, behavioral disorders, and hearing loss.

5.6   Global Monitoring

The Stockholm Convention provides an organizational framework for the collection of comparable monitoring data on the presence of POPs from all regions. It is known as the global monitoring plan (GMP) for POPs, which is carried out in order to monitor POP levels in different matrices around the world and identify changes in their concentrations over time. Moreover, another purpose of this monitoring plan is to know about regional and global environmental transport of POPs. A global coordination group controls the implementation of the global monitoring plan across regions and the development of the global monitoring report.

The goal of the Union Implementation Plan is related to controlling and fulfilling legal obligations, to taking stock of actions done, and to establishing a strategy and action plan for further union measures aiming at elimination or reduction of POP releases.

In order to provide high-quality data to reflect human exposure and environmental concentration of these hazardous chemicals, UNEP Chemicals and Health Branch has implemented several rounds of projects at worldwide (UNEP 2012). The pilot project was undertaken from 2005 to 2007 to assess existing capacity and capacity building needs to analyze POPs. Recently, 2nd Phase UNEP/GEF POPs GMP projects are implementing through four regional projects in 42 countries in Africa, Asia, Pacific Islands, and Latin America and the Caribbean regions from 2016 to 2020.

Therefore, through UNEP-coordinated GMP projects, the experience at country level as well as newly generated results on concentrations of POPs in various matrices contributes to the global assessment of levels of POPs, to the regional capacity building, and finally to the effectiveness evaluation of the Stockholm Convention.

Following all these efforts, and according to Article 16 of the convention, the GMP objective provides *a harmonized organizational framework for the collection of comparable monitoring data on the presence of the POPs listed in Annexes A, B and C of the Convention in order to identify trends in levels over time as well as to provide information on their regional and global environmental transport* (UNEP/POPS/COP.6/INF/31/Add.1 n.d.).

A passive sampler network for monitoring POPs has been established in Latin America since 2010 by deploying a pair of PAS containing one cartridge of XAD-2 resin on each site. A total of 13 countries are involved and are part of the LAPAN – Latin American Passive Atmospheric Sampling Network. The LAPAN has been set up to enable studies of long-term spatial and temporal trends of atmospheric (gas phase) contaminants on a regional scale in a sustainable, cost-effective way.

Moreover, results of this monitoring can be used to verify effectiveness of Stockholm Convention measures implemented for POPs under the UNEP and the POPs Protocol of the Convention on Long-range Transboundary Air Pollution (CLRTAP) under the United Nations Economic Commission for Europe (UNECE) on a local and global scale.

Moreover, other networks at worldwide levels are also implemented in the region, like Global Atmospheric Passive Sampler (GAPS), using PUF samplers to analyze different POPs (old and news) in order to contribute to the global assessment of POPs and to the effectiveness evaluation of the Stockholm Convention (Rauert et al. 2016, 2018). Moreover, the World Health Organization (WHO) uses different matrices like food, maternal milk, and blood for monitoring POP levels at worldwide.

References

Alcock RE, Halsall CJ, Harris CA et al (1994) Contamination of environmental samples prepared for PCB analysis. Environ Sci Technol 28:1838–1842. https://doi.org/10.1021/es00060a013

Anand M, Taneja A (2020) Organochlorine pesticides residue in placenta and their influence on anthropometric measures of infants. Environ Res 182:109106. https://doi.org/10.1016/j.envres.2019.109106

Andrade M, Covelo E, Alonso Vega M (2005) Influencia del manejo agrícola intensivo en la contaminación del suelo. In: Pozzo Ardizzi MC (ed) Revista Pilquen. Sección Agronomía, vol 7. Centro Universitario Regional Zona Atlántica (CURZA) de la Universidad Nacional del Comahue, Río Negro. 17p

Burger J, Gochfeld M (2004) Marine birds as sentinels of environmental pollution. EcoHealth 1:263–274

Camacho M, Boada L, Orós J et al (2014) Monitoring organic and inorganic pollutants in juvenile live sea turtles: results from a study of *Chelonia mydas* and *Eretmochelys imbricata* in Cape Verde. Sci Tot Environ 481:303–310. https://doi.org/10.1016/j.scitotenv.2014.02.051

Chierichetti M, Scenna L, Ondarza P et al (2021) Persistent organic pollutants and chlorpyrifos in the cockfish *Callorhinchus callorynchus* (Holocephali: Callorhynchidae) from Argentine coastal waters influence of sex and maturity. Sci Tot Environ 796:148761. https://doi.org/10.1016/j.scitotenv.2021.148761

Clukey K, Lepczyk C, Balazs G et al (2018) Persistent organic pollutants in fat of three species of Pacific pelagic longline caught sea turtles: accumulation in relation to ingested plastic marine debris. Sci Tot Environ 610–611:402–411. https://doi.org/10.1016/j.scitotenv.2017.07.242

Collins C, Fryer M, Grosso A (2006) Plant uptake of non-ionic organic chemicals. Environ Sci Technol 40:45–52. https://doi.org/10.1021/es0508166

Commendatore M, Yorio P, Scenna L, Ondarza PM, Marinao C, Suarez N, Miglioranza KSB (2018) Persistent organic pollutants in sediments, intertidal crabs, and the threatened Olrog's gull in a northern Patagonia salt marsh, Argentina. Mar Pollut Bull 136:533–546

Covaci A, Gheorghe A, Voorspoels S et al (2005) Polybrominated diphenyl ethers, polychlorinated biphenyls and organochlorine pesticides in sediment cores from the Western Scheldt river (Belgium): analytical aspects and depth profiles. Environ Int 31:367–375. https://doi.org/10.1016/j.envint.2004.08.009

Da Cuña RH, Rey Vázquez G, Piol MN et al (2011) Assessment of the acute toxicity of the organochlorine pesticide endosulfan in *Cichlasoma dimerus* (Teleostei, Perciformes). Ecotox Environ Saf 74(4):1065–1073

de Wit C (2002) An overview of brominated flame retardants in the environment. Chemosphere 46:583–624. https://doi.org/10.1016/S0045-6535(01)00225-9

DeLeon S, Halitschke R, Hames RS et al (2013) The effect of polychlorinated biphenyls on the song of two passerine species. PLoS One 8(9):e73471

Fenster L, Eskenazi B, Anderson B et al (2006) Association of in utero organochlorine pesticide exposure and fetal growth and length of gestation in an agricultural population. Environ Health Perspect 114:597–602. https://doi.org/10.1289/ehp.8423

Fernández-Cruz T, Álvarez-Silvares E, Domínguez-Vigo P et al (2020) Prenatal exposure to organic pollutants in northwestern Spain using non-invasive matrices (placenta and meconium). Sci Total Environ 731:138341. https://doi.org/10.1016/j.scitotenv.2020.138341

Fisk AT, Hobson KA, Norstrom RJ (2001) Influence of chemical and biological factors on trophic transfer of persistent organic pollutants in the northwater polynya marine food web. Environ Sci Technol 35:732–738. https://doi.org/10.1021/es001459w

GESAMP (1990) Review of potentially harmful substances. Choosing priority organochlorines for marine hazard assessment, Joint group of experts on the scientific aspects of marine pollution, vol 42. FAO, Rome, p 10

Gonzalez M, Miglioranza KSB, Aizpún de Moreno JE et al (2005) Evaluation of conventionally and organically produced vegetables for high lipophilic organochlorine pesticide (OCP) residues. Food Chem Toxicol 43:261–269. https://doi.org/10.1016/j.fct.2004.10.002

Gonzalez M, Miglioranza K, Shimabukuro V et al (2012) Surface and groundwater pollution by organochlorine compounds in a typical soybean system from the South Pampa, Argentina. Environ Earth Sci 65:481–491. https://doi.org/10.1007/s12665-011-1328-x

Grondona S, Gonzalez M, Martinez DE et al (2019) Assessment of organochlorine pesticides in phreatic aquifer of pampean region, Argentina. Bull Environ Contam Toxicol 102:544–549. https://doi.org/10.1007/s00128-019-02584-3

Guirlet E, Das K, Thomé J et al (2010) Maternal transfer of chlorinated contaminants in the leatherback turtles, *Dermochelys coriacea*, nesting in French Guiana. Chemosphere 79:720–726. https://doi.org/10.1016/j.chemosphere.2010.02.047

Harmon SM (2015) Chapter 18—The toxicity of persistent organic pollutants to aquatic organisms. In: Comprehensive analytical chemistry. Elsevier, Amsterdam, pp 587–613. https://doi.org/10.1016/B978-0-444-63299-9.00018-1

Hinton DE, Segner H, Au DWT et al (2008) Liver toxicity. In: Di Gulio RT, Hinton DE (eds) The toxicology of fishes. CRC Press, Boca Raton, pp 327–400

Horri K, Alfonso S, Cousin X et al (2018) Fish life-history traits are affected after chronic dietary exposure to an environmentally realistic marine mixture of PCBs and PBDEs. Sci Total Environ 610–611:531–545. https://doi.org/10.1016/j.scitotenv.2017.08.083

IARC (1997) Polychlorinated dibenzo-para-dioxins and polychlorinated dibenzofurans. Summary of data reported and evaluation. IARC monographs on the evaluation of carcinogenic risks humans, vol 69. International Agency for Research on Cancer, Lyon, pp 33–342

Ilyina T, Pohlmann G, Lammel J et al (2006) A fate and transport ocean model for persistent organic pollutants and its application to the North Sea. J Mar Syst 63(1–2):1–19. https://doi.org/10.1016/j.jmarsys.2006.04.007

Iwata H, Tanabe S, Sakai N et al (1993) Distribution of persistent organochlorines in the oceanic air and surface seawater and the role of ocean on their global transport and fate. Environ Sci Technol 27:1080–1098. https://doi.org/10.1021/es00043a007

Jaspers VLB, Covaci A, Van den Steen E et al (2007) Is external contamination with organic pollutants important for concentrations measured in bird feathers? Environ Int 33:766–772

Johnson LL, Anulacion BF, Arkoosh MR et al (2013) Effects of legacy persistent organic pollutants (POPs) in fish—current and future challenges. Fish Physiol 33:53–140. https://doi.org/10.1016/B978-0-12-398254-4.00002-9

Kalyabina V, Esimbekova E, Kopylova K et al (2021) Pesticides: formulants, distribution pathways and effects on human health—a review. Toxicol Rep 8:1179–1192. https://doi.org/10.1016/j.toxrep.2021.06.004

Kelly S, Eisenreich K, Baker J et al (2008) Accumulation and maternal transfer of polychlorinated biphenyls in snapping turtles of the upper Hudson River, New York, USA. Environ Toxicol Chem 27:2565–2574

Ko F-C, Pan W-L, Cheng J-O et al (2018) Persistent organic pollutants in Antarctic notothenioid fish and invertebrates associated with trophic levels. PLoS One 13:e0194147. https://doi.org/10.1371/journal.pone.0194147

Krauthacker B, Romanic SM, Reiser E (2001) Polychlorinated biphenyls and organochlorine pesticides in vegetation samples collected in Croatia. Bull Environ Contam Toxicol 66:334–341. https://doi.org/10.1007/s001280010

Lanfranchi AL, Menone ML, Miglioranza KSB et al (2006) Striped weakfish (*Cynoscion guatucupa*): a biomonitor of organochlorine pesticides in estuarine and near-coastal zones. Mar Pollut Bull 52:74–80. https://doi.org/10.1016/j.marpolbul.2005.08.008

Leppänen MT, Kukkonen JV (2004) Toxicokinetics of sediment-associated polybrominated diphenylethers (flame retardants) in benthic invertebrates (*Lumbriculus variegatus*, Oligochaeta). Environ Toxicol Chem 23:166–172. https://doi.org/10.1897/03-68. PMID: 14768881

Li YF, Macdonald RW, Ma J et al (2004) α-HCH budget in the Arctic Ocean: the arctic mass balance box model (AMBBM). Sci Tot Environ 324:115–139

Lopez-Espinosa MJ, Granada A, Carreno J et al (2007) Organochlorine pesticides in placentas from southern Spain and some related factors. Placenta 28:631–638. https://doi.org/10.1016/j.placenta.2006.09.009

Massone H, Martínez DE, Cionchi JL et al (1998) Suburban areas in developing countries and its relation with groundwater pollution. Mar del Plata (Argentina) as a case study. Environ Manag 22:245–254. https://doi.org/10.1007/s002679900100

Matthiessen P (2003) Endocrine disruption in marine fish. Pure Appl Chem 75:2249–2261. https://doi.org/10.1351/pac200375112249

McFarland VA, Clarke JU (1989) Environmental occurrence, abundance, and potential toxicity of polychlorinated biphenyl congeners: considerations for a congener-specific analysis. Environ Health Perspect 81:225–239. https://doi.org/10.1289/ehp.8981225

Menone ML, Miglioranza KSB, Iribarne O et al (2004) The role of burrowing beds and burrows of the SW Atlantic intertidal crab *Chasmagnathus granulata* in trapping organochlorine pesticides. Mar Pollut Bull 48(3–4):240–247

Menone ML, Miglioranza KSB, Botto F et al (2006) Field accumulative behavior of organochlorine pesticides. The role of crabs and sediment characteristics in coastal environment. Mar Pollut Bull 52:1717–1724

Miglioranza KSB, Aizpún de Moreno JE, Moreno VJ et al (1999) Fate of organochlorine pesticides in soils and terrestrial biota of "Los Padres" pond watershed, Argentina. Environ Pollut 105:91–99. https://doi.org/10.1016/S0269-7491(98)00200-0

Miglioranza KSB, Gonzalez M, Ondarza PM et al (2013) Assessment of Argentinean Patagonia pollution: PBDEs, OCPs and PCBs in different matrices from the Río Negro basin. Sci Total Environ 452–453:275–285. https://doi.org/10.1016/j.scitotenv.2013.02.055

Miglioranza KSB, Ondarza PM, Costa P et al (2021) Spatial and temporal distribution of persistent organic pollutants and current use pesticides in the atmosphere of Argentinean Patagonia. Chemosphere 266:129015. https://doi.org/10.1016/j.chemosphere.2020.129015

Mitton FM, Gonzalez M, Peña A et al (2012) Effects of amendments on soil availability and phytoremediation potential of aged p,p'-DDT, p,p'-DDE and p,p'-DDD residues by willow plants (*Salix* sp.). J Hazard Mater 203:62–68. https://doi.org/10.1016/j.jhazmat.2011.11.080

Mitton F, Gonzalez M, Monserrat J et al (2018) DDTs-induced antioxidant responses in plants and their influence on phytoremediation process. Ecotoxicol Environ Saf 147:151–156. https://doi.org/10.1016/j.ecoenv.2017.08.037

Mocarelli P (2001) Seveso: a teaching story. Chemosphere 43:391–402. https://doi.org/10.1016/S0045-6535(00)00386-6

Mull CG, Lyons K, Blasius ME et al (2013) Evidence of maternal offloading of organic contaminants in white sharks (*Carcharodon carcharias*). PLoS One 8:e62886. https://doi.org/10.1371/journal.pone.0062886

Müller MHB, Polder A, Brynildsr et al (2019) Prenatal exposure to persistent organic pollutants in northern Tanzania and their distribution between breast milk, maternal blood, placenta and cord blood. Environ Res 170:433–442. https://doi.org/10.1016/j.envres.2018.12.026

Muñoz-de-Toro M, Beldoménico H, García SR et al (2006) Organochlorine levels in adipose tissue of women from a littoral region of Argentina. Environ Res 102:107–112. https://doi.org/10. 1016/j.envres.2005.12.017

Mwangi JK, Lee W, Wang L, Sung P, Fang L, Lee Y, Chang-Chien G (2016) Persistent organic pollutants in the Antarctic coastal environment and their bioaccumulation in penguins. Environ Pollut 216:924–934

Narahashi T (1992) Nerve membrane Na+ channels as targets of insecticides. Trends Pharmacol Sci 13:236–241. https://doi.org/10.1016/0165-6147(92)90075-H

Needham LL, Gerthoux PM, Patterson DG Jr et al (1999) Exposure assessment: serum levels of TCDD in Seveso, Italy. Environ Res 80:S200–S206. https://doi.org/10.1006/enrs.1998.3928

Nossen I, Ciesielski TM, Dimmen MV et al (2016) Steroids in house sparrows (Passer domesticus): effects of POPs and male quality signalling. Sci Tot Environ 547:295–304

Ondarza PM, Gonzalez M, Fillmann G et al (2014) PBDEs, PCBs and organochlorine pesticides distribution in edible fish from Negro River basin, Argentinean Patagonia. Chemosphere 94: 135–142. https://doi.org/10.1016/j.chemosphere.2013.09.064

Palmer T, Klein A, Sweet S et al (2022) Using epibenthic fauna as biomonitors of local marine contamination adjacent to McMurdo Station, Antarctica. Mar Pollut Bull 178:113621. https:// doi.org/10.1016/j.marpolbul.2022.113621

Quadri-Adrogué A, Seco Pon JP, García GO et al (2021) Chlorpyrifos and persistent organic pollutants in feathers of the near threatened Olrog's gull in southeastern Buenos Aires Province, Argentina. Environ Pollut 272:115918. https://doi.org/10.1016/j.envpol.2020.115918

Rauert C, Harner T, Schuster JK et al (2016) Towards a regional passive air sampling network and strategy for new POPs in the GRULAC region: perspectives from the GAPS Network and first results for organophosphorus flame retardants. Sci Total Environ 573:1294–1302. https://doi. org/10.1016/j.scitotenv.2016.06.229

Rauert C, Harner T, Schuster J et al (2018) Air monitoring of new and legacy POPs in the Group of Latin America and Caribbean (GRULAC) region. Environ Pollut 243:1252–1262. https://doi. org/10.1016/j.envpol.2018.09.048

Rawn DFK, Sadler AR, Casey VA et al (2017) Dioxins/furans and PCBs in Canadian human milk: 2008–2011. Sci Total Environ 595:269–278. https://doi.org/10.1016/j.scitotenv.2017.03.157

Rodriguez PM, Vera B, Miglioranza KSB et al. (2020) Biomonitoreo de plaguicidas de uso prohibido y uso actual em mujeres embarazadas, Patagonia Norte. Paper presented at XXXVII Jornadas Interdisciplinarias de Toxicología, Asociación Toxicológica Argentina, 16–18 September 2020

Roscales JL, Muñoz-Arnanz J, González-Solís J et al (2010) Geographical PCB and DDT patterns in shearwaters (Calonectris sp.) breeding across the NE Atlantic and the Mediterranean archipelagos. Environ Sci Technol 44:2328–2334

Rudolph I, Chiang G, Galbán-Malagón C et al (2016) Persistent organic pollutants and porphyrins biomarkers in penguin faeces from Kopaitic Island and Antarctic Peninsula. Sci Tot Environ 573:1390–1396

Silva Barni MF, Ondarza PM, Gonzalez M et al (2016) Persistent organic pollutants (POPs) in fish with different feeding habits inhabiting a shallow lake ecosystem. Sci Tot Environ 550:900–909

Silva Barni MF, Gonzalez M, Miglioranza KSB (2019) Comparison of Tillandsia bergeri and XAD-resin based passive samplers for air monitoring of pesticides. Atmos Pollut Res 10:1507–1513. https://doi.org/10.1016/j.apr.2019.04.008

Stapleton H, Brazil B, Holbrook R et al (2006) In vivo and in vitro debromination of decabromodiphenyl ether (BDE 209) by juvenile rainbow trout and common carp. Environ Sci Technol 40:4653–4658. https://doi.org/10.1021/es060573x

Storelli M, Zizzo N (2014) Occurrence of organochlorine contaminants (PCBs, PCDDs and PCDFs) and pathologic findings in loggerhead sea turtles, Caretta caretta, from the Adriatic Sea (Mediterranean Sea). Sci Tot Environ 472:855–861. https://doi.org/10.1016/j.scitotenv. 2013.11.137

Tang J, Zhai J (2017) Distribution of polybrominated diphenyl ethers in breast milk, cord blood and placentas: a systematic review. Environ Sci Pollut Res 24:21548–21573. https://doi.org/10.1007/s11356-017-9821-8

Trapp SI, Köhler A, Larsen LC et al (2001) Phytotoxicity of fresh and weathered diesel and gasoline to willow and poplar trees. J Soils Sediments 1:71–76. https://doi.org/10.1007/BF02987712

UNEP (2002) United Nations Environment Programme—Chemicals. Regionally based assessment of persistent toxic substances (RBA/PTS): North America Regional Report. Global Environment Facility, December, 2002

UNEP (2012) United Nations Environment Programme, 2012. https://www.unep.org/explore-topics/chemicals-waste/what-we-do/persistent-organic-pollutants/global-monitoring. Accessed 12 Aug 2021

UNEP/POPS/COP.6/INF/31/Add.1 (n.d.) Conference of the parties to the Stockholm Convention on persistent organic pollutants sixth meeting, Geneva, 28 April–10 May 2013

Van den Berg M, Birnbaum LS, Denison M et al (2006) The 2005 World Health Organization reevaluation of human and mammalian toxic equivalency factors for dioxins and dioxin-like compounds. Toxicol Sci 93:223–241. https://doi.org/10.1093/toxsci/kfl055

van den Brink NW, van Franeker JA, de Ruiter-Dijkman EM (1998) Fluctuating concentrations of organochlorine pollutants during a breeding season in two Antarctic seabirds: Adélie penguin and southern fulmar. Environ Toxicol Chem 17:702–709

Vazquez N, Chiericchetti M, Acuña F et al (2022) Legacy and current use pesticides in the sea anemone *Bunodosoma zamponii* (Actiniaria: Actiniidae) from Argentina's southeastern coast. Sci Total Environ 806:150824

Vizcaino E, Grimalt J, Fernández-Somoano et al (2014) Transport of persistent organic pollutants across the human placenta. Environ Int 65:107–115. https://doi.org/10.1016/j.envint.2014.01.004

Wania F, Mackay D (1996) Tracking the distribution of persistent organic pollutants. Environ Sci Technol 30:390–396

Weber K, Goerke H (2003) Persistent organic pollutants (POPs) in Antarctic fish: levels, patterns, changes. Chemosphere 53:667–678. https://doi.org/10.1016/S0045-6535(03)00551-4

Weijs L, Briels N, Adams DH et al (2015) Bioaccumulation of organohalogenated compounds in sharks and rays from the southeastern USA. Environ Res 137:199–207. https://doi.org/10.1016/j.envres.2014.12.022

Yamashita R, Takada H, Murakami M et al (2007) Evaluation of noninvasive approach for monitoring PCB pollution of seabirds using preen gland oil. Environ Sci Technol 41:4901–4906

Emergent Organic Contaminants

6

Jesús Alfredo Rodríguez-Hernández,
Saúl Antonio Hernández-Martínez, Rafael G. Araújo, Damià Barceló,
Hafiz M. N. Iqbal, and Roberto Parra-Saldívar

Contents

Abstract

The growth of the human population and anthropogenic activity is strongly related to the increase of contamination of the environment. In marine ecosystems, the growing concentration of EOCs like pesticides, personal care

J. A. Rodríguez-Hernández · S. A. Hernández-Martínez
Tecnologico de Monterrey, School of Engineering and Sciences, Monterrey, Mexico

R. G. Araújo · H. M. N. Iqbal · R. Parra-Saldívar (✉)
Tecnologico de Monterrey, School of Engineering and Sciences, Monterrey, Mexico

Tecnologico de Monterrey, Institute of Advanced Materials for Sustainable Manufacturing, Monterrey, Mexico
e-mail: r.parra@tec.mx

D. Barceló
Department of Environmental Chemistry, Institute of Environmental Assessment and Water Research (IDAEA-CSIC), Barcelona, Spain

Catalan Institute of Water Research (ICRA-CERCA), Parc Científic i Tecnològic de la Universitat de Girona, Girona, Spain

Sustainability Cluster, School of Engineering, UPES, Dehradun, India

J. Blasco, A. Tovar-Sánchez (eds.), *Marine Analytical Chemistry*,
https://doi.org/10.1007/978-3-031-14486-8_6

products, flame retardants, plasticizers, and hormones has become a major problem that affects the flora and fauna altering their feeding behavior and sexual differentiation. Due to this, the development of highly efficient detection and remediation technologies is required to assess the contamination by EOCs. Traditional detection methods and remediation technologies have some major drawbacks, like high-cost and time-intensive methodologies that are needed to overcome to assess the growing contamination of marine environments. The use of electrochemical sensors, especially electrochemical biosensors, is a promising alternative for in situ monitoring the concentration of EOCs. Regarding the degradation of EOCs, the role of technologies employed at WWTPs is crucial to reduce the discharge of contaminants to marine ecosystems. Electrochemical oxidation and photocatalytic degradation are highly effective technologies that could be used soon at WWTPs to improve the actual degradation efficiency of EOCs.

Keywords

WWTPs · Electrochemical biosensors · Electrochemical oxidation · Photocatalyst

6.1   Introduction: Problem Statement and Opportunities

In the past decade, anthropogenic activities and their consequences to the environment have become a growing concern area of study. As mentioned, meaningful consequences of these humans' activities have led to the deterioration of soils (Bhardwaj et al. 2014), land (Mahmoud and Gan 2018), and aquatic systems (Yuan et al. 2016). Specifically, due to exploding population, industrialization, and urbanization, the pollution of aquatic systems, including surface waters, groundwaters, and wastewaters, by anthropogenic pollutants is of crucial concern due to its toxicological effects over aquatic living microorganisms, land animals, humans, plants, and birds (Thompson and Vijayan 2020; Zaid 2017).

The abovementioned concept, anthropogenic pollutant, might be defined as a diverse array of synthetic or natural contaminants released to the environment because of human activities, which are primarily attributed to industrial processes, and human and animal healthcare (Lapworth et al. 2012). Thus, during the past few decades, the contamination of aquatic systems by these contaminants has awaken an increasing interest due to their environmental fate and their potential toxicity. Since the 1970s, quality of water has been improved in developed countries thanks to wastewater treatment plants (WWTPs); however, water contamination is increasing due to the production at large scale of new substances which are commonly recalcitrant to conventional treatments in WWTPs, which provokes their releasement to ecosystems through contaminated effluents (Escudero et al. 2021).

Even though there are different kinds of anthropogenic pollutants, one of those with special concern due to its ubiquity is the "emerging organic contaminants" (EOCs), term which is used to cover already not only known contaminants but also

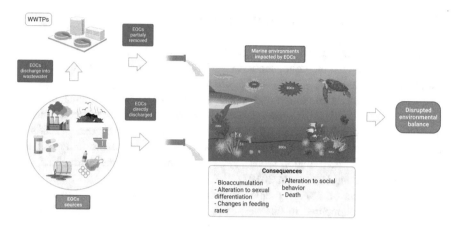

Fig. 6.1 Sources, occurrence, transport, and effects of ECOs in the marine environment. (Created with BioRender.com)

novel developed compounds or newly contaminants discovered in the environment with negative potential effects over ecosystems, Fig. 6.1 (Lapworth et al. 2012).

EOCs include a wide array of different compounds, as well as their metabolites and transformation products, including pharmaceuticals and personal care products (PPCPs), pesticides, hormones, surfactants, flame retardants, plasticizers, veterinary products, industrial additives, and engineered nanomaterials (García et al. 2020; Rojas and Horcajada 2019). As mentioned, anthropogenic activities have led to the increasing presence of these pollutants in the environment. Up to now, it has been established that EOCs enter the environment from different sources and pathways, including WWTPs, septic tanks, hospital effluents, industrial effluents, and livestock activities (Lapworth et al. 2012; Rojas and Horcajada 2019).

To date, the occurrence of EOCs in different ecosystems has been achieved by the application of different technologies of detection, including traditional techniques, such as liquid chromatography/mass spectrometry (LCMS) (Lapworth et al. 2018) and high-pressure liquid chromatography/mass spectrometry (HPLC/MS) (Hossain et al. 2018), and some novel techniques, including electrochemical detection (Feier et al. 2018) and biosensors (Mao et al. 2021; Qian et al. 2021).

Finally, in this chapter, different methodologies are reviewed that have been followed to remove or degrade EOCs from aquatic systems, including procedures used in WWTPs, such as chlorination (Liu et al. 2019), adsorption (Ali et al. 2018), ozonation (Wang and Chen 2020), and UV degradation (Rodríguez et al. 2019), and other lab-scale procedures, such as electrochemical techniques (Feier et al. 2018), photolysis (Tisler et al. 2019), and biodegradation (Bilal et al. 2019).

6.2   Classification and Sources of EOCs

As mentioned, there are different sources of EOCs that lead to the pollution of aquatic environments, including septic tanks, hospital effluents, industrial effluents, livestock activities, and, the most important, effluents of WWTPs, which are characterized to be unable to remove all EOCs that enter through influents of plants (Lapworth et al. 2012). It has been established that wastewaters are the main sources of EOCs in the environment and that surface waters contain the greatest loads of EOCs. Moreover, EOCs have been classified as a function of the nature of the micropollutant in three major types, PPCPs, organic dyes, and herbicides and pesticides (Rojas and Horcajada 2019).

PPCPs are a set of emerging environmental contaminants that are broadly and progressively implemented in the well-being of human and veterinary and which have been demonstrated to have an inherent ability to induce physiological effects in humans (Rojas and Horcajada 2019). In case of pharmaceuticals, these are commonly classified according to their properties and purposes in, for example, antibiotics (e.g., amoxicillin, ampicillin), anticancer (e.g., doxorubicin), anticonvulsant (e.g., carbamazepine), analgesic (e.g., acetaminophen), anti-inflammatory (e.g., ketoprofen), hormones (progesterone), and more, Table 6.1. On the other hand, personal care products are those products that might be or not prescribed, and its main purpose is to treat individual care purposes. Most common personal care products to name are cosmetics, perfumes, shampoos, and UV blockers (Gogoi et al. 2018).

Moreover, due to their universal consumption, low human metabolic capability, and improper disposal, PPCPs have been widely detected at low concentrations (ng/L–μg/L) in aquatic systems, which enter into the environment through several pathways, including WWTPs, industrial services, hospitals, aquaculture facilities, runoff from fields into surface waters, and runoff into soils from animal farming (Ebele et al. 2017).

Organic dyes. It is estimated that a significant proportion of industrial waste is comprised in dye discharges. About 10–15% of the worldwide massive production of dyes (800,000 tons per year) reaches the environment through industrial effluents or due to losses that occur during drying processes (Rojas and Horcajada 2019). Particularly, the release of dyes in aquatic systems is of concern due to their high persistence, toxicity, and potential to bioaccumulate in living organisms (Shanker et al. 2017). In general, dyes are classified according to their solubility, chemical properties, and/or their function in the textile industry; some examples are azo dyes, disperse dyes, fast color bases, ingrain dyes, pigment emulsions, reactive dyes, sulfur dyes, and solubilized dyes, from which azo dyes represent more than 70% of the total production (Rojas and Horcajada 2019; Shanker et al. 2017).

Herbicides and pesticides. These kinds of pollutants are characterized to be effective at the time to eliminate or destroy pests from human being's food. However, it has been widely studied that most of herbicides and pesticides are soluble in water, which is of special concern due to their toxic potential to biological systems

Table 6.1 Groups of EOCs and main adverse effects

Group	Examples	Major adverse effects	References
PPCPs	Sulfamethoxazole Ciprofloxacin Carbamazepine Fluoxetine Triclosan	Bacterial resistance genes, antibiotic resistance, abnormal growth in early stages of development, and alterations to behavior	Valdez-Carrillo et al. (2020), Wang and Wang (2018), Xiang et al. (2021)
Pesticides	Hexachlorocyclohexane Fenitrothion Atrazine	Toxicity, high environmental persistence, and bioaccumulation	Khan et al. (2022)
Hormones	17α-Ethinylestradiol 17β-Estradiol Estriol Estrone	Endocrine disruption, alteration on sexual behavior, and cancer	Azizi-Lalabadi and Pirsaheb (2021), Valdez-Carrillo et al. (2020)
Surfactants	Linear alkylbenzene sulfonates Nonylphenol ethoxylate	Cytotoxicity, bioaccumulation, high environmental persistence, endocrine disruption, and cancer	Smital et al. (2004), Villarreal-Reyes et al. (2021)
Flame retardants	Octabromodiphenyl ether Pentabromodiphenyl ether	Toxic, bioaccumulation, high environmental persistence, alterations to behavior, and abnormal neurological development	Ortega-Olvera et al. (2020)
Plasticizers	Bisphenol A Di-*n*-octyl phthalate Bis(2-ethylhexyl) terephthalate Bis(2-ethylhexyl) phthalate	Endocrine disruption, alteration on sexual and neurological development, and damages to immune system	Barnabé et al. (2008), Ma et al. (2019), Sarkar et al. (2013)

(Rashid et al. 2010). In this case, these micropollutants get into water bodies by leaching, surface drainage, and runoff (Ighalo et al. 2021).

6.3   Environment and Human Health Impact

The high demand in the development of anthropogenic activities generates about 4000 new chemical products daily, in which despite existing regulations, these compounds in the long term can be harmful to the environment and human health, while its properties, toxicological effects, and environmental behavior are little known. This reflects that EOCs of today will be contaminants of the future, highlighting the high importance of controlling the release, detection, and surveillance of EOCs in the environment (Reberski et al. 2022).

The pharmaceutical industry is one of the main sources of new chemical compounds and in turn one of the main sources of emerging pollutants. Hormones and analogs, antibiotics, antidepressants, and many other compounds are part of the list of more than 3000 compounds used in the pharmaceutical industry, and only a small number of compounds are considered for evaluation and study in the field. This is because the release of these compounds is done mainly in drainage systems and a dilution factor is created that decreases the detection efficiency. However, the presence of many EOCs in water, food, and animals has recently been reported concentrations of ng L^{-1}, which may be harmful to humans and wildlife (Khan et al. 2022).

The increase in antibiotic resistance is a worldwide problem derived from the excessive and inappropriate consumption of antibiotics in humans and animals, which has led to greater difficulties in the treatment of common diseases. The release of these compounds and their metabolites into the environment is associated with the development of resistance genes and is a global threat to human health and survival, as well as to the ecological balance. It has been reported that the environment plays an important role as a reservoir and propagation of antibiotic resistance genes (Alderton et al. 2021; Bombaywala et al. 2021).

Herbicides, insecticides, fungicides, and bactericides are compounds with a varied spectrum widely used in modern agroindustry to improve the quality of crops and prevent losses or damage caused by insects or pathogenic microorganisms. However, the constant use of these compounds in farmland has generated a large distribution of these contaminants due to bioaccumulation in living organisms, plants, and fruits and contamination of soils, groundwater, and surface water (Khan et al. 2022; Reshma and Krishna 2017).

6.4 EOCs' Impact in the Marine Environment

The marine environment has also been extensively affected by anthropogenic contamination. The main contaminants of marine ecosystems can be grouped in marine trash or debris (plastic bags, cans, derelict fishing gear, abandoned vessels, and microplastics) and chemical contamination (Barrows et al. 2018; Lu et al. 2020; Rochman et al. 2016; Ying and Kookana 2003). The latter has been related to the development of big cities near the coast, as industrial effluents, wastewater, and runoff may serve as migration pathways of a wide range of contaminants, especially EOCs (Lu et al. 2020; Naik and Dubey 2016).

In recent years, the concentration of several EOCs in marine ecosystems has increased alarmingly, and their negative effects have been well documented in the literature. For example, studies have evaluated effects of chronic exposure to endocrine-disrupting chemicals, like estrogen, on several species, like the zebra fish or swordtails (*Xiphophorus helleri*), and results showed that long-term exposure on males produced demasculinization and feminization and also a higher incidence of intersex (Söffker and Tyler 2012). Furthermore, European perch specimens exposed to psychiatric drugs like oxazepam modified their social and feeding patterns, increasing their feeding rate and promoting a more antisocial conduct (Brodin et al. 2013). A study evaluated the effect of diverse psychiatric drugs like fluoxetine, sertraline, and oxazepam on swimming patterns of larvae from Japanese rice fish (*Oryzias latipes*) (Chiffre et al. 2016). Specimens presented alterations to their swimming behavior, showing a more erratic swimming pattern, and in concentrations above 800 μg/L were lethal.

Another issue related with EOCs in marine environments is the accumulation of several of these groups of chemicals in animal tissue, flora, and sediments (Geyer et al. 2000). Studies have reported that EOCs, like nonylphenol, BPA, 17β-estradiol (E2), and estrone (E1), have been found in marine sediments of the Osaka Bay and are related to the bioaccumulation of EOCs in the tissue of mullets and flounders in the area (Koyama et al. 2013). Another study carried out in Iberian waters from rivers found out that estrone was consistently present in sediments with the highest concentration of 3.5 ng/g (Gorga et al. 2015). The accumulation of diverse EOCs in estuarine sediments from Mumbai was collected and analyzed (Tiwari et al. 2016). Results are a great accumulation of diverse EOCs like BPA and phthalates like di-*n*-butyl phthalate (DBP). This demonstrates that most of these contaminants have a strong potential of accumulation and bioaccumulation, which means that even very low concentrations of EOCs in the environment could cause severe consequences at long term. In this way, the development of highly efficient technologies for the removal/degradation of EOCs is required. Especially attention is needed in the treatment of the main EOC migration pathways toward the sea, like industrial and WWTP effluents.

6.5 Detection and Quantification of EOCs

Most of EOCs decomposed in the water derived from their chemical properties through different water cycles, thus generating potential threats to the environment, specifically aquatic life, and humans once their presence has been reported. The presence of these compounds has increased alarmingly in groundwater, drinking water, wastewater, and surface water (Khan et al. 2022).

EOCs are a multi-class of compounds that can be natural or synthetic compounds produced in urban, agriculture, aquaculture, and pharmaceutical activities (Ismail et al. 2019). Natural and synthetic hormones, which are not metabolized by the human organism, are excreted through the urine or feces reaching and contaminating wastewater and septic tanks and together with industrial effluents, and intensively cultivated agricultural soils are the main sources of EOCs that they contaminate different water systems of consumption, negatively affecting the environmental stability and human and animal health (Vieira et al. 2021; Zwart et al. 2020). EOC concentrations in wastewater are low, on the order of ng/L or μg/L, but mainly are bioaccumulative compounds. The most used techniques for the detection and quantification of EOCs are traditional analysis techniques such as high-performance liquid chromatography tandem mass spectrometer (HPLC-MS/MS), ultra-performance liquid chromatography with tandem mass spectrometer (UPLC-MS/MS), and gas chromatography time-of-flight mass spectrometry (GC-TOF-MS) (Castillo-Zacarías et al. 2021; Ben Chabchoubi et al. 2021). These techniques have a high sensitivity of detection and reproducibility and, however, require many steps of sample preparation, large sample volumes, large times of analysis, expensive consumables, and equipment.

The development of in situ detection instruments for different pollutants is an area of high priority, since it allows real-time monitoring of the presence or release of EOCs into the environment and, in turn, allows the reduction of the consumption of polluting chemical reagents that are used in conventional analysis techniques (Hernandez-Vargas et al. 2018).

Electrochemical sensors have been reported as a new and rapid application procedure for the detection and screening of EOCs and other toxic compounds in wastewater, with a very low detection limit, high sensitivity, and low cost (Chabchoubi et al. 2021). Biosensors have been developed from cells, specific molecules, aptamers, antibodies, DNA, and model organisms for the detection of different EOCs, such as estrogen, BPA, nonylphenol, growth hormone, and cortisol among many others that can be found in food and environment. Transducer sensors have been developed based on carbon nanomaterials such as graphene oxide and carbon nanotubes, metal oxide nanomaterials, noble metal nanomaterials, some polymers, and many others (Jaffrezic-Renault et al. 2020; Kaya et al. 2020; Lu et al. 2020).

6.6 Traditional Technologies for the Degradation of EOC from the Water

6.6.1 The Role of WWTPs in the Degradation of EOCs

Typically, wastewater contains diverse EOCs contained in products used by humans such as disinfectants, pharmaceuticals, personal care products, dyes, and pesticides, among others. In this way, wastewater treatment plays a major role in avoiding the leakage of EOCs into the environment by removing the most possible contaminants from water before discharging it into the environment.

Most common WWTPs use a complex combination of physical, chemical, and biological processes through two main stages to remove contaminants: primary and secondary treatment. The primary treatment consists in physical treatments to remove the most part of suspended solids using screening and grit removal processes and a clarification process carried mainly by sedimentation or flotation.

In the secondary treatment, the biological degradation is carried out using aerobic degradation, for example, by using activated sludge, or through anaerobic degradation, by using anaerobic digestion reactors. In this stage, most of the pathogens and organic matter are degraded; however, some EOCs are very persistent, and low concentrations of these kinds of contaminants are not removed through these treatments.

Due to this, in some plants, a third treatment is required to enhance the removal efficiency of the WWTP. The aim of this stage is to enhance the degradation of some pathogens and other contaminants that could still be present after the secondary treatment. Technologies employed in the tertiary treatment stage depend on target contaminants, but the most employed processes are filtration, adsorption, and chlorination.

Filtration processes use semipermeable membranes with small pore size that can separate molecules of contaminants from water (Sharma et al. 2021). Reverse osmosis is a type of advanced filtration where, due to the osmotic pressure, it is needed to apply pressure into the system, and it can remove molecules and ions with high selectivity (Trishitman et al. 2020). This technology is mainly employed in water purification of drinking water; however, there are some wastewater treatment plants that have incorporated this technology.

The adsorption process consists in the deposition of contaminant molecules on the surface of a solid material (adsorbent). In wastewater treatment plants, the most employed adsorbent is activated carbon, due to its relatively low cost compared to other adsorbents, high surface area, and its porous structure.

Chlorination is an affordable technology that is used in most of WWTPs for disinfection prior to discharging the water back into the environment. This method usually adds Cl_2 or $Ca(OCl)_2$ to the water, and then chlorine interacts with organic compounds either by addition or substitution of atoms of molecules or by oxidizing them (Albolafio et al. 2022).

6.6.2 Main Problems with Current WWTP Technologies Regarding EOCS Degradation

Despite efforts that have been made to improve the removal efficiency of WWTPs, there are still several challenges in the removal of EOCs from wastewater. One major issue is related to the need of more efficient removal/degradation techniques that are economically viable, as costs of some highly effective technologies hinder their implementation at WWTPs, especially in developing countries. For example, reverse osmosis (RO) filtration has been evaluated in several WWTPs with high removal percentage of diverse contaminants; however, costs of operation of the plant can increase significantly, even reaching millions of dollars (Guo et al. 2014).

Another issue with current technologies is the chlorination disinfection by-products (DBPs) generated during the disinfection process. Even though chlorination is very effective against biological pathogens, the reaction of chlorine with organic contaminants may produce toxic compounds such as trihalomethanes and haloacetic acids, among others (Barber 2014; Gad and Pham 2014; Stroheker et al. 2014).

Furthermore, the degradation efficiency for EOCs of traditional WWTPs varies depending on several factors, such as water inlet composition, weather, and technologies employed during treatments, among others. However, in most cases, the efficiency of degradation of EOCs is considerably lower than recommended concentration levels. Due to this, current technologies employed for wastewater treatment are not enough to stop the growing problem of the leakage of EOCs into the environment.

6.6.3 Promising Technologies for the Degradation of EOCs in Water

Due to these issues, it is needed to develop more efficient and cheaper technologies that ensure a higher water quality, thus reducing the leakage of EOCs to the environment. In this way, several approaches have been studied at lab scale with promising results, especially electrochemical advanced oxidation treatments, photocatalytic degradation, and biocatalytic degradation systems.

6.6.3.1 Electrochemical Oxidation

Advanced oxidation processes are based on the addition of an oxidizing agent and transform contaminants into by-products that are not harmful or are easier to remove (Dave and Das 2021). In this way, the electrochemical oxidation is based on the electron transfer to oxidize contaminants by either direct or indirect oxidation. The direct oxidation occurs by the adsorption of contaminants on the anode surface and further electron transfer reaction that breaks down organic molecules. The indirect oxidation is carried out by anodic or cathodic processes that generate strong oxidants like O_3 or H_2O_2 (Kumar and Shah 2021).

Electrochemical systems can be highly efficient, and this mostly depends on materials employed as electrodes. In the literature, there are several reports of the evaluation of carbon-based materials, like graphene or activated carbon, and metallic nanoparticles, with oxidative performances of organic contaminants above 90% (Fajardo et al. 2017). Also, electrochemical systems can be coupled with sustainable energy systems, like fuel cells, that supply the power needed for the operation and make the process more power efficient (Liang et al. 2018).

Even though electrochemical oxidation has major advantages compared to other technologies, such as high removal percentage, energetic efficiency, and a wide range of materials that can be used as electrodes, there are still some challenges to overcome. For example, it is needed to develop low-cost electrocatalysts that are highly efficient for wastewater treatment, as most of the studies are focused on other applications like fuel cells, where usually expensive noble metals, like Pt or Ru, are employed (Du et al. 2021).

6.6.3.2 Photocatalytic Degradation

The use of UV radiation has been reported as a common approach for wastewater remediation, as this technology has some advantages like the capability to use the solar light to carry out the process. However, direct UV-driven photolysis may be a slow process and requires that the contaminant absorbs the UV radiation (Kumar and Shah 2021). In this way, an alternative is photocatalytic degradation, where a photocatalyst is employed to enhance the degradation of contaminants combined with some oxidizing agents.

Photocatalytic degradation has been employed at lab scale for degrading diverse contaminants, especially highly toxic organic compounds where other processes, like biological treatment, are not suitable. This process is based on the excitation of the catalyst due to its interaction with photons, which produce a pair of excited electrons (e^-) and valence hole (h^+) that will migrate to the surface of the photocatalyst (Qi and Yu 2020). Redox species in the solution will react with the photocatalyst and generate radicals like $^\bullet OH$ and $^\bullet O_2^-$, which are strong oxidants and will degrade even highly stable contaminants (Kumar and Shah 2021).

There are several studies that demonstrate that photocatalysis is a very promising technology with high removal efficiency of up to near 100% for diverse organic contaminants. The most employed photocatalysts are metal oxides, and ZnO has been widely reported as one of the most efficient materials for photocatalytic degradation of organic contaminants (Ferreira et al. 2021). Other materials that have been employed successfully as photocatalysts are polymers (Ahmad et al. 2021), metal organic frameworks (Lv et al. 2021), and carbon dots (Wang et al. 2021).

Photocatalysis has shown a great performance for degrading diverse organic contaminants; however, its efficiency is considerably hindered when there is a high content of organic matter (Kumar and Shah 2021). This makes photocatalysis a treatment more suitable to be implemented after a pretreatment that reduces the content of organic contaminants in wastewater.

6.7 Concluding Remarks

The detection and elimination of EOCs are a global environmental problem of great importance, and in recent years, great efforts and advances have been made in the development of technologies for the identification and quantification of different contaminants in situ through high-precision technologies, with rapid response, without the use of polluting reagents, and at low cost. Similarly, the advancement of technology for the treatment of water contaminated with EOCs has made great progress, such as adsorption, ozonation, UV degradation, photolysis, and electrochemical techniques that can be implemented in WWTPs as tertiary treatments.

The implementation of technologies as a single treatment still shows deficiencies, which has deployed strategies of coupled technology methods to increase treatment efficiency. However, there are still inadequate knowledge and few public regulations for the control of EOCs, which continues to be reflected in safety problems in the consumption of reused water or food grown with reused water, which continues to create a high concern about environmental impacts and on human health in the short term.

As a perspective of future studies for the degradation of EOCs, the development of innovative materials and nanocatalysts will allow the creation of highly efficient treatment processes which are friendly to the environment for a broad spectrum of contaminants.

Acknowledgments Authors acknowledge Tecnológico de Monterrey for the financial support for the publication. All listed authors are also grateful to their representative universities/institutes for providing literature facilities. Consejo Nacional de Ciencia y Tecnología (CONACYT) is thankfully acknowledged for partially supporting this work under Sistema Nacional de Investigadores (SNI) program awarded to Rafael G. Araújo (CVU: 714118), Hafiz N. M. Iqbal (CVU: 735340), and Roberto Parra-Saldívar (CVU: 35753).

Funding This work is part of the project entitled "Contaminantes emergentes y prioritarios en las aguas reutilizadas en agricultura: riesgos y efectos en suelos, producción agrícola y entorno ambiental" funded by CSIC-Tecnológico de Monterrey under i-Link + program (LINKB20030).

References

Ahmad N, Anae J, Khan MZ, Sabir S, Yang XJ, Thakur VK, Campo P, Coulon F (2021) Visible light-conducting polymer nanocomposites as efficient photocatalysts for the treatment of organic pollutants in wastewater. J Environ Manag 295:113362. https://doi.org/10.1016/J. JENVMAN.2021.113362

Albolafio S, Marín A, Allende A, García F, Simón-Andreu PJ, Soler MA, Gil MI (2022) Strategies for mitigating chlorinated disinfection byproducts in wastewater treatment plants. Chemosphere 288:132583. https://doi.org/10.1016/J.CHEMOSPHERE.2021.132583

Alderton I, Palmer BR, Heinemann JA, Pattis I, Weaver L, Gutiérrez-Ginés MJ et al (2021) The role of emerging organic contaminants in the development of antimicrobial resistance. Emerg Contam 7:160–171. https://doi.org/10.1016/j.emcon.2021.07.001

Ali MEM, El-aty AMA, Badawy MI, Ali RK (2018) Removal of pharmaceutical pollutants from synthetic wastewater using chemically modified biomass of green alga *Scenedesmus obliquus*. Ecotoxicol Environ Saf 151:144–152. https://doi.org/10.1016/j.ecoenv.2018.01.012

Azizi-Lalabadi M, Pirsaheb M (2021) Investigation of steroid hormone residues in fish: a systematic review. Process Saf Environ Prot 152:14–24. https://doi.org/10.1016/J.PSEP.2021.05.020

Barber LB (2014) Emerging contaminants. Compr Water Qual Purif 1:245–266. https://doi.org/10. 1016/B978-0-12-382182-9.00015-3

Barnabé S, Beauchesne I, Cooper DG, Nicell JA (2008) Plasticizers and their degradation products in the process streams of a large urban physicochemical sewage treatment plant. Water Res 42(1–2):153–162. https://doi.org/10.1016/J.WATRES.2007.07.043

Barrows APW, Cathey SE, Petersen CW (2018) Marine environment microfiber contamination: global patterns and the diversity of microparticle origins. Environ Pollut 237:275–284. https:// doi.org/10.1016/J.ENVPOL.2018.02.062

Ben Chabchoubi I, Belkhamssa N, Ksibi M, Hentati O (2021) Trends in the detection of pharmaceuticals and endocrine-disrupting compounds by field-effect transistors (FETs). Trends Environ Anal Chem 30:e00127. https://doi.org/10.1016/j.teac.2021.e00127

Bhardwaj R, Sharma N, Handa N, Handa H, Kaur R, Kaur G, Sirhindi AK (2014) Chapter 19: Prospects of field crops for phytoremediation of contaminants. In: Emerging technologies and management of crop stress tolerance, vol 2, 2nd edn. Elsevier, Amsterdam. https://doi.org/10. 1016/B978-0-12-800875-1.00019-3

Bilal M, Adeel M, Rasheed T, Zhao Y, Iqbal HMN (2019) Emerging contaminants of high concern and their enzyme-assisted biodegradation—a review. Environ Int 124:336–353. https://doi.org/ 10.1016/j.envint.2019.01.011

Bombaywala S, Mandpe A, Paliya S, Kumar S (2021) Antibiotic resistance in the environment: a critical insight on its occurrence, fate, and eco-toxicity. Environ Sci Pollut Res 28(20): 24889–24916. https://doi.org/10.1007/s11356-021-13143-x

Brodin T, Fick J, Jonsson M, Klaminder J (2013) Dilute concentrations of a psychiatric drug alter behavior of fish from natural populations. Science 339(6121):814–815. https://doi.org/10.1126/ SCIENCE.1226850/SUPPL_FILE/BRODIN.SM.PDF

Castillo-Zacarías C, Barocio ME, Hidalgo-Vázquez E, Sosa-Hernández JE, Parra-Arroyo L, López-Pacheco IY et al (2021) Antidepressant drugs as emerging contaminants: occurrence in urban and non-urban waters and analytical methods for their detection. Sci Total Environ 757:143722. https://doi.org/10.1016/j.scitotenv.2020.143722

Chiffre A, Clérandeau C, Dwoinikoff C, Le Bihanic F, Budzinski H, Geret F, Cachot J (2016) Psychotropic drugs in mixture alter swimming behaviour of Japanese medaka (*Oryzias latipes*) larvae above environmental concentrations. Environ Sci Pollut Res 23(6):4964–4977. https://doi.org/10.1007/S11356-014-3477-4/TABLES/6

Dave S, Das J (2021) Technological model on advanced stages of oxidation of wastewater effluent from food industry. In: Advanced oxidation processes for effluent treatment plants. Elsevier, Amsterdam, pp 33–49. https://doi.org/10.1016/B978-0-12-821011-6.00002-5

Du X, Oturan MA, Zhou M, Belkessa N, Su P, Cai J, Trellu C, Mousset E (2021) Nanostructured electrodes for electrocatalytic advanced oxidation processes: from materials preparation to mechanisms understanding and wastewater treatment applications. Appl Catal B Environ 296:120332. https://doi.org/10.1016/J.APCATB.2021.120332

Ebele AJ, Abou-Elwafa Abdallah M, Harrad S (2017) Pharmaceuticals and personal care products (PPCPs) in the freshwater aquatic environment. Emerg Contam 3(1):1–16. https://doi.org/10.1016/j.emcon.2016.12.004

Escudero J, Muñoz JL, Morera-Herreras T, Hernandez R, Medrano J, Domingo-Echaburu S et al (2021) Antipsychotics as environmental pollutants: an underrated threat? Sci Total Environ 769:144634. https://doi.org/10.1016/j.scitotenv.2020.144634

Fajardo AS, Seca HF, Martins RC, Corceiro VN, Freitas IF, Quinta-Ferreira ME, Quinta-Ferreira RM (2017) Electrochemical oxidation of phenolic wastewaters using a batch-stirred reactor with NaCl electrolyte and Ti/RuO2 anodes. J Electroanal Chem 785:180–189. https://doi.org/10.1016/J.JELECHEM.2016.12.033

Feier B, Florea A, Cristea C, Robert S (2018) Electrochemical detection and removal of pharmaceuticals in waste waters. Curr Opin Electrochem 11:1–11. https://doi.org/10.1016/j.coelec.2018.06.012

Ferreira SH, Morais M, Nunes D, Oliveira MJ, Rovisco A, Pimentel A, Águas H, Fortunato E, Martins R (2021) High UV and sunlight photocatalytic performance of porous ZnO nanostructures synthesized by a facile and fast microwave hydrothermal method. Materials 14(9):2385. https://doi.org/10.3390/MA14092385

Gad SC, Pham T (2014) Trihalomethanes. In: Encyclopedia of toxicology, 3rd edn. Elsevier, Amsterdam, pp 849–851. https://doi.org/10.1016/B978-0-12-386454-3.00956-8

García J, García-galán MJ, Day JW, Boopathy R, White JR, Wallace S, Hunter RG (2020) A review of emerging organic contaminants (EOCs), antibiotic resistant bacteria (ARB), and antibiotic resistance genes (ARGs) in the environment: increasing removal with wetlands and reducing environmental impacts. Bioresour Technol 307:123228. https://doi.org/10.1016/j.biortech.2020.123228

Geyer HJ, Rimkus GG, Scheunert I, Kaune A, Schramm K-W, Kettrup A et al (2000) Bioaccumulation and occurrence of endocrine-disrupting chemicals (EDCs), persistent organic pollutants (POPs), and other organic compounds in fish and other organisms including humans. In: Bioaccumulation—new aspects and developments. Springer, Berlin, pp 1–166. https://doi.org/10.1007/10503050_1

Gogoi A, Mazumder P, Kumar Tyagi V, Tushara Chaminda GG, Kyongjin An A, Kumar M (2018) Occurrence and fate of emerging contaminants in water environment: a review. Groundw Sustain Dev 6:169–180

Gorga M, Insa S, Petrovic M, Barceló D (2015) Occurrence and spatial distribution of EDCs and related compounds in waters and sediments of Iberian rivers. Sci Total Environ 503–504:69–86. https://doi.org/10.1016/J.SCITOTENV.2014.06.037

Guo T, Englehardt J, Wu T (2014) Review of cost versus scale: water and wastewater treatment and reuse processes. Water Sci Technol 69(2):223–234. https://doi.org/10.2166/WST.2013.734

Hernandez-Vargas G, Sosa-Hernández JE, Saldarriaga-Hernandez S, Villalba-Rodríguez AM, Parra-Saldivar R, Iqbal HMN (2018) Electrochemical biosensors: a solution to pollution detection with reference to environmental contaminants. Biosensors 8(2):1–21. https://doi.org/10.3390/bios8020029

Hossain A, Nakamichi S, Habibullah-Al-Mamun M, Tani K, Masunaga S, Matsuda H (2018) Occurrence and ecological risk of pharmaceuticals in river surface water of Bangladesh. Environ Res 165:258–266. https://doi.org/10.1016/j.envres.2018.04.030

Ighalo JO, George A, Adelodun AA (2021) Recent advances on the adsorption of herbicides and pesticides from polluted waters: performance evaluation via physical attributes. J Ind Eng Chem 93:117–137. https://doi.org/10.1016/j.jiec.2020.10.011

Ismail NAH, Wee SY, Kamarulzaman NH, Aris AZ (2019) Quantification of multi-classes of endocrine-disrupting compounds in estuarine water. Environ Pollut 249:1019–1028. https://doi.org/10.1016/j.envpol.2019.03.089

Jaffrezic-Renault N, Kou J, Tan D, Guo Z (2020) New trends in the electrochemical detection of endocrine disruptors in complex media. Anal Bioanal Chem 412(24):5913–5923. https://doi.org/10.1007/s00216-020-02516-9

Kaya SI, Cetinkaya A, Bakirhan NK, Ozkan SA (2020) Trends in sensitive electrochemical sensors for endocrine disruptive compounds. Trends Environ Anal Chem 28:e00106. https://doi.org/10.1016/j.teac.2020.e00106

Khan S, Naushad M, Govarthanan M, Iqbal J, Alfadul SM (2022) Emerging contaminants of high concern for the environment: current trends and future research. Environ Res 207:112609. https://doi.org/10.1016/J.ENVRES.2021.112609

Koyama J, Kitoh A, Nakai M, Kohno K, Tanaka H, Uno S (2013) Relative contribution of endocrine-disrupting chemicals to the estrogenic potency of marine sediments of Osaka Bay, Japan. Water Air Soil Pollut 224(5):1–9. https://doi.org/10.1007/S11270-013-1570-9/FIGURES/5

Kumar V, Shah MP (2021) Chapter 1: Advanced oxidation processes for complex wastewater treatment. In: Advanced oxidation processes for effluent treatment plants. Elsevier, Amsterdam, pp 1–31. https://doi.org/10.1016/B978-0-12-821011-6.00001-3

Lapworth DJ, Baran N, Stuart ME, Ward RS (2012) Emerging organic contaminants in groundwater: a review of sources, fate and occurrence. Environ Pollut 163:287–303. https://doi.org/10.1016/j.envpol.2011.12.034

Lapworth DJ, Das P, Shaw A, Mukherjee A, Civil W, Petersen JO et al (2018) Deep urban groundwater vulnerability in India revealed through the use of emerging organic contaminants and residence time tracers. Environ Pollut 240:938–949. https://doi.org/10.1016/j.envpol.2018.04.053

Liang S, Zhang B, Shi J, Wang T, Zhang L, Wang Z, Chen C (2018) Improved decolorization of dye wastewater in an electrochemical system powered by microbial fuel cells and intensified by micro-electrolysis. Bioelectrochemistry 124:112–118. https://doi.org/10.1016/J.BIOELECHEM.2018.07.008

Liu Y, Liu H, Hu C, Lo S (2019) Simultaneous aqueous chlorination of amine-containing pharmaceuticals. Water Res 155:56–65. https://doi.org/10.1016/j.watres.2019.01.061

Lu J, Zhang C, Wu J, Zhang Y, Lin Y (2020) Seasonal distribution, risks, and sources of endocrine disrupting chemicals in coastal waters: will these emerging contaminants pose potential risks in marine environment at continental-scale? Chemosphere 247:125907. https://doi.org/10.1016/J.CHEMOSPHERE.2020.125907

Lv SW, Cong Y, Chen X, Wang W, Che L (2021) Developing fine-tuned metal–organic frameworks for photocatalytic treatment of wastewater: a review. Chem Eng J 433(2):133605. https://doi.org/10.1016/J.CEJ.2021.133605

Ma Y, Liu H, Wu J, Yuan L, Wang Y, Du X, Wang R, Marwa PW, Petlulu P, Chen X, Zhang H (2019) The adverse health effects of bisphenol A and related toxicity mechanisms. Environ Res 176:108575. https://doi.org/10.1016/J.ENVRES.2019.108575

Mahmoud SH, Gan TY (2018) Impact of anthropogenic climate change and human activities and human activities on environment and ecosystem services in arid regions. Sci Total Environ 633: 1329–1344

Mao K, Zhang H, Pan Y, Yang Z (2021) Biosensors for wastewater-based epidemiology for monitoring public health. Water Res 191:116787. https://doi.org/10.1016/j.watres.2020.116787

Naik MM, Dubey SK (2016) Marine pollution and microbial remediation. Springer, Berlin, pp 1–270. https://doi.org/10.1007/978-981-10-1044-6

Ortega-Olvera JM, Mejía-García A, Islas-Flores H, Hernández-Navarro MD, Gómez-Oliván LM (2020) Ecotoxicity of emerging halogenated flame retardants. Compr Anal Chem 88:71–105. https://doi.org/10.1016/BS.COAC.2019.11.004

Qi K, Yu J (2020) Modification of ZnO-based photocatalysts for enhanced photocatalytic activity. Interface Sci Technol 31:265–284. https://doi.org/10.1016/B978-0-08-102890-2.00008-7

Qian L, Durairaj S, Prins S, Chen A (2021) Nanomaterial-based electrochemical sensors and biosensors for the detection of pharmaceutical compounds. Biosens Bioelectron 175:112836. https://doi.org/10.1016/j.bios.2020.112836

Rashid B, Husnain T, Riazuddin S (2010) Herbicides and pesticides as potential pollutants: a global problem. In: Plant adaptation and phytoremediation. Springer, Berlin, pp 427–447. https://doi.org/10.1007/978-90-481-9370-7

Reberski JL, Terzić J, Maurice LD, Lapworth DJ (2022) Emerging organic contaminants in karst groundwater: a global level assessment. J Hydrol 604:127242. https://doi.org/10.1016/j.jhydrol.2021.127242

Reshma AC, Krishna RR (2017) Contamination of emerging contaminants in Indian aquatic sources: first overview of the situation. J Hazard Toxic Radioactive Waste 21(3):04016026. https://doi.org/10.1061/(ASCE)HZ

Rochman CM, Browne MA, Underwood AJ, Van Franeker JA, Thompson RC, Amaral-Zettler LA (2016) The ecological impacts of marine debris: unraveling the demonstrated evidence from what is perceived. Ecology 97(2):302–312. https://doi.org/10.1890/14-2070.1

Rodríguez EM, Rey A, Mena E, Beltrán FJ (2019) Environmental application of solar photocatalytic ozonation in water treatment using supported TiO2. Appl Catal B Environ 254: 237–245. https://doi.org/10.1016/j.apcatb.2019.04.095

Rojas S, Horcajada P (2019) Metal—organic frameworks for the removal of emerging organic contaminants in water. Chem Rev 120(16):8378–8415. https://doi.org/10.1021/acs.chemrev.9b00797

Sarkar J, Chowdhury PP, Dutta TK (2013) Complete degradation of di-n-octyl phthalate by *Gordonia* sp. strain Dop5. Chemosphere 90(10):2571–2577. https://doi.org/10.1016/J.CHEMOSPHERE.2012.10.101

Shanker U, Rani M, Jassal V (2017) Degradation of hazardous organic dyes in water by nanomaterials. Environ Chem Lett 15(4):623–642. https://doi.org/10.1007/s10311-017-0650-2

Sharma R, Verma N, Lugani Y, Kumar S, Asadnia M (2021) Conventional and advanced techniques of wastewater monitoring and treatment. In: Green sustainable process for chemical and environmental engineering and science. Elsevier, Amsterdam, pp 1–48. https://doi.org/10.1016/B978-0-12-821883-9.00009-6

Smital T, Luckenbach T, Sauerborn R, Hamdoun AM, Vega RL, Epel D (2004) Emerging contaminants—pesticides, PPCPs, microbial degradation products and natural substances as inhibitors of multixenobiotic defense in aquatic organisms. Mutat Res 552(1–2):101–117. https://doi.org/10.1016/J.MRFMMM.2004.06.006

Söffker M, Tyler CR (2012) Endocrine disrupting chemicals and sexual behaviors in fish—a critical review on effects and possible consequences. Crit Rev Toxicol 42(8):653–668. https://doi.org/10.3109/10408444.2012.692114

Stroheker T, Peladan F, Paris M (2014) Safety of food and beverages: water (bottled water, drinking water) and ice. In: Encyclopedia of food safety, vol 3. Elsevier, Amsterdam, pp 349–359. https://doi.org/10.1016/B978-0-12-378612-8.00295-X

Thompson WA, Vijayan MM (2020) Environmental levels of venlafaxine impact larval behavioural performance in fathead minnows. Chemosphere 259:127437. https://doi.org/10.1016/j. chemosphere.2020.127437

Tisler S, Zindler F, Freeling F, Nödler K, Toelgyesi L, Braunbeck T, Zwiener C (2019) Transformation products of fluoxetine formed by photodegradation in water and biodegradation in zebrafish embryos (*Danio rerio*). Environ Sci Technol 53(13):7400–7409. https://doi.org/10. 1021/acs.est.9b00789

Tiwari M, Sahu SK, Pandit GG (2016) Distribution and estrogenic potential of endocrine disrupting chemicals (EDCs) in estuarine sediments from Mumbai, India. Environ Sci Pollut Res 23(18): 18789–18799. https://doi.org/10.1007/S11356-016-7070-X/FIGURES/6

Trishitman D, Cassano A, Basile A, Rastogi NK (2020) Reverse osmosis for industrial wastewater treatment. In: Current trends and future developments on (bio-) membranes: reverse and forward osmosis: principles, applications, advances. Elsevier, Amsterdam, pp 207–228. https://doi.org/ 10.1016/B978-0-12-816777-9.00009-5

Valdez-Carrillo M, Abrell L, Ramírez-Hernández J, Reyes-López JA, Carreón-Diazconti C (2020) Pharmaceuticals as emerging contaminants in the aquatic environment of Latin America: a review. Environ Sci Pollut Res 27(36):44863–44891. https://doi.org/10.1007/S11356-020-10842-9

Vieira WT, De Farias MB, Spaolonzi MP, Da Silva MGC, Vieira MGA (2021) Endocrine-disrupting compounds: occurrence, detection methods, effects and promising treatment pathways—a critical review. J Environ Chem Eng 9(1):104558. https://doi.org/10.1016/j.jece. 2020.104558

Villarreal-Reyes C, Díaz de León-Martínez L, Flores-Ramírez R, González-Lara F, Villareal-Lucio S, Vargas-Berrones KX (2021) Ecotoxicological impacts caused by high demand surfactants in Latin America and a technological and innovative perspective for their substitution. Sci Total Environ 816:151661. https://doi.org/10.1016/J.SCITOTENV.2021.151661

Wang J, Chen H (2020) Catalytic ozonation for water and wastewater treatment: recent advances and perspective. Sci Total Environ 704:135249. https://doi.org/10.1016/j.scitotenv.2019. 135249

Wang J, Wang S (2018) Microbial degradation of sulfamethoxazole in the environment. Appl Microbiol Biotechnol 102(8):3573–3582. https://doi.org/10.1007/S00253-018-8845-4/ FIGURES/1

Wang Y, Li X, Lei W, Zhu B, Yang J (2021) Novel carbon quantum dot modified g-C3N4 nanotubes on carbon cloth for efficient degradation of ciprofloxacin. Appl Surf Sci 559: 149967. https://doi.org/10.1016/J.APSUSC.2021.149967

Xiang L, Xie Z, Guo H, Song J, Li D, Wang Y, Pan S, Lin S, Li Z, Han J, Qiao W (2021) Efficient removal of emerging contaminant sulfamethoxazole in water by ozone coupled with calcium peroxide: mechanism and toxicity assessment. Chemosphere 283:131156. https://doi.org/10. 1016/J.CHEMOSPHERE.2021.131156

Ying GG, Kookana RS (2003) Degradation of five selected endocrine-disrupting chemicals in seawater and marine sediment. Environ Sci Technol 37(7):1256–1260. https://doi.org/10.1021/ ES0262232

Yuan Y, Song D, Wu W, Liang S, Wang Y (2016) The impact of anthropogenic activities on marine environment in Jiaozhou Bay, Qingdao, China: a review and a case study. Reg Stud Mar Sci 8: 287–296. https://doi.org/10.1016/j.rsma.2016.01.004

Zaid AA (2017) Consequences of anthropogenic activities on fish and the aquatic environment. Poult Fish Wildl Sci 3:138. https://doi.org/10.4172/2375-446X.1000138

Zwart N, Jonker W, ten Broek R, de Boer J, Somsen G, Kool J et al (2020) Identification of mutagenic and endocrine disrupting compounds in surface water and wastewater treatment plant effluents using high-resolution effect-directed analysis. Water Res 168:115204. https://doi.org/ 10.1016/j.watres.2019.11520

Nanoparticles in the Marine Environment **7**

Andreas Gondikas, Julian Alberto Gallego-Urrea,
and Karin Mattsson

Contents

Abstract

Our understanding of the role of nanoparticles in biogeochemical processes in aquatic systems is continuously improving along with technological advancements that allow studying matter in the nanoscale. However, environmental nanoscience in the marine environment is lagging behind. This chapter outlines processes that regulate the fate of nanoparticles in the marine environment and highlights parameters that ought to be considered when designing sampling strategies for nanoparticle research. Sample pre-treatment methods are presented, with focus on practical solutions to address challenges of low nanoparticle concentration, high ion concentration, and complex biological processes

A. Gondikas (✉)
Department of Geology and Geoenvironment, National and Kapodistrian University of Athens, Athens, Greece

J. A. Gallego-Urrea
Department of Marine Sciences, Kristineberg Marine Research Station, University of Gothenburg, Fiskebäckskil, Sweden

Present Address: Horiba Europe, Filial Sweden, Gothenburg, Sweden

K. Mattsson
Department of Marine Sciences, Kristineberg Marine Research Station, University of Gothenburg, Fiskebäckskil, Sweden

© The Author(s), under exclusive license to Springer Nature Switzerland AG 2023
J. Blasco, A. Tovar-Sánchez (eds.), *Marine Analytical Chemistry*,
https://doi.org/10.1007/978-3-031-14486-8_7

that are characteristic of marine systems. The most prominent techniques for nanoparticle analysis in environmental samples are outlined with examples, advantages, and disadvantages. Finally, three case studies are discussed, representing different origins, effects, and analytical challenges. Iron oxide is used as an example of naturally occurring nanoparticles with complex and dynamic reactivity, titanium dioxide as an example of natural and intentionally produced anthropogenic nanoparticles, and plastic nanoparticles as an example of incidentally produced anthropogenic nanoparticles. Analytical solutions for studying these materials are discussed, highlighting the scarcity of studies in the marine environment, in comparison with similar studies in land waters, and providing directions for future scientific development.

Keywords

Nanoparticle fate · Sample handling · Analytical techniques · Iron oxide · Titanium dioxide · Nanoplastics

7.1   Introduction

Our understanding of biogeochemical processes and elemental speciation has greatly improved since technological advancements in analytical techniques allowed scientists to study solid matter at the nanoscale. Although there is yet no definition of nanoparticles that is universally adopted across disciplines and geographic regions, it suffices to state that nanoparticles are solids with dimensions between 1 and 100 nanometers that often exhibit different physicochemical properties than their larger counterparts, i.e., materials of larger dimensions with the same composition. Atoms on the surface of a solid are not as thermodynamically stable as in the inner volume, resulting in irregularities of the structure, which give rise to differences in chemical behavior (Banfield and Zhang 2001). At the nanoscale, the ratio of surface to volume of the solid is so high that surface atoms dominate overall particle properties, resulting in novel properties, e.g., unique optical and electronic properties of quantum dots, mechanical strength of metals and alloys, thermal properties of metal oxides, etc. (Auffan et al. 2009). In taking advantage of such phenomena, material scientists are constantly discovering materials with novel characteristics, which are then incorporated in numerous consumer products and often released into the environment after use (Vance et al. 2015). In addition to nanoparticles that are intentionally manufactured, incidental nanoparticles find their way into the environment at a substantially larger scale (Hochella et al. 2019). These are nanoparticles formed as a byproduct of human activities, mostly through friction and combustion, e.g., friction between road and tires, fuel combustion, and natural weathering of waste, among many others. Intentionally and incidentally produced nanoparticles are grouped as anthropogenic nanoparticles, and they are released in the marine environment through direct and indirect routes. Release from antifouling paints on ships, discharge of scrubber water effluent, and other ship wastewater are a few examples of direct release, while atmospheric deposition of exhaust particles and input to estuaries from contaminated rivers are examples of indirect release. Nanoparticles are formed either through fragmentation of larger solids (top-down process, e.g., rock weathering and plastic degradation) or from nucleation and growth of precursor components (bottom-up process, e.g., mineral formation and laboratory-controlled precipitation). Recently, the disposal of plastic waste has attracted scientific interest, and as a result, we are starting to understand the impact of plastics of all sizes, including nanoplastics, in the marine ecosystem. The marine environment is heavily polluted with litter, where the majority of the numerical abundance is plastics (Goldstein et al. 2013). There are two main sources for nanoplastic particles in the environment, primary and secondary particles. Primary particles are particles intentionally manufactured to a fixed size, whereas secondary particles arise from the degradation and fragmentation of larger objects. Most microplastic particles in the marine environment are secondary particles and are in higher abundance of smaller sizes (Ter Halle et al. 2016).

Although much attention is drawn to anthropogenic nanoparticles, naturally occurring nanoparticles are abundantly present in the environment and are often exhibiting similar characteristics and properties as anthropogenic (Wagner et al. 2014). Natural nanoparticles are substantially more abundant than anthropogenic, play a key role in the global cycling of elements, and are thus vital for biological growth and carrying nutrients and essential metals (Hassellöv and von der Kammer 2008). A vast majority of natural nanoparticles are clays, followed by iron oxides, other metal oxides, sulfides, carbonates, and phosphates (Hochella et al. 2019). Main inputs of natural nanoparticles in the marine environment are hydrothermal vents, atmospheric deposition, and riverine flows into estuaries. In the marine environment, natural nanoparticles are formed abiotically, through the nucleation of polyatomic clusters and complexes, which may grow to larger sizes and crystallize under favorable conditions. Biotic production of nanoparticles is also possible inside or outside of cells, either as a defense mechanism to the presence of toxic metals or as a mechanism to store nutrients (Asmathunisha and Kathiresan 2013), while they may also utilize components of nanoparticles with bioavailable geochemical phases, when these are available (Zhang et al. 2012).

Dissolved organic matter (DOM) often regulates the speciation of metals, the formation of dissolved complexes and nanoparticles, and the metal bioavailability. Although DOM was traditionally considered to interact with metal complexes and free metal ions, recent studies have shed light in the ubiquitous role of DOM on metal biogeochemical reactions and metal bioavailability (Fig. 7.1).

Studying the fate and transport of nanoparticles in the marine ecosystem and their role in biogeochemical processes will improve our understanding on pressing environmental challenges, such as global cycling of elements, impacts of climate-related changes, and ocean acidification, and may also help us answer questions related to the origin of life. A distinctive feature of nanoparticles is that although they are solids, they diffuse in water (and air) following a pattern that resembles that of dissolved species. The volume of a nanoparticle is so small that its movement is influenced by the Brownian motion of water molecules continuously colliding on its

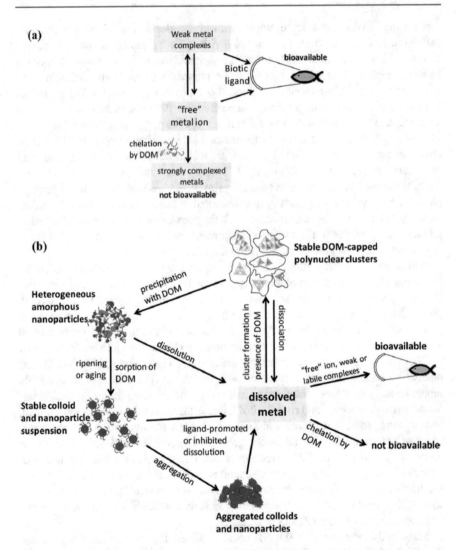

Fig. 7.1 Importance of DOM on metal speciation and bioavailability: (**a**) the traditional approach and (**b**) current understanding (reprinted with permission from Aiken et al. [2011]. Copyright 2011 American Chemical Society)

surface. However, their reactivity is governed by solid surface chemistry. In a sense, nanoparticles *talk like a solid but walk like a liquid*. Nanoparticles may also pass through biological membranes, which makes them a fascinating subject for examining mechanisms of biogeochemical reactions and elaborating their toxicity potential. Their fate and transport in aqueous systems are determined by simultaneous effects of aggregation, attachment, dissolution, growth, and surface reactions (Birdi 2002) as follows:

1. Aggregation between nanoparticles and with other particles is a process that results in larger and heavier particles, which settle on sediments influenced by gravity, thus resulting in the removal of nanoparticles from the water column. The kinetics of aggregation are strongly influenced by the number concentration of particles and the total concentration of ions. As the number of particles per unit volume increases, collisions between particles become more frequent, and as the total ion concentration increases, collisions that result in permanent attachment of the colliding particles become more likely (Chen et al. 2006, 2007).

2. Attachment to solid surfaces such as sediments, plants, and fish is less likely in the open ocean; however, attachment to large organic agglomerates, such as macrogels and marine snow, is an efficient process for the removal of nanoparticles from the water column. Similar to aggregation, the kinetics of nanoparticle attachment to surfaces is dependent on the number of nanoparticles and surface properties (Santschi et al. 2021).

3. Dissolution, when it is thermodynamically favored, results in the reduction of size and disappearance of nanoparticles. Often dissolution is promoted by ligands, i.e., organic or inorganic compounds with specific affinity for atoms on the surface of nanoparticles. Although chloride is a weak ligand for most metals, its abundance in seawater makes it a primary driver of nanoparticle dissolution (Kent and Vikesland 2012), while organic ligands may induce dissolution at trace concentrations (Gondikas et al. 2012).

4. Surface reactions include sorption, complexation, redox, and photocatalytic reactions and often moderate the fate and reactivity of nanoparticles. Alterations of surface chemistry also drive changes in toxicity (Caballero-Díaz et al. 2013). Aggregation, attachment, and dissolution are also surface reactions although they are often regarded separately.

An additional factor regulating the fate of particles is natural organic matter, as it may attach and bind sites on the particles' surface, thus preventing or facilitating further growth, dissolution, attachment to solid surfaces, and aggregation with other particles (Aiken et al. 2011; Deonarine et al. 2011). In marine waters, the regulating effect of organic matter takes several forms; biological activity influenced by seasonal conditions produces a variety of organic matter in dissolved or particulate forms, which strongly interact with nanoparticles. A complex process on a massive scale takes place in the Earth's oceans and seas, when hydrochemical conditions are favorable (algal blooms), resulting in the production of biopolymers and gels from microbial cell exudates, sloppy feeding of zoo plankton, and cell lysis (Tiselius and Kuylenstierna 1996). These biopolymers grow to gradually form nano-, micro-, and macrogels and through various physical and chemical processes and form a complex structure known as marine snow (Fig. 7.2), which is large enough to be influenced by gravity and settles on the ocean floor. Attachment of nanoparticles to all forms of this process results in their removal from the water column and deposition on sediments. Owing to the scale and global spread of marine snow formation and precipitation, this is considered as a major process for particle removal in marine waters. However, in the early stages of this process, biopolymers may act to stabilize nanoparticles,

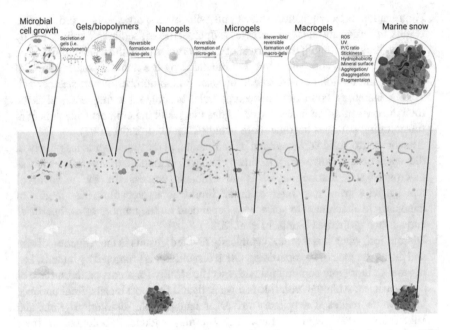

Fig. 7.2 Graphic representation of the marine snow formation process (reprinted under the CCA License from Quigg et al. [2021])

thus extending their presence in the water column (Gondikas et al. 2020). In addition, nanoparticles may enhance or suppress the formation process of nano- and microgels when in sufficiently high concentrations, which is a realistic scenario for plastic nanoparticles, for example (Chen et al. 2011).

7.2   Researching Nanomaterials in Marine Ecosystems

As developments in analytical and sampling equipment improve our ability to study nanoparticles in the natural environment, the importance of nanoscience emerges in several disciplines ranging from geology, evolutionary biology, and environmental science to socioeconomic studies (Hochella 2002). The most prominent example is iron-rich nanoparticles, introduced in seawater through fluxes of hydrothermal fluids or deposition of atmospheric dust that carry essential elements and may interact with organisms through various phases (Fig. 7.3). The multitude of reactions and phases that iron-rich nanoparticles may partake highlight the dynamic nature of nanoparticles and the necessity to develop long-term monitoring activities. Researching behavior and effects of nanoparticles in the marine environment requires reaching remote areas, often in deep seas, applying a sampling procedure that maintains the physicochemical integrity of particles, utilizing state-of-the-art

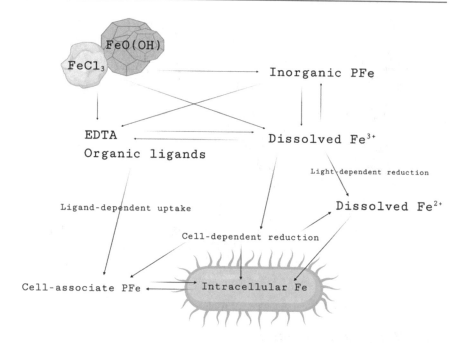

Fig. 7.3 Schematic representation of iron biogeochemical interactions in seawater (reprinted under the CCA License from Hunnestad et al. [2020])

equipment to thoroughly characterize the sample material, and then repeating the whole process in order to obtain multiple snapshots in time and space.

Marine ecosystems pose a challenging environment for conducting research due to the unlimited diversity and often remoteness of sites of interest (e.g., deep-sea hydrothermal vents). It is, therefore, imperative to conduct a thorough research on previous work and of environmental conditions at the site of interest, prior to designing the sampling strategy. In this respect, it is important to extend the literature search beyond one's own discipline and include various others, such as oceanographic, biological, and geochemical but also economical, commercial, and sociological. In addition to the detection and quantification of nanoparticles, it is important to collect information on hydrochemical parameters of the seawater, such as temperature, pH, ion composition, organic matter, as well as water current speed and directions. Biological information, such as seasonal algal blooms that will produce biopolymers and precursors of marine snow, are equally critical. It is also helpful to map socioeconomic activities in the region, such as ports, maritime routes, fisheries, wastewater treatment plant effluents, water sport activities, etc. that may influence water composition. A sampling procedure based on the analysis strategy will then need to be developed. It is also critical to decide in advance what parameters of target nanoparticles are needed to be quantified in order to answer the scientific hypothesis at hand.

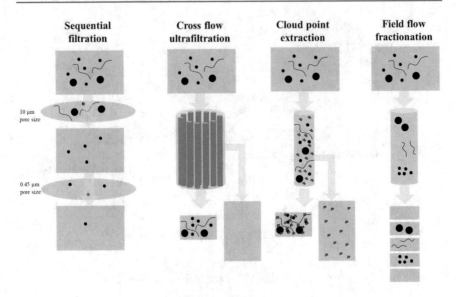

Fig. 7.4 Schematic representation of sample pre-treatment procedures

7.2.1 Sampling and Pre-treatment

Avoiding contamination and sample alterations during sampling is a major challenge in marine sciences and becomes even more important in a relatively new scientific field, such as marine nanosciences, where target materials may be omnipresent (in water, air, surfaces, and tools). In order to avoid contamination, sampling vials should be cleaned ideally under particle-free conditions, e.g., in a laminar flow hood. Washing with a detergent is sufficient for removing particle contamination from vials and should be followed by thorough rinsing with ultrapure water, a final rinse with acetone, and allowing to dry in a particle-free environment. If additional target materials are included, such as trace metal analysis, additional washing procedures ought to be followed (e.g., acid/base treatment). Measures to minimize contamination from the atmosphere should also be taken, particularly from soot, given that in most cases, a ship running on petrol is used, which will produce particle-rich exhaust that will be carried by wind and disperse in the vicinity of the vessel. Finally, sampling equipment ought to be cleaned and rinsed prior to each sample collection. In the laboratory, several approaches are available for isolating and simplifying the particulate fraction of an aqueous sample, including filtration, cross-flow ultrafiltration, cloud point extraction, and field flow fractionation (Fig. 7.4, Wagner et al. 2019).

7.2.1.1 Filtration
Filtration, most likely the oldest sample pre-treatment method, involves passing the aqueous sample through a filter with defined pore sizes, assuming that particles larger than the pore size will be retained on the filter membrane, while smaller

particles will pass through pores. Sequential filtration using filters with various pore sizes may be used, starting with the largest pore size. This setup reduces the risk of filter clogging, by splitting particles on multiple surfaces, and simplifies the analytical process; however, pressure differences across the filtration setup ought to be taken into account, and care should be taken to avoid air bubbles trapped in the tubing or filter casing. Structure and material properties of the membrane and flow rate will influence the interaction between particles and membrane, so it is important to keep in mind that the size cutoff is a nominal value and in reality some particles smaller than this value will be retained and a fraction of larger particles will pass through (e.g., Cai et al. [2020]). This method is advantageous when combined with electron microscopy because the membranes provide an ideal substrate for preserving particles.

7.2.1.2 Cross-Flow Ultrafiltration

Clogging of the filter membranes is a common issue when sampling environmental waters. Gels and large particles quickly form a cake on the filter, blocking its pores and reducing sample flow. In cross-flow ultrafiltration, the water flows through several cylinder-shaped membranes, thus preventing large particles and gels to attach on the membrane surface. A portion of the water is allowed to pass through the filter by applying an additional flow vertical to the main flow, thus increasing the particle to water ratio of the sample. The output may be directed back to the sample container, so the sample becomes more concentrated with time. The membrane material may have such small pore size that is able to reject macromolecules (ultrafiltration), so the technique is also applicable for pre-concentrating organic compounds with a molecular weight cutoff (Guo and Santschi 1996). This method is advantageous for very dilute samples, where the mass or number concentration of particles falls below the detection limit of the analytical equipment.

7.2.1.3 Cloud Point Extraction

When pre-concentration is the treatment of choice, particles in the resulting suspension are more likely to aggregate, because of more frequent collisions between particles. In cloud point extraction, a surfactant is mixed with the sample and heated until the cloud point is reached. The micelles formed by the surfactant are separated from the liquid with centrifugation, and the supernatant is removed. The remaining mix of micelles and particles is diluted and sonicated to recover nanoparticles in a concentrated and stable suspension (El Hadri and Hackley 2017).

7.2.1.4 Field Flow Fractionation

In field flow fractionation, the sample is introduced by an aqueous carrier into a channel whose bottom part is an ultrafiltration membrane. Nanoparticles in samples are carried toward the channel exit by the carrier that flows parallel to the membrane, while a vertical flow is pushing them toward the membrane, where the carrier flow is lower. Smaller particles with higher diffusion coefficient escape toward the center of the channel and into the higher flow profile of the carrier (Giddings 1993). Thus, smaller particles exit the channel faster and size-dependent separation is achieved.

Table 7.1 Parameters influencing the release, fate, and impact of nanoparticles in the marine environment

Hydrobiochemical parameters	Socioeconomic activities	Nanoparticle properties
Temperature	Ports	Elemental composition
pH	Maritime routes	Number concentration
Ionic composition	Fisheries	Size distribution
Total organic carbon	Deep-sea mining	Shape
Dissolved organic carbon	Oil/gas extraction	Compositional homogeneity
Water currents	Water sports	Surface properties
Algal blooms	Hotels/farms on coastal land	

This is a powerful tool for simplifying complex marine water and sediment samples (Hassellöv 2005) (Fig. 7.4).

7.2.2 Analysis

There is variety of analytical techniques for determining properties of nanoparticles, such as the ones listed in Table 7.1, and for a thorough review, the reader is directed to the work by Hassellöv and Kaegi (Hassellöv and Kaegi 2009). Most of these techniques are applicable for nanoparticle suspensions of much higher concentrations than expected in surface waters, even more so in the marine environment where natural and anthropogenic input is diluted in the vast volumes of seas and oceans. Here, we present three techniques that are more appropriate for analysis of marine water samples: electron microscopy, inductively coupled plasma mass spectrometry in single-particle mode, and Raman spectroscopy.

7.2.2.1 Electron Microscopy

Microscopy enables visual inspection of details that are not possible to detect with the naked eye. Electron microscopy is the most powerful tool in microscopy, allowing the observation of nano-objects and even atomic structures. When a beam of electrons is focused and directed on a sample area or spot, electrons interact with the material on the sample producing various signals, such as backscattered electrons, secondary electrons, X-rays, etc. that can be used to infer information about materials' properties. The two-dimensional image produced when capturing scattered electrons can be used to infer dimensions and structure of a nanoparticle, while scattered X-rays can be used to acquire compositional information of the nanoparticle. There are two main operational modes of electron microscopy, namely, transmission electron microscopy (TEM, Fig. 7.5a) and scanning electron microscopy (SEM, Fig. 7.5b). The main differences between the two modes are that in TEM, a high-energy electron beam is focused on a spot and transmitted through the sample where it is captured by a detector located under the sample, while in SEM, a low-energy electron beam is scanning the surface of the sample in a raster motion

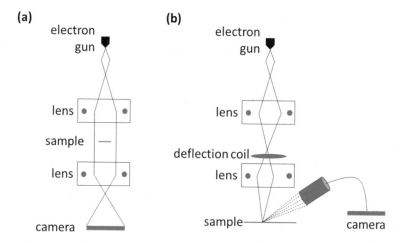

Fig. 7.5 Schematic representation of the transmission electron microscopy (**a**) and scanning electron microscopy (**b**)

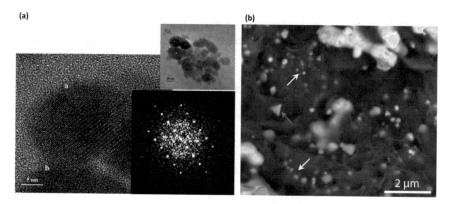

Fig. 7.6 (**a**) High-resolution TEM image of a silver sulfide nanoparticle with the fast Fourier transform pattern showing crystal structure (reprinted with permission from Kim et al. [2010]. Copyright 2010 American Chemical Society) and (**b**) high-resolution SEM image with backscattered electron detector showing gold nanoparticles on clay minerals (reprinted from Shuster et al. [2017] with permission from Elsevier)

and scattered electrons are captured by a detector located above the level of the sample (Hasselöv and Kaegi 2009).

With TEM, the two-dimensional shape of a particle may be observed with precision even for particles of a few nanometers in size. The structure of electrons in a crystalline particle can be observed and compared to library data to identify types of crystal phases present in the nanoparticle (Fig. 7.6a). However, analysis using TEM is time- and effort consuming. The sample needs to be very thin for the electron beam to pass through, so nanoparticles attached on larger objects are impossible to observe with TEM. In addition, scanning the sample area to identify particles of interest is substantially time-consuming. Sample preparation for TEM involves the deposition of nanoparticles on a thin film that is supported on a grid. For aqueous suspensions, the film is immersed in the sample, allowing time for particles to attach on it. Alternatively, a droplet of the sample is deposited on the film and allowed to dry. In order to avoid drying artifacts (e.g., dissolved components solidifying after drying), the film is often rinsed with ultrapure water, thus risking loss of nanoparticles.

With SEM, particles larger than a few tens of nanometers may be observed. The higher limit of detection, compared to TEM, is balanced by the faster analysis and simpler sample preparation. Topography and shape of particles combined with elemental composition offer a powerful tool for identifying individual nanoparticles and clusters of nanoparticles with larger particles (Fig. 7.6b). Sample preparation involves filtration of an aqueous suspension or drying on a flat surface. Conductivity of the substrate on which particles are deposited is crucial to avoid trapping electrons on the sample surface, which may create artifacts and deteriorate signal acquisition.

7.2.2.2 Single-Particle Inductively Coupled Plasma Mass Spectroscopy (spICPMS)

The spICPMS method is developed on a widely used conventional method that is often applied for trace metal analysis in natural waters, among many other applications. In conventional ICPMS, samples are digested prior to analysis using acid combinations to dissolve all solids. The liquid sample is then introduced through a peristaltic pump into a nebulizer, where it is mixed under pressure with

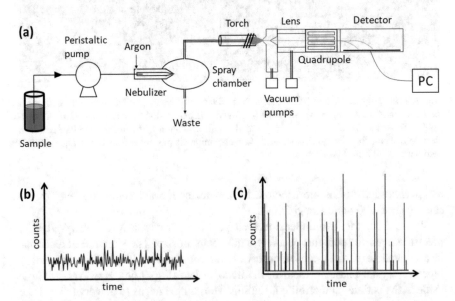

Fig. 7.7 Schematic representation of the ICPMS instrument (**a**), signal recorded in conventional mode (**b**), and signal recorded in single-particle mode (**c**)

argon gas forming a spray of droplets (Fig. 7.7a). Larger droplets crash on the spray chamber walls and exit into the waste, while smaller droplets are directed into the inductively coupled plasma torch. All components of the sample are atomized in the torch and are then separated based on their mass to charge ratio, under an electric and in some cases a magnetic field. The signal from an element is recorded and averaged over a period of time (the dwell time), and the output is reported as counts per second (Fig. 7.7b), which is compared to a calibration curve to derive the mass concentration of the element in the sample. In order to measure nanoparticles, the sample digestion step is obviously omitted. The sample is introduced in the nebulizer and spray chamber, similarly to conventional operation; thus, when a particle reaches the torch, it is atomized into a cloud of atoms. At the detector, the dwell time is reduced drastically to enable recording of single-particle events, which appear as spikes above the background when a cloud of atoms reaches the detector (Fig. 7.7c).

The main disadvantage of the spICPMS technique is that only one element can be recorded in a single particle, with most instruments. In addition, samples with dissolved components of the target element produce background signal overlapping with the nanoparticle signal, thus hindering detection of smaller particles. In a similar manner, polyatomic and isotopic interferences will produce an artificial background signal that cannot be resolved without sacrificing some proportion of instrument sensitivity. Nevertheless, spICPMS is currently spearheading efforts for monitoring nanoparticles in environmental systems, because it offers several key advantages as follows:

1. Minimal sample pre-treatment. In many cases, an aqueous sample may be directly injected in the instrument for analysis, without any pre-treatment. Depending on the type of nebulizer used, large particles (>10 μm) ought to be removed to avoid clogging on the tip of the nebulizer and especially for marine samples; a dilution step may be necessary to reduce the ionic content. The time between dilution and analysis has to be minimal to avoid alterations caused by the drastic change of water composition (Toncelli et al. 2016).
2. Available equipment. Conventional ICPMS instruments are standard equipment for analytical laboratories and can be used with minor alterations for spICPMS. Not needing to purchase new instrumentation or rearrange space is a significant advantage for this new technology.
3. High throughput. Analysis of a sample may last from a few seconds to several minutes, depending on the concentration of particles in the sample, which is relatively fast compared to other methods for nanoparticle analysis. This is partly due to extensively developed sample introduction and detection systems that have already been developed for conventional ICPMS.
4. Relevant parameters. With spICPMS, the number concentration of particles containing the target element can be measured within a reasonable timeframe. The combination of compositional information and number concentration is an attractive attribute of spICPMS. In comparison, obtaining the same output using microscopy would require several hours of analysis for a single sample. However,

compositional information obtained from microscopy is more complete, considering that one element per particle is measured with spICPMS.

5. Ongoing developments. Several technological improvements are addressing key challenges of using spICPMS for nanoparticle detection. For example, reducing the dwell time to such an extent that allows measuring two elements or using a time of flight mass spectroscopy (TOFMS) instrument to measure multiple elements on a single particle provides a holistic set of compositional information. Furthermore, matrix-matching of finely tuned droplets substantially improves instrument calibration for all types of samples (Verboket et al. 2014).

Overall, the incorporation of spICPMS in routine analysis protocols for nanoparticle characterization is gaining ground, supported by ongoing technological advancements. In analyzing marine water samples, care should be taken to appropriately dilute in order to reduce the concentration of dissolved components entering the plasma, which may jeopardize the plasma flame stability, reduce lifetime of parts, and stress the detector. Furthermore, care ought to be taken to avoid clogging of the nebulizer, so a filtration step is often required to remove large particles, e.g., larger than 10 μm.

7.2.2.3 Raman Spectroscopy and Microscopy

Raman spectroscopy is an analytical technique that provides information on vibrational, rotational, and low-frequency transitions in molecules. A monochromatic light such as laser light interacts with the sample, absorbed photons cause molecules to change their dipole moment, and when photons are re-emitted, they exhibit a frequency shift. Micro-Raman or Raman microscopy is a combination of Raman spectroscopy and optical microscopy, which can be used to identify micro-sized particles since the laser light can be focused to a smaller spot (Fig. 7.8). Raman

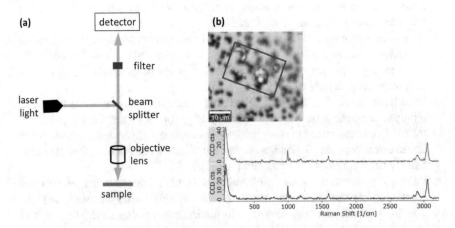

Fig. 7.8 Simplified representation of the Raman microscopy technique (**a**) and example of Raman microscopy analysis of plastic particle allowing identification of chemical transformations (reprinted under the CCA License from Mattsson et al. [2021b])

microscopy can provide chemical information, size, and morphology for single particles of up to several hundreds of nanometers in size. The main advantage of Raman microscopy is the combination of chemical information and imaging, and compared to infrared spectroscopy, there is no interference from water molecules. Its combination with SEM may enhance imaging quality and if needed provide elemental analysis (Mattsson et al. 2021b). Disadvantages include a weak signal that is sensitive to interferences from fluorescence and possibly cosmic rays. Increasing the number of accumulations and consequently the duration of analysis improves signal quality and addresses the issue of interferences from cosmic rays. Fluorescence interferences may be reduced by using a laser with a higher wavelength or thorough cleaning of particles before analysis (Fig. 7.8).

7.3 Case Studies of Natural and Anthropogenic Nanoparticles

Nanoparticles are abundant in the marine environment, with multiple sources—anthropogenic and natural—and exhibit a variety of physicochemical properties. Here, we provide three examples that are in highest abundance as follows:

1. Iron oxide nanoparticles, mostly of natural origin and with a crucial role in providing nutrients and trace elements to biota
2. Titania nanoparticles, both natural and anthropogenic with catalytic properties, most likely to influence biogeochemical reactions on upper layers of the oceans, where exposure to light is maximal, especially on the sea-surface microlayer
3. Plastic nanoparticles, of anthropogenic origin and a byproduct of the degradation of larger plastic objects that continuously accumulate on the marine ecosystem and whose impacts we are beginning to understand

Although this list is not exhaustive, it provides key information about potential sources and mechanisms of transformations that may take place in the marine environment. Most importantly, it highlights the necessity for multidisciplinary approaches in risk assessment.

7.3.1 Iron Oxide Nanoparticles

Iron is regarded as the micronutrient with the largest effect on marine photosynthesis. The main sources of iron to the oceans are freshwater discharges (Hassellöv and von der Kammer 2008), atmospheric deposition (Fig. 7.9), and hydrothermal vents (Yücel et al. 2011). The importance of iron in biota is evidenced when, in large areas of the ocean, phytoplankton growth is limited by the scarcity of bioavailable iron.

The chemistry of iron regulates its fate and bioavailability. The most thermodynamically stable form of iron in the oceans is the oxidised form, iron(III), with very low solubility albeit detectable at low concentrations attributed mainly by strong complexation with organic ligands. In sunlit surface waters, iron (III) may be

Fig. 7.9 Representation of the fate of particles depositing on the surface of the ocean or forming near the surface. At time zero (T0), particles are deposited on the sea-surface microlayer; with time, particles with higher mass and friction coefficient ratio settle faster (i.e., larger and heavier particles represented in red), while smaller and friction-prone particles settle slower (green and black). Notice that heteroaggregation leads to combinations of these particles in this representation

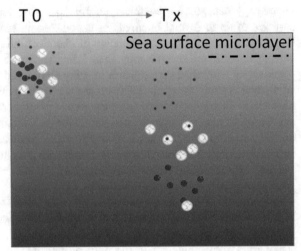

reduced to the more soluble and bioavailable iron (II) form. However, oxidizing agents may transform the iron back to the iron (III) form. The overall status of oxidation, alongside with the ability of iron oxides to form stable solid phases, can play an important role in regulating the availability of iron. These processes depend on salinity, properties and quantity of dissolved organic carbon, presence of another particulate matter, temperature, pH, and oxygen concentration. Adsorbed iron may undergo redox reactions under sunlight or oxidizing agents and transform between iron (II) and (III) states. While in this cycle, ligand complexation and mineral formation of both states may occur (Turner and Hunter 2001). For example, coagulation and sedimentation of particulate iron along salinity gradients—typically present in estuaries—are a major sink of bioavailable iron. As a result, the majority of iron transported by rivers is trapped in estuaries not reaching the oceans.

The formation of iron oxides may follow two routes, (i) precipitation (nucleation) from dissolved forms of iron, i.e., Fe(II) or Fe(III), or (ii) transformation from another phase of iron oxide. There are numerous forms of iron oxides which have different modes of formation (e.g., goethite, akageneite, ferrihydrite, hematite, maghemite, lepidocrocite, magnetite, feroxyhyte, and green rusts). Due to the importance of bioavailable iron on phytoplankton balance and the associated elemental cycles, it is important to follow processes that provide iron from less refractory fractions of these solid phases.

One of the main paths of natural formation of iron oxides is the hydrolysis of Fe (III) followed by the formation of nanosized ferrihydrite. Fe(III) in solution forms the Fe(III) hexa-aquocation which has acidic properties (Raiswell and Canfield 2012):

$$Fe^{3+} + 6\,H_2O - > Fe(H_2O)_6^{3+}$$

Because this complex acts as an acid, it will deprotonate stepwise at neutral to high pH, leading to the formation of the hydroxide species (or hydrolyzed, depending on the way they are presented), and ultimately, all six water molecules will become deprotonated, leading to the formation of a Fe(III) oxide or oxide hydroxide:

$$Fe(H_2O)_6^{3+} - > FeOOH + 3\,H^+ + 4\,H_2O$$

$$2\,Fe(H_2O)_6^{3+} - > Fe_2O_3 + 6\,H^+ + 9\,H_2O$$

It is the latest reactions that are thought to dominate in natural environments during the formation of iron oxides. Bligh and Waite (Bligh and Waite 2011) added pulses of Fe(II) and Fe(III) to oxygenated seawater to follow the formation of iron oxides and discovered that during the first hour of experiments, small iron oxides showed rapid aggregation and Fe(II)- and Fe(III)-derived oxides had the same structure as aged lepidocrocite and ferrihydrite, respectively.

After formation, solid phases of iron interact with dissolved organic carbon which alters surface properties of the solid phase upon adsorption. This process and interactions with further ionic and solid phases determine the fate of this particulate material in water compartments. Some studies have focused on surface processes of iron oxide phases; Garg et al. (Garg et al. 2020) and references therein have investigated the kinetics of dissolution of amorphous iron oxyhydroxide under photoreductive conditions in combination with fulvic acids, although there was no control on the aggregate size distribution. Other studies have investigated the nature of these solid phases of iron in different types of natural waters and the implications

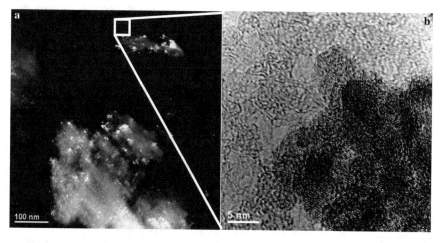

Fig. 7.10 TEM images showing clusters of iron-rich nanoparticles as bright spots (reprinted from Raiswell et al. [2006] with permission from Elsevier)

for the input to oceans. For example, Raiswell and coworkers (Raiswell et al. 2006) found that fine-grain fractions of glacially derived sediment contained individual nanoparticles with sizes <10 nm or assembled into aggregates of several-hundred-nanometer diameter (Fig. 7.10).

These studies showcase the importance of performing analytical studies to follow the dynamics of natural processes that affect iron bioavailability. Manufactured iron-based nanoparticles have been studied to a large extent (Hjorth et al. 2017), but there are not many studies investigating the abundance and dynamics of natural nanoparticles (Von Der Heyden et al. 2019). This is partly due to inherent difficulties in measuring iron in oceanic samples where contamination from outside sources during sampling, transporting, and measuring is historically recognized. The high ionic strength of seawater samples also complicates the use of techniques that are more suitable for freshwater (e.g., ICPMS), and the instability (e.g., transformation of solid phases) of samples after collection and exposure to other conditions (i.e., light exposure or air) can also make it more complicated to obtain reliable results in measuring natural oceanic particles (Fig. 7.10).

7.3.2 Titania Nanoparticles

A distinctive characteristic of titanium dioxide (titania) nanoparticles is that they both are naturally abundant in the environment and are industrially produced in large amounts, thus subsequently released into the environment from consumer products (Williams et al. 2019). In addition to their intentional production, they are also unintentionally produced either from weathering of larger grains or as byproducts of the production of larger particles, e.g., pigments. Although natural production volumes of titania are larger, the industrial and incidental production is substantial and comparable to natural, at least in the proximity of point sources (Kaegi et al. 2017). Titania nanoparticles are insoluble in most surface waters; thus, aggregation and settling are expected to regulate their fate, while surface reactions will regulate their reactivity. Titania forms crystal structures, the most abundant being rutile and anatase with the latter being more stable in the smaller nanoparticle size (less than 10 nanometers). Both phases are photoreactive with anatase exhibiting higher reactivity than rutile and thus are expected to catalyze redox reactions in upper layers of water bodies that are exposed to sunlight. The photocatalytic reactivity of nano-titania is the subject of intense research, due to its potential impact on the degradation of organic compounds and ecotoxicological effects. In the marine environment, the reactivity of nano-titania is influenced by the aggregation state of nanoparticles and surface properties (Fig. 7.11), both of which are regulated by the presence of organic macromolecules and biological activity.

Recent studies demonstrated that nano-titania is accumulating on the water surface of lakes, possibly due to attachment of low density or hydrophobic macromolecules on their surface (Gondikas et al. 2018). It is likely that a similar mechanism is taking place in marine environments; however, the impact of the photoreactivity of nano-titania in marine waters is largely understudied. In fact,

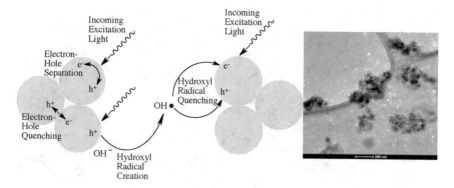

Fig. 7.11 Graphic representation of a photocatalytic kinetic model for nanoparticle aggregates (left) and TEM picture of nano-titania (right, reprinted with permission from Jassby et al. [2012]. Copyright 2012 American Chemical Society)

due to difficulties involved in sample collection, treatment, and analysis of titania nanoparticles in aquatic environments (Gondikas et al. 2018), most studies in seawater correspond to mechanistic investigations of the fate and possible effects. Botta et al. (Botta et al. 2011) demonstrated how the aging of consumer products containing titania nanoparticles would behave in aquatic environments. Similarly, other studies have focused on effects on stability of different organic molecules (Callegaro et al. 2015), interaction with microalgae (Morelli et al. 2018), possible toxic effects to specific marine communities of phytoplankton (Miller et al. 2012), and heteroaggregation with suspended natural particles (Wang et al. 2021). It remains still as an open question to which extent engineered or natural titania nanoparticles disperse and react in the ocean. Data on the release from commercial products into the oceans are lacking; the fate, transformation and behavior under conditions typical of the marine environment and resulting concentrations, mineralogy, and particle-size distribution require further investigation.

7.3.3 Plastic Nanoparticles

In the marine environment, plastics will degrade and fragment through various processes such as hydrolysis, mechanical degradation, thermo-oxidative degradation, photodegradation, and biodegradation. Hydrolysis is introduced by the addition of water and is an environmentally induced bond-breaking reaction. During mechanical degradation, particles are subjected to mechanical stress from waves, rocks, sand, and other forces or substances that the polymer may interact with. Photooxidative degradation is initiated by UV radiation, which is effective on land or air, however, much slower in marine environments (Kalogerakis et al. 2017). UV radiation may break carbon-carbon bonds, but the effectiveness depends on the wavelength of the light and the chemical structure of the polymer, while changes of physical and optical properties are limited to the surface of particles. The process

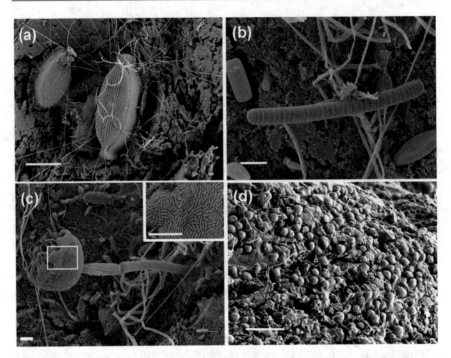

Fig. 7.12 Examples of microorganisms growing on plastic surfaces (reprinted with permission from Zettler et al. [2013]. Copyright 2013 American Chemical Society)

of thermo-oxidation is similar to photooxidation except that the bond-braking process is initiated by heat instead of light. With biodegradation, microorganisms such as bacteria and fungi degrade plastics through extracellular or intracellular enzymes and use plastics as a substrate for growth (Fig. 7.12). However, this process is complex, and not all mechanisms are known. In the marine environment, the most critical processes for the fate of plastics and generation of nanoplastics are UV radiation and mechanical force. For floating particles, photooxidative degradation initiated by UV radiation is essential. Degradation rate increases over time, and the particles' surface is more affected by the creation of cracks and flakes on the surface. These weathering processes of plastics generate nanoplastics; however, reaction rates and byproducts remain largely unknown.

There are three steps one needs to follow to identify nanoplastics in the marine environment, i.e., sampling, extraction/isolation, and identification, and there are no standardized protocols for any of them. There are five major methods for sampling plastic particles in the marine environment, including beachcombing, marine observational surveying, biological sampling, water sampling, and sediment sampling (Cole et al. 2011). Neither beachcombing nor marine observational surveying can be used for nanoplastics since both techniques are based on visual inspection. Biological sampling generally includes marine species and birds, where their gut or tissue is exanimated and to date mostly larger, micro-sized particles have been detected in marine animals. In collecting particles from marine water, the water is filtered through one or several membranes with different pore sizes, and nanoplastics may be retained on several membranes, even with pore sizes larger than their nominal size, due to aggregation with larger particles or random deposition on the membrane material. In isolating plastic particles from sediment, the most commonly used method is high-density liquid separation, where the sediment is mixed with the separation liquid and agitated and allowed to settle. Hence, sediment particles settle at the bottom, while lighter plastic particles float to the surface due to their relative density. Then, the supernatant or buoyant fraction is filtered through one or many membranes with different pore sizes. What polymers will be collected depends on their density in relation to the density separation liquid and the sizes of particles that depend on the membranes' pore size. Due to the high dispersity of nanoplastics, they are likely not efficiently fractionated in the liquid column; however, they are efficiently extracted from the bulk sediment mass.

Raman microscopy has been used to identify microplastic particles in samples from the marine environment including water and sediment. However, to identify nanoplastics in the marine environment is a major challenge mainly due to the multitude and abundance of natural particles present. It has been observed in most environmental systems that particle numbers increase exponentially as size decreases. Therefore, smaller volumes of sample are required to detect nanoparticles; however, it is important to study the whole size continuum; otherwise, it is challenging to apply links with other studies. Sampling larger sizes in combination with a smaller subsample for >1 μm is possible. The number of natural particles will be high so a treatment method to reduce the organic matter needs to be applied. The treatment can be a pre-treatment step or a posttreatment or both (Pfeiffer and Fischer 2020). The goal is to reduce the amount of organic matter but without affecting plastics. Several techniques may be used to quantify the amount and size of nanoplastics; however, very few are able to identify the particle's chemical composition and thus assign a polymer type. The most commonly used techniques to identify microplastic particles are pyrolysis GCMS or spectroscopy such as μ-FTIR and μ-Raman or confocal Raman microscopy (Mattsson et al. 2021a). Pyrolysis provides information regarding the polymer and its mass, meaning that samples need to be first filtered to remove larger particles. Moreover, only the mass of each polymer will be obtained. On the other hand, μ-FTIR and Raman microscopy provide information on chemical composition and size but not mass. For μ-FTIR, the lower detection limit is 10 μm, determined by the microscope diffraction limit. Raman microscopy has a lower detection limit, at several hundreds of nanometers, and confocal Raman microscopy has successfully been used to identify polymers down to 100 nm. To achieve detection of smaller particles, instruments need to be combined with other techniques such as nano-FTIR, a combination of μ-FTIR and atomic force microscopy (Meyns et al. 2019; Huth et al. 2012). Raman can be combined with atomic force microscopy (Schmidt et al. 2005), scattering near-field optical microscope (Webster et al. 1998), and SEM (Zhang et al. 2020). All these techniques are promising and may be applied to environmental nanoplastics.

However, all techniques pose challenges with noise or interferences from the background—mostly organic matter—thus, thorough method development is needed.

7.4 Chapter Summary and Conclusion

Naturally occurring nanoparticles have been present and taking part in biogeochemical reactions in the environment since the formation of the Earth, and recent technological developments have allowed scientific studies to improve our understanding on these processes. Furthermore, industrially manufactured and incidental nanoparticles will continue to appear in the marine environment in increasing numbers, as new nano-enabled technologies and consumer products are developed. The role of nanoparticles in the biogeochemical cycling of elements and impacts on water quality is demonstrated through the multitude and complexity of their transport and reactivity. The scarcity of studies focusing on the marine environment compared to the literature available for land waters, such as lakes and rivers, is not in line with the importance of marine ecosystems in global aspirations for blue growth and ecosystem preservation. Examples of iron oxides, titania, and plastic nanoparticles presented here highlight that there is an urgent need to further improve and harmonize scientific tool kits for studying physicochemical properties of natural and anthropogenic nanoparticles in the environment, their spread, impact on biota, and interactions with organic macromolecules.

One priority is clearly protocols for sample collection, contamination control, pre-concentration, and fractionation. Although such methods are reported in the literature, harmonization and standardization are vital for application in marine waters with varying organic matter content, salinity, and oxygen concentration. A second priority is to improve technological capacity and analytical methods for identifying and characterizing nanoparticles in such complex samples. Electron microscopy, spICPMS, and Raman spectroscopy are widely applicable for environmental samples and produce relevant parameters for adequately characterizing nanoparticles. However, further improvements and method standardization are still lacking. Lastly, it is critical to apply regular monitoring programs for studying the spread and impact of nanomaterials in the marine environment. Since a sampling campaign will offer a snapshot of a dynamic system, it is necessary to conduct multiple campaigns in order to obtain a holistic understanding of mechanisms involved behind the observed phenomena. The task is extremely challenging given that seas and oceans cover more than 70 % of the Earth's surface and may extend to more than 10,000 m in depth. In addition, the multitude of socioeconomic activities directly or indirectly impacting the marine ecosystem complicate physical and chemical characteristics of nanoparticle populations in seawater. Upon building on standardized protocols for sample collection, pre-treatment, and analysis, regular monitoring programs may be envisioned in the near future.

Note: Julián Gallego reports that is currently employed at Horiba Scientific Sweden, but the research reported in the enclosed document is not part of that relationship.

References

Aiken GR, Hsu-Kim H, Ryan JN (2011) Influence of dissolved organic matter on the environmental fate of metals, nanoparticles, and colloids. Environ Sci Technol 45(8):3196–3201

Asmathunisha N, Kathiresan K (2013) A review on biosynthesis of nanoparticles by marine organisms. Colloids Surfaces B Biointerfaces 103:283–287

Auffan M, Rose JJ, Bottero J-YY, Lowry GV, Jolivet J-PP, Wiesner MR (2009) Towards a definition of inorganic nanoparticles from an environmental, health and safety perspective. Nat Nanotechnol 4(10):634–641. Available from: https://www.nature.com/articles/nnano.2009.242

Banfield JF, Zhang H (2001) Nanoparticles in the environment. Rev Mineral Geochemistry 44(1):1–58

Birdi KS (2002) Handbook of surface and colloid chemistry, 2nd edn. CRC Press LLC, Boca Raton

Bligh MW, Waite TD (2011) Formation, reactivity, and aging of ferric oxide particles formed from Fe(II) and Fe(III) sources: implications for iron bioavailability in the marine environment. Geochim Cosmochim Acta 75(24):7741–7758

Botta C, Labille J, Auffan M, Borschneck D, Miche H, Cabié M et al (2011) TiO2-based nanoparticles released in water from commercialized sunscreens in a life-cycle perspective: structures and quantities. Environ Pollut 159(6):1543–1550

Caballero Díaz E, Pfeiffer C, Kastl L, Rivera-Gil P, Simonet B, Valcárcel M et al (2013) The toxicity of silver nanoparticles depends on their uptake by cells and thus on their surface chemistry. Part Part Syst Charact [Internet] 30(12):1079–1085. Available from: https://onlinelibrary.wiley.com/doi/full/10.1002/ppsc.201300215

Cai H, Chen M, Chen Q, Du F, Liu J, Shi H (2020) Microplastic quantification affected by structure and pore size of filters. Chemosphere 257:127198

Callegaro S, Minetto D, Pojana G, Bilanicová D, Libralato G, Volpi Ghirardini A et al (2015) Effects of alginate on stability and ecotoxicity of nano-TiO2 in artificial seawater. Ecotoxicol Environ Saf 117:107–114

Chen CS, Anaya JM, Zhang S, Spurgin J, Chuang CY, Xu C et al (2011) Effects of engineered nanoparticles on the assembly of exopolymeric substances from phytoplankton. PLoS One 6(7): e21865

Chen KL, Mylon SE, Elimelech M (2006) Aggregation kinetics of alginate-coated hematite nanoparticles in monovalent and divalent electrolytes. Environ Sci Technol 40(5):1516–1523. Available from: https://pubs.acs.org/doi/full/10.1021/es0518068

Chen KL, Mylon SE, Elimelech M (2007) Enhanced aggregation of alginate-coated iron oxide (hematite) nanoparticles in the presence of calcium, strontium, and barium cations. Langmuir 23(11):5920–5928. Available from: https://pubs.acs.org/doi/full/10.1021/la063744k

Cole M, Lindeque P, Halsband C, Galloway TS (2011) Microplastics as contaminants in the marine environment: a review. Mar Pollut Bull 62(12):2588–2597

Deonarine A, Lau BLT, Aiken GR, Ryan JN, Hsu-Kim H (2011) Effects of humic substances on precipitation and aggregation of zinc sulfide nanoparticles. Environ Sci Technol 45(8): 3217–3223

El Hadri H, Hackley VA (2017) Investigation of cloud point extraction for the analysis of metallic nanoparticles in a soil matrix. Environ Sci Nano 4(1):105–116. Available from: https://pubs.rsc.org/en/content/articlehtml/2017/en/c6en00322b

Garg S, Xing G, Waite TD (2020) Influence of pH on the kinetics and mechanism of photoreductive dissolution of amorphous iron oxyhydroxide in the presence of natural organic matter:

implications to iron bioavailability in surface waters. Environ Sci Technol 54(11):6771–6780. https://doi.org/10.1021/acs.est.0c01257

Giddings JC (1993) Field-flow fractionation: analysis of macromolecular, colloidal, and particulate materials. Science (80-) 260(5113):1456–1465. Available from: https://www.science.org/doi/abs/10.1126/science.8502990

Goldstein MC, Titmus AJ, Ford M (2013) Scales of spatial heterogeneity of plastic marine debris in the Northeast Pacific Ocean. PLoS One 8(11):e80020. Available from: https://journals.plos.org/plosone/article?id=10.1371/journal.pone.0080020

Gondikas A, Gallego-Urrea J, Halbach M, Derrien N, Hassellöv M (2020) Nanomaterial fate in seawater: a rapid sink or intermittent stabilization? Front Environ Sci 8:151. Available from: https://www.frontiersin.org/article/10.3389/fenvs.2020.00151/full

Gondikas A, Morris A, Reinsch BCBC, Marinakos SMSM, Lowry GVGV, Hsu-Kim H (2012) Cysteine-induced modifications of zero-valent silver nanomaterials: implications for particle surface chemistry, aggregation, dissolution, and silver speciation. Environ Sci Technol 46(13): 7037–7045

Gondikas A, Von Der Kammer F, Kaegi R, Borovinskaya O, Neubauer E, Navratilova J et al (2018) Where is the nano? Analytical approaches for the detection and quantification of TiO2 engineered nanoparticles in surface waters. Environ Sci Nano 5(2):313–326

Guo L, Santschi PH (1996) A critical evaluation of the cross-flow ultrafiltration technique for sampling colloidal organic carbon in seawater. Mar Chem 55(1–2):113–127

Hassellöv M (2005) Relative molar mass distributions of chromophoric colloidal organic matter in coastal seawater determined by flow field-flow fractionation with UV absorbance and fluorescence detection. Mar Chem 94(1–4):111–123

Hassellöv M, Kaegi R (2009) Analysis and characterization of manufactured nanoparticles in aquatic environments. Environ Hum Heal Impacts Nanotechnol:211–266. Available from: https://onlinelibrary.wiley.com/doi/full/10.1002/9781444307504.ch6

Hassellöv M, von der Kammer F (2008) Iron oxides as geochemical nanovectors for metal transport in soil-river systems. Elements 4(6):401–406

Hjorth R, Coutris C, Nguyen NHA, Sevcu A, Gallego-Urrea JA, Baun A et al (2017) Ecotoxicity testing and environmental risk assessment of iron nanomaterials for sub-surface remediation–recommendations from the FP7 project NanoRem. Chemosphere 182:525–531

Hochella MF (2002) There's plenty of room at the bottom: nanoscience in geochemistry. Geochim Cosmochim Acta 66(5):735–743

Hochella MF, Mogk DW, Ranville J, Allen IC, Luther GW, Marr LC et al (2019) Natural, incidental, and engineered nanomaterials and their impacts on the earth system. Science 363(6434):eaau8299

Hunnestad AV, Vogel AIM, Digernes MG, Van Ardelan M, Hohmann-Marriott MF (2020) Iron speciation and physiological analysis indicate that *Synechococcus* sp. Pcc 7002 reduces amorphous and crystalline iron forms in synthetic seawater medium. J Mar Sci Eng 8(12):1–18

Huth F, Govyadinov A, Amarie S, Nuansing W, Keilmann F, Hillenbrand R (2012) Nano-FTIR absorption spectroscopy of molecular fingerprints at 20 nm spatial resolution. Nano Lett 12(8): 3973–3978. Available from: https://pubs.acs.org/sharingguidelines

Jassby D, Budarz JF, Wiesner M (2012) Impact of aggregate size and structure on the photocatalytic properties of TiO 2 and ZnO nanoparticles. Environ Sci Technol 46(13):6934–6941

Kaegi R, Englert A, Gondikas A, Sinnet B, von der Kammer F, Burkhardt M (2017) Release of TiO2–(Nano) particles from construction and demolition landfills. Nano Impact 8:73–79

Kalogerakis N, Karkanorachaki K, Kalogerakis GC, Triantafyllidi EI, Gotsis AD, Partsinevelos P et al (2017) Microplastics generation: onset of fragmentation of polyethylene films in marine environment mesocosms. Front Mar Sci 4(MAR):84

Kent RD, Vikesland PJ (2012) Controlled evaluation of silver nanoparticle dissolution using atomic force microscopy. Environ Sci Technol 46(13):6977–6984. Available from: https://pubs.acs.org/doi/abs/10.1021/es203475a

Kim B, Park CS, Murayama M, Hochella MF (2010) Discovery and characterization of silver sulfide nanoparticles in final sewage sludge products. Environ Sci Technol 44(19):7509–7514

Mattsson K, Björkroth F, Karlsson T, Hassellöv M (2021b) Nanofragmentation of expanded polystyrene under simulated environmental weathering (thermooxidative degradation and hydrodynamic turbulence). Front Mar Sci 7:1252. Available from: https://www.frontiersin.org/articles/10.3389/fmars.2020.578178/full

Mattsson K, da Silva VH, Deonarine A, Louie SM, Gondikas A (2021a) Monitoring anthropogenic particles in the environment: recent developments and remaining challenges at the forefront of analytical methods. Curr Opin Colloid Interface Sci 56:101513

Meyns M, Primpke S, Gerdts G (2019) Library based identification and characterisation of polymers with nano-FTIR and IR-sSNOM imaging. Anal Methods 11(40):5195–5202. Available from: https://pubs.rsc.org/en/content/articlehtml/2019/ay/c9ay01193e

Miller RJ, Bennett S, Keller AA, Pease S, Lenihan HS (2012) TiO2 nanoparticles are phototoxic to marine phytoplankton. PLoS One 7(1):e30321. Available from: https://journals.plos.org/plosone/article?id=10.1371/journal.pone.0030321

Morelli E, Gabellieri E, Bonomini A, Tognotti D, Grassi G, Corsi I (2018) TiO2 nanoparticles in seawater: aggregation and interactions with the green alga *Dunaliella tertiolecta*. Ecotoxicol Environ Saf 148:184–193

Pfeiffer F, Fischer EK (2020) Various digestion protocols within microplastic sample processing– evaluating the resistance of different synthetic polymers and the efficiency of biogenic organic matter destruction. Front Environ Sci 8:263

Quigg A, Santschi PH, Burd A, Chin WC, Kamalanathan M, Xu C et al (2021) From nano-gels to marine snow: a synthesis of gel formation processes and modeling efforts involved with particle flux in the ocean. Gels 7(3):114

Raiswell R, Canfield DE (2012) The iron biogeochemical cycle past and present. Geochemical Perspect 1(1):1–232

Raiswell R, Tranter M, Benning LG, Siegert M, De'ath R, Huybrechts P et al (2006) Contributions from glacially derived sediment to the global iron (oxyhydr)oxide cycle: implications for iron delivery to the oceans. Geochim Cosmochim Acta 70(11):2765–2780

Santschi PH, Chin WC, Quigg A, Xu C, Kamalanathan M, Lin P et al (2021) Marine gel interactions with hydrophilic and hydrophobic pollutants. Gels 7(3):83

Schmidt U, Hild S, Ibach W (2005) Hollricher O. Characterization of thin polymer films on the nanometer scale with confocal Raman AFM. Macromol Symp 230(1):133–143. https://doi.org/10.1002/masy.200551152

Shuster J, Reith F, Cornelis G, Parsons JE, Parsons JM, Southam G (2017) Secondary gold structures: relics of past biogeochemical transformations and implications for colloidal gold dispersion in subtropical environments. Chem Geol 450:154–164

Ter Halle A, Ladirat L, Gendre X, Goudouneche D, Pusineri C, Routaboul C et al (2016) Understanding the fragmentation pattern of marine plastic debris. Environ Sci Technol 50(11):5668–5675. Available from: https://pubs.acs.org/doi/full/10.1021/acs.est.6b00594

Tiselius P, Kuylenstierna M (1996) Growth and decline of a diatom spring bloom: phytoplankton species composition, formation of marine snow and the role of heterotrophic dinoflagellates. J Plankton Res 18(2):133–155

Toncelli C, Mylona K, Tsapakis M, Pergantis SA (2016) Flow injection with on-line dilution and single particle inductively coupled plasma-mass spectrometry for monitoring silver nanoparticles in seawater and in marine microorganisms. J Anal At Spectrom 31(7): 1430–1439. Available from: www.rsc.org/jaas

Turner DR, Hunter KA (2001) The biogeochemistry of iron in seawater. Wiley

Vance ME, Kuiken T, Vejerano EP, SP MG, Hochella MF, Hull DR (2015) Nanotechnology in the real world: redeveloping the nanomaterial consumer products inventory. Beilstein J Nanotechnol 6(1):1769–1780. Available from: https://www.beilstein-journals.org/bjnano/articles/6/181

Verboket PE, Borovinskaya O, Meyer N, Dittrich PS (2014) A new microfluidics-based droplet dispenser for ICPMS. Anal Chem 86:6012–6018

Von Der Heyden B, Roychoudhury A, Myneni S (2019) Iron-rich nanoparticles in natural aquatic environments. Miner 9:287. Available from: https://www.mdpi.com/2075-163X/9/5/287/htm

Wagner S, Gondikas A, Neubauer E, Hofmann T, von der Kammer F (2014) Spot the difference: engineered and natural nanoparticles in the environment-release, behavior, and fate. Angew Chem Int Ed Engl 53:12398–12419

Wagner S, Navratilova J, Gondikas A (2019) Sample preparation for the analysis of nanomaterials in water. Encycl Water:1–10. Available from: https://onlinelibrary.wiley.com/doi/full/10.1002/9781119300762.wsts0078

Wang J, Zhao X, Wu F, Tang Z, Zhao T, Niu L et al (2021) Impact of montmorillonite clay on the homo- and heteroaggregation of titanium dioxide nanoparticles (nTiO2) in synthetic and natural waters. Sci Total Environ 784:147019

Webster S, Smith DA, Batchelder DN (1998) Raman microscopy using a scanning near-field optical probe. Vib Spectrosc 18(1–3):51–59

Williams RJ, Harrison S, Keller V, Kuenen J, Lofts S, Praetorius A et al (2019) Models for assessing engineered nanomaterial fate and behaviour in the aquatic environment. Curr Opin Environ Sustain 36:105–115

Yücel M, Gartman A, Chan CS, Luther GW (2011) Hydrothermal vents as a kinetically stable source of iron-sulphide-bearing nanoparticles to the ocean. Nat Geosci 4(6):367–371. Available from: https://www.nature.com/articles/ngeo1148

Zettler ER, Mincer TJ, Amaral-Zettler LA (2013) Life in the "plastisphere": Microbial communities on plastic marine debris. Environ Sci Technol 47(13):7137–7146. Available from: https://pubs.acs.org/doi/full/10.1021/es401288x

Zhang W, Dong Z, Zhu L, Hou Y, Qiu Y (2020) Direct observation of the release of nanoplastics from commercially recycled plastics with correlative Raman imaging and scanning electron microscopy. ACS Nano 14(7):7920–7926. https://doi.org/10.1021/acsnano.0c02878

Zhang T, Kim B, Levard CC, Reinsch BC, Lowry GV, Deshusses MA et al (2012) Methylation of mercury by bacteria exposed to dissolved, nanoparticulate, and microparticulate mercuric sulfides. Environ Sci Technol 46(13):6950–6958

Microplastics and Nanoplastics

8

Lucia Pittura, Stefania Gorbi, Carola Mazzoli, Alessandro Nardi, Maura Benedetti, and Francesco Regoli

Contents

Abstract

Microplastics have become a constant and ubiquitous component of the marine environment, being found in water surface, along water column, and in sediments, beaches, and organisms worldwide. Assessing microplastics in the environment is necessary to understand sources, distribution, abundance, and ecological consequences for marine ecosystems, especially considering that microplastic contamination is expected to increase in years to come in view of increasing global annual primary plastic production.

This chapter intends to provide tools to approach the issue of microplastics as consciously as possible, including the importance to consider microfibers as a category per se. Definitions of microplastics and microfibers will be provided along with a brief discussion of elements affecting their distribution in the marine environment and how they can cause biological effects. The focus of the chapter is, however, on the most common and suitable sampling strategies, analytical approaches, and standard requirements, for conducting a reliable assessment of microplastics in marine matrices.

L. Pittura · S. Gorbi · C. Mazzoli · A. Nardi · M. Benedetti · F. Regoli (✉)
Department of Life and Environmental Sciences, Polytechnic University of Marche, Ancona, Italy
e-mail: l.pittura@staff.univpm.it; s.gorbi@staff.univpm.it; c.mazzoli@pm.univpm.it; a.nardi@staff.univpm.it; m.benedetti@staff.univpm.it; f.regoli@staff.univpm.it

© The Author(s), under exclusive license to Springer Nature Switzerland AG 2023
J. Blasco, A. Tovar-Sánchez (eds.), *Marine Analytical Chemistry*,
https://doi.org/10.1007/978-3-031-14486-8_8

Keywords

Microplastics · Nanoplastics · Microfibers · Water · Sediment · Biota · Sampling · Characterization · μ-FTIR · Raman spectroscopy

8.1   Introduction

Microplastics have become a constant and ubiquitous component of the marine environment, being found in water surface, along water column, and in sediments and organisms worldwide (Qu et al. 2018). Microplastic contamination is expected to increase in years to come, especially in view of increasing global annual primary plastic production that would reach 1.1 billion tons in 2050 and with a total cumulative production between 1950 and 2050 of 34 billion tons, none of which would be biodegradable (Geyer 2020). Assessing microplastics in the environment is, thus, necessary to understand sources, distribution, abundance, and ecological consequences for marine ecosystems. This problem is also on the agenda of national and international organizations worldwide, asking experts to use their knowledge to compile recommendations and guidelines for routinely monitoring programs (van Bavel et al. 2020).

This chapter intends to provide tools to approach the issue of microplastics as consciously as possible, firstly, describing what constitute a microplastic, what kind of other categories exist, how microplastics originate, and their possible fate in the field and, secondly, focusing on the most suitable sampling strategies, analytical approaches, and standard requirements for a reliable assessment of microplastics in marine matrices.

International consensus has not yet been reached on a definition and categorization of microplastics, often resulting in an ambiguous comparison of data (Hartmann et al. 2019). To date, it is globally accepted that the lower size limit for microplastics is 1 μm (Frias et al. 2018), while, under this value and down to 1 nm, plastics are defined as nanoplastics (Gigault et al. 2018). The upper size limit of microplastics is, instead, still under debate. Frias and Nash (2019) proposed 5 mm, whereas other authors suggest to restrict the definition of microplastics to particles smaller than 1 mm based on the International System of Units definition for "micro" (Cole et al. 2011).

Microplastics and nanoplastics can have a primary origin, when they are intentionally produced in the micro-nanometer size range to be directly used in a wide range of applications (e.g., personal care products and cosmetics, abrasive powders, powders for injection molds, and 3D printing). Otherwise, secondary microplastics represent results of weathering or fragmentation of larger objects either during use or following loss to the environment (UNEP 2021).

Another criterion for defining microplastics is associated to shape. Also in this case, there is no standardized classification. BASEMAN consortium (whose goal was the validation and harmonization of analytical methods for microplastic analysis in environmental matrices) suggested eight categories based on the most common microplastic types described in peer-reviewed publications: (1) pellet, (2) fragment, (3) fiber, (4) film, (5) rope and filament, (6) microbead, (7) sponge/foam, and (8) rubber.

Despite that color is not considered to be crucial to define microplastics, recording this characteristic is considered important for studies concerning aquatic organisms, as some species are thought to potentially ingest microplastics based on a color preference behavior (Frias and Nash 2019).

Although textile fibers are broadly classified as a shape of microplastics, they were suggested as a category of their own. Compared to other types of microplastics, they are much more abundant in the environment, and they are different in source (textile vs. common use), polymer typology (often natural vs. synthetic only), and mitigation actions (industry vs. public awareness, Avio et al. 2020). These differences of microplastics and microfibers are highlighted by definitions given by Bessa et al. (2019) and Liu et al. (2019). Microplastics are synthetic solid particles of polymeric matrix, with size ranging from 1 µm to 5 mm, consisting of either items that are manufactured to be of microscopic dimensions (primary) or that are formed from the weathering and fragmentation of larger plastic waste items, which are insoluble in water at 20 °C (Bessa et al. 2019). Microfibers are natural or artificial fibrous materials of threadlike structure with a diameter lower than 50 µm, length ranging from 1 µm to 5 mm, and length to diameter ratio greater than 100. Microfibers are released or shed to the environment from all kinds of fibrous materials, such as clothes; agricultural, industrial, and home textiles; and some textile products, semimanufactured goods, or accessories used in other fields, during production, use, and end-of-life disposal (Liu et al. 2019).

Distribution and accumulation of microplastics in aquatic ecosystems are widely dependent on particle characteristics such as size, shape, density, and chemical composition along with environmental parameters including wind, temperature, and water current velocity (Gola et al. 2021). Hydrodynamic processes, coastal currents, drift, and river outflow act to disperse microplastics from their sources. Additionally, rotational ocean currents transport surface plastics to convergence zones of oceanic gyres, leading to concentrated areas of accumulation (Coyle et al. 2020). Several factors potentially influence the vertical distribution of plastics along the water column, such as wind-induced mixing, incorporation into marine aggregates or fecal matter, biofouling (Cole et al. 2016), size and shape of materials, and relative density that might vary with additives added during production (Reisser et al. 2015, Table 8.1). For plastic denser than seawater, the shape and the near-bottom current velocity magnitude strongly define the settling (Bagaev et al. 2017).

Due to their small size and widespread occurrence, microplastics can be ingested by wide range of marine organisms causing a range of effects like mechanical damages, attachment of polymers to external surfaces, hindering of mobility and clogging of the digestive tract, inflammation, cellular stress, and decreased growth (Setälä et al. 2016). The chemical impact can be related to additives present in the plastic from manufacturing, as well as to the environmental contaminants which are adsorbed by the hydrophobic nature and high surface-to-volume ratio of microplastics. There is, however, an active debate regarding the toxicological

Table 8.1 Buoyancy of most common polymer in seawater (from Bessa et al. 2019)

Abbreviation	Polymer	Density (g cm^{-3})	Buoyancy of polymer in seawater (1.025 g cm^{-3})
PS	Polystyrene	0.01–1.06	Positive
PP	Polypropylene	0.85–0.92	Positive
LDPE	Low-density polyethylene	0.89–0.93	Positive
EVA	Ethylene vinyl acetate	0.93–0.95	Positive
HDPE	High-density polyethylene	0.94–0.98	Negative
PA	Polyamide	1.12–1.15	Negative
PA 66	Nylon 66	1.13–1.15	Negative
PMMA	Polymethyl methacrylate	1.16–1.20	Negative
PC	Polycarbonate	1.20–1.22	Negative
PU	Polyurethane	1.20–1.26	Negative
PET	Polyethylene terephthalate	1.38–1.41	Negative
PVC	Polyvinyl chloride	1.38–1.41	Negative
PTFE	Polytetrafluoroethylene	2.10–2.30	Negative

relevance of adsorbed pollutants on microplastics and their possible transfer to marine organisms due to the variability of experimental results (Benedetti et al. 2022). Despite that a detailed review of fate, distribution, and biological effects of microplastics is outside the aim of this chapter, these issues are fundamental for a comprehensive risk assessment in the marine environment.

8.2 Sampling the Marine Environment for Microplastic Detection

Monitoring microplastics in the environment is based on sequence of steps starting with the collection of appropriate samples, followed by isolation of particles from the matrix and their physical and chemical characterization.

The sampling strategy always depends on the aim of the research, while sampling devices are related to the matrix to be collected, the size limit of microplastics to be targeted, and available equipment for immediate processing.

Water, sediment, and biota can be collected as bulk or volume-reduced samples (Stock et al. 2019). In the first case, the whole collected volume is taken without any reduction during the sampling process, theoretically, allowing to capture all microplastics regardless of their size and visibility. Volume-reduced samples are subjected, during the sampling, to specific treatments (i.e., filtration, separation, and concentration) to preserve only portions of samples that will be further processed in a laboratory (Wang and Wang 2018). Selective sampling consists of direct collection of microplastics recognizable by the naked eye, and it is applicable only to large microplastics (1–5 mm, e.g., plastic pellet on beaches) (Hidalgo-Ruz et al. 2012).

8.2.1 Sampling of Seawater

Different typologies of devices are used to collect microplastics in seawater, categorized in non-discrete sampling devices for volume-reduced samples and discrete devices for bulk samples (Cutroneo et al. 2020). Depending on the device, it is possible to collect different layers of the water compartment: the upper 30 cm or "surface waters" (Han et al. 2020), including the air-water interface or directly below the interface; the "subsurface layers" still within the upper mix layers affected by winds, surface currents, and vessel movements; and the deeper layers of water column (van Bavel et al. 2020). The choice of device depends on available equipment and characteristics of sampling location, such as turbulence at the surface, hydrodynamic profile, and depth, besides the focus of the research. Irrespective of the sampling method, environmental metadata should always be collected to support data interpretation including bathymetry, water temperature, salinity, water currents, surface wind, and weather conditions (van Bavel et al. 2020).

Nets originally designed for plankton are widely used for collecting microplastics (Fig. 8.1a). They consist of a funnel shape mesh, generally made of nylon and 1.5–4.5 m long, with a circular or a rectangular mouth maintained constantly opened by a rigid frame and connected to a final cylinder called code-end. The main advantage is the possibility to filter large volumes of water concentrating and gathering microplastics in the code-end. There are various mesh sizes for nets possibly ranging from 20 µm to 5000 µm, despite that those of 300 µm are the most commonly used to facilitate comparison among studies. However, 300-µm mesh fails to depict the overall pattern of microplastic pollution, since it does not retain microfibers and smaller microplastics (Tamminga et al. 2019), which are both of great biological relevance. However, nets with lower mesh sizes can easily get clogged up, compromising the sampling process (Löder and Gerdts 2015).

Plankton nets are used for both vertical and horizontal sampling of the water column, allowing the collection of replicate samples in the bongo conformation, with two plankton nets connected through a couple of circular aluminum frames

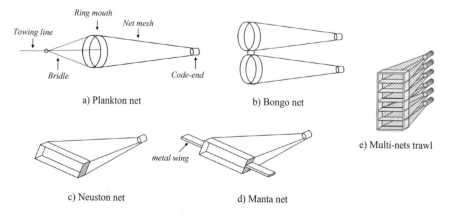

Fig. 8.1 Most common typologies of net devices for sampling microplastics in seawater column (**a**, **b**, and **e**) and on surface and subsurface waters (**c** and **d**, modified from van Bavel et al. 2020)

(Fig. 8.1b). In horizontal sampling, a weight (about 10–20 kg) is fixed to the net, then slowly dropped to the maximum depth (avoiding contact with the bottom), and trawled obliquely at a speed of less than 2 knots to let the water pass through the net with a steady flow. In vertical sampling, the net is lifted toward the surface from a specific depth, thus sampling the entire water column (Campanale et al. 2020).

Manta and neuston nets have frames of rectangular shape (Fig. 8.1c–d) and are the most suitable for surface and subsurface water sampling: nets are deployed from the ship and towed horizontally along horizontal transect for specific periods, varying from a minimum of 10 min (Setälä et al. 2016) to a maximum of 240 min (Pan et al. 2019), at a known speed (between 1 and 5 knots, Cutroneo et al. 2020). Manta nets have two buoys or, more typically, two metal wings equipped on the sides of the frame which give the appearance of a manta ray, to ensure stability and keep the net floating on the surface. Manta nets can thus maintain a constant immersion depth under the sea surface, and the filtered water volume can be estimated accurately. Neuston nets are kept at the surface by floats or suspended beneath the water's surface, filtering the surface layer even in the presence of waves, despite that the volume is difficult to be estimated accurately because the net's immersion depth changes constantly (van Bavel et al. 2020). Neuston nets, mounted one above the other as multi-net trawl, also provide the possibility to synchronously sample various layers of the water column (Fig. 8.1e).

After the sampling, nets must be carefully rinsed from the exterior to assure that plastic debris are washed into the collector and to clean the net before the next sampling. The material within the collector is finally transferred to a sample container for subsequent processing in laboratory (Fig. 8.2): samples can be fixed with plastic-friendly fixatives (e.g., formalin) to preserve the biological component or stored frozen if microplastics are the only target parameters of the study.

During the trawling, it is fundamental to measure the volume of the filtered water and to normalize the concentration of microplastics per volume unit. For this reason, a flow meter is often present on the net opening (Stock et al. 2019); as an alternative, the amount of filtered water can be calculated by the net opening size and trawl distance, the latter easily obtained using smartphone applications for GPS tracking during the sampling.

In addition to nets, pumping systems represent non-discrete sampling devices allowing to filtrate in situ large volumes of water, from surface up to a depth of 100 m (Cutroneo et al. 2020). Pumping systems are equipped with a flow meter to determine the volume of filtered water. Microplastics are collected through a series of filters/sieves, available in different mesh sizes to separate microplastics into different dimensional classes at the time of sampling. Filters can be recovered in petri dishes and preserved until laboratory analysis, whereas sieves need to be rinsed with decontaminated water (e.g., microfiltered water) and microplastics collected and preserved in a glass jar. Some sieves are designed with the possibility to disassemble the frame and mount a clean filter mesh. The first filter/sieve of the filtration unit is often of 5-mm mesh size used to remove larger plastic particles, and a 300-μm mesh filter/sieve is typically included in the battery for comparing results with those obtained with nets (Setälä et al. 2016; Tamminga et al. 2019; Rist et al.

Fig. 8.2 Net-based method for sampling microplastics in surface waters. Activities carried out by the Polytechnic University of Marche (Italy) in the Adriatic Sea using a manta net during activities of RESPONSE project funded by JPI Oceans, 2020–2023 (photo credit, L. Pittura and C. Mazzoli)

2020; Karlsson et al. 2020; Schönlau et al. 2020). Such comparisons revealed that in situ pump filtration methods are more accurate in volume measurement and versatile for point sampling and filter size choice, enabling standardization of sampling (Razeghi et al. 2021); in addition, pump filtration allows to collect smaller microplastics and more significant sampling of microfibers compared to nets (Campanale et al. 2020). On the other side, pump systems are more expensive, and trawling methods can cover larger areas, thus better overcoming some of the problems related to patchiness (Karlsson et al. 2020). Net trawling and pump sampling methods can be considered complementary techniques providing a more comprehensive approach for monitoring microplastics in water compartment (Tamminga et al. 2019; Razeghi et al. 2021).

Diverse pumping systems are commercially available or custom made. They can pump water from a specific depth that is directed to the filtering system outside the water or can work directly submersed. Submersible pumps can be lowered from the vessel using a winch sideways or toward the stern of the ship (Cutroneo et al. 2020). Depending on the technical specification of pumps, different volumes can be sampled, diverse depths can be covered, and filtration time can vary from several hours to a few minutes. For example, a new custom-made plastic-free pump-filter system (UFO system—Universal Filtering Objects system) was applied for collecting microplastics down to 10 μm in the Arctic, filtering approximately 1 m^3 of water from 5-m depth (Rist et al. 2020). The device is composed of a metal hose deployed in the water, a pump controlled by an inverter, and a modular filtering device. The mouth of the hose is equipped with a stainless-steel metal cage of 5-mm mesh to protect the system against large debris. The filtering unit consists of three parts: the water first passes through a filter of 300-μm mesh to retain larger items with the purpose of protecting the finer filtering mesh from clogging. The water is then divided onto two parallel units with filters of 10 μm. Outlets were recombined and connected to a mechanical flowmeter to quantify the filtered volume (Rist et al. 2020, Fig. 8.3). Preston-Whyte et al. (2021) used a Micro Plastic Particle Pump developed by KC Denmark for sampling near surface water (0.5-m depth) in a port environment. Using a crane, the particle pump was deployed from the quayside and held submerged. During deployment, 2000 L of pumped water was directed through stacked sieves (5 mm, 500 μm, 300 μm, and 200 μm). On recovery, sieves were individually removed, and all particulate matter on the sieve was transferred to a 1-L glass jar (Fig. 8.4a). Actually, the pump system of KC Denmark enables to sample down to 40 m, and a flexible range of filter sizes is available (ranging from 1000 μm down to 20 μm). The same model of pump system used by Preston-Whyte et al. (2021) was, in fact, used by the Polytechnic University of Marche (Italy) to sample microplastics in water column up to 30 m along the Adriatic coast (Mediterranean Sea), within activities of the RESPONSE JPI Oceans project: sieves were individually removed and stored in glass petri dishes for subsequent processing in the laboratory (Fig. 8.4b).

Discrete sampling devices can be used to collect defined volumes of water from specific depths. The main advantage of collecting bulk samples is that, theoretically, all present microplastics can be sampled without any size limitation, preventing any loss possibly occurring in volume-reduced samples. This sampling is rapid and reduces the risk of contamination, due to the short handling and sample exposure to the surrounding environment. The main disadvantage consists of the limited amount of samples that can be collected, stored, and processed (Campanale et al. 2020).

Discrete sampling devices can include a plexiglass water sampler, a rosette sampler system (CTD [conductivity-temperature-depth] sampler), and the lander system (Fig. 8.5). The plexiglass water sampler is particularly useful and convenient for microplastic collection in shallow waters with weak currents, while it would not be applicable to deeper aquatic environments because it is usually made of acrylic materials and can be fragile when it is subjected to pressure. The rosette sampler system is typically comprised of a set of Niskin bottles (8–12 L) equipped with CTD sensors, and it can be adopted to collect water samples at various depths in marginal seas and pelagic zones up to 6800 m in depth. Niskin bottles can also be fitted on a lander system to collect bottom water near the seafloor (Liu et al. 2020).

Additional methods can be as simple as a glass bottle of 1 liter, used during various citizen science-driven projects to collect surface waters (van Bavel et al. 2020). Samples obtained with discrete devices are transferred to a jar, and the inside of the device is rinsed with decontaminated water to collect plastic particles that

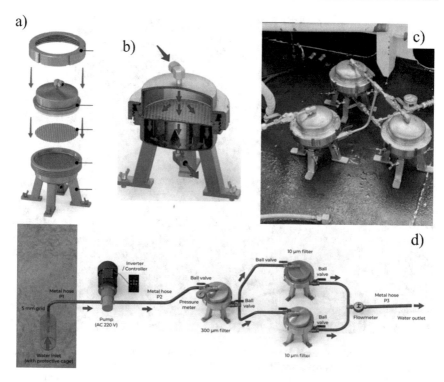

Fig. 8.3 UFO system as an example of custom-made pump-filter device for sampling microplastics in seawater: (**a**) mounting schematic of a single UFO unit, (**b**) cross section of a single UFO (arrows illustrate the water flow), (**c**) picture of the real setup operating on board during the survey, and (**d**) overall schematics of UFO setup (modified from Rist et al. 2020)

remain attached. The sample is preserved until laboratory analysis or filtered or sieved directly on board (Cutroneo et al. 2020).

8.2.2 Sampling of Sediments

Microplastics can be investigated in various typologies of sediments including those of intertidal beaches and subtidal seabed sediments. Different sampling approaches are obviously required, although it is always recommended to collect data associated to the sampling site, such as date of sampling, coordinates of location, morphological and hydrographic characteristics of the area, type of sediment, presence of macroplastics, local point sources (e.g., proximity to urban and/or industrial areas, to river streams and/or estuaries, and to wastewater treatment plants), or any other relevant information for interpretation of results. Exhaustive examples of datasheets to collect data while sampling intertidal and subtidal sediments are provided in the

Fig. 8.4 Micro Plastic Particle Pump (KC Denmark) as an example of submersible device for sampling microplastics in seawater. (**a**) Operation of the pump in a port environment and sample collection (modified from Preston-Whyte et al. 2021). (**b**) Operation of the pump and sample collection carried out by the Polytechnic University of Marche (Italy) along the coasts of Adriatic Sea during activities of RESPONSE project funded by JPI Oceans, 2020–2023 (photo credit, S. Gorbi and F. Regoli)

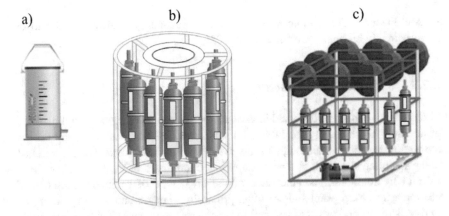

Fig. 8.5 Most common types of discrete sampling devices for sampling microplastics in seawater: (**a**) plexiglass water sampler, (**b**) CTD sampler, and (**c**) lander system (modified from Liu et al. 2020)

protocol for monitoring microplastics in sediments produced by Frias et al. (2018) within the BASEMAN JPI Oceans project.

8.2.2.1 Intertidal Sediments (Beaches)

Sampling sediments on beaches is relatively easy from a technical point of view, and it can be carried out by trained nonscientist operators with unsophisticated equipment: a nonplastic sampling tool (tablespoon, trowel, or small shovel and corers), a frame to define the sampling unit, a 5-mm metal sieve to exclude macro-debris directly at the beach, and a nonplastic container to store the sample (Fig. 8.6). The not trivial aspect is the monitoring design, the correct identification of sampling area (e.g., shoreline and above the strandline), the depth of sediments to be collected, and the frequency of surveys (Löder and Gerdts 2015). Beaches are dynamic environments, and distribution of microplastics can rapidly change according to the depositional regime and environmental characteristics like tides, currents, and winds (Hanke et al. 2013). Other critical issues are the number of replicates and the quantity of sediments to be collected (weight or volume), to ensure a representative sampling and accurate estimation of microplastic concentration (Prata et al. 2019). Sampling strategies for microplastics on beaches are summarized in guidelines of the UNEP (Cheshire et al. 2009), the OSPAR (Wenneker et al. 2010), the NOAA (Lippiatt et al. 2013), and the MSFD Technical Subgroup on Marine Litter (TSGML, Hanke et al. 2013).

Sediment collection is recommended to be performed on the strandline (top of the shore), where litter is more likely to accumulate, defining a transect of 100 m parallel to the water edge (Fig. 8.7); for heavily littered beaches, the transect can be reduced to 50 m. If the survey is extended to the whole beach, at least two transects shall be identified between water edges and above the strandline (back of the beach) with a minimum distance of 50 m. Along transects, sampling is performed within conventional areas of 30×30 cm or 50×50 cm marked through the use of a quadrat (sampling units): collection of the top 5 cm of sediment (total volume of

Fig. 8.6 Tools for sampling sediments on beaches for microplastics and sampling activities by high school students coordinated by researchers of the Polytechnic University of Marche, Italy (photo credit, L. Pittura)

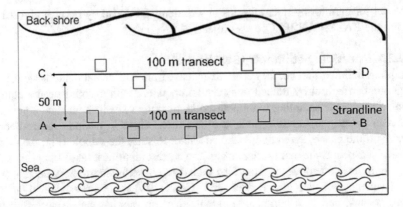

Fig. 8.7 Schematic representation of sampling strategy for microplastics on beach with transects and quadrants (modified from Frias et al. 2018)

approximately 4500 cm^3 = 4.5 L) is a common approach, and a minimum of five replicates, separated by at least 5 m, is recommended. Sampling units should be chosen in a random way to be representative of the sampling area (Fig. 8.7). Four surveys per year in spring (April), summer (mid-June to mid-July), autumn (mid-September to mid-October), and winter (mid-December to mid-January) are suggested, but a higher frequency may be initially necessary to identify significant seasonal patterns.

8.2.2.2 Subtidal Sediments (Seabed Sediments)

The most common devices to collect submerged sediments for microplastic analysis are grabs and corers (Fig. 8.8a–e), the choice of which depends on purpose of the survey, sediment typology, and various environmental and logistical constraints, such as water depth and vessel characteristics.

Grabs are easy to use and allow to collect samples in a relatively short time. The most common are van Veen, Ekman-Birge, and Shipek grabs. The van Veen grab is constituted by two jaws which close when these arrive at the bottom holding the sediment inside. Some models have upper doors to collect the upper centimeters of sediment directly inside the device. The Ekman-Birge grab has a box shape with upper windows and lower jaws; the smaller type can be manually operated by a rod that allows the insertion into the sediment. The Shipek grab has a bucket that rotates into the sediment when it reaches the sea bottom (Romano et al. 2018), facilitating the collection of difficult substrates (gravel or compacted sands).

Compared to grabs which retrieve disturbed sediments, core samplers collect cylindrical sections maintaining the sediment integrity from the surface to the deeper layers. This allows to determine historical profiles and inputs of microplastics as indicators of anthropogenic activity. As advantages, corers produce small volumes of sample and potentially enable to sample at a depth of more than 5000 m (Löder and Gerdts 2015). Among these devices, box corer (e.g., Reineck box corer) consists of a stainless-steel square box of variable size and a support frame that stabilizes the

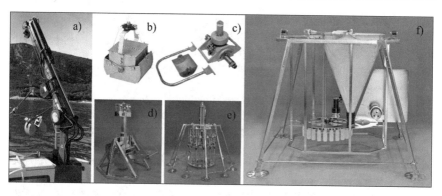

Fig. 8.8 Devices for sampling interdital sediments for microplastics: (**a**) van Veen grab (photo credit, S. Gorbi), (**b**) Ekman-Birge grab, (**c**) Shipek grab, (**d**) box corer, (**e**) multiple corer, and (**f**) sediment trap for microplastic vertical fluxes

sampler, ensuring the vertical penetration in the sea bottom. The box allows the recovery of nearly undisturbed samples of around 30 cm thick, and samples can be collected in a single device or a multiple corer constituted by 4 up to 12 core barrels (Romano et al. 2018).

Independently on the device, subsamples of ideally 250 ml of sediment should be collected to be representative of sampling location (Hanke et al. 2013). Samples are then placed into containers (preferably of nonplastic material) and stored frozen at $-20\,^{\circ}\text{C}$ in the dark if these are not immediately analyzed (Frias et al. 2018).

While analysis of microplastics in sediments can answer questions related to their presence, accumulation, and characteristics, the sinking process of microplastics along the water column can be monitored through the deployment of sediment traps (Saarni et al. 2021; Fig. 8.8f). These instruments, commonly used in oceanography to measure settling of organic and inorganic particles, consist of an upward-facing funnel that directs sinking particles toward a collection vessel. Traps typically operate over an extended period of time (weeks to months), and their series can be cycled to follow temporal changes in sinking flux with time, for instance, across a seasonal cycle.

8.2.3 Sampling of Biota

The selection of appropriate species is crucial to assess the ingestion of microplastics and their potential biological effects. A sampling design that includes multispecies, covering different trophic positions, habitat, and feeding strategies is preferable to a single-species approach. This strategy provides a more ecologically relevant assessment of microplastics in the marine environment facilitating comparisons among areas with different species but with similar trophic web structures (Avio et al. 2020). In general, the selected species should meet as many as possible of the following criteria:

(i) They occur naturally with high abundance and wide geographic distribution.
(ii) The biology and ecology are well known.
(iii) They are easy to sample and to process in laboratory.
(iv) They are already used as bioindicators in monitoring programs for other contaminants.
(v) The ingestion of microplastics is documented.
(vi) They have ecological relevance (i.e., key species in maintaining ecosystem functions) and commercial value (Fossi et al. 2018; Bray et al. 2019).

Bivalves fulfill most of such criteria and are abundant in intertidal and coastal locations worldwide. Adults are sessile organisms relatively easy to collect and process in laboratory conditions, having been used as bioindicators to monitor contaminants and marine environmental status for several years (e.g., Mussel Watch Programme): mussels have already been suggested as suitable sentinel species for microplastic pollution (Bessa et al. 2019). Also fish were highlighted as valuable indicators of the occurrence of microplastics in the marine environment, since they exploit almost every kind of habitat, occupy many ecological niches, and are an important food source for human populations worldwide (Sbrana et al. 2020). Other candidate species include benthic sediment-dwelling organisms (including marine worms) that are abundant and are widely used for biomonitoring contaminants in aquatic systems (Bessa et al. 2019). Based on such considerations, Fossi and collaborators (2018) proposed some species for monitoring different habitats (from coastal areas to offshore and from benthic environments to pelagic waters) at different spatial scales in the Mediterranean Sea. Mussels (*Mytilus galloprovincialis*), polychaetes (*Arenicola marina*), and crabs (*Carcinus* spp.) were highlighted as small-scale bioindicators of microplastics along the coastline, while red mullet (*Mullus barbatus*), sole (*Solea* spp.), European hake (*Merluccius merluccius*), and catshark (*Galeus melastomus*) as small-scale sea-bottom indicators being demersal fish living in close connection with sediments and depending on benthic prey for feeding. The bogue *Boops boops* and the pompano *Trachinotus ovatus* can be sentinel species of microplastics in coastal waters, while, for monitoring open waters at small scale, mesopelagic and pelagic fish (the European anchovy *Engraulis encrasicolus* and the European pilchard *Sardina pilchardus*) were suggested. Large pelagic predators *Thunnus alalunga* and *Coryphaena hippurus* were proposed as medium-scale bioindicators of microplastics in open waters, while, for basin-scale studies, large filter feeding whales (e.g., *Balaenoptera physalus*) and sharks (e.g., *Cetorhinus maximus*) were proposed as the most suitable for their migratory behavior.

Despite that the MSFD-TSGML recommends at least 50 individuals per species as a suitable sample size, the majority of published studies differs considerably (Wesch et al. 2016). Sample size could be lowered in those species or population showing high frequency of microplastic ingestion. Even though larger sample size provides more reliable results (Hermsen et al. 2018), as a practical suggestion, the number of individuals analyzed per single species could be reduced in favor of more species to be included in the study (Avio et al. 2020).

Due to the diversity of habitats and species, a large variety of techniques have been used for sampling biota. Organisms can be collected in grasps, traps, creels, or bottom trawling (benthic species), by manta or bongo nets (planktonic and nektonic invertebrates), by hand (e.g., bivalves or crustaceans), or by electrofishing (Stock et al. 2019). In addition to scientific sampling campaign, sources of samples for microplastic monitoring may include sportfishing events, farmed organisms, or animals bought at fish markets. All ethical requirements must be followed, i.e., avoiding protected/endangered species or invasive sampling methods (Bessa et al. 2019). The analysis of dead animals is particularly useful for obtaining data on microplastic ingestion by large marine vertebrates (e.g., seabird, turtles, and cetaceans) without killing individuals for scientific purposes (Hanke et al. 2013; Wesch et al. 2016).

Target tissues for analyses are mainly those of the digestive tracts (the esophagus, the stomach, and the gut) for larger biota, while whole specimens are usually analyzed for smaller species. Depending on the research question, additional tissues can be selected, for example, muscle (fillet) of commercial fish to evaluate the potential exposure to microplastics for humans. After collection, samples (whole organisms or dissected tissues) can be stored at −20 °C until further processing in laboratory: fixatives, like formalin, ethanol, or formaldehyde, have also been used to preserve microplastics. The time between organism collection and their dissection/storage should be as short as possible to avoid gut clearance (Lusher et al. 2017). During sampling/dissection of biota, it is advisable to record the following information: date and time of activity; type of equipment used for sampling; sampling site or location data (GPS coordinates, site name, depth, and environmental conditions); number and name of species collected from the same site; number of individuals for each species; length, weight, and sex of individuals; and additional observations (Bessa et al. 2019).

8.3 Sample Processing for Microplastic Isolation

In following sample collection, microplastics must be isolated from water, sediment, suspended matter, and organic material (Stock et al. 2019). Isolating microplastics in an appropriate manner is fundamental to achieve high extraction efficiencies, preservation of particles, and accurate data generation (Lusher et al. 2020). Sample processing includes three main steps: (1) organic matter digestion, (2) density separation, and (3) filtration. There are several methods to perform each of these steps that vary in complexity, time, and cost of materials, and the more appropriate choice often depends on the typology of samples.

8.3.1 Organic Matter Digestion

This digestion step is crucial for facilitating the extraction of microplastics from biological samples, and it is often applied to sediments and water samples as pre-treatment before density separation and/or filtration (Frias et al. 2018; Gago et al. 2018). The removal of organic matter must be performed without altering microplastics in terms of number, shape, polymer characteristics, and color (Bessa et al. 2019). Thus, the overall recommendation is the use of a previously tested digestion protocol, and the choice should be driven considering digestion efficiency, polymer resistance, dangerousness of reagents, and costs (Campanale et al. 2020). There are different types of digestion, including acid, alkaline, oxidative, and enzymatic, and all these methods have advantages and limitations (Lusher et al. 2020).

Nitric acid (HNO_3) is widely used in acid treatments: it is very effective in digesting the organic material present in the sample with >98 % weight loss for biological tissue. However, it causes degradation of polystyrene, polyamide, and polyethylene or change of color (polyvinyl chloride), especially if it is used in high concentrations and at high temperatures. Hydrochloric acid (HCl) has also been suggested, but it seems to be inefficient in treating large quantities of biologic material, and it causes alteration of some polymers. Therefore, acid digestion may be avoided or used with caution since it may lead to underestimation of microplastics in environmental samples (Lusher et al. 2020).

Alkaline digestion is an alternative with great potential, but it may also damage or discolor plastics, leave oily residues and bone fragments, or redeposit tissue debris on plastic surfaces, complicating the subsequent characterization (Prata et al. 2019). Nevertheless, digestion with 10 % potassium hydroxide (KOH) at a maximum temperature of 40 °C was suggested as a cost-effective method for processing biota samples (Bessa et al. 2019). It was recommended to not exceed the temperature of 40 °C since higher temperatures could degrade and reduce the recovery rate of some polymers. The limitation of proposed approach is that it is not particularly suitable for tissues with a considerable amount of fat, as the digestion may take several days and can partially damage some polymers. Combining KOH with a detergent (e.g., Tween 20) may accelerate the degradation process (Bessa et al. 2019). Digestion of 10 % KOH at 40 °C was also proposed as pre-treatment for seawater samples with medium and high organic matter content (water with eggs and larvae, plus zoo- and phytoplankton). In following this step, an additional step was recommended using hydrogen peroxide (H_2O_2) if all organic matter was not digested: H_2O_2 (15 % solution) at 1:1 volume sample/solution ratio is added to oxidize and digest the biological material (Gago et al. 2018).

H_2O_2 is indeed a popular oxidizing agent to efficiently remove organic matter with little effect on microplastic integrity if it is used at less than 20 %: polymeric changes have been identified in terms of transparency and shrinking in size when a 30–35 % solution is applied (Prata et al. 2019).

In contrast to the chemical digestion, the use of enzymes (e.g., proteinase K, trypsin, collagenase, papain, and cellulase) does not affect the polymer's structure and has an excellent removal efficiency of the organic fraction, but it takes more time and higher costs, especially for samples that contain a considerable amount of organic matter to digest and thus require a significant quantity of enzymes

(Campanale et al. 2020): in this respect, it is less affordable during extracting microplastics from large tissues or organisms.

8.3.2 Density Separation

Differences in density can be used to separate plastics from inorganic solid materials (e.g., sand, shells and carapace of invertebrates, and frustule of algae). The density separation step is necessary for sediments to discriminate microplastics from other particles, but it can be applied also to biota samples after digestion and to water samples for minimizing the filtration time and facilitating sorting and characterization steps (Bessa et al. 2019; Gago et al. 2018). The process is based on floating properties of microplastics in denser salt solution. In simple terms, a saturated salt solution with a known density can be carefully mixed with the sample containing microplastics and left to settle: plastic particles will float to the surface, and the supernatant is then collected and filtered for further investigation, while nonplastic debris heavier than microplastics are deposited on the bottom (Ribeiro-Claro et al. 2017; Fig. 8.9). The duration of mixing, as well as the floating time, can vary considerably depending on the sample volume and type, ranging from minutes to several hours or days. For example, coarse sediments settle out relatively quickly, while samples with fine particulate matter require a longer period. The recovery of supernatant can be carried out using tubes, volumetric flasks, separating funnels, or specific devices such as the Sediment-Microplastic Isolation unit (Lusher et al. 2020).

A range of salts differing for their density can be used to separate and extract microplastics: salts also vary in cost and toxicity (Bessa et al. 2019). Sodium chloride (NaCl) and sodium tungstate dihydrate (STD) are both cost-effective and nontoxic salts: for these characteristics, NaCl is recommended by the MSFD-TSML, the NOAA, and the BASEMAN consortium for monitoring programs (Lusher et al. 2020). However, these salts are not effective for the density separation of heavier polymers. Sodium iodide (NaI) is a higher-density salt that allows the separation of most polymers, but it is quite expensive. Zinc chloride ($ZnCl_2$), despite being considered the most effective and least expensive method, is a highly dangerous and corrosive substance: consequently, careful handling, disposal, and recycling of this reagent are required. Sodium polytungstate (SPT) and its derivatives are extremely expensive (although these are recyclable), can be hazardous, and, therefore, are not a first choice for routine monitoring. Furthermore, despite that salts like $ZnCl_2$, NaI, and $ZnBr_2$ allow the density separation of heavier polymers, they are highly soluble in water, and therefore, larger quantities are required in respect to NaCl, SPT, or NaBr (Campanale et al. 2020).

Since some polymers may be lost in separation more frequently than others depending on the salt solution applied (Table 8.2), it is necessary to consider this limitation when the density separation step is included in the processing of sample for microplastic isolation. The way to perform the procedure (e.g., time of mixing and settling and the way to recover the supernatant) can also influence extraction

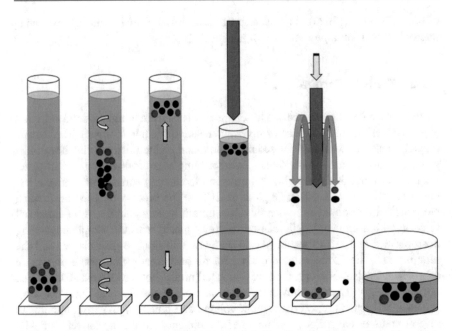

Fig. 8.9 Representation of a density separation for isolating microplastics from digested samples, including mixing, settling, and recovery of supernatant containing microplastics for subsequent filtration. Dark dots, microplastics, and red dots, inorganic or biological material (credit, CG Avio)

results. It is, thus, advisable to evaluate the extraction yield of the density separation method, especially if it was never experienced on the matrix to be processed. A general recommendation is to perform sequential extractions to increase the efficiency of procedure: on average only, the 30.2 % of microplastics were recovered after the first extraction, while reaching between 88.7 % and 100 % following sequential extractions (Lusher et al. 2020).

The effectiveness of a procedure to isolate microplastics from a matrix can be tested using recovery experiments with spiked microplastics (Brander et al. 2020). These tests consist of the addition to the sample of a known number of microplastics, for which also size, shape, color, and polymer are known. The extraction yield is then measured calculating the percentage of spiked microplastics recovered at the end of sample processing, also verifying if changes in physical and chemical characteristics of spiked microplastics have occurred. It is suggested to prepare a heterogeneous mixture of microplastics for these experiments, since the extraction efficiency depends on the method used but also on the characteristics of particles (Löder and Gerdts 2015): commercial microplastics or handmade particles obtained from plastic objects can be used. Quinn et al. (2017) prepared microparticles of eight typologies of plastic polymers and colors from 11 different post-consumer products, cutting them by various physical methods including a coffee grinder, food processor, and liquid nitrogen. These handmade particles were used along with commercial microspheres, to test the effectiveness of NaCl, NaBr, NaI, and ZnBr2 salts in the

Table 8.2 Separation abilities of different density solutions compared to some of common polymers. +, separation; ±, possible separation; and −, not separated (modified from Lusher et al. 2020)

Polymer (abbr.)	Buoyancy of polymers in density solution (g cm^{-3})						
	NaCl (1.0–1.2)	STD (1.4)	NaBr (1.37–1.4)	NaI (1.8)	ZnCl$_2$ (1.6–1.8)	ZnBr$_2$ (1.71)	SPT (2.94–3.1)
PS	+	+	+	+	+	+	+
PP	+	+	+	+	+	+	+
LDPE	+	+	+	+	+	+	+
EVA	+	+	+	+	+	+	+
HDPE	+	+	+	+	+	+	+
PA	+	+	+	+	+	+	+
PA 66	+	+	+	+	+	+	+
PMMA	+	+	+	+	+	+	+
PC	+−	+	+	+	+	+	+
PU	+−	+	+	+	+	+	+
PET	−	+−	+	+	+	+	+
PVC	−	+−	+−	+	+	+	+
PTFE	−	−	−	−	−	−	+

density separation of microplastics from sediments. Miller et al. (2021) produced, instead, high-density polyethylene microparticles of irregular shapes and of various sizes from yellow lids sourced from single-use sterile containers milled in a blender: recovery rates and effects on particles were evaluated after testing four isolation methods from seawater samples. A recovery experiment was performed by Avio et al. (2017) to compare the efficiency of six methods for the extraction of microplastics from fish tissues, including five already published and a new one developed by authors. Four different size classes of polyethylene and polystyrene particles, obtained by sieving a commercial stock powder, were spiked in the gastrointestinal tracts dissected from mullets. The new protocol was validated, showing an extraction yield of 95 % with a density separation step carried out twice, and applied to extract microplastics from fish exposed under laboratory conditions (Avio et al. 2017).

8.3.3 Filtration

Direct filtration of seawater with low organic matter content (clear water sample), or filtration of supernatants obtained by density separation, is the most frequent method to isolate microplastics from environmental samples. Among different filtering systems used, the vacuum filtration is by far the most common (Lusher et al. 2020; Fig. 8.10a). Types of filters include polytetrafluoroethylene, polycarbonate, nylon, glass fiber, cellulose (nitrocellulose, cellulose acetate, or mixed cellulose), stainless steel, silicon, and Anodisc filters (aluminum oxide). The type of filter

Fig. 8.10 (**a**) Example of a vacuum filtration apparatus used for isolate microplastics from pre-treated biota samples, (**b**) example of a cellulose filter resulting from the filtration step, (**c**) visual examination of the filter under a stereomicroscope and sorting of particles using tweezers, and (**d**) example of a clean support, on which microparticles are transferred for a single-point analysis for chemical identification by micro Fourier Transform Infrared (μ-FTIR) spectroscopy (photo credit, L Pittura)

should be chosen based on availability, porosity, and structure and on suitability for analytical techniques used for the subsequent characterization of retained particles (Martellone et al. 2021).

Pore size will have a significant effect on the overall number of particles collected as they determine the lower size of microplastics detected ranging from 0.3 μm to 200 μm. Larger pore size facilitates rapid filtration but will result in the loss of smaller plastics that are those preferentially ingested by marine organisms (Avio et al. 2020), while small pore or mesh size may result in quick obstruction by organic and mineral matter in the absence of adequate pre-treatment of sample (Prata et al. 2019). Structure of filter will have, instead, a direct effect on the dominant shape of retained microplastics. For example, nylon and cellulose filters can retain more fibers than polycarbonate filters. In fact, nylon and cellulose filters have deep and curvy pore canals organized in a lattice and in a multilayer, respectively, in which fibers are more likely to get stuck; pores on polycarbonate filters are, instead, circular; and canals are shallow and straight allowing fibers to go through more easily (Cai et al. 2020). A standardized pore size, as well as type of filters, should be defined to allow comparison between different studies.

Once filtration of samples is completed, filters are preserved in cleaned glass petri dishes, and remaining solutions can be removed by an oven or a drier at room temperature (Cutroneo et al. 2020). Dried filters are subsequently visually examined through microscopy techniques for the physical characterization of retained particles, which must be followed by the chemical identification of polymers. If the chemical characterization cannot be carried out directly on filters through an automated analysis with spectroscopy techniques (i.e., micro-FTIR and micro-Raman), a manual sorting is previously necessary, consisting of isolation of single particles using tweezers and their transfer onto a clean support (Fig. 8.10b–d).

Polytetrafluoroethylene filters have proved to be a better choice compared to aluminum oxide, glass fiber, and polycarbonate filters, when physical characterization is performed through scanning electron microscopy coupled with energy dispersion spectrometry (SEM-EDX), because they do not pose any interference to the analysis (Pivokonsky et al. 2018). Glass fiber, cellulose, and stainless-steel filters are instead the most used for filtration and subsequent examination of particles under optical microscopy; however, glass fiber filters, in particular, are not suitable for spectroscopic analysis because they absorb in the infrared area (Martellone et al. 2021).

Aluminum oxide filters are widely used for spectroscopic techniques: they are suitable for analyses by Infrared (IR) spectroscopy in transmittance mode but not in reflectance being transparent to infrared light; they can also be successfully used for Raman spectroscopy. However, Anodisc filters have a self-absorption in the mid-infrared fingerprint range ($1400–600$ cm^{-1}), hampering a distinct identification of potential microplastic particles and an accurate classification of the polymer type. In addition, aluminum oxide filters are also among the most fragile, because they are thin although they are being rigid.

Silicon filters can represent an alternative, since they guarantee sufficient transparency for the broad mid-infrared region of $4000–600$ cm^{-1} and offer good mechanical stability during analysis of microplastic samples using both transmission FTIR microscopy and FTIR imaging as well as Raman microscopy (Käppler et al. 2015). Gold-coated polycarbonate filters are also used for analysis of particles using micro-FTIR spectroscopy in reflectance mode (Martellone et al. 2021).

8.3.4 Quality Assurance and Quality Control (QA/QC) of Analysis

Contamination issues are a challenge in quality assurance and quality control during microplastic studies: the risk of external contamination is extremely important, particularly for microfibers (Prata et al. 2021). QA includes a series of systematic steps or activities to ensure that generated data are accurate and reliable. QC is the process of verifying and checking all data, results, and reported methods to ensure their validity and prevent erroneous conclusions (Brander et al. 2020). QA/QC practices should be considered in advance and carefully followed throughout the whole study process, including sampling and collection, extraction, and analysis.

Regardless of the environmental matrix to collect, the use of plastic devices and materials for sampling and store samples should be eliminated, replacing plastics with glass or metal. If plastic materials cannot be avoided, as the case of nylon nets for seawater or biota sampling, they must be characterized and compared with microplastics extracted from samples: if they match, they must be removed from results. Similarly, certain sampling characteristics intrinsically contain multiple potential sources of microplastics that sometimes cannot be removed. For example, sampling activities carried out on a vessel might expose to potential external contamination derived from hull paints, life vests, ropes, and sails, all materials that should be characterized (Brander et al. 2020). Controls for air contamination should also be performed: leaving a wet filter paper or an open container with filtered water during sample collection is possible to register the deposition of microfibers or microplastics from the surrounding environment.

Another important measure is to control the release of fibers from clothes. Providing protection to operators, including adequate cotton lab coats, can prevent the release of synthetic textile fibers from clothes to some extent. Nonetheless, natural textiles can also release fibers, while cellulosic fibers, such as cotton, can be abundant in indoor air. In this respect, the use of coats, gloves, and paper towels of recognizable colors may help to identify accidental contaminations, thus improving control and prevention procedures (Prata et al. 2021): this recommendation should be applied also during the processing of samples in laboratory.

The processing of samples in laboratory is suggested to be carried out under fume hoods or laminar flow hood. However, fume hoods are poorly efficient in controlling for air contamination since unfiltered air from the laboratory is drawn inside the hood and then expelled outside or filtered. On the other hand, laminar flow hoods draw air in through HEPA filters and create a laminar flow toward the front, preventing the entrance of uncontrolled air: the use of laminar flow hoods is thus preferred. Alternatives include the use of rooms with controlled air flow and access; minimal personnel circulation is always recommended. Metal or glass materials and aluminum foils are also required for all laboratory activities, as plastic objects can release particles and should be fully avoided. All materials must be cleaned with pure water or ethanol, and solutions (especially salt solutions) should always be filtered (0.22- or 0.45-μm pore size) to avoid external contamination with microplastics. During laboratory preparative steps, air deposition controls and running procedural blanks allow to check the background contamination along with the influence of sample processing. Procedural blanks typically include pure reagents (or water) treated with the same procedures as samples.

Results on analysis of both air deposition controls and blank samples should be reported and considered for the final quantification of microplastics in analyzed samples.

8.3.5 Operative Protocol for Isolation of Microplastics from the Gastrointestinal Tract of Marine Species

This paragraph provides a step-by-step description of an operative protocol, including necessary materials, reagents, and equipment, that students can easily perform to extract microplastics from tissues of marine organisms. The protocol is based on methods tested, validated, and directly experienced on a range of marine species by authors of the present chapter and intercalibrated in a joint exercise within several partners of three JPI Oceans projects (EPHEMARE, BASEMAN, and RESPONSE; Vital et al. 2021; Avio et al. 2015, 2017, 2020; Cau et al. 2019, 2020; Bour et al. 2018; Bessa et al. 2019).

The gastrointestinal tract of the red mullet will be the target of the procedure as a practical example, for assessing the ingestion of microplastics by marine biota. However, the same protocol and recommendations can be applied to other species and tissues and to the whole specimens for smaller species (e.g., mussels).

Materials

- Forceps
- Metal tweezer
- Metal spatula
- Glass petri dishes
- Glass flask (250 ml)
- Glass beakers (5 L, 1 L, and 250 ml)
- Glass cylinders (100 ml and 250 ml)
- Glass tube
- Mortar and pestle
- Nitrate cellulose filters (8-μm pore size)
- Acetate cellulose membrane (0.45-μm pore size)
- Magnetic stirrer plate and cylindrical stirrer
- Graduated cylinders (100 ml and 250 ml)
- Stirrer plate and cylindrical stirrer

Reagents

- 10 % KOH solution
- 15 % H2O2 solution
- Saturated solution of NaCl salt (1.2 g/cm3)
- Ultrapure/distilled water

Equipment

- Oven (work temperature 40–50 °C).
- Hood.
- Vacuum filtration system schematically represents in Fig. 8.11 the following:
  - 1 = filtration ramp (Speedflow)
  - 2 = glass filter holder (diam 47 mm) + max volume 500 ml
  - 3 = vacuum tube HW/55 diam mm 8 × 15
  - 4 = non-return PP valve
  - 5 = vacuum trap 2 L
  - 6 = vacuum tube HW/55 diam mm 8 × 15
  - 7 = second vacuum trap 2 L (to protect the pump)
  - 8 = additional filter to protect the pump (optional)
  - 9 = vacuum generator RCK400 34 L/min

Preparation of Solution

- 10 % KOH solution. Dissolve 10 g of KOH for each 100 ml of ultrapure/distilled water.
- 15 % H2O2 solution. Dilute 30 % of hydrogen peroxide in ultrapure/distilled water (1:1, volume/volume).
- Saturated solution of NaCl salt (1.2 g/cm3). Dissolve 10 g of NaCl in 100 ml of ultrapure/distilled water reaching a density of 1.2 g/cm3. To check the density, it is possible to weight 1 ml of solution: if 1 ml of solution weights at least 1.2 g, the

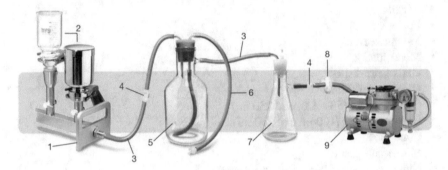

Fig. 8.11 Schematic representation of vacuum filtration apparatus

right density of 1.2 g/cm3 has been obtained. If NaCl salt of pure grade is not available, the commercial cooking salt can be used.

Sample Dissection

Organisms are dissected to obtain the gastrointestinal tract: open the fish from the anus to the mouth using forceps and isolate the gastrointestinal tract including the esophagus, the stomach, and the intestine, with the help of forceps and tweezer (Fig. 8.12). Register the main morphological parameters of dissected specimens: total body weight, total length (from the mouth to the caudal fin), and weight of the gastrointestinal tract.

Digestion of the Gastrointestinal Tract

Put the gastrointestinal tract in a 250-ml glass flask or in a glass beaker and add 10 % KOH solution until the sample is covered or according to the minimum ratio of 5:1, volume/gastrointestinal tract weight. Leave the sample in oven at a maximum temperature of 50 °C until the digestion process is completed (Fig. 8.13).

Density Separation of Digested Sample

Put the digested sample in a graduated glass cylinder (100 ml or 250 ml depending on the volume of digested sample) and add the saturated NaCl salt solution and a cylindrical stirrer. Leave to mix the solution on a magnetic stirrer plate for 10 min and leave to settle for additional 10 min. After this step, a first aliquot of supernatant is collected from the glass tube into a graduated glass beaker (1 L or 5 L depending on processed volumes). The density separation step is carried out twice for a better extraction performance: rinse and refill the cylinder with saturated NaCl salt solution, before repeating the mixing, settling, and recovery of supernatant. All materials used for the collection of supernatant need to be washed with ultrapure/distilled water to collect particles potentially attached to the wall of glass tube and cylinder. The supernatant, obtained from sequential extractions, is ready to be filtered (Fig. 8.14).

Fig. 8.12 Red mullets and European hakes and dissection of the gastrointestinal tract from the red mullet

Fig. 8.13 Digestion of the gastrointestinal tract of the red mullet using 10 % KOH solution

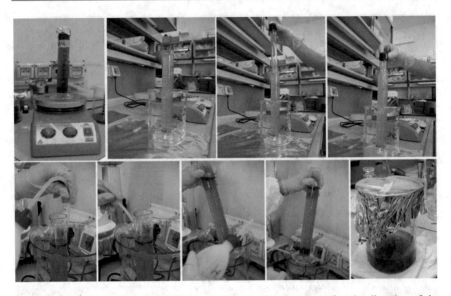

Fig. 8.14 Density separation procedure and recovery of supernatant after the digestion of the gastrointestinal tract of red mullet

Processing of Fatty Tissues

When the tissues to be analyzed are particularly fatty, the digestion with KOH is not suitable, because formation of oil droplets in the solution hampers the density separation and the filtration steps. In these conditions, it is suggested to dry the tissues, instead of digesting, before the density separation and subsequent filtration. The procedure consists of placing the sample in a petri dish and leaving it to dry in oven at a maximum temperature of 50 °C. Once it is dried, sample is gently triturated using a mortar to obtain a powder (Fig. 8.15). The powder is then put in a 100-ml glass cylinder to carry out the density separation according to procedure described above (Figs. 8.14 and 8.15).

Filtration of Supernatant

Filter under vacuum the supernatant through a nitrate cellulose. The pore size of filter should be chosen according to the detection limit of the analytical method used for the subsequent chemical identification of microplastics: 8-μm pore size is suitable for the μ-FTIR spectroscopy. Filtration can be speeded up mixing the supernatant with a metal spatula during procedure.

At the end of filtration, recover the filter and put it in a petri dish adding 15 % H_2O_2 solution, useful to digest the organic material eventually remained as residue, irrespective of digestion and density separation steps. Petri dish is left in oven at a maximum temperature of 50 °C until the end of digestion and the drying of filter. To avoid the formation of salt crystals on dried filter, ultrapure/distilled water can be used to rinse the filter before its recovery at the end of filtration (Fig. 8.16).

Fig. 8.15 Drying of gastrointestinal tract and trituration before density separation and filtration as an alternative procedure to KOH digestion method for processing fatty tissues

Dried filters are observed under a stereomicroscope for the manual isolation of all particles resembling microplastics and fibers (Fig. 8.17a–e), which are classified based on shape and color, and then measured using an image analysis software for categorization in size classes (Fig. 8.17 and Table 8.3). The chemical identification will be performed later on single particles and fibers, for example, using μ-FTIR or μ-Raman spectrometry (Fig. 8.20a).

QA/QC Procedures
To reduce and monitor the potential contamination of sample by external microplastics and microfibers during all phases' protocol, be sure to follow recommendations as follows:

- Perform dissection and various processing steps in a dedicated room with closed windows and restricted access.
- Clean all laboratory surfaces and materials used for processing samples with ultrapure water and ethanol.
- Dress only cotton wear. It is also recommended to register colors of clothes worn underneath the lab coats.

Fig. 8.16 Filtration of supernatant after the density separation step and digestion of recovered filters with 15 % H$_2$O$_2$ in oven to eliminate residual organic material

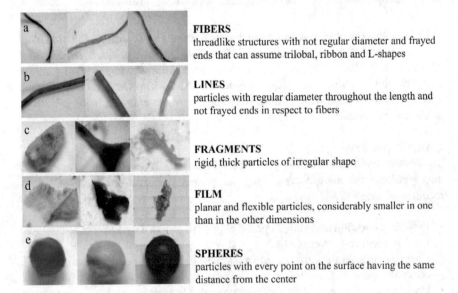

FIBERS
threadlike structures with not regular diameter and frayed ends that can assume trilobal, ribbon and L-shapes

LINES
particles with regular diameter throughout the length and not frayed ends in respect to fibers

FRAGMENTS
rigid, thick particles of irregular shape

FILM
planar and flexible particles, considerably smaller in one than in the other dimensions

SPHERES
particles with every point on the surface having the same distance from the center

Fig. 8.17 Example of visual guidelines for the identification of particles based on shape

Table 8.3 Example of datasheet showing classification of each particle extracted per sample into the main categories of physical characteristics

Sample code	Particle ID	Shape					Color					Size classes (mm)			
		Fiber	Line	Fragment	Film	Sphere	Black	Blue	Red	Transparent	Others	5–1	1–0.5	0.5–0.1	0.1–0.01
S1	S1.1	x					x					x			
	S1.2			x			x					x			
	S1.3				x		x					x			

- Cover the top of glass containers with aluminum foil, especially during the filtration step.
- Prefilter working solution before their use with 0.45-µm pore size filter.
- Include a blank sample starting from dissection: hold an open beaker with distilled water on the workbench while dissecting and process it according to the same procedures applied to samples.

8.4   Analytical Techniques for Microplastic Characterization

After extraction and isolation, particles need to be accurately characterized in terms of number, size, shape, surface texture, color, and chemical typology of polymers. This information is relevant to trace microplastic sources, origins, weathering, and residence time in the field, as well as to highlight those typologies more available for biota. Chemical analysis can also allow to identify additives and associated contaminants or impurities on microplastic surface, like organic and inorganic material. Various techniques can be used in sequence or in association with their own advantages and limitations, often dependent on dimensions of microplastics to be detected: the actual challenge is to implement existing tools or develop new approaches to overcome the characterization of smallest microplastics and nanoplastics.

8.4.1   Physical Characterization

8.4.1.1 Microscopy Techniques

Physical characterization through microscopy techniques is primarily used to identify and classify microplastics preserved on a filter or in petri dishes or jars (Cutroneo et al. 2020).

Optical microscopy (OM) is suitable to visually examine particles of submillimeter size retaining the 3D shape and color of suspected microplastics. OM allows to distinguish between plastics and other organic/inorganic compounds by analyzing detailed surface textures and structural information (Jung et al. 2021). Visual guidelines can help the operator in the identification of suspected microplastics, including bright and unnaturally colored particles, fragments with sharp geometrical shapes, shiny surfaces, and featureless fibers with a consistent width (Fig. 8.17). Physical and tactile guidelines include the particle holding its shape or stretched when poked and resistance to easy breakage (Primpke et al. 2020). Once particles were identified, they are measured using an image analysis software and categorized by shape, color, and size classes (Table 8.3). The main advantage of light microscopy in microplastic characterization is that it is a relatively cheap and easy approach. However, visual sorting under a stereomicroscope can be difficult for microplastics with no specific color, and it requires considerable time and resources in terms of researchers involved in counting hundreds of particles (Campanale et al. 2020). Since this technique does not provide information on the chemical composition of objects, further characterization is necessary to confirm the plastic nature of particles.

Scanning electron microscopy (SEM) can also be used, allowing to visualize nanometer-sized particles. Discrimination of surface structures of plastics and other materials can be integrated with an energy-dispersive X-ray probe (SEM-EDX) to provide further information on the elemental composition of organic and inorganic species, particularly useful for environmental samples. However, SEM-EDX is expensive and requires substantial time and effort for sample preparation and examination, which limits the number of samples that can be handled in routinary analyses (Jung et al. 2021).

8.4.1.2 Light-Scattering Technique

Multiple methods apply the scattering of laser light on particles to obtain information on physical properties like particle size and particle-size distribution. Dynamic light scattering (DLS), the most widely used, measures particle sizes in the range from 1 nm to 3 mm based on the fluctuation of intensity of a laser beam that passes the suspension. These particle size analyzers calculate total particle-size distribution without distinguishing microplastics and other particles. Therefore, microplastic particle size and distribution can be measured using light diffraction and/or dynamic light scattering only when these particles have been previously isolated and represent the only present in the solution matrix: in this respect, a rigid sample pre-treatment is fundamental to completely remove all other organic/inorganic particles. DLS may provide different results from those obtained by visual inspection and might not detect small-size particles that are masked by the effect of larger particles on strongly scattered light (Lee and Chae 2021).

Nanoparticle tracking analysis (NTA) can represent another approach on characterization of environmental matrices. NTA gives information on size profile, recording scattered laser light with a microscope and a digital camera. NTA visualizes nanoparticles and particle concentration in the solution and derives size of particles correlating their hydrodynamic diameter due to its Brownian motion (Schwaferts et al. 2019). The size distribution obtained with NTA may be less sensitive to the presence of large particles and aggregates than DLS. NTA can detect particles up to 30 nm but not particles larger than 2 μm: analysis using NTA is more time-consuming (up to 1 h) than DLS (several minutes; Lee and Chae 2021).

8.4.2 Chemical Characterization

8.4.2.1 Spectroscopy Methods: FTIR and Raman

The spectroscopic methods, including Fourier-transform infrared (FTIR) spectroscopy and Raman spectroscopy, are the most common approaches in the chemical identification of microplastics, being also recommended by the MSFD-TSGML. These methods are based on the energy absorption by characteristic functional groups of polymer particles, resulting in a vibrational spectrum which is unique for every polymer type. The chemical identification of particles is obtained by

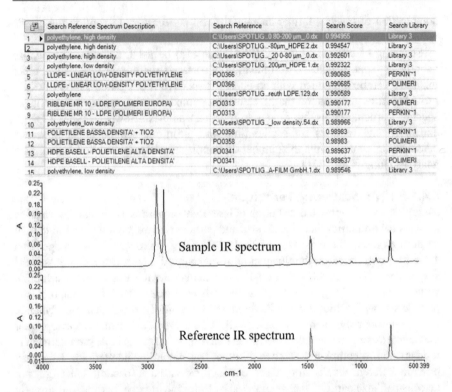

Fig. 8.18 Result of acquisition of an infrared spectrum of a microparticle extracted from sample (sample IR spectrum) and of matching with a reference polymer in a database (reference IR spectrum): the match factor threshold of 0.99 (search score) validates the polymer identification as polyethylene (photo credit, L. Pittura)

comparing the spectrum of the investigated sample with spectra of known polymers by matching them to spectral libraries through database comparison algorithms (Fig. 8.18). In coupling the spectrometer (FTIR or Raman) to a microscope, small microplastics are measurable through the "micro"-spectroscopy (μ-FTIR and μ-Raman): μ-Raman spectroscopy can characterize microplastic samples higher than 1 μm, while μ-FTIR spectroscopy could identify microparticles higher than 10–20 μm (Silva et al. 2018). These techniques are nondestructive and can be coupled with other methodologies to obtain additional and complementary information on the composition of plastic polymers (Campanale et al. 2020). Atomic force microscopy (AFM) combined with either FTIR or Raman spectroscopy is a potential candidate for nanoplastic analysis: AFM probes can be operated in both contact and noncontact modes with objects providing images at nanometer resolutions, while FTIR or Raman spectroscopy determines the chemical composition of the object (Shim et al. 2017).

Transmission, reflectance, and attenuated total reflectance (ATR) are acquisition modes available in FTIR analysis (Fig. 8.19). In transmission mode, the FTIR

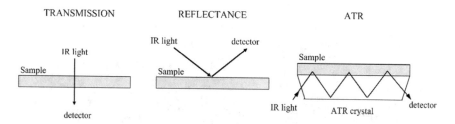

Fig. 8.19 Acquisition modes of FTIR analysis

spectrometer records the IR light that passes through the sample. Working in the transmission mode makes the characterization of microplastics difficult for two main reasons: (i) if the particle is too thick, the IR beam does not pass making the characterization impossible; (ii) if the particle is not clean, there is disturbance, and the IR spectrum is difficult to interpret. In addition, the transmittance mode needs IR transparent filters (e.g., aluminum oxide), and it is limited, owing to total absorption patterns, by a certain thickness of microplastic samples. The reflectance mode records the IR signal that is reflected from the sample: the disadvantage is that measurements of irregularly shaped microplastics may result in non-interpretable spectra due to refractive error. The use of micro-ATR accessory in combination with microscopy can prevent these problems since IR spectra are collected at the surface of a particle (Löder and Gerdts 2015): the sample is in contact with a crystal of high refractive index, and the IR light passes throughout the crystal hitting the sample several times and finally reaches the detector obtaining the IR spectrum. The pressure produced by the ATR probe may, however, damage highly weathered or fragile microplastics, and tiny plastic particles can be pulled from the filter paper by adhesion to or electrostatic interaction with the probe tip. On the other side, an ATR probe made of germanium can be easily damaged by contact analysis with hard and sharp inorganic particles like those possibly remained on a filter paper from a sandy sample (Shim et al. 2017).

Instruments available on the market differ mainly by the type of microscope coupled to the spectrometer and the mode of particle acquisition, being manual or automated. A manual sample placement means that there is a single-point acquisition and particles must be positioned singularly. More expensive instruments have the possibility of fully automated measurements of multiple particles in a sample and to map or generate spatial chemical images of whole-membrane filters through the motorized movement of the sample table of the microscope (Fig. 8.20a–b). Micro-Raman imaging theoretically allows for the spectral analysis of whole-membrane filters at a spatial resolution below 1 μm. Focal plane array (FPA)-based FTIR imaging allows for detailed and unbiased high-throughput analysis of total microplastics on a sample filter. This technique enables the simultaneous recording of several thousand spectra within an area with a single measurement and thus the generation of chemical images (Löder and Gerdts 2015). The main disadvantages of these automated methods are extended processing time to map an entire filter (9 h to

a b

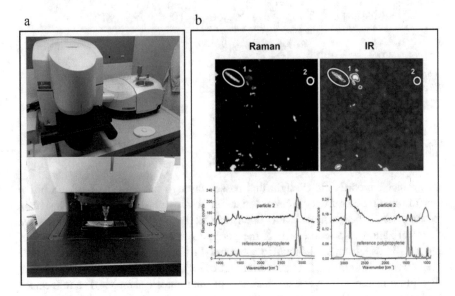

Fig. 8.20 (a) Single-point analysis of particles through μATR-FTIR spectroscopy (photo credit, L. Pittura) and (b) Raman and IR chemical images with false coloring denoting the spectral intensity of particles and spectra identification in comparison with a reference (modified from Araujo et al. 2018)

scan one filter paper; Shim et al. 2017), refractive errors during measurement of irregularly shaped microplastic particles, lack of information on associated organic additives to MPPs (microplastics), and overlap of polymer bands given by organic and inorganic contaminations that can disturb identification of particles (Campanale et al. 2020).

8.4.2.2 Thermoanalytical Methods: Py-GC-MS

Pyrolytic gas chromatography in combination with mass spectrometry (Py-GC-MS) can be used to assess the chemical composition of potential microplastic particles by analyzing their thermal degradation products. In following a pyrolytic process, decomposition products characteristic of each polymer are trapped on a solid-phase adsorbent and thermally desorbed. Volatile compounds are then separated by gas chromatography and identified by mass spectrometry (Campanale et al. 2020).

The pyrolysis of plastic polymers results in characteristic pyrograms, which facilitate the polymer identification by comparing combustion products with reference pyrograms of known virgin-polymer samples (Löder and Gerdts 2015). Contrarily to Raman or FTIR technique, which only investigates the surface of a particle, Py-GC-MS allows the analysis of the whole particle, enabling to simultaneously identify polymer types and associated organic plastic additives. Although Py-GC-MS has the advantage that individual sorting of particles is not needed, the limit is the amount of sample (e.g., 0.35–7 mg) that can be analyzed (Shim et al. 2017). This quantity may compromise the representativeness of the sample composition when complex environmental samples are analyzed, as it may not be homogenous on a small scale. Variants of this technique have been used to develop new methods, such as thermo-extraction and desorption coupled with GC-MS which combines thermogravimetric analysis (TGA) and thermal desorption-gas chromatography-mass spectrometry (TD-GC-MS; Silva et al. 2018). Thermal analysis provides an alternative method to spectroscopy for chemical identification of polymer types but, as a destructive method, prevents the possibility of additional characterization of particles.

8.5 Data Expression

The expression and normalization of obtained results should be harmonized for comparison among studies.

For water samples, data are usually provided as the number of microplastics (MPPs) per unit of volume (MPPs L^{-1} or MPPs m^{-3}) or per unit area (MPPs m^{-2}; Gago et al. 2018).

For both intertidal and subtidal sediments, it is recommended reporting the number of microplastics (MPPs) per unit area (MPPs m^{-2}) and per unit dry mass (MPPs kg^{-1} dry). In addition, since samples collected are a function of the length, breadth, and thickness of the collected area, also the number of particles per cubic cm or m should be provided (MPPs cm^{-3}/m^{-3}; Uddin et al. 2020; Frias et al. 2018).

Data on ingestion of microplastics by biota should be presented containing, at least, the following information for each investigated species:

(a) The frequency of ingestion given as the percentage of specimens containing one or more particles on the total of analyzed specimens.
(b) The average number of particles calculated on organisms positive to ingestion (reporting only the average number of particles per individual might be misleading if it is not specified whether organisms without particles [i.e., 0 values] were included in the average).
(c) The number of analyzed specimens.

Frequency of ingestion reflects the probability for organisms to interact with microplastics in their own habitat, appearing to be a more appropriate index than the number of ingested items to monitor microplastics in natural population and to better highlight differences among sampling areas and species (Avio et al. 2020). Expression of data as number of particles per weight of tissues is not recommended since microplastics are not homogeneously distributed among and within tissues and because tissues are often subjected to marked weight variations (Bessa et al. 2019).

In most cases, microplastic numbers are not sufficient to make a mass determination. However, if a mass determination is possible, mass of MPPs is also provided: for water and sediment, it is normalized on abovementioned units; for biota, the mass of MPPs is usually given per unit mass of tissue (MP g^{-1} tissue; Uddin et al. 2020).

As previously mentioned, it would be appropriate to represent separately quantification data on microplastic particles (MPPs) and those of textile microfibers (MFs): for the latter, data on natural microfibers should also be provided using the same reporting units for MPPs.

In addition to data on total quantification, it is important to provide a detailed representation of all physical and chemical characteristics of particles extracted from environmental samples. Each typology of size classes, shape, color, and polymer can be easily provided as relative contribution to the total number of microplastics extracted from sample. Presentation of data on extracted microplastics as percentage distribution into size classes is particularly important, as this improves the understanding of the size distribution of microplastics in the marine environment and highlights dimensions more available for biota. Regarding the shape, inter-study comparison is often hampered by the absence of standardized definitions and categories for microplastic characteristics; in this respect, it is suggested to always provide the applied definition of shapes and to provide a corresponding photo of extracted microplastics. Results of studies reporting microplastics without a chemical characterization should not be considered reliable.

It is important to stress that the adequate presentation of obtained results is of key importance to trace origin, distribution and fate of microplastics, as well as their biological impact and risk for the marine environment.

Acknowledgments This chapter has been written as part of activities of the projects "RESPONSE, Toward a risk-based assessment of microplastic pollution in marine ecosystems" funded by JPI Oceans (2020–2023) and "EMME, Exploring the fate of Mediterranean microplastics: from distribution pathways to biological effects" (PRIN-2017) funded by MUR Italian Agency (Ministero dell'Università e della Ricerca).

References

Araujo CF, Nolasco MM, Ribeiro AM, Ribeiro-Claro PJ (2018) Identification of microplastics using Raman spectroscopy: latest developments and future prospects. Water Res 142:426–440

Avio CG, Cardelli LR, Gorbi S, Pellegrini D, Regoli F (2017) Microplastics pollution after the removal of the Costa Concordia wreck: first evidences from a biomonitoring case study. Env Poll 227:207–214

Avio CG, Gorbi S, Regoli F (2015) Experimental development of a new protocol for extraction and characterization of microplastics in fish tissues: first observations in commercial species from Adriatic Sea. Mar Environ Res 111:18–26

Avio CG, Pittura L, d'Errico G, Abel S, Amorello S, Marino G, Gorbi S, Regoli F (2020) Distribution and characterization of microplastic particles and textile microfibers in Adriatic food webs: general insights for biomonitoring strategies. Environ Pollut 258:113766

Bagaev A, Mizyuk A, Khatmullina L, Isachenko I, Chubarenko I (2017) Anthropogenic fibres in the Baltic Sea water column: field data, laboratory and numerical testing of their motion. Sci Total Environ 599:560–571

Benedetti M, Giuliani ME, Mezzelani M, Nardi A, Pittura L, Gorbi S, Regoli F (2022) Emerging environmental stressors and oxidative pathways in marine organisms: current knowledge on regulation mechanisms and functional effects. Biocell 46(1):37

Bessa F et al. (2019) Harmonized protocol for monitoring microplastics in biota. Deliverable 4.3. JPI-Oceans BASEMAN project

Bour A, Avio CG, Gorbi S, Regoli F, Hylland K (2018) Presence of microplastics in benthic and epibenthic organisms: influence of habitat, feeding mode and trophic level. Env Pollut 243: 1217–1225

Brander SM, Renick VC, Foley MM, Steele C, Woo M, Lusher A, Carr S, Helm P, Box C, Cherniak S, Andrews RC, Rochman CM (2020) Sampling and quality assurance and quality control: a guide for scientists investigating the occurrence of microplastics across matrices. Appl Spectrosc 74(9):1099–1125

Bray L, Digka N, Tsangaris C, Camedda A, Gambaiani D, de Lucia GA, Matiddi M, Miaud C, Palazzo L, Pérez-del-Olmo A, Raga JA, Silvestri C, Rag JA (2019) Determining suitable fish to monitor plastic ingestion trends in the Mediterranean Sea. Environ Pollut 247:1071–1077

Cai H, Chen M, Chen Q, Du F, Liu J, Shi H (2020) Microplastic quantification affected by structure and pore size of filters. Chemosphere 257:127198

Campanale C, Savino I, Pojar I, Massarelli C, Uricchio VF (2020) A practical overview of methodologies for sampling and analysis of microplastics in riverine environments. Sustainability 12(17):6755

Cau A, Avio CG, Dessì C, Follesa MC, Moccia D, Regoli F, Pusceddu A (2019) Microplastics in the crustaceans *Nephrops norvegicus* and *Aristeus antennatus*: flagship species for deep-sea environments? Env Pollut 255:113107

Cau A, Avio CG, Dessì C, Moccia D, Pusceddu A, Regoli F, Follesa MC (2020) Benthic crustacean digestion can modulate the environmental fate of microplastics in the deep sea. Environ Sci Technol 54(8):4886–4892

Cheshire AC, Adler E, Barbière J, Cohen Y, Evans S, Jarayabhand S, Jeftic L, Jung RT, Kinsey S, Kusui ET, Lavine I, Manyara P, Oosterbaan L, Pereira MA, Sheavly S, Tkalin A, Varadarajan S, Wenneker B, Westphalen G (2009) UNEP/IOC guidelines on survey and monitoring of marine litter. UNEP Regional Seas Reports and Studies. No. 186

Cole M, Lindeque PK, Fileman E, Clark J, Lewis C, Halsband C, Galloway TS (2016) Microplastics alter the properties and sinking rates of zooplankton faecal pellets. Environ Sci Technol 50(6):3239–3246

Cole M, Lindeque P, Halsband C, Galloway TS (2011) Microplastics as contaminants in the marine environment: a review. Mar Pollut Bull 62(12):2588–2597

Coyle R, Hardiman G, O'Driscoll K (2020) Microplastics in the marine environment: a review of their sources, distribution processes, uptake and exchange in ecosystems. Case Stud Chem Environ Eng 2:100010

Cutroneo L, Reboa A, Besio G, Borgogno F, Canesi L, Canuto S, Dara M, Enrile F, Forioso I, Greco G, Lenoble V, Malatesta A, Mounier S, Petrillo M, Rovetta R, Stocchino A, Tesan J, Vagge G, Capello M (2020) Microplastics in seawater: sampling strategies, laboratory methodologies, and identification techniques applied to port environment. Environ Sci Pollut Res 27(9):8938–8952

Fossi MC, Peda C, Compa M, Tsangaris C, Alomar C, Claro F, Ioakeimidis C, Galgani F, Hema T, Deudero S, Romeo T, Battaglia P, Andaloro F, Caliani I, Casini S, Panti C, Baini M (2018) Bioindicators for monitoring marine litter ingestion and its impacts on Mediterranean biodiversity. Environ Pollut 237:1023e1040

Frias JPGL, Nash R (2019) Microplastics: finding a consensus on the definition. Mar Pollut Bull 138:145–147

Frias JPGL et al. (2018) Standardised protocol for monitoring microplastics in sediments. Deliverable 4.2. JPI-Oceans BASEMAN project

Gago J et al. (2018) Standardised protocol for monitoring microplastics in seawater. Deliverable D4.1. JPI-Oceans BASEMAN project

Geyer R (2020) Production, use and fate of synthetic polymers in plastic waste and recycling. In: Letcher TM (ed) Plastic Waste and Recycling: Environmental Impact, Societal Issues, Prevention, and Solutions. Academic Press, Cambridge, MA, pp 13–32

Gigault J, ter Halle A, Baudrimont M, Pascal PY, Gauffre F, Phi TL, El Hadri H, Grassl B, Reynaud S (2018) Current opinion: what is a nanoplastic? Environ Pollut 235:1030–1034

Gola D, Tyagi PK, Arya A, Chauhan N, Agarwal M, Singh SK, Gola S (2021) The impact of microplastics on marine environment: a review. Environ Nanotechnol Monit Manag 16:100552

Han M, Niu X, Tang M, Zhang BT, Wang G, Yue W, Kong X, Zhu J (2020) Distribution of microplastics in surface water of the lower Yellow River near estuary. Sci Total Environ 707: 135601

Hanke G, Galgani F, Werner S, Oosterbaan L, Nilsson P, Fleet D, Kinsey S, Thompson R, Van Franeker JA, Vlachogianni T, Palatinus A, Scoullos M, Veiga JM, Matiddi M, Alcaro L, Maes T, Korpinen S, Budziak A, Leslie H, Gago J, Liebezeit G (2013) Guidance on monitoring of marine litter in euro-pean seas. European Commission

Hartmann NB, Huffer T, Thompson RC, Hassellöv M, Verschoor A, Daugaard AE, Rist S, Karlsson T, Brennholt N, Cole M, Herrling MP, Hess MC, Ivleva NP, Lusher AL, Wagner M (2019) Are we speaking the same language? Recommendations for a definition and categorization framework for plastic debris. Environ Sci Technol 53:1039–1047

Hermsen E, Mintenig SM, Besseling E, Koelmans AA (2018) Quality criteria for the analysis of microplastic in biota samples: a critical review. Environ Sci Technol 52(18):10230–10240

Hidalgo-Ruz V, Gutow L, Thompson RC, Thiel M (2012) Microplastics in the marine environment: a review of the methods used for identification and quantification. Environ Sci Technol 46(6): 3060–3075

Jung S, Cho SH, Kim KH, Kwon EE (2021) Progress in quantitative analysis of microplastics in the environment. A Review Chem Eng J 130154

Käppler A, Windrich F, Löder MG, Malanin M, Fischer D, Labrenz M, Eichhorn KJ, Voit B (2015) Identification of microplastics by FTIR and Raman microscopy: a novel silicon filter substrate opens the important spectral range below 1300 cm−1 for FTIR transmission measurements. Anal Bioanal Chem 407(22):6791–6801

Karlsson TM, Kärrman A, Rotander A, Hassellöv M (2020) Comparison between manta trawl and in situ pump filtration methods, and guidance for visual identification of microplastics in surface waters. Environ Sci Pollut Res 27(5):5559–5571

Lee J, Chae KJ (2021) A systematic protocol of microplastics analysis from their identification to quantification in water environment: a comprehensive review. J Hazard Mater 403:124049

Lippiatt S, Opfer S, Arthur C (2013) Marine Debris Monitoring and Assessment. NOAA Technical Memorandum NOS-OR & R-46

Liu K, Courtene-Jones W, Wang X, Song Z, Wei N, Li D (2020) Elucidating the vertical transport of microplastics in the water column: a review of sampling methodologies and distributions. Water Res 116403

Liu J, Yang Y, Ding J, Zhu B, Gao W (2019) Microfibers: a preliminary discussion on their definition and sources. Environ Sci Pollut Res 26(28):29497–29501

Löder MGJ, Gerdts G (2015) Methodology used for the detection and identification of microplastics-a critical appraisal. In: Bergmann M, Gutow L, Klages M (eds) Marine anthropogenic litter. Springer, Cham

Lusher AL, Munno K, Hermabessiere L, Carr S (2020) Isolation and extraction of microplastics from environmental samples: an evaluation of practical approaches and recommendations for further harmonization. Appl Spectrosc 74(9):1049–1065

Lusher AL, Welden NA, Sobral P, Cole M (2017) Sampling, isolating and identifying microplastics ingested by fish and invertebrates. Anal Methods 9:1346–1360

Martellone L, Lucentini L, Mattei D, De Vincenzo M, Favero G, Bogialli S, Litti L, Meneghetti M, Corami F, Rosso B. (2021) Strategie di campionamento di microplastiche negli ambienti acquatici e metodi di pretrattamento. Roma: Istituto Superiore di Sanità (Rapporti ISTISAN 21/2)

Miller ME, Motti CA, Menendez P, Kroon F (2021) Efficacy of microplastic separation techniques on seawater samples: testing accuracy using high-density polyethylene. Biol Bull 240(1):52–66

Pan Z, Guo H, Chen H, Wang S, Sun X, Zou Q, Zhang Y, Lin H, Cai S, Huang J (2019) Microplastics in the northwestern Pacific: abundance, distribution, and characteristics. Sci Total Environ 650:1913–1922

Pivokonsky M, Cermakova L, Novotna K, Peer P, Cajthaml T, Janda V (2018) Occurrence of microplastics in raw and treated drinking water. Sci Total Environ 643:1644–1651

Prata JC, da Costa JP, Duarte AC, Rocha-Santos T (2019) Methods for sampling and detection of microplastics in water and sediment: a critical review. TrAC–Trends Anal Chem 110:150–159

Prata JC, Reis V, da Costa JP, Mouneyrac C, Duarte AC, Rocha-Santos T (2021) Contamination issues as a challenge in quality control and quality assurance in microplastics analytics. J Hazard Mater 403:123660

Preston-Whyte F, Silburn B, Meakins B, Bakir A, Pillay K, Worship M, Paruk S, Mdazuka Y, Mooi G, Harmer R, Doran D, Tooley F, Maes T (2021) Meso- and microplastics monitoring in harbour environments: a case study for the port of Durban, South Africa. Mar Pollut Bull 163: 111948

Primpke S, Christiansen SH, Cowger W, De Frond H, Deshpande A, Fischer M et al (2020) Critical assessment of analytical methods for the harmonized and cost-efficient analysis of microplastics. Appl Spectrosc 74(9):1012–1047

Qu X, Su L, Li H, Liang M, Shi H (2018) Assessing the relationship between the abundance and properties of microplastics in water and in mussels. Sci Total Environ 621:679–686

Quinn B, Murphy F, Ewins C (2017) Validation of density separation for the rapid recovery of microplastics from sediment. Anal Methods 9(9):1491–1498

Razeghi N, Hamidian AH, Wu C, Zhang Y, Yang M (2021) Microplastic sampling techniques in freshwaters and sediments: a review. Environ Chem Lett:1–28

Reisser JW, Slat B, Noble KD, Plessis KD, Epp M, Proietti M, de Sonneville J, Becker T, Pattiaratchi C (2015) The vertical distribution of buoyant plastics at sea: an observational study in the North Atlantic Gyre. Biogeosciences 12(4):1249–1256

Ribeiro-Claro P, Nolasco MM, Araújo C (2017) Characterization of microplastics by Raman spectroscopy. In: Rocha-Santos TAP, Duarte AC Characterization and Analysis of Microplastics.. Compr Anal Chem 75

Rist S, Vianello A, Winding MHS, Nielsen TG, Almeda R, Torres RR, Vollertsen J (2020) Quantification of plankton-sized microplastics in a productive coastal Arctic marine ecosystem. Environ Pollut 266:115248

Romano E, Celia Magno M, Bergamin L (2018) Grain size of marine sediments in the environmental studies, from sampling to measuring and classifying. A critical review of the most used procedures. Acta IMEKO 7(2):10–15

Saarni S, Hartikainen S, Meronen S, Uurasjärvi E, Kalliokoski M, Koistinen A (2021) Sediment trapping–an attempt to monitor temporal variation of microplastic flux rates in aquatic systems. Environ Pollut 274:116568

Sbrana A, Valente T, Scacco U, Bianchi J, Silvestri C, Palazzo L, de Lucia GA, Valerani C, Ardizzone G, Matiddi M (2020) Spatial variability and influence of biological parameters on microplastic ingestion by *Boops boops* (L.) along the Italian coasts (Western Mediterranean Sea). Environ Pollut:114429

Schönlau C, Karlsson TM, Rotander A, Nilsson H, Engwall M, van Bavel B, Kärrman A (2020) Microplastics in sea-surface waters surrounding Sweden sampled by manta trawl and in-situ pump. Mar Pollut Bull 153:111019

Schwaferts C, Niessner R, Elsner M, Ivleva NP (2019) Methods for the analysis of submicrometer- and nanoplastic particles in the environment. TrAC–Trends Anal Chem 112:52–65

Setälä O, Magnusson K, Lehtiniemi M, Norén F (2016) Distribution and abundance of surface water microlitter in the Baltic Sea: a comparison of two sampling methods. Mar Pollut Bull 110(1):177–183

Shim WJ, Hong SH, Eo SE (2017) Identification methods in microplastic analysis: a review. Anal Methods 9(9):1384–1391

Silva AB, Bastos AS, Justino CI, da Costa JP, Duarte AC, Rocha-Santos TA (2018) Microplastics in the environment: challenges in analytical chemistry–a review. Anal Chim Acta 1017:1–19

Stock F, Kochleus C, Bänsch-Baltruschat B, Brennholt N, Reifferscheid G (2019) Sampling techniques and preparation methods for microplastic analyses in the aquatic environment–a review. TrAC–Trends Anal Chem 113:84–92

Tamminga M, Stoewer SC, Fischer EK (2019) On the representativeness of pump water samples versus manta sampling in microplastic analysis. Environ Pollut 254:112970

Uddin S, Fowler SW, Saeed T, Naji A, Al-Jandal N (2020) Standardized protocols for microplastics determinations in environmental samples from the Gulf and marginal seas. Mar Pollut Bull 158: 111374

United Nations Environment Programme (UNEP) (2021) From Pollution to Solution. A global assessment of marine litter and plastic pollution Nairobi

van Bavel B, Lusher A, Jaccard PF, Pakhomova S, Singdahl-Larsen C, Andersen JH, Murray CJ (2020) Monitoring of microplastics in Danish marine waters using the Oslo-Kiel ferry as a ship-of-opportunity. NIVA-rapport

Vital SA, Cardoso C, Avio CG, Pittura L, Regoli F, Bebianno MJ (2021) Do microplastic contaminated seafood consumption pose a potential risk to human health? Mar Pollut Bull 171:112769

Wenneker B, Oosterbaan L, Intersessional Correspondence Group on Marine Litter (ICGML) (2010) Guideline for Monitoring Marine Litter on the Beaches in the OSPAR Maritime Area. Edition 1.0. London, UK, OSPAR Commission, 15pp & Annexes

Wang W, Wang J (2018) Investigation of microplastics in aquatic environments: an overview of the methods used, from field sampling to laboratory analysis. TrAC–Trends Anal Chem 108:195–202

Wesch C, Bredimus K, Paulus M, Klein R (2016) Towards the suitable monitoring of ingestion of microplastics by marine biota: a review. Environ Pollut 218:1200–1208

Remote Sensing: Satellite and RPAS (Remotely Piloted Aircraft System)

9

Martha Bonnet Dunbar, Isabel Caballero, Alejandro Román, and Gabriel Navarro

Contents

Abstract

Remote sensing is the science of detecting, monitoring, and acquiring information about the physical characteristics of a certain area without actually coming into contact with it, contrary to in situ observations. This chapter on satellites and RPASs (remotely piloted aircraft systems) provides an introduction to remote sensing, including a brief historical background; an overview of elements involved in remote sensing, for example, how radiation of different wavelengths reflected by or emitted from objects or materials in the distance is detected and measured; and the difference between spatial, temporal, and spectral ranges and resolutions. Advantages and limitations of remote sensing compared to in situ measurements are discussed as well as three main remote sensing platforms, i.e., ground level (towers and cranes), aerial (RPAS), and spaceborne (space shuttles, polar-orbiting satellites, and geostationary satellites), and their respective sensors, with a focus on satellite and RPAS sensors. The final two sections of the chapter

M. B. Dunbar · I. Caballero · A. Román · G. Navarro (✉)
Department of Ecology and Coastal Management, Institute of Marine Sciences of Andalusia (ICMAN), Spanish National Research Council (CSIC), Puerto Real, Cádiz, Spain
e-mail: Gabriel.navarro@icman.csic.es

J. Blasco, A. Tovar-Sánchez (eds.), *Marine Analytical Chemistry*,
https://doi.org/10.1007/978-3-031-14486-8_9

take the reader from theory to practice, discussing the role that remote sensing plays in the development of many technologies and applications in major scientific fields, in particular, its application in marine analytical and environmental chemistry. The chapter concludes with a series of real-world practical case studies by authors of this chapter, in which different remote sensing techniques have been used around the world to monitor and assess areas of ecological concern, following extreme environmental events, such as volcanic eruptions, hurricanes, typhoons, phytoplankton blooms and catastrophic floods.

Keywords

Remote sensing · Satellite · RPAS · Environmental hazard

9.1 Introduction

Remote sensing is the science of detecting, monitoring, and acquiring information about the physical characteristics of a certain area without actually coming into contact with it, contrary to in situ or on-site observation. The process of remote sensing includes the detection and measurement of radiation of different wavelengths reflected by or emitted from objects or materials in the distance. These different wavelengths are key to their identification and form the basis for subsequent classification by class/type, substance, and spatial distribution.

By measuring the area's reflected and emitted radiation at a distance, typically from satellite, aircraft, or remotely piloted aircraft system (RPAS), special sensors can capture images of much larger areas than can be observed from the ground.

The term "remote sensing" was first used in the USA in the 1950s by Evelyn L. Pruitt, a research geographer at the US Office of Naval Research who, prompted by the need to define emerging imaging capabilities, came up with a dedicated term, which is now used ubiquitously. The concept of remote sensing itself, however, began much earlier with aerial photography as the first form of remotely capturing information about the Earth. Deriving from the Greek word photo, photography literally means light and graph, i.e., to draw. Going back to 500 BC, the philosopher Aristotle discovered that by directing sunlight through a pinhole, an upside-down image of the sun appeared on the ground. He used this method to view an eclipse without having to look directly at the sun.

After the French inventor Joseph Nicéphore Niépce successfully produced the first photographs in the 1800s, in 1850, Gaspard-Félix Tournachon ("Nadar") captured the first aerial photograph from a hot-air balloon of a French village in 1858, although his earliest surviving photograph was taken above Paris, France, in 1866. The oldest aerial photograph that has survived was taken in Boston, USA, in 1860 by James Wallace Black. In 1909, Wilbur Wright who, together with his brother Orville Wright—the famous Wright Brothers—developed the world's first airplane, was the first to successfully capture aerial images from an airplane. By World War I, airplanes equipped with cameras proved invaluable in military reconnaissance by providing aerial images of large surface areas, and by World War II, this became almost standard practice. In fact, the allied forces had a dedicated team of experts detecting hidden Nazi rocket bases with the help of millions of stereoscopic aerial images. During the Cold War between the Soviet Union and the USA, the Lockheed U-2 ("Dragon Lady") was able to fly at the ultrahigh altitude of 21,300 meters, resulting in an increased use of aerial reconnaissance. Shortly afterward in an attempt to conquer space, the Soviet Union launched Sputnik 1 and Sputnik 2, the first artificial satellites in space. The USA followed close behind with the successful launch of Explorer 1 in 1960, beginning the so-called Cold War Space Race.

These pioneers were followed by a long series of military and civilian space missions, and interest in using these platforms to obtain data from the Earth's atmosphere and surface soon became apparent. In 1960, the US National Aeronautics and Space Administration (NASA) launched the first meteorological satellite TIROS 1. The next in the series—TIROS 2—was the first to detect differences in sea-surface temperature, marking the beginning of ocean observation via satellite. Realizing this potential, NASA sponsored an initial meeting at the Woods Hole Oceanographic Institution (WHOI), in Falmouth, MA, in 1964 to discuss ways in which space-based remote sensing could contribute to oceanography, resulting in a road map of remote sensing toward ocean exploration.

The first Earth-observing satellite was launched in 1972, then known as the Earth Resources Technology Satellite (ERTS), and subsequently renamed Landsat 1, the first in a series of Landsat missions jointly managed by NASA and the U.S. Geological Survey (USGS), with the latest Landsat 9- in orbit since 2021. Landsat 1 was the first satellite launched with the specific aim to study and monitor Earth's landmasses. The Landsat program's continuous archive since 1972—the world's longest continuously acquired collection in fact—provides important land change data and current information that would otherwise not be available.

> I'd go to meetings and people were just jumping up and down because they had discovered another use for the data.
>
> Virginia T. Norwood, talking about early Landsat data, published in the article "The woman who brought us the world", MIT Technology Review, June 29, 2021
>> That is one cool thing about Landsat… people are always finding new applications.
>> Jeff Masek, Landsat 9 Project Scientist, NOVA Now podcast, Dec 17, 2020.

Remotely piloted aircraft systems (RPASs), also commonly referred to as unmanned aerial vehicles (UAVs), unmanned aircraft systems (UASs), and drones, are aircrafts that operate without any humans onboard, with the additional ground-based controller and a component that communicates with the RPAS. These systems are operated via remote control by a human operator, but there are also various degrees of autonomy, including autopilot assistance and even aircrafts that function completely autonomously without any human intervention. Although RPASs were originally developed for military purposes, in order to carry out missions that were too dangerous for human intervention, and continued doing so throughout the

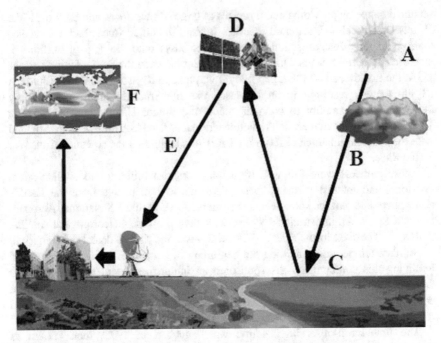

Fig. 9.1 Elements involved in remote sensing, mainly from satellite

twentieth century, the potential for their use in mapping was already appreciated in the late 1970s by different research groups (Przybilla and Wester-Ebbinghaus 1979; Wester-Ebbinghaus 1980). Navigation and mapping sensors were mounted onto radio-controlled platforms that could acquire low-altitude, high-resolution imagery, an idea that at first did not gain much attention in academia. However, in the technology and service industries, visionary companies as well as open-minded civil aviation authorities that anticipated the social and business benefits of unmanned aircrafts did start developing, applying, and regulating the technology (Petrie 2013).

The following elements of remote sensing are graphically presented in Fig. 9.1:

1. Illumination or source of energy (A). The basic requirement for remote sensing is a source of electromagnetic radiation that reaches the object or is directly emitted by the object itself.
2. Interaction between the radiation and the atmosphere (B). The electromagnetic radiation interacts with the atmosphere as it travels from the source to the object and from the object to the remote sensor. Therefore, the atmospheric interference contained in the signal that is received by the sensor must be eliminated.
3. Interaction with the object (C). In cases where the electromagnetic radiation comes from an external source and is not emitted by the object itself, it interacts with the object. This interaction depends on the properties of the land or ocean area and the electromagnetic radiation.

4. Detection of the electromagnetic radiation by the sensor (D). After emission from the object or reflection after interaction, the electromagnetic radiation is detected and registered by the sensor mounted on the remote sensing platform.
5. Transmission, reception, and processing (E). The signal registered by the sensor is transferred to a data reception station, stored in a specific format, and made available to end users. For RPASs, data are collected directly by the operator.
6. Interpretation and analysis (F). Using suitable processing methods, unwanted data are eliminated or corrected and made a parameter of interest.

9.2   Advantages and Limitations of Remote Sensing

In stark contrast to remote sensing, another way of obtaining information from the Earth is by means of field or in situ measurements, carried out with instruments in direct contact with the land or water. This intimate contact means that these measurements are usually more accurate than those made from space, which rarely match the former in precision. However, despite this, remotely sensed data have a number of advantages that compensate for this limitation. The first major advantage is related to the panoramic perspective and synoptic character of the data taken from space, as well as the possibility of obtaining information from remote or difficult to access areas, as is the case in some parts of the ocean, for example. The privileged position of remote sensors in space, hundreds of kilometers away from the Earth's surface, allows them to observe large areas, as opposed to the punctual measurements of in situ sampling. Moreover, this view of large portions of the Earth is almost instantaneous, contrasting with the long times needed for in situ sampling campaigns, especially larger areas or regions. In the specific case of the marine environment, its dynamic and changing nature, coupled with the characteristic scales of some oceanographic phenomena, gives remote sensing data a clear advantage when studying certain oceanographic phenomena. Moreover, certain aspects of the ocean, such as the ubiquity of mesoscale ocean eddies or the large extent of rapid phytoplankton blooms, were first revealed in space-based data. Another example in the marine environment concerns the global and repetitive coverage of the oceans. The orbits described by space platforms allow remote sensors to acquire data from almost the entire surface of the Earth in a continuous and repeatable manner. This global dimension is important as the ocean covers more than 70 % of the planet's surface, supports important ecosystems, and plays an important role in climate regulation. The repeatability of measurements allows multi-temporal studies to be carried out and the evolution of certain ocean processes to be monitored. Long-term time series of satellite data are now available, such as those obtained with the US National Oceanic and Atmospheric Administration's (NOAA) Geostationary Operational Environmental Satellites (GOES), which have been continuously collecting data since 1975. Another important advantage is the immediate data transmission from satellites to receiving stations and from there in near real-time to end users. This is possible thanks to the presentation of remotely sensed data in digital format, which also facilitates their visualization, interpretation, and integration with other data and models.

9.3   Spatial, Temporal, and Spectral Resolutions and Ranges

Generally speaking, resolution is defined as the ability of the sensor to discriminate detailed information about the object under study and is related to the smallest features or group of features that can be observed. The range, on the other hand, is analogous to the total coverage of the sensor and refers to the largest features or group of features that can be detected.

Spatial resolution refers to the size of the smallest portion of the land or ocean that the sensor is able to differentiate. In electro-optical sensors, it is often related to the concept of the instantaneous field of view (IFOV), which is the cone of angular visibility, measured in radians and observed at a given time by the sensor. It is precisely the intersection of the IFOV with the Earth's surface that determines the portion of the land or ocean that contributes to the measurement made by the sensor. The remote sensor detects the average electromagnetic radiation preceding this portion. It is usually assumed, but not always correct, that this resolution cell is square at least with regard to the nadir, and the side of the cell is taken as a measure of the spatial resolution. The smaller the size of this side, the better the spatial resolution of the data. This distance corresponds to the size of the smallest unit of information included in the satellite image, which is called a pixel (from *picture element*). When the sensor points the surface away from the nadir, the size of the resolution cell increases and becomes distorted, since the distance between the sensor and the object also increases, and effects of overlapping in the sampling of the land or sea surface and of the curvature of the Earth's surface that intensify these facts begin to be important. In any case, in electro-optical sensors, the spatial resolution mainly depends on the IFOV value and the orbital height, and to a lesser extent on other factors, such as the scanning speed and the number of detectors, whereas the spatial range is the total area covered by the remote sensor when acquiring the data. This range in electro-optical sensors is related to the concept of field of view (FOV), which is the total angle of view, measured in radians, observed by the sensor. The intersection of the FOV with the surface of the land or the ocean, which in turn depends on the orbital height, determines the spatial range, which is determined by the lateral extent of the observed surface, i.e., the swath. In general, resolution and spatial extent are inversely related, and each application requires remotely sensed data with a specific resolution and spatial extent.

Another important feature of satellite remote sensing acquisition systems is their ability to periodically observe the same area of the Earth's surface. This allows users to study dynamic processes and phenomena that occur near the land or sea surface. Therefore, *temporal resolution* refers to the periodicity (or frequency) with which the remote sensor can obtain data from the same area. The shorter this time period (or the higher the frequency), the better the temporal resolution of the sensor. This is also referred to as the revisit period of a satellite sensor and is usually several days, and the absolute temporal resolution of the remote sensing system to recapture images from the exact same area at the same viewing angle is equal to this revisit period.

While one of the strengths of remote sensing is the spatial detail and coverage of its measurements, the frequency with which these measurements are obtained is one of its weaknesses, especially when compared to in situ methods that allow data to be obtained almost continuously in time. The temporal resolution is primarily a function of the orbital characteristics of the platform, as well as the design of the sensor, especially its FOV, which determines the swath of land or ocean observed. Geostationary orbiting satellites that are directed toward the same region of the Earth's surface have the highest data acquisition frequency. In contrast, non-imaging sensors, such as the altimeter, which are located on polar-orbiting satellites, have the lowest temporal resolution. In these cases, the temporal resolution would be exactly the time it takes the satellite to complete a full orbital cycle, repeating the trace of its orbit over the Earth's surface. This orbital repeat cycle would be shortened to a duration of 1 to 3 days, albeit at the expense of very poor spatial coverage. If the orbital cycle is longer than 10 days, the spatial sampling density is greatly improved, but a temporal resolution longer than 10 days can be a significant disadvantage for many applications. Imaging remote sensors aboard polar-orbiting satellites can overcome problems associated with the orbital cycle depending on how large an area of the Earth's surface they observe as they travel.

The area of the electromagnetic spectrum in which the sensor operates should also be taken into account when considering temporal resolution. Sensors that detect visible radiation can only obtain data from the surface of the Earth during the day, while those that operate in the infrared can do so during both day and night, increasing their sampling frequency. Cloud coverage is an impediment to obtaining data for sensors that record visible and infrared radiation, affecting the temporal resolution in areas with frequent cloud cover. However, clouds are not an obstacle for sensors operating in the microwave region.

The temporal scale refers to the total period for which data from a certain region recorded by a sensor are available. This temporal range is limited in land and ocean observation from space by the lifetime of remote sensors and satellites, which is usually a few years. Once in orbit, it is not feasible to carry out direct repairs or maintenance of the satellite and its sensors, which deteriorate with the passage of time. Therefore, sometimes, in order to extend the time range, when a satellite is no longer operational, another satellite with similar characteristics is launched, increasing the time for which data are obtained with a sensor in a specific area. Through the previously mentioned Landsat series, for example, remote sensing data have been collected continuously this way since 1972. For aircraft or UAV, the temporal resolution is based on when flights are performed and accordingly scheduled by the pilot or the agency funding the flight.

Remote sensors do not detect electromagnetic radiation for an isolated wavelength but rather average radiation from continuous wavelength intervals centered on the wavelengths of interest. These ranges are referred to as spectral bands or sensor channels. The narrower the wavelength range for a particular band or channel, the finer the *spectral resolution*; thus, the spectral resolution refers to the sensor's ability to differentiate the electromagnetic radiation received in different wavelengths of the spectrum and depends on the spectral width of the sensor channels. The smaller the spectral width of these channels, the better the spectral resolution. Multispectral sensors record energy over several separate wavelength ranges at various spectral resolutions, and hyperspectral sensors, which are advanced multispectral sensors, detect hundreds of very narrow spectral bands throughout the visible, near-infrared, and mid-infrared portions of the electromagnetic spectrum. Their very high spectral resolution facilitates fine discrimination between different targets based on their spectral response in each of the narrow bands. Spectral range refers to the total number of spectral bands and the range of the electromagnetic spectrum covered by these bands. Therefore, the spectral range improves with increasing number of spectral bands and increasing size of inspected region. Radars are usually the acquisition systems with the lowest resolution and spectral range, many of which work with a single channel, although some of them are capable of distinguishing the polarization state of electromagnetic radiation. In contrast, electro-optical sensors have the highest number of narrow spectral bands that occupy a substantial region of the electromagnetic spectrum, in particular, hyperspectral systems, as is the case of the Hyperion sensor, which has 220 spectral bands. It can, therefore, be argued that the greater the number of bands in a sensor and the narrower the bands, the better the sensor's ability to reproduce the spectral response of an observed object. However, the choice of the number, width, and location of spectral bands included in a sensor is related to the objectives for which it is designed. Therefore, meteorological sensors such as Meteosat or AVHRR can operate correctly with only one band in the visible, since clouds do not offer significant chromatic variations. On the other hand, sensors designed to obtain information on ocean color, such as the SeaWiFS (Sea-viewing Wide Field-of-view Sensor) or MODIS, need several bands that allow them to distinguish different components that absorb and reflect light and estimate their concentration.

The different types of resolution and range described above/previously are not independent and are closely related to each other. As an obvious example, if the spatial resolution of a sensor is increased, then the spatial range decreases, and consequently, the temporal resolution also decreases. Similarly, by increasing the spatial resolution, the observation time and the amount of radiation arriving from the inspected area are reduced by decreasing the IFOV dimensions, and this can reduce the radiometric resolution and also the spectral resolution if the size of the sensor bands is increased to receive a greater amount of electromagnetic radiation.

Another important aspect to keep in mind is that any increase in resolution and/or range implies an increase in the volume of data to be processed. Therefore, it is not possible to increase all resolutions for a system simultaneously, and in designing a remote sensor, specific objectives should be the deciding factor for which resolutions and/or ranges should be improved.

9.4 Platforms

A platform is defined as the carrier for remote sensing sensors. There are three major remote sensing platforms: ground-level platforms (towers and cranes), aerial platforms (UAV, helicopters, and aircraft), and spaceborne platforms (space shuttles, polar-orbiting satellites, and geostationary satellites) (Fig 9.2). In this chapter, we focus on satellite and UAV sensors.

9.4.1 Satellite Agencies

9.4.1.1 National Aeronautics and Space Administration (NASA)
Established in 1958, NASA is the USA's civil space program. NASA studies the Earth, including climate, our sun, and our solar system, at its 20 centers and facilities located across the USA and in the only national laboratory in space.

Although satellite technology continues improving throughout the years, the mission stays the same: monitor Earth's land and coastal regions to help people manage natural resources.

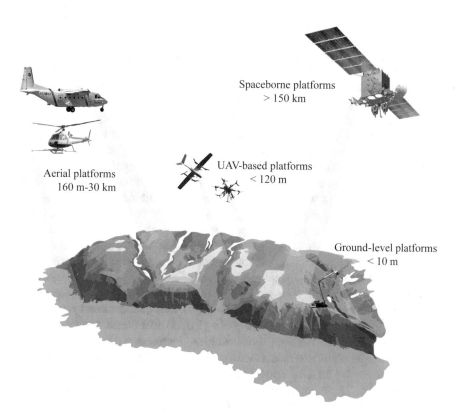

Spaceborne platforms
> 150 km

Aerial platforms
160 m-30 km

UAV-based platforms
< 120 m

Ground-level platforms
< 10 m

Fig. 9.2 Remote sensing platforms

NASA was responsible for the development and launch of a number of satellites with Earth applications, such as the Landsat series mentioned previously in the introduction. Landsat satellites give us a global perspective of how the Earth is changing because of natural causes, such as earthquakes, or because of human-caused drivers, such as greenhouse gas emissions that lead to warming temperatures on a global scale. The latest Landsat mission, Landsat 9, was launched on September 27, 2021, aboard a United Launch Alliance Atlas V rocket. Together in orbit, Landsat 9 and its sister satellite, Landsat 8, join forces to collect images from the planet with an 8-day revisit time. Although satellite technology continues improving throughout the years, the mission aims remain the same: monitor Earth's land and coastal regions to help people manage natural resources.

Additionally, MODIS (or Moderate Resolution Imaging Spectroradiometer) is a key instrument aboard the Terra and Aqua satellites launched by NASA. Terra MODIS and Aqua MODIS acquire multispectral data by viewing the entire Earth's surface on a daily basis. These data improve our understanding of global dynamics and processes that occur on land, in the oceans, and in the lower atmosphere, providing earth and climate measurements. MODIS plays a vital role in the development of validated, global, interactive Earth system models able to predict global change accurately enough to assist policymakers in making sound decisions concerning the protection of our environment.

9.4.1.2 National Oceanic and Atmospheric Administration (NOAA)

The NOAA Satellite and Information Service provides timely access to global environmental data from satellites and other sources to monitor the Earth. NOAA has a long history of geostationary and polar-orbiting environmental satellites in operational programs, such as the Joint Polar Satellite System (JPSS), the Geostationary Operational Environmental Satellite (GOES) program, and the Polar Operational Environmental Satellite (POES) program. The organization's satellite program has been essential for life-saving weather and climate forecasts for the USA and beyond. However, these satellites have also evolved to gather environmental data used for a wide range of applications in the ocean, coastal regions, agriculture, and the atmosphere, among many others, as well as in space. Geostationary satellites help monitor and predict weather and environmental events, including tropical systems, tornadoes, flash floods, dust storms, volcanic eruptions, and forest fires. Polar-orbiting satellites collect data for weather, climate, and environmental monitoring applications including precipitation, sea-surface temperatures, atmospheric temperature and humidity, sea ice extent, forest fires, volcanic eruptions, global vegetation analysis, as well as search and rescue. The Visible Infrared Imaging Radiometer Suite (VIIRS) instrument aboard the joint NASA/NOAA Suomi National Polar-orbiting Partnership (Suomi NPP) and NOAA-20 satellites produce reflectance imagery in near real-time (NRT), providing continuity from the MODIS corrected reflectance imagery that provides natural looking images. These continuous global environmental observations are subsequently derived to produce various geophysical variables that help to describe the Earth's oceanic, atmospheric, and terrestrial systems. NOAA's satellite data improve resilience to climate variability, maintain economic vitality, and improve the security and well-being of society.

9.4.1.3 The European Space Agency (ESA)

The ESA was founded in 1975 resulting from the merger of the European Launcher Development Organisation (ELDO) and the European Space Research Organisation (ESRO), both established in 1964. It is an international organization with 22 member states pushing the boundaries of science and technology and promoting economic growth in Europe. The ESA's mission is to shape the development of Europe's space capability and ensure that investment in space continues to deliver benefits to the citizens of Europe and the world. Therefore, it is dedicated to the peaceful exploration and the use of space for the benefit of humankind. By coordinating financial and intellectual resources of its members, it can undertake programs and activities far beyond the scope of any single European country. The ESA's programs are designed to discover and understand more about Earth, its immediate space environment, our solar system, and the universe, as well as to develop satellite-based technologies and services, and to promote European industries. The ESA also works closely with space organizations outside of Europe and has cooperated with NASA on many projects. The ESA also established a system of meteorological satellites known as Meteosat. The Medium Resolution Imaging Spectrometer (MERIS) was one of the main instruments on board the ESA's Envisat platform, a programmable spectrometer operating in the solar-reflective spectral range. Although it is primarily dedicated to ocean-color observations, MERIS broadened its scope of objectives to atmospheric and land surface-related studies, being operational throughout the Envisat mission lifetime, from 2002 to 2012. MERIS had a high spectral and radiometric resolution and a dual spatial resolution, within a global mission covering open ocean and coastal zone waters and a regional mission covering land surfaces. The primary objective of MERIS is to observe the color of the ocean, both in the open ocean (clear or case I waters) and in coastal zones (turbid or case II waters). These observations are used to derive estimates of the concentration of chlorophyll and sediments in suspension in the water, for example. It has also been used for monitoring and mapping *Posidonia oceanica* deposits. These measurements are useful for studying the oceanic component of the global carbon cycle and the productivity of these regions, among other applications.

Copernicus Program: "Europe's Eyes on Earth"

Copernicus is the European Union's Earth observation program, looking at the Earth and its environment to benefit all European citizens. It offers information services that draw from satellite Earth observation and in situ (non-space) data. The European Commission manages the program, and it is implemented in partnership with the member states, the ESA, the European Organization for the Exploitation of Meteorological Satellites (EUMETSAT), the European Centre for Medium-Range Weather Forecasts (ECMWF), EU agencies, and Mercator Ocean. Global information from satellites and ground-based, airborne, and seaborne measurement systems provide information to help service providers, public authorities, and other international organizations improve European citizens' quality of life and beyond. The ESA is developing a series of next-generation Earth observation missions called Sentinels specifically for operational needs of the Copernicus program, on behalf of the joint ESA/European Commission initiative Copernicus. The objective of the Sentinel program is to replace retired Earth observation missions, such as the ERS and Envisat missions, or those that are currently nearing the end of their operational life span. This will ensure a continuity of data so that there are no gaps in ongoing studies. Each mission focuses on a different aspect of Earth observation; atmospheric, oceanic, and land monitoring and the data can be used in many applications. The information services provided are free and openly accessible to users, reaching different degrees of maturity. Some of the services were already declared operational several years ago: in 2012 for the Land Monitoring Service and the Emergency Management Service—Mapping—and in 2015 for the Atmosphere Monitoring Service and the Marine Environment Monitoring Service. Others were declared operational more recently: in 2016 for the Border Surveillance and Maritime Surveillance components of the security service and in May 2017 for the Support to External Action component and in July 2018 for the Climate Change Service.

Each Sentinel mission is based on a constellation of two satellites to fulfill revisit and coverage requirements, providing robust datasets for Copernicus services. The missions correspond to the following:

(a) *Sentinel-1* is a polar-orbiting, all-weather, day-and-night radar imaging mission for land and ocean services. With objectives of land and ocean monitoring, Sentinel-1 will be composed of two polar-orbiting satellites operating day and night and will perform radar imaging, enabling them to acquire imagery regardless of the weather. Sentinel-1A was launched on 3 April 2014 and Sentinel-1B on 25 April 2016.

(b) *Sentinel-2* is a polar-orbiting, multispectral high-resolution imaging mission. The objective of Sentinel-2 is land monitoring to provide, for example, imagery of vegetation, soil and water cover, inland waterways, and coastal areas. Sentinel-2 can also deliver information for emergency services. Sentinel-2A was launched on 23 June 2015, and Sentinel-2B followed on 7 March 2017.

(c) *Sentinel-3* is a multi-instrument mission, in which the primary objective is marine observation to measure sea-surface topography, sea- and land-surface temperature, ocean color, and land color with high-end accuracy and reliability. The mission supports ocean forecasting systems, as well as environmental and climate monitoring. Composed of three satellites, the mission's primary instrument is a radar altimeter, but polar-orbiting satellites will carry multiple instruments, including optical imagers. Sentinel-3A was launched on 16 February 2016, and Sentinel-3B joined its twin in orbit on 25 April 2018.

(d) *Sentinel-5* Precursor—also known as Sentinel-5P—is the forerunner of Sentinel-5 to provide timely data with high spatiotemporal resolution on a multitude of trace gases and aerosols affecting air quality, ozone, UV radiation, and climate. It has been developed to reduce data gaps between the Envisat satellite and the launch of Sentinel-5. Sentinel-5P was taken into orbit on 13 October 2017.

(e) *Sentinel-4* is a payload devoted to atmospheric monitoring that will be embarked upon a Meteosat Third Generation-Sounder (MTG-S) satellite in geostationary orbit. The mission aims to provide continuous monitoring of the composition of the Earth's atmosphere at high temporal and spatial resolution, and the data will be used to support monitoring and forecasting over Europe.

(f) *Sentinel-5* is a payload that will monitor the atmosphere from polar orbit aboard a MetOp Second Generation satellite.

(g) *Sentinel-6* Michael Freilich is the next radar altimetry reference mission to extend the legacy of sea-surface height measurements, until at least 2030. It is an Earth observation satellite mission developed to provide enhanced continuity to the very stable time series of mean sea-level measurements and ocean sea state that started in 1992, with the TOPEX/Poseidon mission. It carries a radar altimeter to measure global sea-surface height, primarily for operational ocean-ography and for climate studies. The first satellite was launched into orbit on 21 November 2020.

Did you know?

In 1972, Virginia Tower Norwood (also known as "the mother of Landsat") invented the first multispectral scanner to image Earth from space. Landsat 1 and its successors have been scanning the planet continuously ever since. Norwood was a pioneer inventor in the field of microwave antenna design. She designed the transmitter for a reconnaissance mission to the moon that cleared the way for the Apollo landings. And she conceived and led the development of the first multispectral scanner to image Earth from space—the first in a series of satellite-based scanners that have been continuously imaging the world for nearly half a century.

9.4.2 UAV

An unmanned aerial vehicle (UAV), also referred to as a "drone," is a relatively small and mobile flying robot that can operate autonomously or is controlled telemetrically over opened and confined areas, obtaining ultrahigh-resolution images at centimeter spatial resolution (Boukoberine et al. 2019; Hassanalian and Abdelkefi 2017; Townsend et al. 2020; Valavanis and Vachtsevanos 2015). The growing interest among the scientific community in the use of UAVs for research has led to the development of different types of drones in many shapes and sizes to be applied in a variety of activities. There are many metrics for UAV classification, such as size, mean takeoff weight (MTOW), operational altitude, or flight range (Valavanis and Vachtsevanos 2015; Watts et al. 2012). One of the most commonly accepted classifications is by wing type, and although multi-rotors are the most popular in the drone world, there are other options that work better in other applications. This section shows four categories of UAVs classified by wing type in single rotor, multi-rotor, fixed wing, and hybrid, each with their own characteristics (Table 9.1), and briefly described as follows.

Table 9.1 Advantages and disadvantages of different types of UAVs

UAV Type	Advantages	Disadvantages
Single rotor	• VTOL and hover flight • Long endurance (higher if the drone is gas powered) • Greater area coverage • Higher payload capacity	• More dangerous and harder to fly • More expensive • Higher complexity
Multi-rotor	• VTOL and hover flight • Easy control and maneuver • Very stable and good camera control • Accessibility (indoors and outdoors) • Simple design	• Short flight time • Limited area coverage • Small payload capabilities
Fixed wing	• Long-endurance flight • Greater area coverage • Fast flight speed • Heavier payload capacity • Higher flight altitude and flight time	• Space is necessary for launching and recovery • No VTOL and hover flight • Harder to fly • Only forward movement • More expensive
Hybrid	• VTOL flight • Long-endurance flight • Large area coverage	• Under development • Not perfect transition between hovering and forward flight

9.4.2.1 Single Rotor

This platform consists of a single main rotor and a tail rotor for flight heading. It is commonly used in manned aviation, whereas single rotor drones are not commonly used for research purposes by the scientific community. Its main feature is greater endurance compared to multi-rotors (which can be even higher with gas engines), which also allows it to cover large areas. Its main advantages are its high load capacity (being able to equip, e.g., an aerial lidar device) and the ability to combine hovering with fast-forward flights. Its drawbacks are its complexity, cost, danger, and flight difficulty.

9.4.2.2 Multi-rotor

This type of UAV, the most commonly used in scientific research, consists of multiple rotors; therefore, it is classified according to the number of rotor blades. These devices are easily controllable and highly maneuverable and can monitor hard-to-reach places indoors and outdoors. However, they are very limited by the flight time since batteries have little autonomy and by the payload capacity, which is very reduced compared to other types of UAVs. In addition, these drawbacks limit the spatial coverage of these devices, requiring more than one maneuver to cover a wide flight area.

9.4.2.3 Fixed Wing

These drones use a single wing for their lift instead of vertical lift rotors, which makes them more efficient. The fact that they only have to use energy for moving, and not for staying in the air, allows them to cover much greater distances and to have a longer flight time (which can be even higher with gas engines). However, their main disadvantage lies in the difficulty of working in geographically limited areas, since they require space for takeoff and landing maneuvers. In addition, they are expensive and difficult to handle and can only fly forward, which rules out the vast majority of photogrammetric work.

9.4.2.4 Hybrid

This platform merges benefits of fixed-wing and multi-rotor drones, thus assuming a transition between two modes during flight. This means greater flight autonomy and, therefore, greater area coverage, which makes it the ideal platform for scientific research in almost any location. However, despite its many advantages, this device is still under development, especially in terms of optimization of transition between flight modes.

9.5 Types of Sensors

As described previously, remote sensing sensors receive electromagnetic radiation and convert it into a signal that can be recorded and displayed as either numerical data or images. Imaging sensors are the core component of any remote sensing system, and they come with a wide range of spatial, temporal, and spectral resolutions. Sensor system implementation is equally varied, including single- and multiple-sensor configurations, active and passive sensors, and completely solid or optomechanical sensors. In general, sensors are commonly grouped by their spectral sensitivity, and there are several regions of the electromagnetic spectrum that are

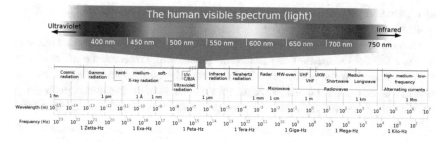

Fig. 9.3 Electromagnetic spectrum

useful for remote sensing. Figure 9.3 shows the electromagnetic spectrum with the most important bands used in remote sensing.

9.5.1 Satellite Sensors

As briefly mentioned in the introduction, remote sensing instruments can be either active or passive. While active instruments have their own illumination or source of energy, passive instruments detect natural energy that is reflected off of or emitted directly by the observed target, with reflected sunlight being the most common external source of radiation sensed by passive instruments (optical sensors). These sensors work in several bands including spectra within and beyond what humans can see (visible, IR, NIR, TIR, and microwave).

Optical remote sensing is a passive technique for land and ocean observation that relies on the sun as a source of illumination. These systems are classified into different types according to the number of spectral bands used in image acquisition: panchromatic sensor (1 single band that combines Red, Green, and Blue bands), multispectral imaging sensors (3–50 spectral bands), and hyperspectral imaging sensors (50–300 spectral bands). These sensors use the following types of technologies: (a) multispectral imaging using discrete detectors and scanning mirrors (e.g., Landsat multispectral scanner and Landsat thematic mapper), (b) multispectral imaging using linear arrays (e.g., SPOT and IKONOS), and (c) imaging spectrometry using linear and area arrays (e.g., AVIRIS and MODIS). Optical sensor data enable the assessment of sea-surface properties, such as phytoplankton concentration, chromophoric dissolved organic matter (CDOM), suspended matter (SPM), type of benthic substrate, vegetation composition, and bathymetry in shallow waters.

Active instruments, which have their own source of energy (electromagnetic radiation) to illuminate the target object, send a pulse of energy from the sensor to the object and subsequently receive the radiation that is reflected back from the target object. There are many different types of active remote sensors, including the following examples:

Radars (radio detection and ranging) use a transmitter operating at either radio or microwave frequencies to emit electromagnetic radiation, and a directional antenna or receiver to measure the time of arrival of reflected or backscattered pulses of radiation from target objects. Distance to the object is determined since electromagnetic radiation propagates at the speed of light. These sensors are widely used for monitoring oil spills.

Scatterometers are a type of active high-frequency microwave radars that transmit microwave pulses to the Earth's surface and subsequently measure the radiation that is backscattered to the instrument. These backscattered coefficients are related to surface roughness. Over ocean surfaces, measurements of backscattered radiation in the microwave spectral region can be used to derive maps of surface wind speed and direction.

Lidar (light detection and ranging) is a remote sensing method that uses light in the form of a pulsed laser to measure variable distances to the Earth. These light pulses—combined with other data recorded by the airborne system—generate precise, three-dimensional information about the shape of the Earth and its surface characteristics.

A lidar instrument primarily comprises a laser, a scanner, and a specialized GPS receiver. Airplanes and helicopters are the most commonly used platforms for acquiring lidar data over broad areas, with two main types of lidar being topographic and bathymetric. Topographic lidar typically uses a near-infrared laser to map the land, while bathymetric lidar uses water-penetrating green light to also measure seafloor and riverbed elevations. Distance to the target object is determined by recording the time between transmitted and backscattered pulses and using the speed of light to calculate the distance traveled.

9.5.2 UAV Sensors

Unmanned aerial vehicles (UAVs) can be equipped with a wide range of sensors, ranging from the visible to the infrared spectrum (Fig. 9.3), and are in continuous technological evolution in order to guarantee the success of marine and coastal monitoring missions. Most of the existing satellite or aircraft sensors have been adapted to the size and load capacity of UAV platforms, resulting in a wide variety of low-cost models of RGB, multispectral, hyperspectral, thermal, and lidar cameras (Ren et al. 2019; Valavanis and Vachtscvanos 2015). The discussed sensors and their characteristics as well as some case studies are summarized in Table 9.2, and details of these sensors are highlighted in the following subsections.

RGB sensor. This is the most common type of sensor mounted on a UAV platform and allows images of true color of red, green, and blue bands from the visible spectrum (between 400 and 760 nm) to be acquired. High-quality RGB cameras ensure quality photogrammetric products and low-signal/noise-ratio data for data analysis (Ren et al. 2019; Valavanis and Vachtsevanos 2015). The vast majority of drones used for entertainment or television are manufactured with their own integrated RGB sensor (e.g., DJI drones). However, in scientific research, mountable RGB sensors are more commonly used (e.g., Sony Nex5n RGB, Zenmuse Z30, and Sony A6000), since they offer the option of parameter modification, such as focal length, lens distortion, or pixel size, and are successfully used to monitor coastal environmental issues, such as plastics/debris accumulation (Garcia-Garin et al. 2020), invasive species proliferation (Marzialetti et al. 2021), or harmful algal blooms (HABs, Xu et al. 2018).

Thermal sensor. Lightweight thermal sensors that can acquire information in the mid- and long-wave infrared (wavelengths from 5000 nm to 35,000 nm) are broadly used in surface temperature and thermal emission measurements. Unlike satellite imagery, atmospheric effects are negligible, the data processing is less complex, and temperature values are theoretically more precise. However, the lower structural complexity of UAV platforms means that lightweight thermal sensors generally do not come with cooled detectors, thus resulting in lower capture rates, lower spatial resolution, and lower sensitivity due to a reduced signal-to-noise ratio (Valavanis

Table 9.2 Most commonly used UAV sensor types with examples

Sensor Type		Electromagnetic Spectrum	Examples	Case Studies
RGB		Visible light	Sony Nex5n RGB, Zenmuse Z30, Sony A6000, and Yuneec E90	Plastic/Debris Detection (Garcia-Garin et al. 2020), Invasive Species Detection (Marzialetti et al. 2021), and HABs monitoring (Xu et al. 2018)
Thermal		Mid-wave infrared and long-wave infrared	FLIR Vue Pro 19 mm and Workswell Wiris Pro	Plastic/Debris Detection (Goddijn-Murphy and Williamson 2019, Topouzelis et al. 2019) and HAB monitoring (Berni et al. 2009)
Multispectral		Visible light and near infrared	MicaSense RedEdge-MX, MicaSense RedEdge-P, and Sentera 6X	Plastic/Debris Detection (Biermann et al. 2020), Invasive Species Detection (Marzialetti et al. 2021), and HAB monitoring (Goldberg et al. 2016)
Hyperspectral		Visible light, near infrared, and short-wave infrared	HySpex SWIR-640 and Headwall VNIR-SWIR	Plastic/Debris Detection (Balsi et al. 2021), Invasive Species Detection (Bolch et al. 2021), and HAB monitoring (Young et al. 2019)
Lidar		Visible light and infrared	LeddarTech Vu8, LeddarOne and LiDAR's Puck Lite	Plastic/Debris Detection (Ge et al. 2016)

and Vachtsevanos 2015). Some typical thermal camera models equipped on drones (e.g., FLIR Vue Pro 19 mm and Workswell Wiris Pro) have been successfully used for the detection of macroplastics in coastal areas (Goddijn-Murphy and Williamson 2019; Topouzelis et al. 2019) or for the monitoring of HABs (Berni et al. 2009), among others.

Multispectral sensor. This sensor can collect spectral information in more than two channels, in addition to standard RGB bands. Its small size and low price, in addition to the high precision in the medium-scale monitoring of the surface, make it a fundamental device for scientific research (Ren et al. 2019). Multispectral cameras are usually employed in precision agriculture, although they are increasingly used to monitor marine/coastal environmental phenomena, such as detection and identification of marine litter (Biermann et al. 2020), detection of invasive species (Marzialetti et al. 2021), or monitoring of HABs (Goldberg et al. 2016). Examples include the MicaSense RedEdge-MX, the MicaSense RedEdge-P, the Sentera 6X, or the Parrot Sequoia.

Hyperspectral sensor. Compared with the limited number of bands that multispectral sensors have, hyperspectral sensors acquire spectral information in greater abundance (hundreds of narrow bands) and higher resolution. In addition, the vast majority are linear-array cameras, which capture large volumes of extremely useful information to address coastal events such as the spectral identification of marine litter (Balsi et al. 2021) or the monitoring of the proliferation of invasive species based on the enhanced spectral detectability (Bolch et al. 2021). However, these sensors are less accessible due to their high cost and because they have compatibility limitations with drones, since they require a large payload capacity to equip them. Storage is another limiting factor given the multidimensional nature of hyperspectral datasets (Ren et al. 2019; Valavanis and Vachtsevanos 2015). Some of the most used models are the HySpex SWIR-640 sensor or the VNIR-SWIR Headwall sensor.

Lidar sensor. As described in the previous section on satellite sensors, lidar sensors work with a laser that is transmitted to the ground surface and subsequently received by the sensor after surface reflection, operating as a radar and assuming one of the most precise ways to obtain real-time geometric data. These sensors allow information to be obtained in places where there are not enough ground control points (GCPs), and they also allow the generation of precise 3D models of surface information. Although they require higher payloads to be mounted on UAVs, they are increasingly used in marine research due to their relative low cost and the precision of the data obtained (Ren et al. 2019; Valavanis and Vachtsevanos 2015). Some examples of lidar sensors that have been equipped on drones are LeddarTech Vu8, LeddarOne, or LiDAR's Puck Lite, which have allowed, among other applications, the detection and identification of macroplastics in coastal regions (Ge et al. 2016).

9.6 Application in Marine Analytical and Environmental Chemistry

Remote sensing plays a huge role in the development of many technologies and applications in major scientific fields, being one of the most innovative research areas for Earth observation. Remote sensing presents a cost-effective complementary approach for a comprehensive assessment of marine environments and aquatic resources, providing several advantages compared to traditional in situ approaches. Remote sensing observations are useful in obtaining up-to-date water quality patterns, biogeochemical parameters, and indicators of environmental health of large areas at any given time and also monitor their dynamic patterns of change, especially in ocean observation. Remote sensing imagery can be used to study degradation of inland and marine environments, pollution, and eutrophication of water bodies. In addition, remote sensing is extensively used to track and monitor the impact of natural hazards, such as hurricanes, floods, active volcanoes, and typhoons on coastal regions in order to detect possible damage and plan appropriate response and management, preventing damage as far as possible. These events have profound effects on marine life and the surrounding environment, requiring a prompt response in order to avoid negative impacts. Some unique remote sensing technologies are also applied for mapping out underwater reefs and bathymetry in strategic locations. Moreover, satellite information is helpful for monitoring and identifying algae with the rapid preparation of synoptic maps. A lot can be learned about a marine ecosystem's health by studying algae or harmful algal blooms (HABs), since they are an indicator of the amount of nitrogen and phosphorous that is leaking into a certain water body. Through long-term monitoring, biochemical seawater properties can be understood, supporting the evaluation of environmental problems and potential health risks and preventing negative impacts on the environment and biodiversity. Therefore, remote sensing provides real-time and historical data for studies of water quality trends and potential impacts of land use and land cover change on biogeochemical parameters over coastal and marine ecosystems. This added-value information is used by regional planners and administrators to conserve and frame policy matters for sustainable development at local, regional, and national scales in the context of a changing global climate.

9.7   Case Studies

A series of case studies are presented as follows, in which different remote sensing techniques are used with different sensors in the context of marine analytical and environmental chemistry:

9.7.1. Volcanic eruption in La Palma, Spain

9.7.2. Monitoring marine macrophytes, Spain

9.7.3. Antarctic environmental research

9.7.4. Impact of a hurricane in the USA

9.7.5. Phytoplankton blooms in Chile

9.7.6. Impact of typhoons in the Philippines

9.7.7. Dredging operations in an estuary in Spain

9.7.8. Catastrophic floods in Spain

Volcanic eruption in La Palma

Case study

In September 19th 2021, the Cumbre Vieja volcanic eruption took place on La Palma Island, Spain. The lava flow and the material expelled by the volcano did not only affect human infrastructure and the general economy of the island, but also reached the ocean near the municipality of Tazacorte forming a lava delta on the west cliff of the island, with significant physical-chemical and biological alterations of seawater. This study highlights the value of Unmanned Aerial Vehicles (UAVs) as a feasible, precise, rapid and safe tool for real time monitoring of the impacts of a volcanic event. In addition, UAVs substantially contributed helping scientists and managers in the emergency assessment. Different areas affected by the volcanic eruption were assessed with optical RGB, thermal and multispectral sensors, and a water sampling device, equipped on board drones.

More info: Román et al. (2021). Unmanned aerial vehicles (UAVs) as a tool for hazard assessment: The 2021 eruption of Cumbre Vieja volcano, La Palma Island (Spain). Science of the Total Environment, 843: 157092.

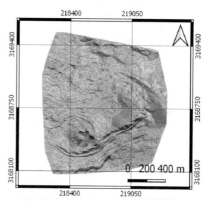

Slope map generated with the SfM workflow followed for the DJI Mavic 2 Pro flight performed on October 2021 over the volcanic crater.

Acknowledgements: This work was financially supported by funds from the Interdisciplinary Thematic Platforms (PTI) WATER:iOS and TELEDETECT granted by the Spanish National Research Council (CSIC) and the grants/projects EQC2018-004275-P, EQC2019-005721, RTI2018-098784-J-I00 and IJC2019-039382-I funded by MCIN/AEI/10.13039/501100011033 and by "ERDF A way of making Europe". Data at sea were collected in the context of the VULCANA-III (IEO-2021-2023) project funded by the IEO-CSIC.

Monitoring Marine Macrophytes

Case study

Marine macrophytes constitute one of the most productive ecosystems on the planet due to the large amount of ecosystem services they provide, since they act as atmospheric CO_2 sinks, provide refuge and food for numerous animal species or perform coastal protection tasks, among others. However, they are seriously affected by climate change and anthropogenic activities, and that is why their monitoring is essential and has evolved over the years until the development of a methodology based on the use of sensors embedded in UAVs, which make it possible to accurately identify the different species and to quantify the extent of these meadows in coastal areas. In this study, a 10-band multispectral sensor (MicaSense RedEdge-MX) is used on a hexacopter drone for the detection and subsequent supervised classification of marine macrophyte meadows in Santibañez (Cádiz Bay, Spain). The results obtained confirm the suitability of this technique to study and monitor marine macrophytes in a range between 0-2 meters depth in coastal areas.

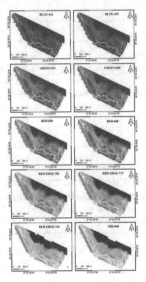

At-sensor reflectance orthomosaics of multispectral bands in the study area.

More info: Román et al. (2021). Using a UAV-Mounted Multispectral Camera for the Monitoring of Marine Macrophytes. Frontiers in Marine Science, 8: 722698.

Acknowledgements: This research was supported by the project EQC2018-004175-and PY20_00244 funded by the National Government and Regional Government of Andalusia, respectively.

Antarctic Environmental Research

Case study

The importance of the Antarctic continent in the Earth's climate, in global ocean circulation and in the global ecosystem is well known. That is why, given the impact of climate change on the polar regions, the need arises among the scientific community to know the functioning and the different responses of the Antarctic ecosystem to extreme environmental conditions, seasonality and isolation. This study presents a novel, fast and accurate methodology based on the use of Unmanned Aerial Vehicles (UAVs) to achieve a comprehensive understanding of the processes taken place in Antarctica, using Deception Island (South Shetland Islands) as a case study. UAV surveys with visible, thermal and multispectral sensors, and a water sampling device, allow the elaboration of precise thematic ecological maps, 3D models of geological structures, anomalous thermal zones mapping, or the sample of dissolved chemicals waters from inaccessible areas.

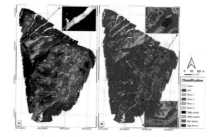

Photomosaics of Vapour Col Chinstrap penguin colony on Deception Island composed of 3800 pictures taken at 100 m altitude with a 10 bands multispectral camera onboard a hexacopter, achieving 6 cm/pixel size. Panel (A): visible RGB mosaic (Red-668, Green-560 and Blue-475) with a zoom capture showing red snow patch; Panel (B): thematic map generated through non-supervised classification method.

More info: *Tovar-Sánchez et al. (2021). Applications of unmanned aerial vehicles in Antarctic environmental research. Scientific Reports, 11: 21717.*

Acknowledgements*: This research has been funded by the Spanish Government projects PiMetAn (ref. RTI2018-098048-BI00), EQC2018-004275-P and EQC2019-005721. This research is part of the POLARCSIC research initiatives.*

Impact of a hurricane in USA

Case study

Mapping bathymetric change is a core task for a wide range of navigation, research, monitoring, and design applications. Satellite-derived bathymetry (SDB) can support this activity, particularly when using data from a platform, like the Sentinel-2A/B twin mission of the Copernicus programme, which provides routine and repetitive image acquisition at 10 m spatial resolution. In this study, we use SDB, in comparison with high-resolution airborne lidar bathymetry (ALB), to quantify bathymetric changes at two inlets in North Carolina following the impacts of the devastating Hurricane Florence in September 2018. We identify bathymetric changes in shallow areas with navigation channels in two of the most dynamic inlets in the Outer Banks, Oregon and Hatteras. Multiple lidar surveys are used to validate the SDB method and for an assessment of accuracy and vertical uncertainty. The multi-temporal SDB products and ALB both show similar results in the erosion/accretion patterns. Comparing the change determined from the two methods, gives a median absolute error of ~0.5 m of SDB compared with ALB, with bias of ±0.2 m for depths ≤7 m; errors that are equivalent to those associated with the SDB estimated absolute depths. By implementing the multi-temporal turbidity correction, SDB based on Sentinel-2 may substantially enhance existing survey methods for change detection and support operational and recursive coastal monitoring on local to regional scales.

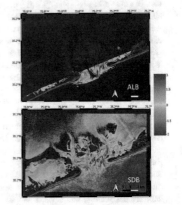

a) Erosion/deposition map after Hurricane Florence in Hatteras Inlet for high-resolution lidar surveys (ALB 2017-ALB 2019), and for b) SDB (SDB 2017-SDB 2019). Green colors indicate erosion (deepening) and brown colors accretion (shallowing). The base map corresponds to a Sentinel-2 image at 10 m spatial resolution.

More information: *Caballero, I., & Stumpf, R. P. (2021). On the use of Sentinel-2 satellites and lidar surveys for the change detection of shallow bathymetry: The case study of North Carolina inlets. Coastal Engineering, 103936.*

Acknowledgements: *This research was supported by the Spanish Ministry of Science, Innovation and Universities (MCIU), the Spanish State Research Agency (AEI) and the European Regional Development Fund (ERDF) (Sen2Coast Project; RTI2018-098784-J-I00). The research was also supported by the National Oceanic and Atmospheric Administration (NOAA).*

Phytoplankton Blooms in Chile

Case study

During the southern summer of 2020, large phytoplankton blooms were detected using satellite technology in Chile (western Patagonia), where intensive salmonid aquaculture is carried out. Some harvesting sites recorded massive fish mortalities, which were associated with the presence of the dinoflagellate species *Cochlodinium sp.* The bloom included other phytoplankton species, such as *Lepidodinium chlorophorum*, which persistently changed the colour of the ocean to green. These blooms coincided with the government-managed emergency lockdown due to the COVID-19 pandemic. Local in situ sampling was reduced. However, imagery from the Copernicus programme allowed operational monitoring. This study shows the benefits of both Sentinel-3 and Sentinel-2 satellites in terms of their spectral, spatial and temporal capabilities for improved algal bloom monitoring. These novel tools, which can foster optimal decision-making, are available for delivering early warning during natural catastrophes and blockages, such as those that occurred during the global COVID-19 lockdown.

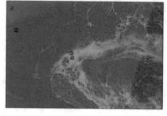

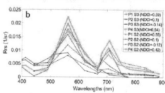

a) RGB (red-green-blue) composite from a Sentinel-3 scene acquired on 8 April 2020, b) Spectral signature in some pixels (P1, P2, P3, and P4) of the scene from Sentinel-3 (300m) and Sentinel-2 on 8 April 2020. The normalized difference chlorophyll index (NDCI) values are indicated for each pixel and satellite (Sentinel-3 and Sentinel-2).

More information: *Rodríquez-Benito et al. (2020). Using Copernicus Sentinel-2 and Sentinel-3 data to monitor harmful algal blooms in Southern Chile during the COVID-19 lockdown. Marine Pollution Bulletin, 161, 111722.*

Acknowledgements*: This research was supported by Mariscope companies, the Spanish Ministry of Science, Innovation and Universities (MCIU), the Spanish State Research Agency (AEI) and the European Regional Development Fund (ERDF) in the frame of the Sen2Coast Project (RTI2018-098784-J-I00).*

Impact of typhoons in Philippines

Case study

Laguna Lake, the largest freshwater lake in the Philippines, is permanently subject to nutrient-driven eutrophication and pollution and experiences harmful algal blooms (cyanoHABs) periodically with serious socio-economic implications. The aim of this study is to evaluate the suitability of the Sentinel-2 imagery of the European Commission's Copernicus Earth Observation programme for lake monitoring during the 2020 Pacific typhoon season (September-November 2020). The Case-2 Regional CoastColour processor is used to atmospherically correct Level 1 data and generate water quality parameters, such as chlorophyll-a (Chl-a) and total suspended matter (TSM) at 10 m. Results show that Super Typhoon Goni and Typhoon Vamco delivered high suspended sediment loads to the reservoir at concentrations above 170 g/m^3 compared to pre-storm situations (0-35 g/m^3). The typhoons also affect Chl-a, with a mean concentration of 10 mg/m^3 and 30 mg/m^3 for pre- and post-typhoons, respectively. In addition, the normalized difference chlorophyll index (NDCI) is used in the Google Earth Engine platform for near-real time monitoring of cyanoHABs at 20 m spatial resolution. Satellite maps are key for detecting the distribution of the blooms due to the patchiness of the green algae species, which usually form scum and elongated slicks in the lake. Maximum records of bloom detection during the study period occur in the Central Bay, one of the lake sections with major aquaculture and fisheries activities. The Sentinel-2 mission improves synoptic mapping of cyanoHABs and enables trends in their extent and severity to be documented, which will assist and benefit the cost-effective management of Laguna Lake.

a) Normalized difference chlorophyll index-NDCI (dll) and b) false composite (NIR-Red-Green) on 3 November 2020 after Typhoon Goni; c) NDCI (dll) and d) false composite (NIR-Red-Green) on 13 November 2020 after Typhoon Vamco.

More information: *Caballero, I., & Navarro, G. (2021). Monitoring cyanoHABs and water quality in Laguna Lake (Philippines) with Sentinel-2 satellites during the 2020 Pacific typhoon season. Science of The Total Environment, 788, 147700.*

Acknowledgements*: This research was supported by the Spanish Ministry of Science, Innovation and Universities (MCIU), the State Research Agency (AEI), and the European Regional Development Fund (ERDF) in the frame of the Sen2Coast Project (RTI2018-098784-J-I00). We also thank Brian Skerry (https://brianskerry.com/) and Rhoy Cobilla for their photographs of the algal blooms.*

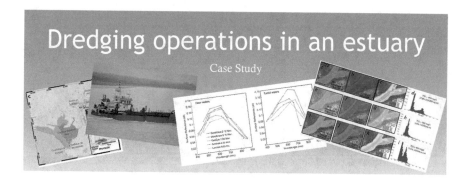

Dredging activities in estuaries frequently cause deleterious environmental effects on the water quality, which can impact flora, fauna, and hydrodynamics, among others. A medium- and high-resolution satellite-based procedure is used in this study to monitor turbidity plumes generated during the dredging operations in the Guadalquivir estuary, a major estuarine system providing important ecosystem services in southwestern Europe. A multi-sensor scheme is evaluated using a combination of five public and commercial medium- and high-resolution satellites, including Landsat-8, Sentinel-2A, WorldView-2, WorldView-3, and GeoEye-1, with pixel sizes ranging from 30 m to 0.3 m. Applying a multi-conditional algorithm after the atmospheric correction of the optical imagery with ACOLITE, Sen2Cor and QUAC processors, the feasibility of monitoring suspended solids during dredging operations is demonstrated at a spatial resolution unachievable with traditional satellite-based ocean color sensors (>300m). The frame work can be used to map on-going, post and pre-dredging activities and assess Total Suspended Solids (TSS) anomalies caused by natural and anthropogenic processes in coastal and inland waters. These promising results are suitable for effectively improving the assessment of features relevant to environmental policies for the challenging coastal management, and may serve as a notable contribution to the Earth Observation Program.

More information: Caballero, I., Navarro, G., & Ruiz, J. (2018). Multi-platform assessment of turbidity plumes during dredging operations in a major estuarine system. International journal of applied earth observation and geoinformation, 68, 31-41.

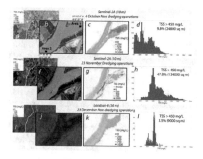

a) Map showing the Guadalquivir estuary RGB (bands 4-3-2) composite and TSS concentration of Sentinel-2 image at 10 m spatial resolution on 4 October 2016 with red rectangle delimiting the Region Of Interest (ROI) at the entrance of the Port of Seville, b) RGB composite, c) TSS concentration within the ROI, and d) Histogram of TSS within the ROI; e-h) the same for Sentinel-2 image on 23 November 2016. The hopper dredger and its shadow are masked in this scene; i-l) the same for Landsat-8 image at 30 m spatial resolution on 23 December 2016. Differences in TSS concentration between the main river channel (> 250 mg/L) and the parallel channel crossing the city of Seville (< 200 mg/L) are evident in the three images (a, e, i).

Acknowledgements: The commercial imagery from WordlView-2/-3 and GeoEye-1 were provided by the European Space Agency in the framework of a Third Party Mission Project (id35116). This work was partially supported by the Projects P09-RNM-4853, PR11-RNM-7722, PIE201530I012 and CTM2014-58181-R.

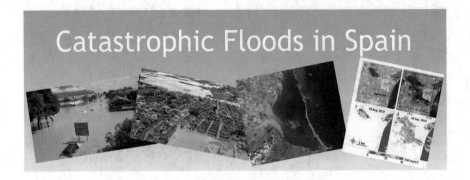

Catastrophic Floods in Spain

Flooding is among the most common natural disasters on our planet and one of the main causes of economic and human life loss worldwide. The devastating event in the Western Mediterranean during the second week of September 2019 is a clear case of this risk crystallization, when a record-breaking flood turned into a catastrophe in southeastern Spain surpassing previous all-time records. Using a straightforward approach with the Sentinel-2 twin satellites from the Copernicus Programme and the ACOLITE atmospheric correction processor, an initial approximation of the delineated flooded zones, including agriculture and urban areas, was accomplished in quasi-real time. The robust and flexible approach does not require ancillary data for rapid implementation. A composite of pre- and post-flood images was obtained to identify change detection and mask water pixels. Sentinel-2 identifies not only impacts on land but also on water ecosystem services, providing information on water quality deterioration and concentration of suspended matter in highly sensitive environments. Subsequent water quality deterioration occurred in large portions of the Mar Menor, the largest coastal lagoon in the Mediterranean. The present study demonstrates the potentials brought by the free and open-data policy of Sentinel-2, a valuable source of rapid synoptic spatio-temporal information at the local or regional scale to support scientists, managers, stakeholders, and society in general during and after emergency.

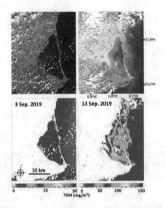

Sentinel-2 RGB composites (red–green–blue) at 10 m spatial resolution showing the (a) before (3 September 2019) and (b) after (13 September 2019) of the flooding event in Murcia province (Spain). The high concentration of suspended material (TSM) in Mar Menor can be observed in the map (d) on 13 September with TSS > 200 mg/m³ compared to the normal situation (c) on 3 September 2019 with minimum concentration (10 mg/m³). White areas correspond to land, clouds, or cloud shadows masked after ACOLITE.

More information: Caballero, I., Ruiz, J., & Navarro, G. (2019). Sentinel-2 satellites provide near-real time evaluation of catastrophic floods in the west Mediterranean. Water, 11(12), 2499.

Acknowledgements: This research was funded by the Spanish Ministry of Science, Innovation and Universities (MCIU), the State Research Agency (AEI), and the European Regional Development Fund (ERDF) in the frame of the Sen2Coast Project (RTI2018-098784-J-I00).

References

Balsi M, Moroni M, Chiarabini V, Tanda G (2021) High-resolution aerial detection of marine plastic litter by hyperspectral sensing. Remote Sens 13(8):1557

Berni JAJ, Zarco-Tejada PJ, Suárez-Barranco MD, Fereres-Castiel E (2009) Thermal and narrow-band multispectral remote sensing for vegetation monitoring from an unmanned aerial vehicle. IEEE Trans Geosci Remote Sens 47:722–738

Biermann L, Clewley D, Martinez-Vicente V, Topouzelis K (2020) Finding plastic patches in coastal waters using optical satellite data. Sci Rep 10:5364

Bolch EA, Hestir EL, Khanna S (2021) Performance and feasibility of drone-mounted imaging spectroscopy for invasive aquatic vegetation detection. Remote Sens 13(4):582

Boukoberine MN, Zhou Z, Benbouzid M (2019) A critical review on unmanned aerial vehicles power supply and energy management: solutions, strategies, and prospects. Applied Energy 255

Garcia-Garin O, Borrell A, Aguilar A, Cardona L, Vighi M (2020) Floating marine macro-litter in the North Western Mediterranean Sea: results from a combined monitoring approach. Mar Pollut Bull 159:111467

Ge Z, Shi H, Mei X, Dai Z, Li D (2016) Semi-automatic recognition of marine debris on beaches. Sci Rep 6:25759

Goddijn-Murphy L, Williamson B (2019) On thermal infrared remote sensing of plastic pollution in natural waters. Remote Sens 11:2159

Goldberg SJ, Kirby JT, Licht SC (2016) Applications of aerial multi-spectral imagery for algal bloom monitoring in Rhode Island. SURFO technical report no. 16-01. University of Rhode Island, South Kingstown, RI, USA, p 28

Hassanalian M, Abdelkefi A (2017) Classifications, applications, and design challenges of drones: a review. Prog Aerosp Sci 91:99–131

Marzialetti F, Frate L, De Simone W, Frattaroli AR, Acosta ATR, Carranza ML (2021) Unmanned aerial vehicle (UAV)-based mapping of *Acacia saligna* invasion in the Mediterranean coast. Remote Sens 13:3361

Petrie G (2013) Commercial operation of lightweight UAVs for aerial imaging and mapping. GEO Informatics 16:28–39

Przybilla H, Wester-Ebbinghaus W (1979) Bildflug mit ferngelenktem Kleinflugzeug. Bildmessung und Luftbildwesen 47:137–142

Ren H, Zhao Y, Xiao W, Hu Z (2019) A review of UAV monitoring in mining areas: current status and future perspectives. Int J Coal Sci Tech 6(5)

Topouzelis K, Papakonstantinou A, Garaba SP (2019) Detection of floating plastics from satellite and unmanned aerial systems (plastic litter project 2018). Int J App Earth Observ 79:175–183

Townsend A, Jiya IN, Martinson C, Bessarabov D, Gouws R (2020) A comprehensive review of energy sources for unmanned aerial vehicles, their shortfalls and opportunities for improvements. Heliyon 6:e05285

Valavanis KP, Vachtsevanos GJ (2015) Handbook of unmanned aerial vehicles. Springer, New York, pp 1–3022

Watts AC, Ambrosia VG, Hinkley EA (2012) Unmanned aircraft systems in remote sensing and scientific research: classification and considerations of use. Remote Sens 4:1671–1692

Wester-Ebbinghaus W (1980) Aerial photography by radio controlled model helicopter. Photogramm. Rec 10:85–92

Xu F, Gao Z, Jiang X, Shang W, Ning J, Song D, Ai J (2018) A UAV and S2A data-based estimation of the initial biomass of green algae in the South Yellow Sea. Mar Pollut Bull 128:408–414

Young K, Hyams H, Provanzano J, Williams M, Nguyen K, Amejecor K, Nikulin A, Chiu K, Cleckner L, Yokota K, De Smet TS, Wigdahl-Perry C (2019) Hyperspectral drone detection of harmful algal blooms: ground trothing new approaches for water quality assessment. American Geophysical Union, Fall Meeting, p 2019

In Situ Sensing: Ocean Gliders

Nikolaos D. Zarokanellos, Miguel Charcos, Albert Miralles,
Matteo Marasco, Mélanie Juza, Benjamin Casas,
Juan Gabriel Fernández, Manuel Rubio, and Joaquin Tintoré

Contents

Abstract

The chapter begins with a description of ocean gliders, their purpose, and applications in operational oceanography and discussion about the contribution to the global ocean observing system with a particular focus on the western Mediterranean. Ocean gliders can resolve scales of kilometers and hours and have the seasonal to yearly endurance needed to describe climatic variability and capture episodic events. Sampling different spatial-temporal scales has played an essential role in resolving societal challenges and economic applications, particularly in ocean boundaries, by understanding the impact of human activities in coastal areas. The growing number of ocean glider observations emphasizes the importance of developing, improving, and sharing best practices for ocean glider operation, data collecting, and analysis to get high-quality in situ observations.

N. D. Zarokanellos (✉) · M. Charcos · A. Miralles · M. Marasco · M. Juza · B. Casas ·
J. G. Fernández · M. Rubio
SOCIB, Balearic Islands Coastal Observing and Forecasting System, Palma, Spain
e-mail: nzarokanellos@socib.es

J. Tintoré
SOCIB, Balearic Islands Coastal Observing and Forecasting System, Palma, Spain

IMEDEA (CSIC-UIB), Mediterranean Institute of Advanced Studies, Esporles, Spain

Although ocean gliders are a mature platform with proven scientific output, improvements in dependability, ease of use, and range would significantly influence platform efficiency, allowing for more extensive adoption and application to a broader range of scientific and operational activities. The chapter aims to enable the ocean community to develop improved methodologies for all ocean glider activities, from research to operations to widely accepted applications.

Keywords

Ocean gliders · Ocean observing system · Data management · Best practices · Ocean glider operations

10.1   Ocean Gliders

10.1.1 Description

Ocean gliders (hereafter gliders) are small autonomous underwater vehicles (AUVs) that can migrate vertically by changing their buoyancy and steer horizontally by gliding on wings (Stommel 1989; Eriksen et al. 2001; Sherman et al. 2001). They can be deployed at sea with various physical and biogeochemical sensors that allow sustained observations at high spatial and temporal resolutions. Since gliders require little human assistance while traveling, these small AUVs are uniquely suited for safely collecting data in local and remote locations at a relatively low cost. They allow sampling the ocean and collecting data where it is impractical for human access, such as in the middle of a hurricane or under sea ice. While many glider designs use different techniques to balance and drive through the water, all gliders share the ability to travel far distances over long periods without the need for maintenance.

Gliders use buoyancy to ascend or descend the water column in the ocean surface to 1000 m in the same way as profiling floats (Stommel 1989; Eriksen et al. 2001). Whereas profilers can only move vertically up and down, gliders fitted with wings that allow them to glide forward with a horizontal velocity of 0.27 m s^{-1} and vertical velocity of 0.09 m s^{-1} as they ascend or descend (Rudnick 2016). In contrast with traditional propellers or thrusters, this low cost energy technique allows the gliders to perform missions of long-duration. Although, gliders are not as fast as conventional AUVs, they have the capability to cover greater range and longer endurance than traditional AUVs, extending ocean sampling missions from hours to weeks or even months, covering several hundreds of nautical miles. They achieve these trajectories by performing sawtooth-like profiles that result from repeated cycles consisting of altering the buoyancy at edges of the profile and modifying the center of mass by moving the internal mass (usually by pitching the battery).

Gliders navigate with the assistance of periodic surface GPS fixes, pressure sensors, altimeter, tilt sensors, and magnetic compasses. Gliders' ability to dive from weeks to months contingent on the type of batteries (alkaline, lithium, or rechargeable), mission configuration, and sensor payload, following a programmed trajectory, allows scientists to observe the same area repeatedly or to monitor a

Source:https://www.alseamar-
alcen.com/products/underwater-glider/seaexplorer

Photo by Ben Allsup, Teledyne Webb
Research

Source:https://www.whoi.edu/what-we-
do/explore/underwater-vehicles/auvs/spray-glider/

Source:https://apl.uw.edu/project/project
.php?id=seaglider

Fig. 10.1 Major types of gliders. Source, https://www.alseamar-alcen.com/products/underwater-glider/seaexplorer. Photo by Ben Allsup, Teledyne Webb Research. Source, https://www.whoi.edu/what-we-do/explore/underwater-vehicles/auvs/spray-glider/. Source, https://apl.uw.edu/project/project.php?id=seaglider

specific physical process, such as a mesoscale eddy providing data at temporal and spatial scales unattainable by powered AUVs and less costly sampling using traditional shipboard techniques (Davis et al. 2002).

Gliders typically collect measurements such as conductivity-temperature-depth (CTD, conductivity to calculate salinity), currents, chlorophyll fluorescence, optical backscatter, bottom depth, and sometimes acoustic backscatter or ambient sound. There is also the possibility of mounting new commercial and custom sensors that enhance their potential and promising future. Glider observations help us better understand the ocean state and variability, complementing satellites or research vessels. Navies and research organizations use a wide variety of glider designs and sensors for their research.

There are four major glider platforms available nowadays: Slocum, SeaExplorer, Spray, and Seaglider (Fig. 10.1). They use different methods to balance and steer throughout the water column. The vehicle pitch is controllable by movable internal ballast (usually battery packs). However, steering and buoyancy methods differ between Slocum gliders and other types of vehicles. Slocum steering is accomplished with a rudder, while other gliders steer by moving internal ballast to control roll. Regarding the buoyancy, Slocum gliders use a piston to flood or evacuate a compartment with seawater. In contrast, Spray, Seaglider, and SeaExplorer gliders move oil in/out from an external bladder. In all cases, because buoyancy adjustments are relatively small, the glider's ballast must typically be adjusted before the start of the mission to achieve an overall vehicle density close to the density of the water where the glider will be deployed.

One of the most significant features of AUVs is their near-real-time capability through iridium satellite communication at every surface (Testor et al. 2010; Ruiz et al. 2012). When the glider is surfacing (usually every 3–12 h), every parameter can be configured, such as trajectory or sampling ratio, to adapt the glider to changing conditions of the sea to respond to scientific or operational needs. Additionally, a fraction of the data set can be sent in real time during the surface time. Real-time glider observations already benefit ocean numerical modeling and forecasting activities. They can significantly improve regional and coastal models by providing physical and biogeochemical observations in areas with high socio-economical interest.

10.1.2 Contribution to the Global Ocean Observing System

In the last decade, the need for ocean monitoring has significantly increased, making it necessary to maintain and expand monitoring capabilities for better understanding ocean state, variability and changes, and impact on the marine ecosystem, particularly in the context of climate change (Tintoré et al. 2013). Specifically, gliders have a significant role in addressing societal issues and economic applications, especially in areas that connect the open basin with the coastal environment (Davis et al. 2002; Testor et al. 2010; Rudnick 2016). The understanding of ocean boundaries will help us to study the impact of human activities in coastal areas.

The glider's ability to monitor the ocean variability at different temporal and spatial scales allowed sampling the ocean at scales of km and hours while maintaining a persistent presence over periods of seasons to decades (Lee and Rudnick 2018). They can also resolve physical and biogeochemical processes that range from extreme events to climate signals (Glenn et al. 2008; Todd et al. 2011; Zaba and Rudnick 2016; Rudnick et al. 2017). In addition, they are capable of sampling across string lateral gradients (e.g., boundary currents and eddies), capturing small-scale and episodic processes (e.g., phytoplankton blooms and carbon export events), and quantifying climate variability.

In recent years, several studies have used glider observations to understand mesoscale processes with a spatial scale of 10–100 km and a temporal scale of a few weeks (Ruiz et al. 2009; Bouffard et al. 2010; Rudnick et al. 2015; Thomsen et al. 2016b; Pascual et al. 2017; Zarokanellos et al. 2017). Mesoscale processes consist of a mixture of geostrophic and ageostrophic dynamics that induce significant vertical fluxes of carbon and biogeochemical tracers from the surface to the interior ocean. These motions can substantially affect the marine ecosystem by significantly modulating phytoplankton biomass and productivity (Briggs et al. 2011; Olita et al. 2014; Thomalla et al. 2015; Zarokanellos and Jones 2021). Mesoscale processes are characterized by a small Rossby number scale ($Ro = V/fL$, where $V \approx 10^{-1}$ m/s represents the horizontal velocity, f is the Coriolis parameter, and $L \approx 10^4 \sim 10^5$ m is the length scale, and the reported vertical velocity (w) can range between 10 and 50 m day^{-1} (Tintoré et al. 1991; Rudnick 1996; Allen and Smeed 1996).

Mesoscale processes can also link with fine-scale (1–10 km) processes like instabilities of ocean currents or frontogenesis (Pietri et al. 2013; Mahadevan 2016; Bosse et al. 2016; du Plessis et al. 2017) and filamentary processes (McWilliams et al. 2009). The importance of vertical velocity has been identified in several global ocean areas, including the Mediterranean (Tintore et al. 1988; Allen et al. 2001; Ruiz et al. 2019). Several studies have shown that fine-scale (1–10 km) processes can be linked to connectivity pathways (Thomsen et al. 2016a; Mahadevan et al. 2020). Understanding the role of fine-scale and mesoscale processes and tracing large-scale connectivity pathways in marine systems will bring new insights into the control of phytoplankton communities and ecosystem health: these are essential steps for long-term goals to connect with modeling jellyfish invasions and bluefin tuna sprawling areas and distribution of crustacean populations in the future decade.

10.1.3 Contribution to the Mediterranean Sea Observing System

The Mediterranean Sea contains one of the most diverse marine ecosystems (almost 18 % of the total marine species worldwide), covering only 0.82 % of the ocean surface and 0.32 % of the global ocean volume (Bianchi and Morri 2000). Consequently, the Mediterranean Sea can be used as an ideal basin to understand global processes on a regional scale (Bethoux et al. 1999). Mesoscale and sub-mesoscale features play a significant role in distributing properties such as heat, salt, and biochemical tracers, which significantly impact the ocean's primary productivity (McGillicuddy et al. 2007; Lévy et al. 2010, 2012; Mahadevan 2014). Understanding the net role of mesoscale (10–100 km) and their instability processes leading to sub-mesoscale features (0.1–10 km) is essential for predicting the behavior of marine ecosystems in response to changes over a basin scale. Front, meanders, eddies, and filaments have a fundamental role in the vertical exchange as they contribute to the redistribution of heat, freshwater, carbon, dissolved oxygen, and nutrients throughout the water column. One of the most significant scientific challenges is understanding how energy and mass transfer between different spatial and temporal scales. Dynamics associated with these mesoscale and sub-mesoscale structures result in enhanced vertical velocities and mixing and modified stratification on temporal scales that range from a few days to several months and from a few to hundreds of km.

The Mediterranean Sea has always featured and still has a rich biodiversity. However, human activity has stressed coastal and ocean areas (e.g., maritime traffic, fishing, tourism, pollution, and invasive species [Millot and Taupier-Letage 2005; Testor et al. 2010]). Moreover, consequences of climate change are noticeable, increasing extreme events (e.g., storms and marine heat waves) that strongly impact society (Tintoré et al. 2019; Schroeder et al. 2020). Climate change will likely worsen problems that coastal areas already face from the intense human activity.

The western Mediterranean coasts are very sensitive to sea-level rise and changes in the intensity and frequency of these extreme events. Confronting existing challenges that affect manufactured infrastructure and coastal ecosystems (e.g., coastal flooding, water pollution, and shoreline erosion) has already been an issue in several regions (Roca et al. 2008; Drius et al. 2019). Furthermore, the increase of atmospheric concentrations of carbon dioxide (CO_2) negatively affects our ocean as it absorbs relatively more gas and increases the level of pH, which dramatically impacts the marine ecosystem (Flecha et al. 2015; Marcellin Yao et al. 2016). Addressing additional stress factors that enhance climate change may require the development of new state-of-the-art sensors.

Many physical and biogeochemical processes in the Mediterranean Sea have many similarities with the global ocean (such as thermohaline circulation and water mass formation), making the Mediterranean an ideal basin for evaluating effects of these processes at different temporal scales on the marine ecosystem due to its proximity and size. Despite the importance of understanding these processes to mitigate the continuous pressure of its environment, only a few studies have been done to provide a complete view of thermohaline changes within the basin (Rixen et al. 2005; Schroeder et al. 2010) due to the lack of comprehensive hydrographic data set.

The capability to sustain monitoring programs between open and coastal areas is crucial to guarantee the preservation of the marine ecosystem for the next generations. Currently, glider observations play an important role in understanding and evaluating the status of such ecosystems. An observational network of gliders is vital to better understand the impact that mesoscale and sub-mesoscale processes have on the marine ecosystem from local to global along a large timescale. Improving the sampling coverage in marginal seas such as the Mediterranean Sea will help respond to worldwide challenges and, thus, meet societal needs.

10.1.4 Monitoring Programs in the Western Mediterranean

ICTS SOCIB has developed four monitoring programs (one endurance glider line and three quasi-endurance glider lines) in the western Mediterranean Sea. Monitoring the western Mediterranean with gliders for long periods at key "choke" points will help us evaluate local and subregional effects of climate change in thermohaline-driven circulation on marine ecosystems (Heslop et al. 2012; Juza and Tintoré 2021). In addition, the very dynamic character of the western Mediterranean allows studying different physical and biogeochemical processes at different temporal and temporal scales that strongly interact with the thermohaline circulation.

Among its long-term monitoring programs (Fig. 10.2), the ICTS SOCIB has developed the *CANALES* glider endurance line since 2011 (Heslop et al. 2012). This program aims to study physical and biogeochemical characteristics and changes of different water masses in the Mallorca and Ibiza Channels. The *CANALES* monitoring program provides long-term observations for studying physical and biogeochemical variability changes from weekly to seasonal and interannual temporal

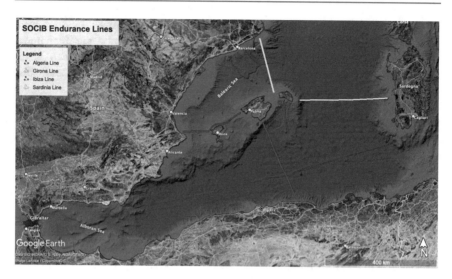

Fig. 10.2 Endurance and quasi-endurance SOCIB glider lines in the western Mediterranean Sea. *CANALES* (red), ABACUS (magenta), SMART (green), and CALYPSO Girona (yellow)

scales in the western Mediterranean. The Ibiza Channel is a key area of water mass interactions between the recent Atlantic water, which has recently entered through the Strait of Gibraltar, and the modified Atlantic water, which has circulated in the Mediterranean basin, and consequently an important area of biodiversity. These interactions trigger various physical and biogeochemical processes that have consequences on the richness of the marine ecosystem of the Balearic Sea. Increasing the number of glider missions and monitoring days in the *CANALES* endurance line will help understand and identify better physical processes (e.g., eddies, mesoscale instabilities, and filaments) that modulate the biogeochemistry of the region and impact the marine ecosystem.

The Algeria quasi-endurance line (called "Abacus") aims to investigate the characteristic of the Algerian basin (AB) circulation. The Algeria basin dominates by mesoscale eddies that usually develop and detach by the meandering of the Algerian current (Escudier et al. 2016). By using glider observations, we understand the dynamic characteristic of the mesoscale activity within the basin. Since 2014, several glider missions have taken place within the AB, bringing new insights into the three-dimensional structure of mesoscale features like eddies and filaments that could not be observed to this extent using ship-based observations (Cotroneo et al. 2016, 2019; Aulicino et al. 2018). Previous studies on the Algerian basin showed that biological processes in the upper layer are strongly influenced by the main flow of Atlantic water and by the presence of eddies (Taupier-Letage et al. 2003; Cotroneo et al. 2016). In addition, monitoring the AB will allow us to distinguish and trace the modification of physical and biogeochemical properties of the surface and intermediate water masses between the Balearic Islands and Algerian coasts.

The Sardinia quasi-endurance line (called "Smart") provides long-term observations between the Mallorca and the Sardinia for studying the interannual variability of physical and biogeochemical properties of Mediterranean water masses. By understanding the propagation of these water masses and their perturbations from the east to the west of the Mediterranean Sea, we can examine their influence on the dense water formation process. Glider observations will help reveal the abrupt shift in terms of temperature, salinity, and density that has been observed in the past (Schroeder et al. 2016) for the deep western Mediterranean water (DWMW). In addition, significant changes also have been observed at the intermediate waters, where warming and salinification take place at a faster rate compared with the past (Schroeder et al. 2017; Margirier et al. 2020).

The Girona quasi-endurance line (called "Calypso) aims to understand and predict three-dimensional pathways by which water from the surface subducts into the interior and vice-versa. Discovering the mechanisms of vertical transport in meso - and sub-mesoscale processes will help can help us to trace substances, phytoplankton and dissolved gases like oxygen. A better understanding and prediction of the three-dimensional circulation will help us evaluate the role of sub-mesoscale processes in large-scale circulation and the biogeochemical effects in the marine ecosystem. Process studies targeting these scales have already shown that the high sub-mesoscale dynamics are linked with connectivity pathways (Zarokanellos 2022; Ruiz et al. 2009, 2019).

10.2 Glider Operations at SOCIB

10.2.1 Infrastructure

SOCIB is a marine research infrastructure that implements and brings state-of-the-art multiplatform systems and technology to monitor and sustain high-resolution observations in coastal and open-ocean regions. The SOCIB glider facility (SOCIB-GLF) is part of the research infrastructure and has been fully active since 2012. It has accomplished more than 120 missions, including 2942 days in water, 32.735 nm navigated, and 108.174 vertical profiles collected.

The SOCIB-GLF is a multidisciplinary team composed of two full-time glider technicians (for glider operations), two part-time field technicians (for sea operations), one part-time data engineer (for glider data management), one part-time TA access scientist, and one full-time glider scientist. The glider fleet in 2021 consists of five gliders (two G2 and three G3 versions), equipped with a suite of sensors that can collect both physical (temperature and conductivity) and biogeochemical observations (chlorophyll fluorescence, oxygen, Colored Dissolved Organic Matter (CDOM), PAR, backscatter at 700 nm, and turbidity) at high spatial resolutions ($\sim$ 2 km). SOCIB-GLF installations include a pressure chamber, a ballasting tank, a lab, and an office space.

The SOCIB-GLF also has strongly collaborated with other SOCIB facilities such as the ETD (Engineering and Technology Development) team and the data center regarding the data management, the public repository, and the online Web-based platform tracker for mission monitoring and development and also for the development of tools such as the glider processing toolbox (Troupin et al. 2015; available through GitHub at https://github.com/socib/glider_toolbox).

Part of the SOCIB's success is the well-trained scientific and technical personnel. SOCIB-GLF, in the last decade, has developed best practices for glider operations, data acquisition, and data quality by leading and following international practices. Recent achievements were possible to accomplish by the availability of sufficient equipment and well-trained personnel. As this platform has a circle of life of a few years and highly depends on technological advances and sensor development, keeping our glider fleet updated is highly associated with our scientific and monitoring needs.

10.2.2 Transnational Access and Marine Services

Besides deployments (endurance lines and projects) relative to SOCIB needs, competitive access programs (in particular, the SOCIB competitive access, the JERICO-S3, and the JERICO-NEXT transnational access) have offered scientists and engineers the opportunity to use the glider fleet for research, monitoring, and/or testing activities in the coastal, shelf, and open waters. Submitted proposals will be evaluated based on scientific excellence, enhancement of technology development, and response to critical societal challenges. The access granted to a specific experiment generally includes support from SOCIB on mission preparation and setup; training and logistical, technical, and data management; and scientific advice.

The requested competitive access to SOCIB-GLF and their available infrastructure can follow three modes: (a) remote, the experiment will be implemented by SOCIB, and the presence of the user group is not required; (b) partially remote, the presence of the user group will be required at some stage of the experiment; and (c) in-person/hands-on, the presence of the user group will be required during the whole access period of the experiment.

Unless otherwise stated, the observation program will be provided by the user group. However, in all instances, potential users are encouraged to consult with SOCIB-GLF at any stage in proposal development. Careful scheduling and synchronizing events by SOCIB-GLF will minimize costs and maximize the return from the overall glider program.

Furthermore, users access to several specific complimentary services in addition to the access of the equipment as described as follows:

1. Access to glider platforms prepared and ready for operation in line with the highest international standards
2. Qualified personnel for the management of gliders (platforms and sensors), including logistics for deployment and recovery
3. A 24/7 monitoring system to pilot gliders at sea—access to a collaborative piloting system

4. Access to a collaborative data management system
5. Quasi-real-time (usually less than eight h) reception of data and visualization system
6. Near-real-time and delayed-time data in NetCDF format
7. Post-mission glider report

10.2.3 International Framework

In recent years, SOCIB has developed strong collaborations on a national and international level. Through the competitive access, monitoring, and processes, studies have carried out in the western Mediterranean Sea in collaboration with Spanish (e.g., IMEDEA, the University of the Balearic Islands, and IEO) and European (e.g., ISMAR-CNR research center, Instituto Hidrográfico in Portugal, Parthenope University of Naples, and the University of Liege) institutions.

SOCIB-GLF has also been participating in the following initiatives:

1. The Boundary Ocean Observing Network (BOON) project focuses on understanding how society experiences change in the global ocean through effects on ocean boundaries.
2. The Global Ocean Observing System (GOOS) is led by the Intergovernmental Oceanographic Commission (IOC) of the UNESCO, and SOCIB glider observations contribute in the improvement of the global observing system and forecasting modeling.
3. SOCIB glider activities are additionally aligned with the CoastPredict and JERICO activities for monitoring physical and biogeochemical essential ocean variables (EOVs).

Additionally, in recent years, SOCIB-GLF, in collaboration with IMEDEA and ICMAN research institute, has started to build and develop the capacity to monitor ocean acidification in the western Mediterranean Sea. Furthermore, SOCIB-GLF with the University of Liege has integrated hydrophones on gliders to better monitor the marine ecosystem at higher trophic levels.

10.3   SOCIB Glider Data Management System

10.3.1 Data Management Plan

SOCIB data management plan (DMP) template was established by analyzing existing DMP documents in the international framework. SOCIB DMP needs to be compatible with CSIC standards, creating a more transparent, collaborative, and sustainable scientific communication system. This approach allows SOCIB's data to be integrated into DIGITAL CSIC, pursuing a more significant impact of research results to address future challenges.

The International Oceanographic Data and Information Exchange (IODE) template was considered, as it outlines different steps of data flows for different platforms for each facility. On the other hand, the open research data (ORD) pilot aims to improve and maximize access to the reuse of research generated data with the primary purpose of making research data FAIR.

In recent years, the Australian National Data Service (ANDS) helped improve data discoverability and connectivity of the Integrated Marine Observing System (IMOS). The IMOS is Australia's premier marine infrastructure system, and it provides the all-important ocean component for Australia's national research data landscape. Therefore, ANDS guidelines have been a critical part of managing such a significant volume of data. In order to complement the study of guidelines mentioned above, our analysis also considered the previous internal DMPs relevant to SOCIB and Northwest Association of Networked Ocean Observing Systems (NANOOS).

By harmonizing these guidelines, it is, therefore, identified as a series of recommendations to be summarized in a new template that helps with the day-to-day operation, which improves the data management life cycle. The use of the template gives us the flexibility to be adapted to different projects and platforms. Description of the operational steps includes additional valuable information for end users, improving the transparency of the data life cycle.

10.3.2 Observation Flow

Glider observations are managed through a scientific and engineering detail record of the metadata (e.g., deployment and recovery day of the glider, glider platform, and sensor type and configuration). SOCIB data management plan (DMP) template includes steps and relations of processes during planning, deployment, data collection, data processing, data delivery, and data dissemination to external partners (e.g., aggregators). The DMP frame describes different roles of persons involved in the glider observation workflow and data life cycles, such as the glider facility, data center facility, data managers, and scientists (Fig. 10.3). Glider observations have processed along three data flows during each mission:

1. Real time (RT)—data processed during the deployment
2. Delayed-time (DT) or recovery data—data processed after the instrument is retrieved
3. Delayed mode (DM)—data post-processed, which includes scientific corrections

Most of data management processes are performed automatically. The metadata catalog plays a critical role in automatic processes and supporting hand operation. RT data is fully automated. Recovery data follows a similar flow to the real-time cycle when the mission is complete. In this case, the data are collected by glider operators and uploaded to an FTP server manually. Delay-mode data flow involves

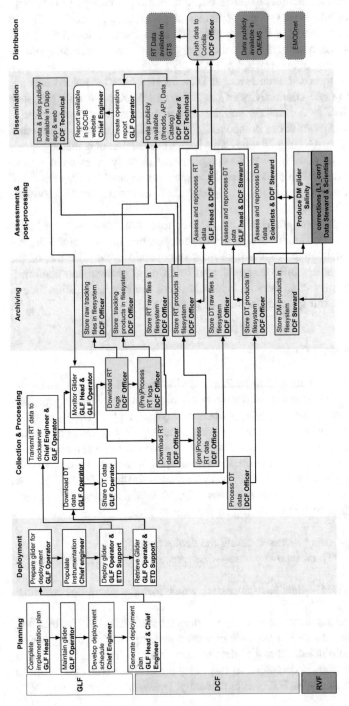

Fig. 10.3 Schematic of SOCIB glider facility data management plan along the data life cycle

higher human intervention in the loop to tackle data intercomparison and occurs once recovered data outputs are available.

Glider quality control (QC) in SOCIB considers historical glider observations (based on 10 years of glider observations) and ship-based observations in the study area. The application for QC is currently performing manually to support and optimize glider operations. Automatic real-time quality control is in the development phase for the following physical and biogeochemical parameters: temperature, salinity, chlorophyll fluorescence, optical backscatter, and oxygen.

10.3.3 Data Outputs

During a glider mission, three types of data outputs have been created: *data* files (NetCDF), *plots* (png), and *trajectories* (kmz). RT data is collected automatically from the dock server to create the glider trajectory and visualize science parameters in the dapp application. Simultaneously, the SOCIB glider toolbox produces NetCDF files and plots the real-time and recovery data. Leading institutions have used the SOCIB glider toolbox to process their glider data, such as NOAA fisheries (Reiss et al 2021) and DFO-MPO (mission report). Plots are used to support the operation and to facilitate the visualization of key measured variables in RT. Glider operators and principal investigators (PI) use the NetCDF files to assess the RT and recovery data. In addition, an automatic tool monitors the engineering data from the glider and detects potential failures of the glider during the mission. This tool creates png figures and provides glider operators with email alerts when data engineering or science issues are detected. The SOCIB salinity correction toolbox is used to perform delay-mode data. The output of the DM process consists of NetCDF files, figures, and calibration coefficients stored in a database.

10.3.4 Data Levels and Distribution

NetCDF files are created for various levels of data: L0, L1, and L2. The L0 NetCDF contains scientific and engineering information as provided by the glider. L1 NetCDF files contain calibrations, unit corrections, and derived variables such as salinity. The L1 processing also automatically calculates some delayed-mode corrections, including the thermal lag. In addition to L0 and L1, L2 NetCDF files are also created (gridded data of the glider profiles). Besides, delayed-mode processing (DM) has been carried out to calibrate the salinity data and included in the L1 NetCDF files.

All data outputs at all levels (L0, L1, and L2) and data modes (real time, recovery, and delayed mode) are shared using the same dissemination mechanisms: SOCIB Thredds data server, SOCIB data catalog, and SOCIB Data API. In addition to this, SOCIB also shares L1 NetCDF files (real time, recovery, and delayed mode) in EGO format with Coriolis via FTP syncing. This additional exchange ensures the final integration of observations into the Global Telecommunication System (by way of

Meteo-France) and the Copernicus Marine Service (by way of Ifremer). A set of memorandum of understandings (MOUs) between CMEMS (In Situ Thematic Assembly Center) and other major European projects (SeaDataNet) and networks (EMODnet) enable the dissemination of observations even further. Mission metadata is also uploaded to SeaDataNet by operators that create and upload cruise summary reports (CSR) for our glider missions.

Technical debts regarding the present workflow include the link with Coriolis, which ideally should be readdressed in order to better use some of the machine-to-machine tools available at SOCIB, such as the SOCIB Data API. This mechanism will be applied to all SOCIB glider data to respond to the European Commission to use the "real" source as much as possible to make users closer to producers.

10.3.5 Metadata Catalog

Figure 10.4 highlights the importance of the metadata database in the data life cycle. Combining the metadata catalog and the processing mechanisms allows us to perform the flow automatically. It also allows us to incorporate in workflow glider operations from institutions outside SOCIB. The same processing chain has been applied to various external Trans-National Access Programme (TNA) calls and projects. Our metadata catalog record is related to the equipment (platform, instrument, and sensor level), deployments, and processes used to handle the observations from each instrument. It is flexible enough to allow the registration of different observation platforms, including gliders, research vessels, mooring, HFR, fix stations, and weather stations. This catalog has linked with a user graphical interface that facilitates glider operators to input their metadata. The information in this catalog has been partially exposed through SOCIB Data API, which provides information on the data and data products and gives access to users to the equipment information in the database.

10.4   Best Practices

During the last decade, the need for continuous monitoring of our ocean has significantly increased. In this context, gliders have played an essential role in GOOS. The increase of glider observations highlights the necessity to create, improve, and share best practices regarding glider operation, data collection, and analysis to achieve high-quality in situ observations (Pearlman et al. 2019). It is also essential to monitor the glider activity and support the real-time transmission of disseminating observations through regional and global databases for better marine forecasting.

We have developed SOCIB best practices for glider operations (Fig. 10.5). Annual planning is highly recommended to provide the necessary time for the facility to get supplies, get prepared, and avoid tight schedules regarding glider operations. The annual plan should include the glider availability, the number of

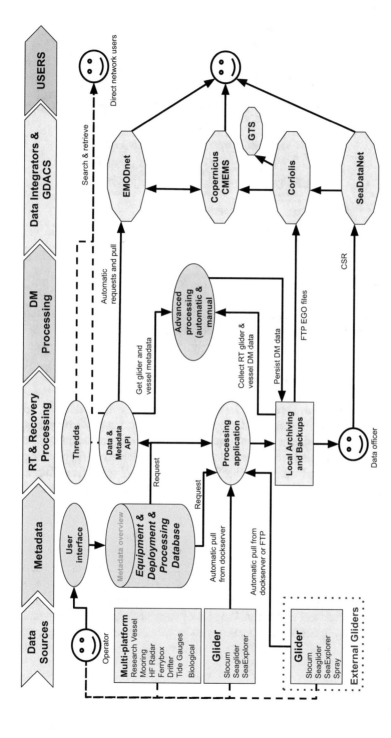

Fig. 10.4 SOCIB metadata support (instrumentation database) along the data life cycle

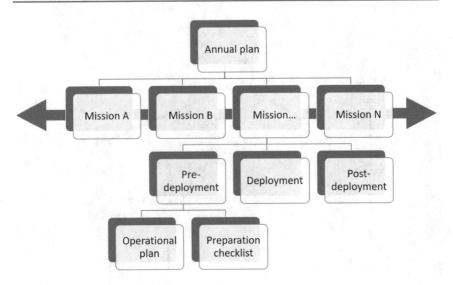

Fig. 10.5 SOCIB best practice overview

planned missions based on monitoring observation programs and projects, the infrastructure maintenance needs, and the personnel availability.

10.4.1 Pre-deployment and Preparation Phases

Glider operations can be separated into three categories that referred to pre-deployment, deployment, and post-deployment operations: the pre-deployment can be split into two independent phases—operational and glider preparation.

During the pre-deployment phase, the following tasks are included: (1) planning the availability of the desired vessel, (2) checking with the vessel's captain and crew members (3) taking into account other campaigns that could interfere, (4) checking the forecasting in the study area, (5) analyzing previous and current observations to have an estimate of expected currents in the study area, and (6) analyzing marine traffic and fishing activities in the study area and emergency plan for glider recovery.

Pre-deployment phase operations include a list of decisions and actions that the scientific and engineer team must make before each deployment. Decisions include the project's design, sampling strategy for each sensor, and study area. In addition, it is crucial to set up a communication channel between different divisions that participate in the experiment and define the role of each participant. Furthermore, a checklist of actions is needed to check the glider's hardware and software, as summarized in Table 10.1.

Table 10.1 Summarize steps of a glider preparation (hardware and software)

Hardware check	Software configuration
Picking glider(s) and checking for any uncalibrated sensors (out of date)	Mission definition: • Expected water days • Expected consumption • Deployment point • Recovery point • Waypoint list • Drive (glider density range). Related with expected currents, endurance, and season • Satellite communication. How often and why the glider will surface. It is important to keep in mind to not waste time at the surface; it adds risk and cost
Communication check	Definition of sampling strategy: • Sensor selection • Sampling rate • Target depths • Profile rate
Battery selection and check	Near-real-time strategy or glider data decimation, which and how much data (density of points) will be transmitted in real time. During the first stage of the mission, it is important to have a high density to not miss any issue
Sensor test—check sensors in the lab	Surface dialog configuration, which parameters will be displayed during the surface time
Ballasting in a glider tank (Fig. 10.6)	Operational or scientific changes during the mission. For example, pre-defined or adaptative target waypoints
Final sealing	Sampling priority (regarding energy consumption)
Pressure test (Fig. 10.7)	CEM (compass error measurement) and calibration
Harbor check (right before the deployment day)	

10.4.2 Deployment Phase

During the deployment phase, the engineering, scientific, and data managing team evaluates the glider's performance, data quality, and data flow during the mission. A checklist has been developed following the next steps:

- Deployment check.
- Launching point (it depends on the research vessel; check for the specific protocol launching procedure).
- Web page visualization. Check if the glider-sampled data appear correctly in digital platforms.
- Check the metadata (mission name, mission start (UTC), mission end (UTC), and short description of the glider deployment).
- Check if the glider toolbox creates NetCDF data (L0, L1, and L2).
- Check L0, L1, and L2 values for each sensor/variable, depending on needs.

Fig. 10.6 Glider in a ballasting tank

10.4.3 Post-Deployment Phase and Data Calibration

During the post-deployment phase, the delayed-time data include both the engineer and scientific data collected from gliders and stored internally in the vehicle during the mission. The data needed to be downloaded from the platform before the processing, quality control, and delay-mode analysis. Gliders' observations can be calibrated using a range of other data sources gathered and analyzed over an extended period. Nowadays, this approach is more common, and intercalibration procedures are included between observations from different platforms (such as cruises, floats, and occasionally moorings [Bosse et al. 2015; Zarokanellos et al. 2017]). Cross-calibration and validation are needed to ensure high-quality observations. To perform intercalibrations, oceanographers compare their CTD data by examining the stability of deep-water masses and their mixing lines from the same study area. The Argo community has already put procedures for intercalibrating the float data as best as possible, which provides an excellent opportunity for the glider community to take it further. At the *CANALES* monitoring program, the calibration of glider data is performed primarily using data from a 3-day seasonal cruise from missions in the same transects as the glider close to dates of the same mission. Using seasonal ship-based cruises, we can provide the

Fig. 10.7 Pressure chamber for pressure test for the glider

necessary laboratory analysis for the quality of glider observations. The cross-calibration might be taken as consideration of the well-calibrated CTD reference data to be implemented in another processing study or monitoring line taking into account the dynamic characteristics of the region. In addition, SOCIB has developed a semiautomatic intercalibration correction pack (available through GitHub at github.com/socib/salinity-correction-toolbox) to calibrate glider observations against ship-based observations by using white maximization image analysis (Allen et al. 2020).

10.4.4 From Data to Products

The Balearic Channels, which have been monitored through the SOCIB *CANALES* endurance line since 2011 (Sect. 10.1.4), are of significant interest since they are the place where water mass exchanges between the northern and southern sub-basins occur (Heslop et al. 2012; Juza et al. 2013, 2019; Barceló-Llull et al. 2019). The monitoring of water mass transport allows addressing key science processes such as coastal shelf-open seawater mass formation and changes, ocean circulation and variability, changes in marine ecosystem activity, and ocean response to climate changes (Heslop et al. 2012; Juza and Tintoré 2021).

The delayed-time L1 product of temperature and salinity (Heslop et al. 2012; Troupin et al. 2015) is used to compute geostrophic velocity and monitor

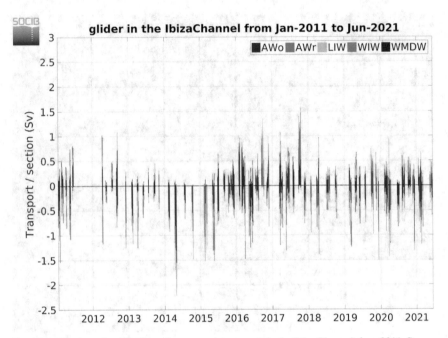

Fig. 10.8 Northward and southward water mass transports in the Ibiza Channel since 2011. Source https://apps.socib.es/subregmed-indicators

geostrophic (total and water mass) transports in the Mallorca and Ibiza Channels. Typical water masses of the western Mediterranean are identified: the recent and modified AW, the Western Intermediate Water (WIW), the Levantine Intermediate Water (LIW), and the Western Mediterranean Deep Water (WMDW). The geometry-based criterion (Juza et al. 2019), which has been shown to detect more properly WIW than the "classical" fixed-range method, is used for the WIW detection.

The "sub-regional Mediterranean Sea Indicators" visualization tool (https://apps. socib.es/subregmed-indicators [Juza and Tintoré 2021]) has been recently implemented to easily follow the evolution of subregional ocean indicators in the Mediterranean Sea and the Balearic Islands region by marine scientists, environmental agencies, ocean managers, and the general public. In particular, this user-friendly website displays the time series of northward and southward geostrophic transports of water masses in the Balearic Channels since January 2011, when the SOCIB endurance line started (https://apps.socib.es/subregmed-indicators/ transports.htm). They are automatically updated once a new delayed-time mission is available and distributed in the SOCIB Thredds server.

SOCIB glider water mass transports (from data processing to figures; Fig. 10.8) are also part of the JERICO-S3 "Data to Product Thematic Service."

10.5   Conclusions

Traditionally, physical and biogeochemical measurements have been carried out from research vessels. However, the rapid expansion and need for ocean observations made them have a fundamental role in observing systems. Challenges to obtaining sustainable observations in areas with high socio-economical interest have helped to be developed several long-term glider monitoring programs world-wide, especially in ocean boundaries.

Nowadays, the glider allows us a higher degree of adaptation for society. Sustainable regional glider lines successfully provide data to regional models of coastal ocean circulation, which help evaluate effects of climate change in the marine ecosystem. Gliders primarily have been focused on physical parameters; however, different biogeochemical sensors have been integrated in the last years, and glider measurements have expanded to several biogeochemical essential ocean variables (EOVs).

The SOCIB DMP and best practices have aimed to ensure operations and data delivery. By improving and harmonizing procedures and guidelines, we will improve the data management life cycle and the flexibility to be adapted to different projects and platforms. The latter will help maximize the access and reuse of glider observations by making the observations FAIR. Furthermore, the DMP will allow the glider facility to design a customized workflow based on their need.

SOCIB best practice procedures will also enhance and standardize quality-controlled processes by providing the user with high-quality glider observations and services. The description of operation steps during the preparation, deployment, and post-deployment mission period includes valuable information for end users.

Marine platforms like gliders enable the Digital Ocean by unlocking new approaches for environmental monitoring with lower costs and new capabilities. Gliders are autonomous underwater vehicles that operate individually or in fleets, delivering near-real-time data for several months with no fuel. By breaking through traditional cost and risk barriers, they enable long-duration missions compared with the previously too costly ship-based marine operations. Gliders can now provide the necessary data to evaluate the environmental impact of human activities in the ocean.

References

Allen JT, Munoz C, Gardiner J et al (2020) Near-automatic routine field calibration/correction of glider salinity data using whitespace maximization image analysis of theta/S data. Front Mar Sci 7:1–14. https://doi.org/10.3389/fmars.2020.00398

Allen JT, Smeed DA (1996) Potential vorticity and vertical velocity at the Iceland-Færœs front. J Phys Oceanogr 26:2611–2634. https://doi.org/10.1175/1520-0485(1996)026<2611:PVAVVA>2.0.CO;2

Allen JT, Smeed DA, Tintoré J, Ruiz S (2001) Mesoscale subduction at the Almeria–Oran front. J Mar Syst 30:263–285. https://doi.org/10.1016/S0924-7963(01)00062-8

Aulicino G, Cotroneo Y, Ruiz S et al (2018) Monitoring the Algerian basin through glider observations, satellite altimetry and numerical simulations along a SARAL/AltiKa track. J Mar Syst 179:55–71. https://doi.org/10.1016/j.jmarsys.2017.11.006

Barceló-Llull B, Pascual A, Ruiz S et al (2019) Temporal and spatial hydrodynamic variability in the Mallorca Channel (western Mediterranean Sea) from 8 years of underwater glider data. J Geophys Res Ocean 124:2769–2786. https://doi.org/10.1029/2018JC014636

Bethoux JP, Gentili B, Morin P et al (1999) The Mediterranean Sea: a miniature ocean for climatic and environmental studies and a key for the climatic functioning of the North Atlantic. Prog Oceanogr 44:131–146. https://doi.org/10.1016/S0079-6611(99)00023-3

Bianchi CN, Morri C (2000) Marine biodiversity of the Mediterranean Sea: situation, problems and prospects for future research. Mar Pollut Bull 40:367–376. https://doi.org/10.1016/S0025-326X (00)00027-8

Bosse A, Testor P, Houpert L et al (2016) Scales and dynamics of submesoscale coherent vortices formed by deep convection in the northwestern Mediterranean Sea. J Geophys Res Ocean 121: 7716–7742. https://doi.org/10.1002/2016JC012144

Bosse A, Testor P, Mortier L et al (2015) Spreading of Levantine intermediate waters by submesoscale coherent vortices in the northwestern Mediterranean Sea as observed with gliders. J Geophys Res Ocean 120:1599–1622. https://doi.org/10.1002/2014JC010263

Bouffard J, Pascual A, Ruiz S et al (2010) Coastal and mesoscale dynamics characterization using altimetry and gliders: a case study in the Balearic Sea. J Geophys Res Ocean 115:1–17. https:// doi.org/10.1029/2009JC006087

Briggs N, Perry MJ, Cetinić I et al (2011) High-resolution observations of aggregate flux during a sub-polar North Atlantic spring bloom. Deep Sea Res Part I Oceanogr Res Pap 58:1031–1039. https://doi.org/10.1016/j.dsr.2011.07.007

Cotroneo Y, Aulicino G, Ruiz S et al (2019) Glider data collected during the Algerian basin circulation unmanned survey. Earth Syst Sci Data 11:147–161. https://doi.org/10.5194/essd-11-147-2019

Cotroneo Y, Aulicino G, Ruiz S et al (2016) Glider and satellite high resolution monitoring of a mesoscale eddy in the Algerian basin: effects on the mixed layer depth and biochemistry. J Mar Syst 162:73–88. https://doi.org/10.1016/j.jmarsys.2015.12.004

Davis R, Eriksen C, Jones C (2002) Autonomous buoyancy-driven underwater gliders. . . . Appl Auton Underw . . .:1–11. https://doi.org/10.1201/9780203522301.ch3

Drius M, Bongiorni L, Depellegrin D et al (2019) Tackling challenges for Mediterranean sustainable coastal tourism: an ecosystem service perspective. Sci Total Environ 652:1302–1317. https://doi.org/10.1016/j.scitotenv.2018.10.121

du Plessis M, Swart S, Ansorge IJ, Mahadevan A (2017) Submesoscale processes promote seasonal restratification in the Subantarctic Ocean. J Geophys Res Ocean 122:2960–2975. https://doi.org/ 10.1002/2016JC012494

Eriksen CC, Osse TJ, Light RD et al (2001) Seaglider: a long-range autonomous underwater vehicle for oceanographic research. IEEE J Ocean Eng 26:424–436. https://doi.org/10.1109/48.972073

Escudier R, Renault L, Pascual A et al (2016) Eddy properties in the western Mediterranean Sea from satellite altimetry and a numerical simulation. J Geophys Res Ocean 121:3990–4006. https://doi.org/10.1002/2015JC011371

Flecha S, Pérez FF, García-Lafuente J et al (2015) Trends of pH decrease in the Mediterranean Sea through high frequency observational data: indication of ocean acidification in the basin. Sci Rep 5:1–8. https://doi.org/10.1038/srep16770

Glenn S, Jones C, Twardowski M et al (2008) Glider observations of sediment resuspension in a middle Atlantic bight fall transition storm. Limnol Oceanogr 53:2180–2196. https://doi.org/10. 4319/lo.2008.53.5_part_2.2180

Heslop EE, Ruiz S, Allen J et al (2012) Autonomous underwater gliders monitoring variability at choke points in our ocean system: a case study in the western Mediterranean Sea. Geophys Res Lett 39:1–6. https://doi.org/10.1029/2012GL053717

Juza M, Escudier R, Vargas-Yáñez M et al (2019) Characterization of changes in western intermediate water properties enabled by an innovative geometry-based detection approach. J Mar Syst
191:1–12. https://doi.org/10.1016/j.jmarsys.2018.11.003

Juza M, Renault L, Ruiz S, Tintoré J (2013) Origin and pathways of winter intermediate water in the
northwestern Mediterranean Sea using observations and numerical simulation. J Geophys Res
Ocean 118:6621–6633. https://doi.org/10.1002/2013JC009231

Juza M, Tintoré J (2021) Multivariate sub-regional ocean indicators in the Mediterranean Sea: from
event detection to climate change estimations. Front Mar Sci 8:1–20. https://doi.org/10.3389/
fmars.2021.610589

Lee CM, Rudnick DL (2018) Underwater Gliders. In: Venkatesan R, Tandon A, D'Asaro E,
Atmanand M (eds) Observing the oceans in real time. Springer Oceanography. Springer,
Cham. https://doi.org/10.1007/978-3-319-66493-4_7

Lévy M, Ferrari R, Franks PJS et al (2012) Bringing physics to life at the submesoscale. Geophys
Res Lett 39:1–13. https://doi.org/10.1029/2012GL052756

Lévy M, Klein P, Tréguier A-M et al (2010) Modifications of gyre circulation by sub-mesoscale
physics. Ocean Model 34:1–15. https://doi.org/10.1016/j.ocemod.2010.04.001

Mahadevan A (2016) The impact of submesoscale physics on primary productivity of plankton.
Annu Rev Mar Sci 8:161–184. https://doi.org/10.1146/annurev-marine-010814-015912

Mahadevan A (2014) Ocean science: eddy effects on biogeochemistry. Nature 506:168–169.
https://doi.org/10.1038/nature13048

Mahadevan A, Pascual A, Rudnick DL et al (2020) Coherent pathways for vertical transport from
the surface ocean to interior. Bull Am Meteorol Soc 101:E1996–E2004. https://doi.org/10.1175/
bams-d-19-0305.1

Marcellin Yao K, Marcou O, Goyet C et al (2016) Time variability of the north-western Mediterranean Sea pH over 1995–2011. Mar Environ Res 116:51–60. https://doi.org/10.1016/j.
marenvres.2016.02.016

Margirier F, Testor P, Heslop E et al (2020) Abrupt warming and salinification of intermediate
waters interplays with decline of deep convection in the northwestern Mediterranean Sea. Sci
Rep 10:1–11. https://doi.org/10.1038/s41598-020-77859-5

McGillicuddy DJ, Anderson LA, Bates NR et al (2007) Eddy/wind interactions stimulate extraordinary mid-ocean plankton blooms. Science (80-) 316:1021–1026. https://doi.org/10.1126/
science.1136256

McWilliams JC, Colas F, Molemaker MJ (2009) Cold filamentary intensification and oceanic
surface convergence lines. Geophys Res Lett 36:1–5. https://doi.org/10.1029/2009GL039402

Millot C, Taupier-Letage I (2005) Circulation in the Mediterranean Sea. In: Life in the
Mediterranean Sea: A Look at Habitat Changes, pp 29–66

Olita A, Sparnocchia S, Cusí S et al (2014) Observations of a phytoplankton spring bloom onset
triggered by a density front in NW Mediterranean. Ocean Sci. https://doi.org/10.5194/os-10-
657-2014

Pascual A, Ruiz S, Olita A et al (2017) A multiplatform experiment to unravel meso- and
submesoscale processes in an intense front (AlborEx). Front Mar Sci 4:1–16. https://doi.org/
10.3389/fmars.2017.00039

Pearlman J, Bushnell M, Coppola L et al (2019) Evolving and sustaining ocean best practices and
standards for the next decade. Front Mar Sci 6:1–19. https://doi.org/10.3389/fmars.2019.00277

Pietri A, Testor P, Echevin V et al (2013) Finescale vertical structure of the upwelling system off
southern Peru as observed from glider data. J Phys Oceanogr 43:631–646. https://doi.org/10.
1175/JPO-D-12-035.1

Rixen M, Beckers JM, Levitus S et al (2005) The western Mediterranean deep water: a proxy for
climate change. Geophys Res Lett 32:1–4. https://doi.org/10.1029/2005GL022702

Roca E, Gamboa G, Tàbara JD (2008) Assessing the multidimensionality of coastal erosion risks:
public participation and multicriteria analysis in a Mediterranean coastal system. Risk Anal 28:
399–412. https://doi.org/10.1111/j.1539-6924.2008.01026.x

Rudnick DL (2016) Ocean research enabled by underwater gliders. Annu Rev Mar Sci 8:519–541. https://doi.org/10.1146/annurev-marine-122414-033913

Rudnick DL (1996) Intensive surveys of the Azores front: 2. Inferring the geostrophic and vertical velocity fields. J Geophys Res Ocean 101:16291–16303. https://doi.org/10.1029/96JC01144

Rudnick DL, Gopalakrishnan G, Cornuelle BD (2015) Cyclonic eddies in the Gulf of Mexico: observations by underwater gliders and simulations by numerical model. J Phys Oceanogr 45: 313–326. https://doi.org/10.1175/JPO-D-14-0138.1

Rudnick DL, Zaba KD, Todd RE, Davis RE (2017) A climatology of the California current system from a network of underwater gliders. Prog Oceanogr 154:64–106. https://doi.org/10.1016/j.pocean.2017.03.002

Ruiz S, Claret M, Pascual A et al (2019) Effects of oceanic mesoscale and submesoscale frontal processes on the vertical transport of phytoplankton. J Geophys Res Ocean 124:5999–6014. https://doi.org/10.1029/2019JC015034

Ruiz S, Garau B, Martínez-Ledesma M et al (2012) Nuevas tecnologías para la investigación marina: 5 años de actividades de gliders en el IMEDEA. Sci Mar 76:261–270. https://doi.org/10.3989/scimar.03622.19L

Ruiz S, Pascual A, Garau B et al (2009) Mesoscale dynamics of the Balearic front, integrating glider, ship and satellite data. J Mar Syst 78:S3–S16. https://doi.org/10.1016/j.jmarsys.2009.01.007

Schroeder K, Chiggiato J, Bryden HL et al (2016) Abrupt climate shift in the western Mediterranean Sea. Sci Rep 6:1–7. https://doi.org/10.1038/srep23009

Schroeder K, Chiggiato J, Josey SA et al (2017) Rapid response to climate change in a marginal sea. Sci Rep 7:1–7. https://doi.org/10.1038/s41598-017-04455-5

Schroeder K, Cozzi S, Belgacem M, Borghini M (2020) Along-path evolution of biogeochemical and carbonate system properties in the intermediate water of the western Mediterranean. Front Mar Sci 7:1–19. https://doi.org/10.3389/fmars.2020.00375

Schroeder K, Josey SA, Herrmann M et al (2010) Abrupt warming and salting of the western Mediterranean deep water after 2005: atmospheric forcings and lateral advection. J Geophys Res 115:C08029. https://doi.org/10.1029/2009JC005749

Sherman J, Davis RE, Owens WB, Valdes J (2001) The autonomous underwater glider "Spray.". IEEE J Ocean Eng 26:437–446. https://doi.org/10.1109/48.972076

Stommel H (1989) The Slocum mission. Oceanography 2:22–25. https://doi.org/10.5670/oceanog.1989.26

Taupier-Letage I, Puillat I, Millot C, Raimbault P (2003) Biological response to mesoscale eddies in the Algerian basin. J Geophys Res Ocean 108. https://doi.org/10.1029/1999jc000117

Testor P, Meyers G, Pattiaratchi C et al (2010) Gliders as a component of future observing systems. Proc Ocean Sustain Ocean Obs Inf Soc 961–978. https://doi.org/10.5270/OceanObs09.cwp.89

Thomalla SJ, Racault M-F, Swart S, Monteiro PMS (2015) High-resolution view of the spring bloom initiation and net community production in the Subantarctic Southern Ocean using glider data. ICES J Mar Sci J du Cons 72:1999–2020. https://doi.org/10.1093/icesjms/fsv105

Thomsen S, Kanzow T, Colas F et al (2016a) Do submesoscale frontal processes ventilate the oxygen minimum zone off Peru? Geophys Res Lett 43:8133–8142. https://doi.org/10.1002/2016GL070548

Thomsen S, Kanzow T, Krahmann G et al (2016b) The formation of a subsurface anticyclonic eddy in the Peru- Chile undercurrent and its impact on the near-coastal salinity, oxygen, and nutrient distributions. J Geophys Res Ocean 121:476–501. https://doi.org/10.1002/2015JC010878

Tintoré J, Gomis D, Alonso S, Parrilla G (1991) Mesoscale dynamics and vertical motion in the Alborán Sea. J Phys Oceanogr 21:811–823

Tintore J, La Violette PE, Blade I, Cruzado A (1988) A study of an intense density front in the eastern Alboran Sea: the Almeria–Oran front. J Phys Oceanogr 18:1384–1397

Tintoré J, Pinardi N, Álvarez-Fanjul E et al (2019) Challenges for sustained observing and forecasting systems in the Mediterranean Sea. Front Mar Sci 6. https://doi.org/10.3389/fmars.2019.00568

Tintoré J, Vizoso G, Casas B et al (2013) SOCIB: the Balearic Islands Coastal Ocean observing and forecasting system responding to science, technology and society needs. Mar Technol Soc J 47:101–117. https://doi.org/10.4031/MTSJ.47.1.10

Todd RE, Rudnick DL, Davis RE, Ohman MD (2011) Underwater gliders reveal rapid arrival of El Niño effects off California's coast. Geophys Res Lett 38:n/a-n/a. https://doi.org/10.1029/2010GL046376

Troupin C, Beltran JP, Heslop E et al (2015) A toolbox for glider data processing and management. Methods Oceanogr 13–14:13–23. https://doi.org/10.1016/j.mio.2016.01.001

Zaba KD, Rudnick DL (2016) The 2014–2015 warming anomaly in the Southern California current system observed by underwater gliders. Geophys Res Lett 43:1241–1248. https://doi.org/10.1002/2015GL067550

Zarokanellos ND, Jones BH (2021) Influences of physical and biogeochemical variability of the central Red Sea during winter. J Geophys Res Ocean 126:1–23. https://doi.org/10.1029/2020jc016714

Zarokanellos ND, Papadopoulos VP, Sofianos SS, Jones BH (2017) Physical and biological characteristics of the winter-summer transition in the central Red Sea. J Geophys Res Ocean 122:6355–6370. https://doi.org/10.1002/2017JC012882

Zarokanellos ND, Rudnick DL, Garcia-Jove M, Mourre B, Ruiz S, Pascual A, Tintoré J (2022) Frontal dynamics in the Alboran Sea: 1. Coherent 3D pathways at the Almeria-Oran front using underwater glider observations. J Geophys Res Oceans 127:e2021JC017405. https://doi.org/10.1029/2021JC017405

Marine Chemical Metadata and Data Management

11

Mohamed Adjou and Gwenaëlle Moncoiffé

Contents

Abstract

Marine chemical data obtained from various marine environments, like other scientific data, should be accompanied with the context describing what they are in a consistent manner. This context is known as metadata. Providing the (meta-) data consistently, following existing best practices, will enable better readability and improve its reusability by peers. These best practices are provided for suggestion, in a non-exhaustive way, yet it remains very important to document marine chemical data. Furthermore, such methodical (meta-)data preparation before sharing will facilitate its 'cataloguing' in specialised data centres/units to promote their discoverability, accessibility and interoperability.

Keywords

Data · Metadata · Data management · Quality assurance · Best practices · Parameters · Data sharing · Data legacy

M. Adjou (✉)
KLaIM, L@bISEN, YNCREA OUEST, Brest, France
e-mail: mohamed.adjou@isen-ouest.yncrea.fr

G. Moncoiffé
National Oceanography Centre, British Oceanographic Data Centre, Liverpool, UK

11.1   Introduction

The acquisition of marine samples and their chemical analysis are often costly. Chemical data collected in the environment are also extremely valuable when these are placed in a different context to the one associated with their original collection. This is critical to improving our understanding of temporal and spatial dynamics so that ensuring their high-quality reusability and long-lasting availability has benefit beyond objectives of the research or monitoring project that delivered these data in the first instance.

The measurement quality made by analysts in the laboratory, together with the description of the context of the measurement, is intimately linked to the quality of the data available for reuse. However, one does not imply the other as the quality assessment (QA) of laboratory measurements, and the data management belongs to two different domains (i.e. analytical chemistry vs. information science). It is important to decouple these two domain-specific QA and emphasise the bottom-up needs, provided by data generators, to better value marine chemical data measurement at the level of nowadays and future information technology.

With the increasing capability of digital facilities and technologies and information accessibility via the World Wide Web, there is a growing trend towards what is known as Open Data. In addition to the goodwill of sharing research data that proved its multiple benefits, this could be optimised by a consistent data management (Popkin 2019). To access such a big hub, there are prerequisites: best practices (BP) and standards need to be followed to allow a given dataset to be findable, accessible and reusable.

In this chapter, inspired from the GEOTRACES[1] summer school 2019, the aim is mainly to provide BP in preparing marine chemical (meta-)data to facilitate their dissemination in long-lasting term through national data centres or equivalent suitable repositories. The focus will be given on discrete sample measurements as opposed to data generated from in situ chemical sensors. This is because many GEOTRACES target measurements are from discrete samples, and these have greater challenges when it comes to producing data that are readily accessible and interoperable due to human-driven variations in procedure and reporting practices. They also tend to be more diverse, complex and challenging to curate for long-term preservation and data reuse.

> **Why Manage Research Data**
> Most scientists collect data to answer specific scientific questions, and the output is to write reports and articles to present results in their own words and

(continued)

---

[1] GEOTRACES is an international research programme that aims to improve the understanding of the *Marine Biogeochemical Cycles of Trace Elements and Isotopes*, initiated in 2006 and formally launched its seagoing effort in January 2010.

add their interpretation. This is how science progresses; however, in a more pragmatic view, the dataset generated is the only objective legacy of a scientific work, as it is not subject to bias and it could be used for other scientific interpretations, on its own, or when it is grouped with other datasets from other sources.

11.2 The Metadata

The metadata of marine chemical measurements are all the information describing the context under which a given dataset was obtained. Without this set of information, the data reusability could be compromised even if it is only used by its own generator (e.g. using temporary undefined abbreviations to label measured parameters). Therefore, as a BP for each dataset, essential metadata should be recorded and gathered in the same file, once the measured data is in its final version.

Here, the idea is to provide essential information about a dataset content in order to support its reusability and facilitate potential linking and merging with other datasets that share a similar context.

11.2.1 Data Narrative

Each dataset should have a *title* consisting of one sentence summarising concisely the key content of the dataset. For a marine chemistry dataset, this should include at the minimum what was measured (i.e. chemical substance or group of substances in a given matrix) and where it was sampled (i.e. name of geographical location and/or cruise name).

With these two kinds of information, a title cannot be exhaustive, but it responds to what data users look for in general: what was measured and where. Additional important information related to the sampling or analytical method or any targeted sampling feature (e.g. 'surface' or 'air-sea interface' or 'euphotic zone' etc.) may be included if it is key to quickly identify (and gauge) the content of the dataset.

A more extensive description should be provided in a short *summary* of the dataset describing as concisely and as clearly as possible the content of the dataset and expanding on the information already provided in the title. It should also summarise information provided in the other sections. This includes the main purpose of the data collection and key contextual information about the fieldwork (e.g. name of the cruise and/or research programme), the temporal and spatial coverage including vertical extent, the sampling method(s) and gear types and the variables contained in the dataset alongside key aspect of methodology. In this summary, only objective descriptions of the data are required, and any data interpretation should be limited to aspect of data reusability that helps the reader understand the data's fitness for purpose.

11.2.2 Data Credit

Names of the data originator(s) should be listed in a contribution order and with their organisational affiliation(s) as for any other scientific publication. This enables that credit be given to the people and the organisation who were responsible for the creation of the dataset. For similar aims, the project(s) funding the work that generated the dataset should also be clearly identified with its acronyms, the grant ID and the funding organisation.

11.2.3 Activity

Most discrete marine chemical data sampling takes place during research cruises on board dedicated research vessels; hence, the attention here will be on this major kind of deployment. Research cruises are attributed a unique ID which needs to be provided, as well as the name of the chief scientist(s) and start and end dates and ports including countries. It is of high importance to have this information and/or a cruise report when it is published (Pollard et al. 2011) or potentially link it to any existing persistent Web link to the cruise description (e.g. cruise summary reports, https://csr.seadatanet.org/).

11.2.4 Material and Methods Used to Sample and Acquire the Data

This section should report the sampling gear, the sample preparation and the analytical method including the instrumentation used. To keep this section short and coherent, it is acceptable to reference existing method(s) to publications that contain the information; however, authors must ensure that any changes operated by data samplers and/or analysts are highlighted. Publication references should also be listed in reference lists with full journal reference and hyperlink to the publication if available.

11.2.5 Measurement Descriptors

In this section, the measured parameters, the associated error, the units and the data quality flag scheme used should be explained in plain text. The labelling system used for column headers by the data creator be it personal or common usage should also be explained. Links to controlled list of acronyms or codes should be provided. This information is crucial to allow a clear understanding of abbreviations used in the data template.

11.2.6 The Sampling and the Logs

Upon collecting samples for marine chemistry studies, each *cruise* will have at least one sampling *station (or site)*. At each station, at least one sampling *event* occurs, and as part of each event, at least one *sample* is collected. It is important to keep this hierarchy in mind, to better visualise the path to identify a given sample and the utility of logs for a data management purpose (Fig. 11.1).

A sample is a defined quantity of a matrix (e.g. seawater, sediment etc.) that is considered representative of the environment from which it is collected an environment that is believed to be a homogeneous matrix ensemble. Seawater samples are identified by their spatial position (e.g. geographical coordinates, water depth or pressure) and sampling date and time. All analysed subsamples (i.e. aliquots) are linked to the same sample. In the case of water samples, all aliquots are linked to the bottle used to sample the water (e.g. a Niskin bottle).

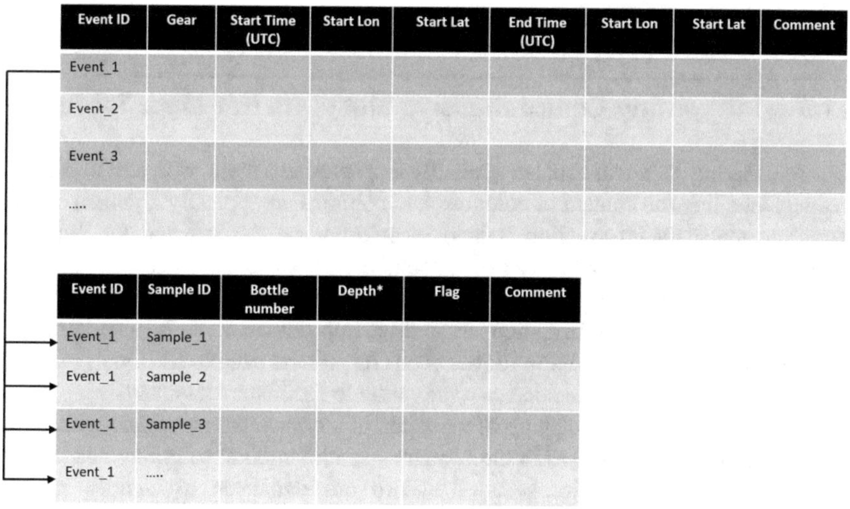

Event ID	Gear	Start Time (UTC)	Start Lon	Start Lat	End Time (UTC)	Start Lon	Start Lat	Comment
Event_1								
Event_2								
Event_3								
.....								

Event ID	Sample ID	Bottle number	Depth*	Flag	Comment
Event_1	Sample_1				
Event_1	Sample_2				
Event_1	Sample_3				
Event_1	.....				

Fig. 11.1 Event and sample logs. This diagram elucidates the connection between event (upper section) and sample (lower section) information records. For an illustration purpose, only essential columns are suggested. *Event logs—gear* represents the sampling device and could be identified as an acronym or abbreviation referenced in details in the cruise report or hyperlinked to a controlled vocabulary. Date/time should be in Universal Time Coordinated (UTC) to standardize against a fixed time reference. Event end time and end spatial coordinates are recommended for static or quasi-static sampling events, but mandatory for 'dynamic' sampling events. *Sample logs—bottle number* applies mainly to Conductivity, Temperature, and Depth (CTD) rosette water sampling events, as it is an essential identification reference for the sample within a given cast). *Depth*, at early log creation stage, could be a nominal one and could help to check significant depth issues. *Flag* could be applied if using a known sample qualifier flagging scheme (e.g. World Ocean Circulation Experiment (WOCE) quality flags for water samples) as an alternative to free text in the *comment* field to log potential complementary information including sampling issues like e.g. leaks

'A station' might have different definitions for different use cases; however, for our purpose, we define a station or site as an intemporal concept with a nominal longitude and latitude; it is a 'cross on the map' where spatial coordinates fluctuate, over 200 m in some cases due to the ship relative drift or sometimes to a tolerance distance radian decided by the chief scientist. It is in fact a 'rough' location identified as an important sampling site as part of the research question and could be resampled in future independent cruises. On the other hand, an event happening at a given sampling station is temporally defined with accurate longitude and latitude at the start and the end of the event. The sampling gear used to collect samples (e.g. the CTD, a pump and a corer) also defines the event.

Chief scientist should have direct, or indirect, responsibility in ensuring that an event log is maintained with clear unique event IDs. This should be the reference to link samples analysed by individuals. With the exception of major sampling events, like CTD rosette casts, the log at sample level is not always maintained centrally, so it is recommended that individual scientists maintain their own individual sample log but reference samples to identifiers used in the central log. Every sample should be linked to an entry in the central event log.

11.3   Integrating Comprehensive Metadata in a Data Table

Discrete marine chemical data are generally displayed in a table, with text in column headers labelling the content of columns, and columns are typically arranged so that sampling metadata information is held in columns on the left and the value of analytical results and any derived parameters are added to the right. Best practice makes it mandatory to have, at a minimum, precise date, time and time zone, latitude, longitude and depth information in individual columns. It is also strongly recommended to have a column for the event ID and the sample ID. Other variable metadata parameter can be added as appropriate to facilitate data analysis.

Figure 11.2 suggests a display of metadata in a typical data template with three different metadata zones, two in the headers—spatiotemporal sampling descriptors and measurement descriptors—and the third one dedicated to identify sample position in space and time in rows below the spatiotemporal sampling headers.

The focus of this chapter is on water column samples; however, by analogy, other sampling types could be represented (e.g. sediment cores and plankton nets), adapting metadata fields to the ones specifically required by this kind of sampling methods.

11.3.1 Sampling and Spatiotemporal Information

Analytical chemistry results generally require to be merged with their essential metadata manually by the analyst or researcher. For this reason, they are easily affected by metadata errors (e.g. typos, coordinate imprecision, missing or different date/time zone and other discrepancies when these are compared to the

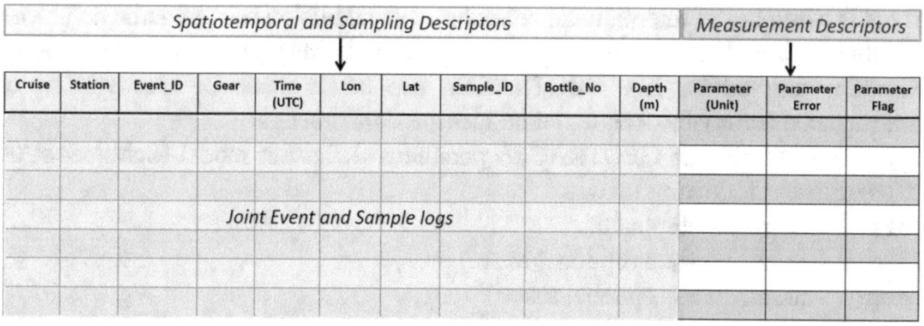

Cruise	Station	Event_ID	Gear	Time (UTC)	Lon	Lat	Sample_ID	Bottle_No	Depth (m)	Parameter (Unit)	Parameter Error	Parameter Flag

Fig. 11.2 Integrating comprehensive metadata in a data table. *Spatiotemporal and sampling descriptors*—the corresponding headers are the most important but non-exhaustive descriptors of the sample spatial/time position and acquisition from the water column. The data creator could add other important descriptors, as needed (e.g. different sample types). When a descriptor requires units, it could be added to the descriptor. *Joint event and sample logs*—the information provided in the rows under *spatiotemporal and sampling descriptors'* headers should be aligned with the event log as presented in Fig. 11.1. Additional sample descriptors (e.g. Sample_id, Bottle_No and Depth) need to be aligned with any existing sample log or alternatively with the processed bottle hydrography in the case of CTD sampling. *Measurement descriptors*—a triplet of the measurement description: the parameter, its associated error and a measurand qualifier flag to qualify the parameter value. The measurand qualifier flag is different from sample quality flag and should follow a proper flagging schema. This figure illustrates a case of only one parameter. Additional parameters could be added as a succession of similar triplet of columns (i.e. parameter, error and flag)

main event and sample logs). To address this issue, it is wise to check systematically potential inconsistencies with central logs before further distribution or data analysis. Having well-maintained centralised logs of events and samples, in addition to using unique identifiers to label samples, is a key method to decrease risks of sample identification errors. However, the manual process of transferring information across (be it time/location or sample ID) will always be prone to human error; hence, it is recommended to always err on the side of caution and report mandatory metadata fields of time and space plus the event and sample identifiers.

11.3.2 Measurement Descriptors

Measurement descriptors are crucial to identify and understand what was measured. This includes variable names but also reporting units and any quality statement. As explained in the metadata *Measurement Descriptors'* section, abbreviated descriptors here should link to a plain text description of what it is.

11.3.2.1 Parameter and its Associated Error

Scientists creating the data are the best persons to accurately describe parameter variables held in their dataset. *Parameters* are frequently provided using shorthand text or acronyms which are sometimes personal to the data creator. These are not always well defined, and they can be a source of confusion in future reuse of the data.

There is a number of international community standards on how to name parameter variables although consensus as yet to be reached (Magagna et al. 2022). Some research communities like GEOTRACES use labels made of acronyms or of concatenated acronym-token to define measured parameters.

As an example, the GEOTRACES parameter-definition model (Schlitzer et al. 2018) uses the following tokens:

Token 1—element/compound.
Token 2—oxidation state (optional token).
Token 3—atomic mass (optional token).
Token 4—phase.
Token 5—data type (i.e. the measurement property).
Token 6—sampling system.

To create a parameter four tokens are concatenated using an underscore character. Token 2 and Token 3 are optional. As an example, concentration of dissolved Fe (II) in seawater sampled by trace metal clean CT would be displayed as *Fe_II_D_CONC_BOTTLE* where *Fe* represents the measured element at its oxidation state *II*, in the dissolved phase (*D*) expressed as a concentration property (*CONC*), for a sample taken from the water column in a **bottle**. The parameter error descriptor in this same GEOTRACES example derives directly from the measured parameter by adding an extra mandatory *error token*: *SD_Fe_II_D_CONC_BOTTLE* (where SD stands for standard deviation).

The parameter definition used in the GEOTRACES programme covers well the use case of the GEOTRACE data product that focusses on marine trace elements and their isotopes in the marine environment. However, for other marine chemical measurements and applications that include complex chemical substances or complex matrices, other parameter labelling schemes need to be adopted.

The best known and used parameter labelling scheme in usage in the marine domain is the British Oceanographic Data Centre (BODC) Parameter Usage Vocabulary (PUV, also known as P01). This vocabulary is based on a parameter-definition model that is machine-readable and applicable to the full range of marine data types, not just chemical ones. P01 is used extensively by the marine community in Europe as part of the European SeaDataNet network. P01 is compatible with the recently proposed framework and decomposition model for describing variables across multiple domains (Magagna et al. 2022). It is based on a conceptual model that defines a measurement as the *property* of an *object* in relation to a *matrix* by a *method*. Its code is uniquely identified by its persistent Uniform Resource Identifier (URI). It is published as a controlled vocabulary that is findable, reusable and accessible online (https://vocab.nerc.ac.uk/collection/P01/current/ or https://vocab.nerc.ac.uk/search_nvs/P01/).

11.3.2.2 Units

It is clear that a dataset containing parameter values, and associated error values, without a unit cannot be reusable. Although a few parameter-defining models have the unit embedded in the parameter definition, the recommended best practice is that the unit should be specified in the data template, as suggested in Fig. 11.2, using consistent units, derived units and prefixes as suggested in the International Union of Pure and Applied Chemistry (IUPAC) terminologies (https://goldbook.iupac.org/) when applicable.

Alternatively, it is recommended to reference the unit to a well-managed controlled vocabulary such as the one maintained by QUDT.org or BODC (https://vocab.nerc.ac.uk/collection/P06/current/ or https://vocab.nerc.ac.uk/search_nvs/P06/).

11.3.2.3 Flags

Measurement qualifier flags are usually expressed in an existing flagging schema. For chemical data, it is important to include a schema with an extensive list of adequate flags (e.g. including below the limit of detection or value in excess flag, rather than bad value) in order to provide accurate information to data users.

Table 11.1 displays SeaDataNet qualifier flags with a range of flags suitable for chemical data measurements.

11.4 The Metadata and Data File Format

In the previous section, the accent was put on important practices to generate an explicit content of (meta-)data, and the selection of a sure 'container' is as essential in the process of producing and safeguarding this information.

A variety of file formats exist to hold information; however, it is recommended to share (meta-)data in a long-lasting and interoperable format relying on open standards. While spreadsheets are widely used to create data files, allowing calculations and customised font, it is a good practice to save data in an open file format (i.e. CSV or plain text) to avoid a situation where a closed file format readability will not be guaranteed over time (e.g. non-available software version).

Table 11.1 SeaDataNet measurement qualifier flags (http://vocab.nerc.ac. uk/collection/L20/current/)

Flag	Label
0	No quality control
1	Good value
2	Probably good value
3	Probably bad value
4	Bad value
5	Changed value
6	Value below detection
7	Value in excess
8	Interpolated value
9	Missing value
A	Value phenomenon uncertain
B	Nominal value
Q	Value below limit of quantification

11.5   Dedicated Data Centres to Preserve and Share Data

Submitting data to national oceanographic data centres (NODC) or an appropriate Associate Data Units (ADU), as identified by IODE,[2] is a guarantee that the data will be safe in the long term and made available to other users with better discoverability.

11.5.1 Importance of Depositing your Dataset in a Data Centre

In general, specialised data centres like an oceanographic data centre or an associated data unit, in addition to being a safe deposit for data with a secure backup policy that neither a research laboratory nor a university can guarantee, add much value to your data by providing greater findability and accessibility. More and more data centres are now offering a DOI minting service and therefore guaranteeing a better discoverability, traceability and online availability of your dataset as well as enabling credit to be made whenever the data are reused.

11.5.2 Rich Metadata: The Key to Boost the Visibility of your Data

When a dataset is submitted to a NODC or an ADU, the metadata are scrutinised in order to ensure that the new data are consistent with and can easily be linked to existing datasets. All metadata fields are harmonised and quality controlled using standard best practice methods and automated checks. This information is then packaged in metadata records according to international exchange and World Wide Web standards, therefore increasing its discoverability by peers.

11.5.3 Policies

In general, researchers and data generators are reluctant to submit their data, to dedicated data NODC/ADU, before publishing their own analyses and interpretations in scientific articles. There is a well-defined framework when a dataset is submitted to NODC/ADU where a release policy is applied. These policies deal mainly with the release date and who can access the data. They are predefined in the agreement made with the funding organisation. In some cases, they can be modified if a data generator requests an extension of the date of release due to valid reasons (e.g. delay in sample analysis after the cruise). Note that requesting a DOI for a given dataset implies that data originators agree on Open Data policy. This is in line with the majority of research funding organisations.

---

[2]International Oceanographic Data and Information Exchange (https://www.iode.org/)

References

Magagna B, Moncoiffé G, Devaraju A, Stoica M, Schindler S, Pamment A, RDA I-ADOPT WG (2022) InteroperAble Descriptions of Observable Property Terminologies (I-ADOPT) WG Outputs and Recommendations (1.1.0). https://doi.org/10.15497/RDA00071

Pollard RT, Moncoiffé G, O'Brien TD (2011) The IMBER data management cookbook–a project guide to good data practices. IMBER Report No. 3, IPO Secretariat, Plouzané, France. 16pp

Popkin G (2019) Data sharing and how it can benefit your scientific career. Nature 569(7756): 445–447

Schlitzer R, Anderson RF, Dodas EM, Lohan M, Geibert W, Tagliabue A et al (2018) The GEOTRACES intermediate data product 2017. Chem Geol 493:210–223

Printed in the United States
by Baker & Taylor Publisher Services